LIBRARY
UNION CARBIDE
TARRYTOWN TECHNICAL CENTER

Handbook of Plastics Testing Technology

SPE MONOGRAPHS

Injection Molding

Irvin I. Rubin

Nylon Plastics

Melvin I. Kohan

Introduction to Polymer Science and Technology: An SPE Textbook

Edited by Herman S. Kaufman and Joseph J. Falcetta

Principles of Polymer Processing

Zehev Tadmor and Costas G. Gogos

Coloring of Plastics

Edited by Thomas G. Webber

Analysis and Performance of Fiber Composites

Bhagwan D. Agarwal and Lawrence J. Broutman

The Technology of Plasticizers

J. Kern Sears and Joseph R. Darby

Fundamental Principles of Polymeric Materials

Stephen L. Rosen

Plastics Polymer Science and Technology

Edited by M. D. Baijal

Plastics vs. Corrosives

Raymond B. Seymour

Handbook of Plastics Testing Technology

Vishu Shah

Handbook of Plastics Testing Technology

VISHU SHAH
Performance Engineered Products, Inc.
Pomona, California

A WILEY-INTERSCIENCE PUBLICATION

JOHN WILEY & SONS

New York · Chichester · Brisbane · Toronto · Singapore

Copyright © 1984 by John Wiley & Sons, Inc.

All rights reserved. Published simultaneously in Canada.

Reproduction or translation of any part of this work beyond that permitted by Section 107 or 108 of the 1976 United States Copyright Act without the permission of the copyright owner is unlawful. Requests for permission or further information should be addressed to the Permissions Department, John Wiley & Sons, Inc.

Library of Congress Cataloging in Publication Data:

Shah, Vishu, 1951–
Handbook of plastics testing technology.

(SPE monographs, ISSN 0195-4288)
"A Wiley-Interscience publication."
Includes bibliographical references and index.
1. Plastics—Testing—Handbooks, manuals, etc.
I. Title. II. Series

TA455.P5S457 1983 620.1′923′0287 83-12434
ISBN 0-471-07871-9

Printed in the United States of America

10 9 8 7 6 5 4 3

Series Preface

The Society of Plastics Engineers is dedicated to the promotion of scientific and engineering knowledge of plastics and to the initiation and continuation of educational programs for the plastics industry. Publications, both books and periodicals, are major means of promoting this technical knowledge and of providing educational materials.

New books, such as this volume, have been sponsored by the SPE for many years. These books are commissioned by the Society's Technical Volumes Committee and, most importantly, the final manuscripts are reviewed by the Committee to ensure accuracy of technical content. Members of this Committee are selected for outstanding technical competence and include prominent engineers, scientists, and educators.

In addition, the Society publishes *Plastics Engineering, Polymer Engineering and Science (PE & S), Journal of Vinly Technology, Polymer Composites,* proceedings of its Annual, National, and Regional Technical Conferences (ANTEC, NATEC, RETEC), and other selected publications. Additional information can be obtained from the Society of Plastics Engineers, 14 Fairfield Drive, Brookfield Center, Connecticut 06805.

Technical Volumes Committee
James L. Throne, Chairman
Lee L. Blyler, Jr.
A. Wayne Cagle
Jerome L. Dunn
Tom W. Haas
John L. Kardos
Richard R. Kraybill
Timothy Lim
Eldridge M. Mount, III
Nikolaos A. Peppas
B. J. Sexton

ROBERT D. FORGER

Executive Director
Society of Plastics Engineers

Preface

The desire to compile this book was initiated mainly because of the virtual non-existence of a comprehensive work on testing of plastic materials. The majority of the literature concerning the testing of plastics is scattered in the form of sales and technical brochures, private organizations' internal test procedures, or a very brief and oversimplified explanation of the test procedures in plastics literature. The main objective of the present book is to provide a general purpose practical text on the subject with the main emphasis on the significance of the test or *why* and not so much on *how* without being extremely technical.

Over the years ASTM (American Society for Testing and Materials) has done an excellent job in providing the industry with standard testing procedures. However, the test procedures discussed in ASTM books lack the theoretical aspects of testing. The full emphasis is not on *significance* of testing but on *procedures* of testing. The ASTM books are also deficient in showing the diagrams and photographs of actual, commercial testing equipment. In this book I have tried to bridge the gap between the oversimplified and less explained tests described in ASTM books and the highly technical and less practical books in existence today.

This handbook is not intended primarily for specialists and experts in the area of plastics testing but for the neophyte desiring to acquire a basic knowledge of the testing of plastics. It is for this reason that detailed discussions and excessive technical jargon have been avoided. The text is aimed at anyone involved in manufacturing, testing, studying, or developing plastics. It is my intention to appeal to a broad segment of people involved in the plastic industry.

In Chapter 1 the basic concepts of testing are discussed along with the purpose of specifications and standards. Also discussed is the basic specification format and classification system. The subsequent chapters deal with the testing of five basic properties: mechanical, thermal, electrical, weathering, and optical properties of plastics. The chapter on mechanical properties discusses in detail the basic stress–strain behavior of the plastic materials so that a clear understanding of testing procedures is obtained. Chapter 7 on material characterization is intended to present a general overview on the latest in characterization techniques in existence today. A brief explanation of the polymer combustion process along with various testing procedures are discussed in Chapter 8. An attempt is made to briefly explain the importance of conditioning procedures. A table summarizing the most common conditioning procedures should be valuable. Several tests that are difficult to incorporate into a specific category were placed in the chapter on miscellaneous tests. End-product testing, an area generally neglected by the majority of processors of plastic products, is discussed along with some useful suggestions on common end-product tests.

Chapter 13 on identification analysis should be important to everyone involved in plastics and particularly useful to plastic converters and reprocessors. The flowchart summarizes the entire identification technique. Since there are so many different tests in existence on the testing of foam plastics, only a brief explanation of each test is given. The chapter on failure analysis is a compilation of methods commonly used by material suppliers. A step-by-step procedure for analyzing product failure should prove valuable to anyone involved in failure analysis. Quality control, although not part of the testing, is included in order to explain quality control as it relates to plastics. The section on visual standard, mold control, and workmanship standard is a good example. In this increasing world of product liability, the chapter on product liability and testing should be of value to everyone.

In order to increase the versatility of this book and meet the goal of providing a ready reference on the subject of testing, a large appendix section is given. One will find very useful data: names and addresses of equipment manufacturers, a glossary, names and addresses of trade publications, information on independent testing laboratories, and a guide to plastics specifications. Many useful charts and tables are included in the appendix. Throughout the book, wherever possible, numerous diagrams, sketches, and actual photographs of equipment are given.

A handbook of this magnitude must make inevitable compromises. Depending on the need of the individual user, there is bound to be a varying degree of excess and shortage. In spite of every effort made to minimize mistakes and other shortcomings in this book, some may still exist. For the sake of future refinement and improvements, all constructive comments will be welcomed and greatly appreciated.

VISHU H. SHAH

Pomona, California
October 1983

Acknowledgments

Although it is practically impossible to acknowledge the help and guidance of everyone who has assisted me in preparing this book, there are a number of people, companies, and organizations to whom credit must be given. I am primarily indebted to the Society of Plastic Engineers who encouraged me to undertake the task of writing this book. I am extremely grateful to George Epstein for reviewing the manuscript, making numerous suggestions, and for being a constant source of guidance and inspiration. I am also very thankful to Penton Publication for allowing me to reprint tables and charts from the latest issue of *Materials Engineering* and also to the publishers of *Modern Plastics Encyclopedia, Plastics Technology, Plastics World,* and *Plastics Design and Processing* for allowing me to reprint several diagrams and tables. Special thanks to Cordura Publications for their help in providing important data for reprinting. I wish to thank Instron Corporation, Tinius Olsen Testing Machine Company, Pacific Scientific–Gardner Instrument Division, DuPont Company, Celanese Plastics, Atlas Electric Devices Company, Macbeth, Waters Associates, Perkin–Elmer Corporation and many others for providing technical assistance and photographs for reproduction. Chapter 5 on Weathering Properties was reviewed by Ray Kinmonth of Atlas Electric Devices Company. His help and guidance is certainly appreciated. The assistance and encouragement from Bill Yahne, Director of quality assurance at Rain Bird has been very important in making this book possible. I wish to especially thank my good friend and business partner, Carl Dispenziere, for his guidance and encouragement.

This book would never have been possible without the help, encouragement, and patience of my lovely wife, Charlene. I am forever indebted to her for her understanding during the many evenings and weekends when I was wrapped up in the preparation of this book.

V. S.

Contents

1. BASIC CONCEPTS **1**

1.1. Basic Concepts, 1
1.2. Specifications and Standards, 2
1.3. Purpose of Specifications, 3
1.4. Basic Specification Format, 4
References, 6

2. MECHANICAL PROPERTIES **7**

2.1. Introduction, 7
2.2. Tensile Tests, 14
2.3. Flexural Properties, 23
2.4. Compressive Properties, 28
2.5. Creep Properties, 30
2.6. Stress Relaxation, 46
2.7. Impact Properties, 49
2.8. Shear Strength, 71
2.9. Abrasion, 73
2.10. Fatigue Resistance, 74
2.11. Hardness Tests, 79
References, 84
Suggested Reading, 86

3. THERMAL PROPERTIES **89**

3.1. Introduction, 89
3.2. Tests for Elevated Temperature Performance, 89
3.3. Thermal Conductivity, 102
3.4. Thermal Expansion, 106
3.5. Brittleness Temperature, 110
References, 112
Suggested Reading, 112

4. ELECTRICAL PROPERTIES **113**

4.1. Introduction, 113
4.2. Dielectric Strength, 114
4.3. Dielectric Constant and Dissipation Factor, 117
4.4. Electrical Resistance Tests, 119
4.5. Arc Resistance, 120

4.6. UL Requirements, 122
References, 125
Suggested Reading, 126

5. WEATHERING PROPERTIES 127

5.1. Introduction, 127
5.2. Accelerated Weathering Tests, 129
5.3. Outdoor Weathering of Plastics, 140
5.4. Resistance of Plastic Materials to Fungi, 142
5.5. Resistance of Plastic Materials to Bacteria, 142
5.6. Limitations of Accelerated Microbial Growth Resistance Testing, 143
5.7. Outdoor Exposure Test for Studying the Resistance of Plastic Materials to Fungi and Bacteria and its Limitations, 143
References, 144
Suggested Reading, 144

6. OPTICAL PROPERTIES 147

6.1. Introduction, 147
6.2. Refractive Index, 148
6.3. Luminous Transmittance and Haze, 150
6.4. Photoelastic Properties, 152
6.5. Color, 154
6.6. Specular Gloss, 162
References, 164
Suggested Reading, 164

7. MATERIAL CHARACTERIZATION TESTS 165

7.1. Introduction, 165
7.2. Melt Index Test, 166
7.3. Capillary Rheometer Tests, 170
7.4. Viscosity Tests, 174
7.5. Gel Permeation Chromatography, 178
7.6. Thermal Analysis Techniques, 183
7.7. Material Characterization Tests for Thermosets, 189
References, 197
General References, 197
Suggested Reading, 198

8. FLAMMABILITY 199

8.1. Introduction, 199
8.2. Flammability Test (Flexible Plastics), 205
8.3. Flammability Test (Self-Supporting Plastics), 206
8.4. Incandescence Resistance Test, 206
8.5. Ignition Properties of Plastics, 208
8.6. Oxygen Index Test, 211
8.7. Surface Burning Characteristics of Materials, 214

8.8. Flammability of Cellular Plastics—Vertical Position, 216
8.9. Flammability of Cellular Plastics—Horizontal Position, 216
8.10. Flame Resistance of Difficult-to-Ignite Plastics, 216
8.11. Smoke Generation Tests, 218
8.12. UL 94 Flammability Testing, 222
8.13. Meeting Flammability Requirements, 227
References, 229
Suggested Reading, 230

9. CHEMICAL PROPERTIES **231**

9.1. Introduction, 231
9.2. Immersion Test, 232
9.3. Stain-Resistance of Plastics, 232
9.4. Solvent Stress-Cracking Resistance, 233
9.5. Environmental Stress-Cracking Resistance, 235
References, 237

10. ANALYTICAL TESTS **239**

10.1. Introduction, 239
10.2. Specific Gravity, 239
10.3. Density by Density Gradient Technique, 241
10.4. Bulk (Apparent) Density Test, 243
10.5. Water Absorption, 243
10.6. Moisture Analysis, 244
10.7. Sieve Analysis (Particle Size) Test, 246
References, 247

11. CONDITIONING PROCEDURES **249**

11.1. Conditioning, 249
11.2. Designation for Conditioning, 249
References, 251
Suggested Reading, 251

12. MISCELLANEOUS TESTS **253**

12.1. Torque Rheometer Test, 253
12.2. Plasticizer Absorption Tests, 254
12.3. Cup Viscosity Test, 259
12.4. Burst Strength Tests, 260
12.5. Crush Test, 264
12.6. Acetone Immersion Tests, 265
12.7. Acetic Acid Immersion Test, 266
12.8. End-Product Testing, 266
References, 268
General References, 269

13. IDENTIFICATION ANALYSIS OF PLASTIC MATERIALS 271

13.1. Introduction, 271
13.2. Chemical and Thermal Analysis for Identification of Polymers, 276
13.3. Identification of Plastic Materials, 277
References, 281
General References, 281

14. TESTING OF FOAM PLASTICS 283

14.1. Introduction, 283
14.2. Rigid Foam Test Methods, 283
14.3. Flexible Foam Test Methods, 291
14.4. Foam Properties, 304
General References, 304

15. FAILURE ANALYSIS 307

15.1. Introduction, 307
15.2. Types of Failure, 308
15.3. Analyzing Failures, 308
References, 315
General References, 316

16. QUALITY CONTROL 317

16.1. Introduction, 317
16.2. Statistical Quality Control, 318
16.3. Quality Control System, 334
16.4. General, 340
References, 344
General References, 345

17. PRODUCT LIABILITIES AND TESTING 347

17.1. Introduction, 347
17.2. Product/Equipment Design Considerations, 348
17.3. Packaging Considerations, 348
17.4. Instructions, Warning Labels, and Training, 349
17.5. Testing and Record-Keeping, 349
17.6. Safety Standards Organizations, 350
References, 350

18. NONDESTRUCTIVE TESTING 351

18.1. Introduction, 351
18.2. Ultrasonic Testing, 351
18.3. Application of Ultrasonics NDT in Plastics, 356

References, 357
Suggested Reading, 358

19. PROFESSIONAL AND TESTING ORGANIZATIONS **359**

19.1. American National Standards Institute, 359
19.2. American Society for Testing and Materials, 360
19.3. Food and Drug Administration, 360
19.4. National Bureau of Standards, 361
19.5. National Electrical Manufacturers Association, 361
19.6. National Fire Protection Association, 361
19.7. National Sanitation Foundation, 362
19.8. Plastics Technical Evaluation Center, 362
19.9. Society of Plastics Engineers, 363
19.10. Society of Plastics Industry, 363
19.11. Underwriters Laboratories, 364

APPENDIXES

A. Index of Test Equipment Manufacturers, 367
B. Abbreviations, 375
C. Glossary, 377
D. Trade Names, 389
E. Safety Standards Organizations, 391
F. Trade Publications, 395
G. Independent Testing Laboratories, 397
H. Specifications, 399
I. Charts and Tables, 429

Index **489**

References [illegible]
Suggested Readings [illegible]

18 [illegible]

18.1 [illegible] Standards [illegible]
18.2 American Society for Testing and Materials [illegible]
18.3 Food and Drug Administration [illegible]
18.4 National Bureau of Standards [illegible]
18.5 National Electrical Manufacturers Association [illegible]
18.6 National Fire Protection Association [illegible]
18.7 National Sanitation Foundation 362
18.8 Plastics Technical Evaluation Center 363
18.9 Society of Plastics Engineers [illegible]
18.10 Society of Plastics Industry [illegible]
18.11 Underwriters Laboratories [illegible]

APPENDICES

A Index of Testing Equipment Manufacturers [illegible]
B Trade Names [illegible]
C Glossary [illegible]
D Abbreviations [illegible]
E Safety Standards [illegible]
F [illegible]
G [illegible]
H [illegible]
I Conversion Tables [illegible]

Index

Handbook of Plastics Testing Technology

CHAPTER 1

Basic Concepts

1.1. BASIC CONCEPTS

Not too long ago, the concept of testing was merely an afterthought of the procurement process. But now, with the advent of science and technology, the concept of testing is an integral part of research and development, product design, and manufacturing. The question that is often asked is "why test?" The answer is simple. Times have changed. The manner in which we do things today is different. The emphasis is on automation, high production, and cost reduction. There is a growing demand for intricately shaped, high-tolerance parts. Consumer awareness, a subject totally ignored by the manufacturers once upon a time, is now a major area of concern. Along with the requirements, our priorities have also changed. While designing a machine or a product, the first order of priority in most cases is safety and health. Manufacturers and suppliers are now required to meet a variety of standards and specifications. Obviously, relying merely on past experience and quality of workmanship is simply not enough. The following are some of the major reasons for testing:

1. To prove design concepts.
2. To provide a basis for reliability.
3. Safety.
4. Protection against product liability suits.
5. Quality control.
6. To meet standards and specifications.
7. To verify the manufacturing process.
8. To evaluate competitors' products.
9. To establish a history for new materials.

In the last two decades, just about every manufacturer has turned to plastics to achieve cost reduction, automation, and high yield. The lack of history along with the explosive growth and diversity of polymeric materials has forced the plastics industry into placing extra emphasis on testing, and developing a wide variety of testing procedures. Through the painstaking efforts of various standards organizations, material suppliers, and mainly the numerous committees of the

American Society for Testing and Materials (ASTM) over 2000 different test methods have been developed.

The need for developing standard test methods specifically designed for plastic materials originated for two main reasons. Initially, the properties of plastic materials were determined by duplicating the test methods developed for testing metals and other similar materials. The Izod impact test, for example, was derived from the manual for testing metals. Because of the drastically different nature of plastic materials, the test methods often had to be modified. As a result, a large number of nonstandard tests were written by various parties. As many as eight to ten distinct and separate test methods were written to determine the same property. Such practice created total chaos among the developers of the raw materials, suppliers, design engineers, and ultimate end-users. It became increasingly difficult to keep up with various test methods as well as to comprehend the real meaning of reported test values. The standardization of test methods acceptable to everyone solved the problem of communication between developers, designers, and end-users, allowing them to speak a common language when comparing the test data and results.

In spite of the standardization of various test methods, we still face the problem of comprehension and interpretation of the test data by an average person in the plastics industry. This is due to the complex nature of the test procedures and the number of tests and testing organizations. The key in overcoming this problem is to develop a thorough understanding of what the various tests mean and the significance of the result to the application being considered (1). Unfortunately, the plastics industry has placed more emphasis on *how* and not enough on *why*, which obviously is more important from the standpoint of comprehension of the test results and understanding the true meaning of the values. The lack of understanding of the real meaning of heat deflection temperature, which is often interpreted as the temperature at which a plastic material will sustain static or dynamic load for a long period, is one such classic example of misinterpretation. In the chapters to follow, we concentrate on the significance, interpretation, and limitations of physical property data and test procedures. Finally, a word of caution: it is extremely important to understand that the majority of physical property tests are subject to rather large errors. As a general rule, the error of testing should be considered ± 5 percent. Some tests are more precise than others. Such testing errors come about from three major areas: one from the basic test itself, the second from the operators conducting the tests, and the third from the variations in the test specimens. While evaluating the test data and making decisions based on test data, one must consider the error factor to make certain that a valid difference in the test data exists (2).

1.2. SPECIFICATION AND STANDARDS

A specification is a detailed description of requirements, dimensions, materials, etc. A standard is something established for use as a rule or a basis of comparison in measuring or judging capacity, quantity, content, extent, value, quality, etc.

A specification for a plastic material involves defining particular requirements in terms of density, tensile strength, thermal conductivity, and other related properties. The specification also relates standard test methods to be used to determine such properties. Thus, standard methods of test and evaluation commonly provide the bases of measurement required in the specification for needed or desired properties (3).

As discussed earlier, the ultimate purpose of a standard is to develop a common language, so that there can be no confusion or communication problems among developers, designers, fabricators, end-users, and other concerned parties. The benefits of standards are innumerable. Standardization has provided the industry with benefits such as improved efficiency, mass production, superior quality goods through uniformity, and new challenges. Standardization has opened the doors for international trade, technical exchanges, and establishment of common markets. One can only imagine the confusion the industry would suffer without the specific definition of fundamental units of distance, mass, and time, and without the standards of weights and measures fixed by the government (4).

Standards originate from a variety of sources. The majority of standards originate from industry. The industry standards are generally established by voluntary organizations that make every effort to see that the standards are freely adopted and represent a general agreement. Some of the most common voluntary standards organizations are the American Society for Testing and Materials, the National Sanitation Foundation, the Underwriters Laboratories, the National Electrical Manufacturers Association, and the Society of Automotive Engineers. Quite often, the industry standards do not provide adequate information or are not suitable for certain applications, in which case, private companies are forced into developing their own company standards. The company standards are generally adapted from modified industry standards.

The Federal Government is yet another major source of standardization activities. The standards and specifications related to plastics are developed by the U.S. Department of Defense and the General Services Administration under the common heading of Military Standards and Federal Standards, respectively.

After the Second World War, there was a tremendous increase in international trade. The International Standards Organization (ISO) was established for the sole purpose of international standardization. ISO consists of the national standards bodies of over 50 countries from around the world. The standardization work of ISO is conducted by technical committees established by agreement of five or more countries. ISO's Technical Committee 61 on plastics is among the most productive of all ISO committees.

1.3. PURPOSE OF SPECIFICATIONS

There are many reasons for writing specifications, but the major reason is to help the purchasing department purchase equipment, materials, and products on an equal basis. The specifications, generally written by engineering departments, allow the purchasing agent to meet his requirements and ensure that the material received at different times is within the specified limits (5). The specification is intended to ensure batch-to-batch uniformity, as well as remove confusion between the purchaser and supplier, as we all known that more often than not what is provided by the supplier is not what is expected by the purchaser.

1.4. BASIC SPECIFICATION FORMAT

Many guidelines and directives have been written for writing specifications. The specifications for materials should include the following:

1. A descriptive title and designation.
2. A brief but all inclusive statement of scope.
3. Applicable documents.
4. A classification system.
5. Definition of related terms.
6. Materials and manufacturing requirements.
7. Physical (property) requirements.
8. A sampling procedure.
9. Specimen preparation and conditioning requirements.
10. Reference to, or descriptions of, methods of test acceptable for determining conformance.
11. Inspection requirements.
12. Instructions for retest and rejection.
13. Packaging and marking requirements.

1.4.1. Classification System

Since plastic materials are seldom supplied without the addition of certain additives and fillers, a classification system must be used to avoid confusion. For

Table 1-1. Detailed Requirement for Acetal Resin Injection Molding and Extrusion Materials[a]

	Type I	Type II
Specific gravity, 23/23°C (73.4/73.4°F):		
min	1.39	1.39
max	1.44	1.44
Melting point, min, °C	158	158
Flow rate, g/10 min:		
load 1050 g:		
min	3.5	—
max	—	2.0
or		
load 2160 g:		
min	7.5	—
max	—	4.3

[a] Reprinted by permission of ASTM.

Table 1-2. Detailed Requirements for Molded Specimen Properties[a]

	Type I			Type II
	Class A	Class B	Class C	Class A
Tensile yield strength, 23°C (73.4°F), min				
MPa	55.2	55.2	55.2	55.2
psi	8000	8000	8000	8000
Elongation, 23°C (73.4°F), min, percent	10	10	10	30
Modulus (tensile) of elasticity, 23°C (73.4°F), min				
MPa	2415	2415	2415	2415
psi	350 000	350 000	350 000	350 000
Deflection temperature under load, 1.82 MPa (264 psi), °C, min	100	100	100	100
Weather resistance, retention of original elongation, min, percent	—	65	50	—

[a] Reprinted by permission of ASTM.

example, the specification for acetal resin covers the two types of acetal resins classified according to melt flow. Type I is the material for injection molding and Type II is the material for extrusion. The two types are subdivided into classes according to applications. Class A is natural color only, containing an antioxidant for general purpose use. Class B is black, containing carbon black pigment for weather-resistant uses. Class C is natural, containing a light stabilizer for weather-resistant use. Table 1-1 lists the detailed requirements for acetal resin.

1.4.2. Requirements

Physical requirements should include specific quantitative information according to the classification. Table 1-2 lists the detailed requirements for molded specimen properties.

1.4.3. Sampling and Conditioning

In order to reproduce the test results time after time, sampling and conditioning procedures must be religiously followed. The procedures for sampling and conditioning must be clearly specified.

1.4.4. Test Methods

The specifications, without adequate information regarding test methods, are useless. The standard and generally accepted test methods must be specified. If no standard test methods exist, a detailed description and requirements of the test should be an integral part of the specification.

REFERENCES

1. Lamond, L., "Right Perspective on Test Data Matches Material to Application Criteria," *Plastics Design and Processing* (Mar 1976), pp. 6–10.
2. "A Guide to Standard Physical Test for Plastics," *Dupont Tech. Rept. No. TR 91.*
3. Schmitz, J. V. (Ed.), *Testing of Polymers,* Interscience, New York, 1965, p. 3.
4. *Ibid*, p. 4.
5. *Ibid*, p. 22.

CHAPTER 2

Mechanical Properties

2.1. INTRODUCTION

The mechanical properties, among all the properties of plastic materials, are often the most important properties because virtually all service conditions and the majority of end-use applications involve some degree of mechanical loading. Nevertheless, these properties are the least understood by most design engineers. The material selection for a variety of applications is quite often based on mechanical properties such as tensile strength, modulus, elongation, and impact strength. These values are normally derived from the technical literature provided by material suppliers. More often than not, too much emphasis is placed on comparing the published values of different types and grades of plastics and not enough on determining the true meaning of the mechanical properties and their relation to end-use requirements. In practical applications, plastics are seldom, if ever, subjected to a single, steady deformation without the presence of other adverse factors such as environment and temperature. Since the published values of the mechanical properties of plastics are generated from tests conducted in a laboratory under standard test conditions, the danger of selecting and specifying a material from these values is obvious. A thorough understanding of mechanical properties, tests employed to determine such properties, and the effect of adverse conditions on mechanical properties over a long period is extremely important.

The basic understanding of stress–strain behavior of plastic materials is of utmost importance to design engineers. One such typical stress–strain (load–deformation) diagram is illustrated in Figure 2-1. For a better understanding of the stress–strain curve, it is necessary to define a few basic terms that are associated with the stress–strain diagram.

STRESS: The force applied to produce deformation in a unit area of a test specimen. Stress is a ratio of applied load to the original cross-sectional area expressed in $lbs/in.^2$

STRAIN: The ratio of the elongation to the gauge length of the test specimen, or simply stated, change in length per unit of the original length ($l/\Delta l$). It is expressed as a dimensionless ratio.

ELONGATION: The increase in the length of a test specimen produced by a tensile load.

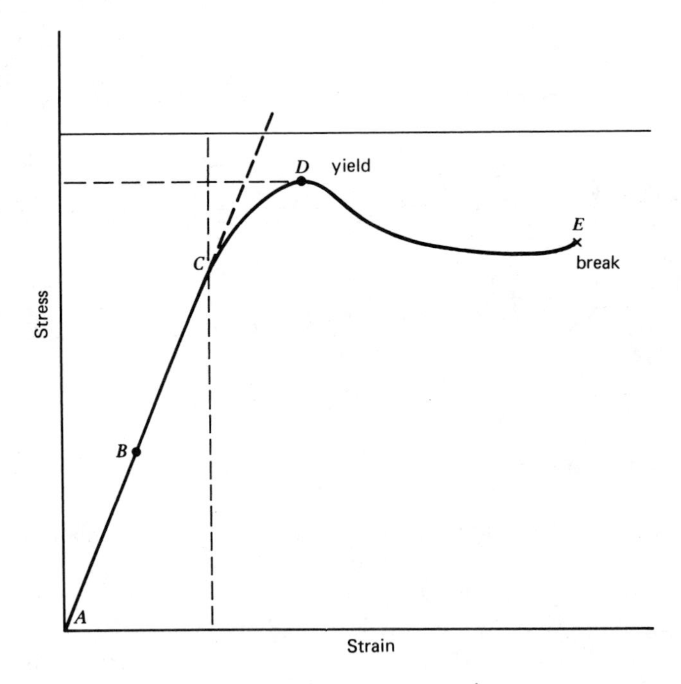

Figure 2-1. A typical stress–strain curve.

YIELD POINT: The first point on the stress–strain curve at which an increase in strain occurs without the increase in stress.

YIELD STRENGTH: The stress at which a material exhibits a specified limiting deviation from the proportionality of stress to strain. Unless otherwise specified, this stress will be at the yield point.

PROPORTIONAL LIMIT: The greatest stress at which a material is capable of sustaining the applied load without any deviation from proportionality of stress to strain. (Hooke's Law). This is expressed in lb/in.^{2}

MODULUS OF ELASTICITY: The ratio of stress to corresponding strain below the proportional limit of a material. It is expressed in F/A, usually lb/in.^{2} This is also known as Young's modulus. A modulus is a measure of material's stiffness.

ULTIMATE STRENGTH: The maximum unit stress a material will withstand when subjected to an applied load in compression, tension, or shear. This is expressed in lb/in.^{2}

SECANT MODULUS: The ratio of the total stress to corresponding strain at any specific point on the stress–strain curve. It is also expressed in F/A or lb/in.^{2}

The stress–strain diagram illustrated in Figure 2-1 is typical of that obtained in tension for a constant rate of loading. However, the curves obtained from other loading conditions, such as compression or shear are quite similar in appearance.

The initial portion of the stress–strain curve between points A and C is linear and it follows Hooke's Law which states that for an elastic material the stress is proportional to the strain. The point C at which the actual curve deviates from the straight line is called the proportional limit meaning that only up to this point is stress proportional to strain. The behavior of plastic material below the proportional limit is elastic in nature and therefore the deformations are recoverable. The deformations up to point B in Figure 2-1 are relatively small and have been associated with the bending and stretching of the interatomic bonds between atoms of plastic molecules as shown in Figure 2-2(*a*). This type of deformation is instantaneous and recoverable. There is no permanent displacement of the molecules relative to each other. The deformation that occurs beyond point C in Figure 2-1 is similar to a straightening out of a coiled portion of the molecular chains [Figure 2-2(*b*)]. There is no intermolecular slippage and the deformations may be recoverable ultimately but not instantaneously. The extensions that occur beyond the yield point or the elastic limit of the material are not recoverable [Figure 2-2(*c*)]. These deformations occur due to the actual displacement of the molecules with respect to each other. The displaced molecules cannot slip back to their original positions, and therefore, a permanent deformation or set occurs. These three types of deformations as shown in Figure 2-2 do not occur separately but are superimposed on each other. The bonding and the stretching of the interatomic bonds are almost instantaneous. However, the molecular uncoiling is relatively slow. Molecular slippage effects are the slowest of all three deformations (1).

These deformations can be further explained by using a mechanical model that duplicates the behavior of plastics under various external conditions. One such spring and dashpot mechanical model known as the Maxwell Model is illustrated in Figure 2-3. The spring is perfectly elastic or Hookean and accounts for the normal elastic behavior. The spring extensions are analogous to the deformations that occur due to the bending and stretching of interatomic bonds. If a nonlinear

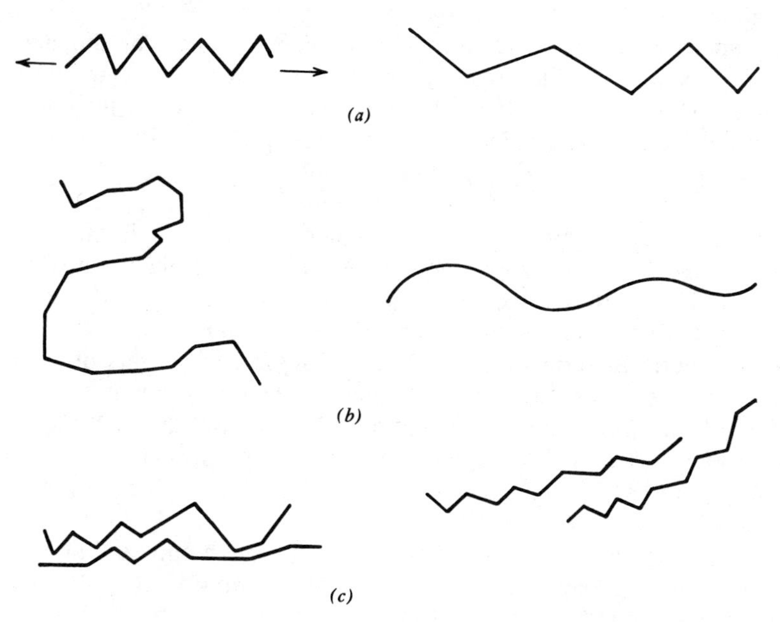

Figure 2-2. Extension types (*a*) bond bending, (*b*) uncoiling, (*c*) slippage.

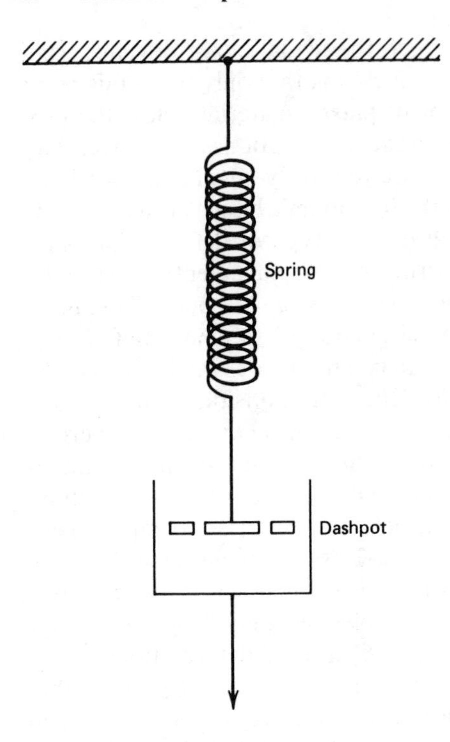

Figure 2-3. Maxwell model.

spring is substituted for a linear one, the deformations are similar to the ones occurring due to uncoiling of portions of molecular chains. The dashpot is filled with a viscous fluid that must leak through holes in the plunger disc before the disc can move. Extensions in the dashpot are not recoverable and correspond to the permanent set. They represent the result of intermolecular slippage.

There are many other interesting correlations that can be made with the mechanical model such as the effect of temperature on the mechanical properties of polymers. For example, at higher temperatures, the viscosity of the fluid in the dashpot decreases and the plunger movement is relatively smoother, resulting in greater extensions. Conversely, at lower temperatures, the liquid becomes more viscous and failures occur before appreciable extensions, similar to a brittle fracture. Other correlations involve models with different leakage rates and rate of loading (2). There are many other variations of mechanical models that are discussed in detail by Rodriguez (3), Baer (4), and Williams (5).

The polymeric materials can be broadly classified in terms of their relative softness, brittleness, hardness, and toughness. The tensile stress–strain diagrams serve as a basis for such a classification (6). The area under the stress–strain curve is considered as the toughness of the polymeric material. Figure 2-4(*a*) illustrates typical tensile stress–strain curves for several types of polymeric materials.

A soft and weak material is characterized by low modulus, low yield stress, and a moderate elongation at break point. Polytetrafluoroethylene (PTFE) is a good example of one such type of plastic material.

A soft but tough material shows low modulus, low yield stress, but very high

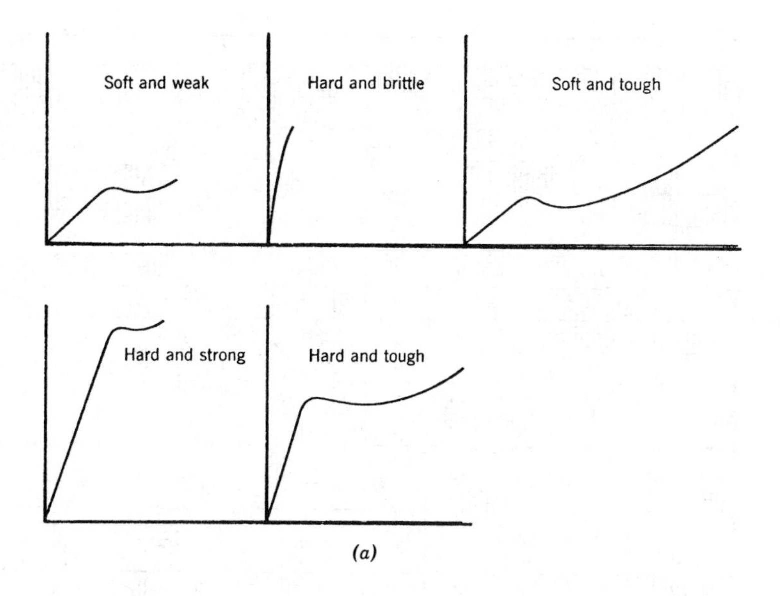

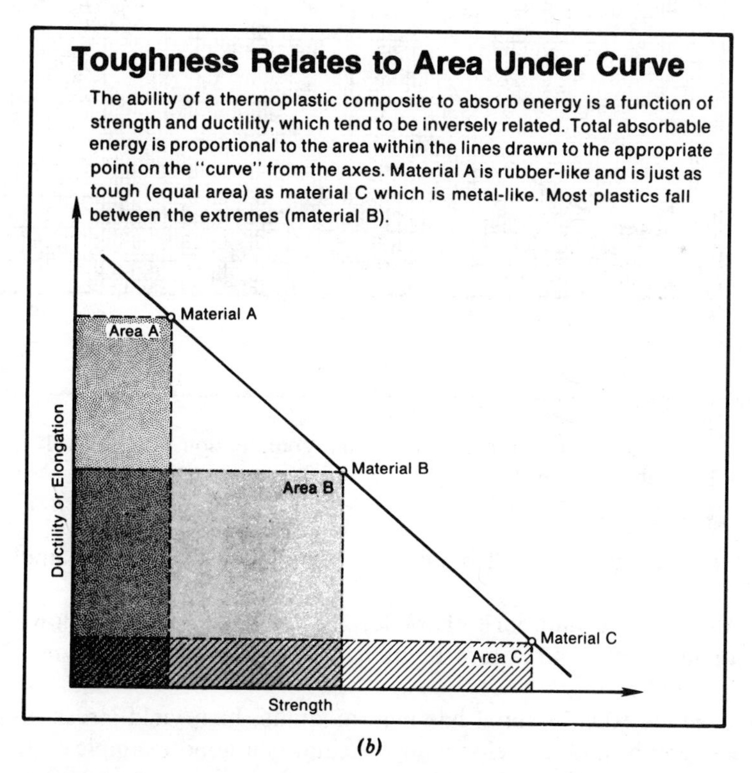

Toughness Relates to Area Under Curve
The ability of a thermoplastic composite to absorb energy is a function of strength and ductility, which tend to be inversely related. Total absorbable energy is proportional to the area within the lines drawn to the appropriate point on the "curve" from the axes. Material A is rubber-like and is just as tough (equal area) as material C which is metal-like. Most plastics fall between the extremes (material B).
Material A
Area A
Material B
Area B
Material C
Area C
Ductility or Elongation
Strength
(b)

Figure 2-4. (*a*) Types of stress–strain curves. (Reprinted by permission of Wiley–Interscience.) (*b*) Relation between ductility and strength. (Courtesy LNP Corporation.)

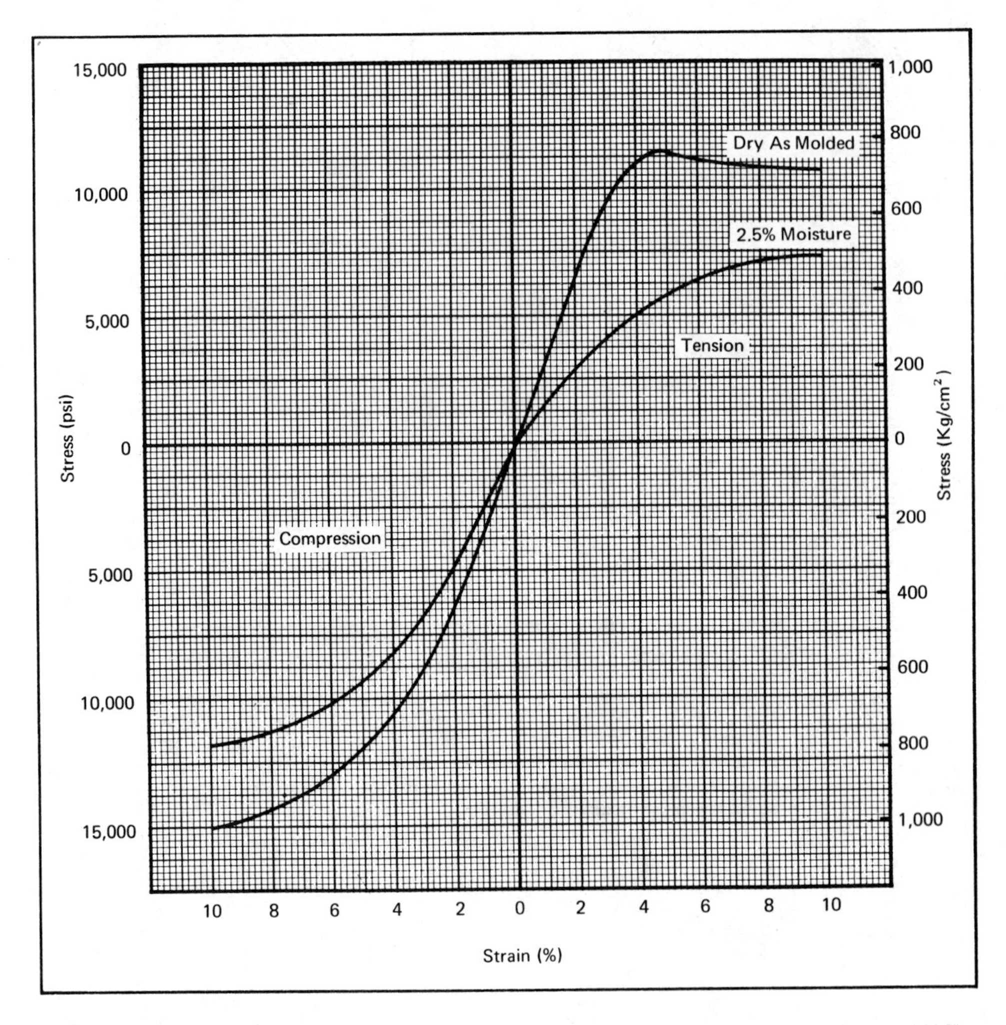

Figure 2-5. Stress–strain curve in tension and compression of Zytel 101, 73°F (23°C). (Courtesy DuPont Company.)

elongation and high stress at break. Polyethylene is a classic example of these types of plastics.

A hard and brittle material is characterized by high modulus and low elongation. It may or may not yield before breaking. One such type of polymer is general purpose phenolic.

A hard and strong material has high modulus, high yield stress, usually high ultimate strength, and low elongation. Acetal is a good example of this class of materials.

A hard and tough material is characterized by high modulus, high yield stress, high elongation at break, and high ultimate strength. Polycarbonate is considered hard and tough material. Figure 2-4(b) illustrates the relation between ductility and strength.

Table 2-1. Characteristic Features of Stress–Strain Curves as Related to Polymer Properties[a]

Description of polymer	Modulus	Yield stress	Ultimate strength	Elongation at break
Soft, weak	Low	Low	Low	Moderate
Soft, tough	Low	Low	Yield Stress	High
Hard, brittle	High	None	Moderate	Low
Hard, strong	High	High	High	Moderate
Hard, tough	High	High	High	High

[a] From F. Billmeyer, *Textbook of Polymer Science*, John Wiley & Sons, Inc. Reprinted by permission of John Wiley & Sons.

Table 2-1 lists the characteristic features of stress–strain curves as they relate to the polymer properties (7). In some applications it is important for a designer to know the stress–strain behavior of a particular plastic material in both tension and compression. At relatively lower strains, the tensile and compressive stress–strain curves are almost identical. Therefore, at low strain, compressive modulus is equal to tensile modulus. However, at a higher strain level, the compressive stress is significantly higher than the corresponding tensile stress. This effect is illustrated in Figure 2-5.

Stress–strain tests are considered short-term tests which means that the mechanical loading is applied within a relatively short period of time. This limits the

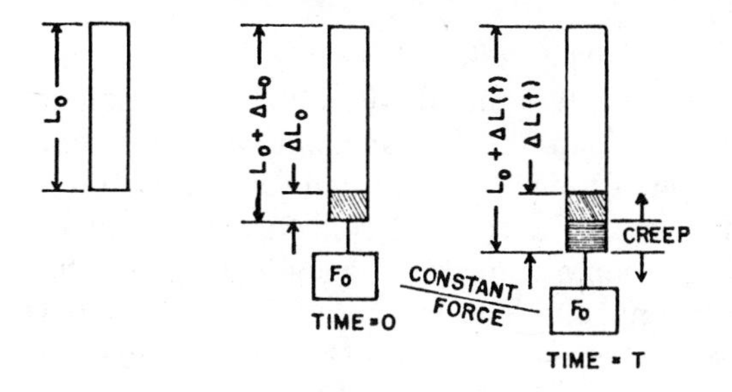

RELAXATION

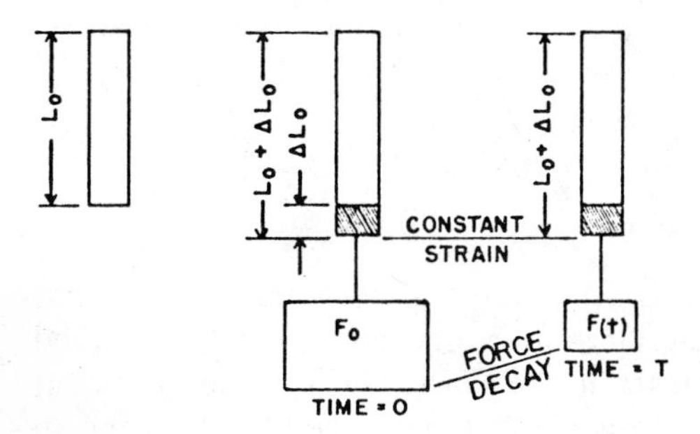

Figure 2-6. Diagram illustrating creep and stress relaxation. (Reprinted by permission of Van Nostrand–Reinhold Company.)

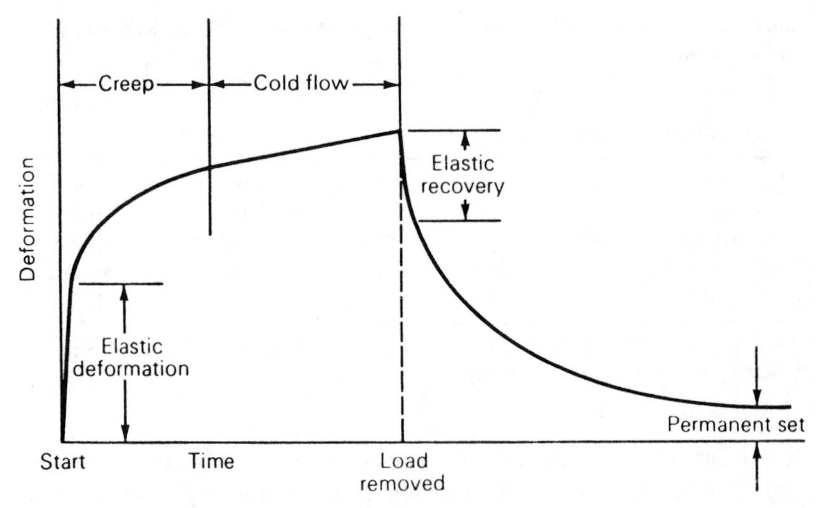

Figure 2-7. Diagram illustrating creep and cold flow. (Reprinted by permission of McGraw–Hill Book Company.)

usefulness of stress–strain tests in the actual design of a plastic part. Stress–strain tests fail to take into account the dependence of rigidity and strength of plastics on time. This serious limitation can be overcome with the use of creep and stress relaxation data while designing a part.

Stress relaxation is the application of a fixed deformation to a specimen and the measurement of the load required to maintain that deformation as a function of time. Creep is the application of a fixed load to a specimen and measurement of the resulting deformation as a function of time (8). Creep and stress relaxation effects are shown schematically in Figure 2-6. Creep and cold flow behavior of plastics are illustrated in Figure 2-7. As soon as the stress is applied, the elastic deformation takes place and continues until the load is removed. A substantial portion of the elastic recovery is not immediate and the material continues to return to its original size or length. The material is considered to have achieved permanent set if the recovery is not complete. The magnitude of permanent set is dependent on the stress applied, the length of time, and the temperature of the material (9). This type of behavior of a polymer in creep can also be represented by a mechanical model that combines a Maxwell and a Voigt element in series (10,11).

2.2. TENSILE TESTS (ASTM D 638)

Tensile elongation and tensile modulus measurements are among the most important indications of strength in a material and are the most widely specified properties of plastic materials. Tensile test, in a broad sense, is a measurement of the ability of a material to withstand forces that tend to pull it apart and to determine to what extent the material stretches before breaking. Tensile modulus, an indication of the relative stiffness of a material, can be determined from a stress–strain diagram. Different types of plastic materials are often compared on the basis of tensile strength, elongation, and tensile modulus data. Many plastics are very sensitive to the rate of straining and environmental conditions. Therefore, the data obtained by this method cannot be considered valid for applications involving load–time scales or environments widely different from this method. The tensile property data are more useful in preferential selection of a particular type of plastic from a large group of plastic materials and such data are of limited use in actual design of the product. This is because the test does not take into account the time-dependent behavior of plastic materials.

2.2.1. Apparatus

The tensile testing machine of a constant-rate-of-crosshead movement is used. It has a fixed or essentially stationary member, carrying one grip and a movable member carrying a second grip. Self-aligning grips employed for holding the test specimen between the fixed member and the movable member prevent alignment problems. A controlled velocity drive mechanism is used. Some of the commercially available machines use a closed loop servo-controlled drive mechanism to provide a high degree of speed accuracy. A load-indicating mechanism capable of indicating total tensile load with an accuracy of ±1 percent of the indicated value or better is used. Lately, the inclination is towards using digital-type load indicators which are easier to read than the analog-type indicators. An extension indicator, commonly known as the extensometer, is used to determine the distance between two designated points located within the gauge length of the test specimen as the specimen is stretched. Figure 2-8 shows a commercially available tensile testing machine. The advent of new microprocessor technology has virtually eliminated time-consuming manual calculations. Stress, elongation, modulus, energy, and statistical calculations are performed automatically and presented on a visual display or hard copy printout at the end of the test.

2.2.2. Test Specimens and Conditioning

Test specimens for tensile tests are prepared many different ways. Most often, they are either injection molded or compression molded. The specimens may also be prepared by machining operations from materials in sheet, plate, slab, or similar form. Test specimen dimensions vary considerably depending upon the requirements and are described in detail in the ASTM book of standards. Figure 2-9 shows ASTM D 638 Type I tensile test specimen most commonly used for testing rigid and semirigid plastics.

The specimens are conditioned using standard conditioning procedures. Since the tensile properties of some plastics change rapidly with small changes in temperature, it is recommended that tests be conducted in the standard laboratory atmosphere of 23 ± 2°C and 50 ± 5 percent relative humidity. The Procedure A of ASTM methods D 618 (as explained in Chapter 11) is recommended for this test.

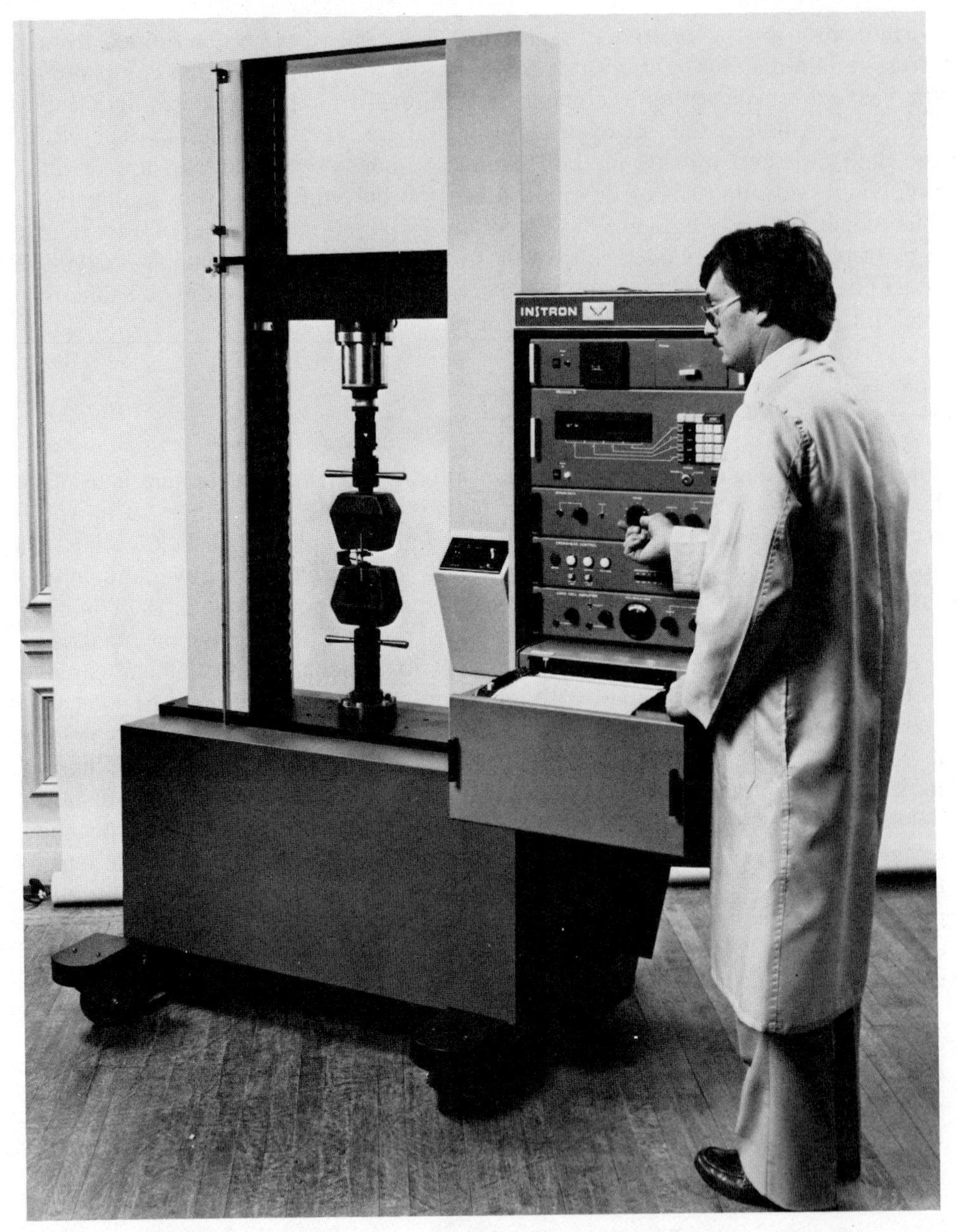

Figure 2-8. Tensile testing machine. (Courtesy Instron Corporation.)

2.2.3. Test Procedures

A. *Tensile Strength*

The speed of testing is the relative rate of motion of the grips or test fixtures during the test. There are basically four different testing speeds specified in the ASTM D 638 Standard. The most frequently employed speed of testing is 0.2 in./min. The test specimen is positioned vertically in the grips of the testing ma-

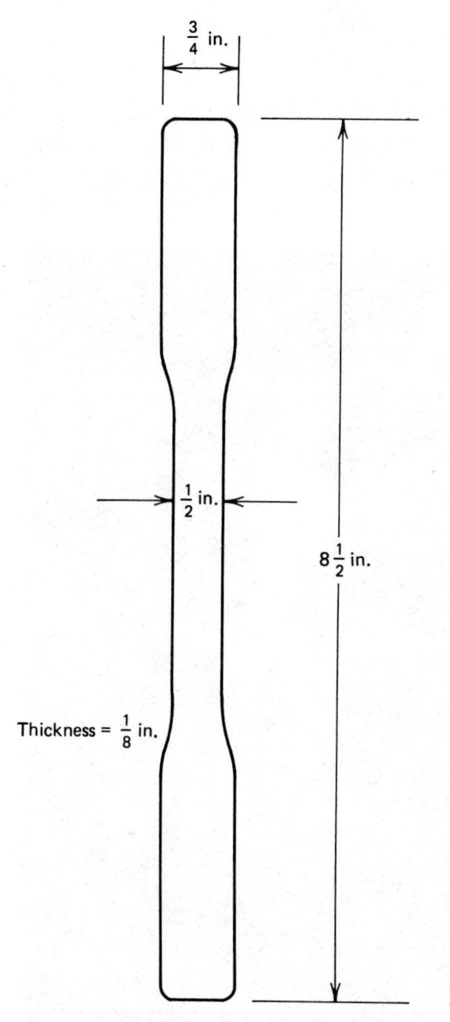

Figure 2-9. Tensile test specimen (Type I).

chine. The grips are tightened evenly and firmly to prevent any slippage. The speed of testing is set at the proper rate and the machine is started. As the specimen elongates, the resistance of the specimen increases and is detected by a load cell. This load value (Force) is recorded by the instrument. Some machines also have a provision to record the maximum load obtained by the specimen which can be recalled after the completion of the test. The elongation of the specimen is continued until a rupture of the specimen is observed. Load value at break is also recorded. The tensile strength at yield and tensile strength at break (ultimate tensile strength) is calculated.

$$\text{Tensile strength} = \frac{\text{Force (load) (lb)}}{\text{Cross section area (sq. in.)}}$$

$$\text{Tensile strength at yield (psi)} = \frac{\text{Maximum load recorded (lb)}}{\text{Cross section area (sq. in.)}}$$

$$\text{Tensile strength at break (psi)} = \frac{\text{Load recorded at break (lb)}}{\text{Cross section area (sq. in.)}}$$

B. Tensile Modulus and Elongation

Tensile modulus and elongation values are derived from a stress–strain curve. An extensometer is attached to the test specimen as shown in Figure 2-10. The extensometer is a strain gauge type of device that magnifies the actual stretch of

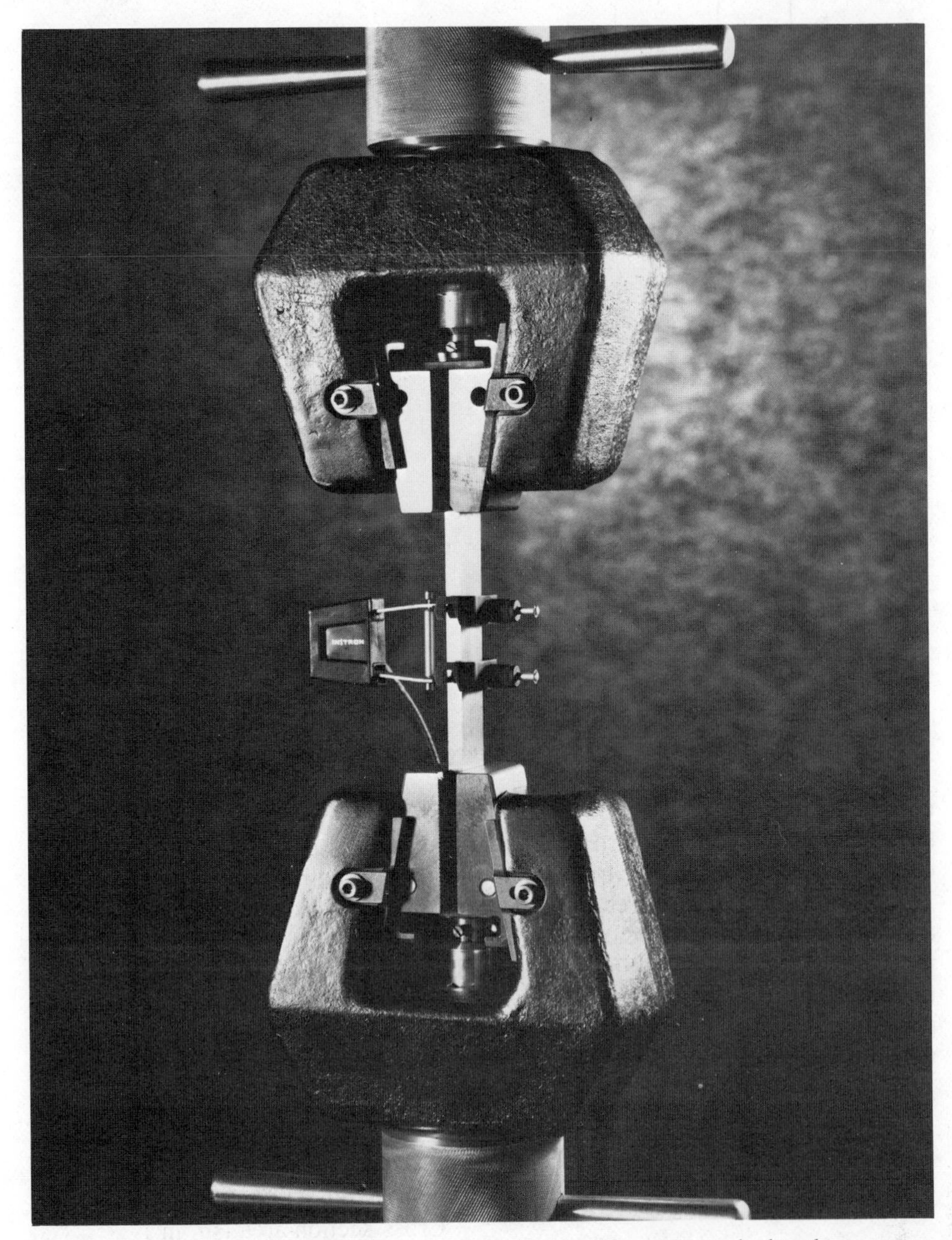

Figure 2-10. Diagram illustrating an extensometer (strain gauge) attached to the test specimen. (Courtesy Instron Corporation.)

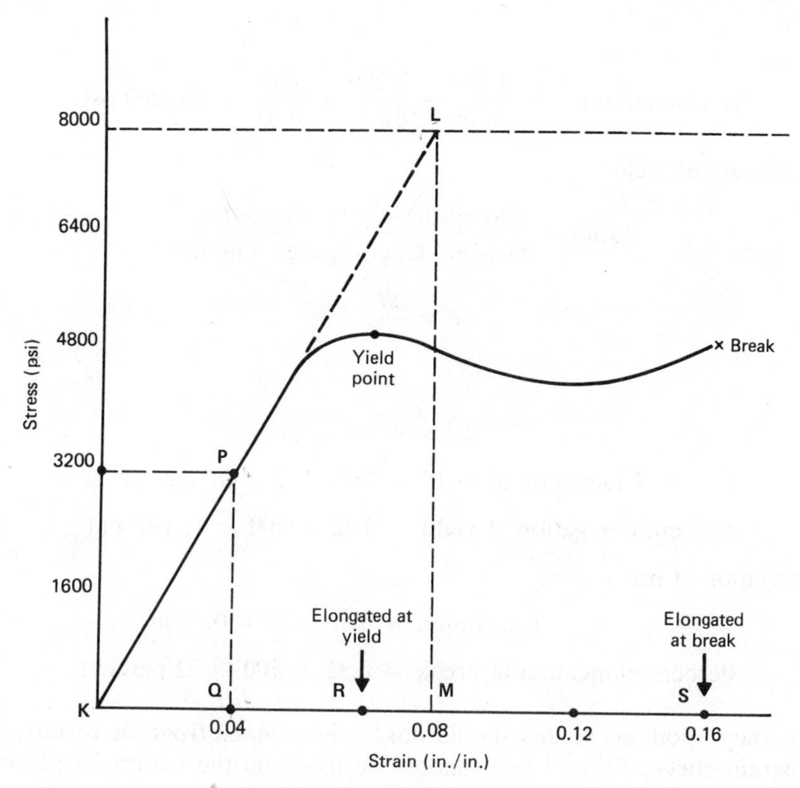

Figure 2-11. Diagram illustrating stress–strain curve from which modulus and elongation values are derived.

the specimen considerably. The simultaneous stress–strain curve is plotted on graph paper. The following stepwise procedure is generally used to carry out the calculations. (Refer to Figure 2-11).

1. Mark off the units of stress in lb/in.^{2} on the x axis of the chart. This is done by dividing the force by cross-section area of the specimen.
2. Mark off the units of strain in in./in. on the y axis. These values are obtained by dividing the chart value by the magnification selected.
3. Carefully draw a tangent KL to the initial straight line portion of the stress–strain curve.
4. Select any two convenient points on the tangent. (Points P and L are selected in this case.)
5. Draw a straight line PQ and LM connecting points P and L with y axis of the chart.
6. Stress value at L = 8000 psi, corresponding strain value at M = 0.08 in./in. Stress value at P = 3200 psi, corresponding strain value at Q = 0.04.

$$\text{Tensile modulus} = \frac{\text{Difference in stress}}{\text{Difference in corresponding strain}}$$

or

$$\text{Tensile modulus} = \frac{8{,}000 - 3{,}200}{0.08 - 0.04} = \frac{4800}{0.04} = 120{,}000 \text{ psi}$$

7. Elongation at yield

$$\text{Strain} = \frac{\text{Change in length (elongation)}}{\text{Original length (gauge length)}}$$

$$\epsilon = \frac{\Delta l}{l}$$

or

$$\Delta l = \epsilon \times l$$

$$\text{Elongation at yield} = 0.06 \times 2 = 0.12 \text{ in.}$$

$$\text{Percent elongation at yield} = 0.12 \times 100 = 12 \text{ percent}$$

8. Elongation at break

$$\text{Elongation} = 0.16 \times 2 = 0.32 \text{ in.}$$

$$\text{Percent elongation at break} = 0.32 \times 100 = 32 \text{ percent}$$

For accuracy, modulus values should not be determined from the results of one stress–strain curve. Several tests should be made on the material and average tensile modulus should be calculated.

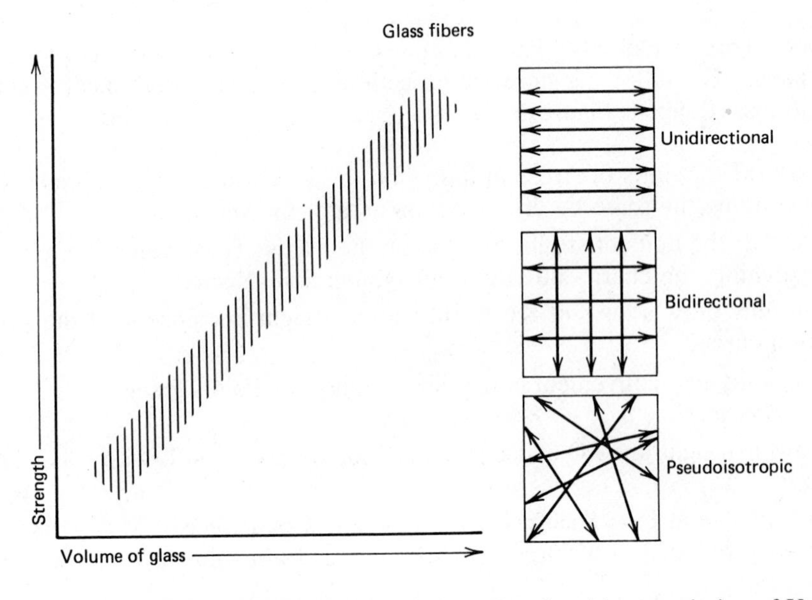

Figure 2-12. The effect of fiberglass orientation. (Reprinted by permission of Van Nostrand Reinhold Company.)

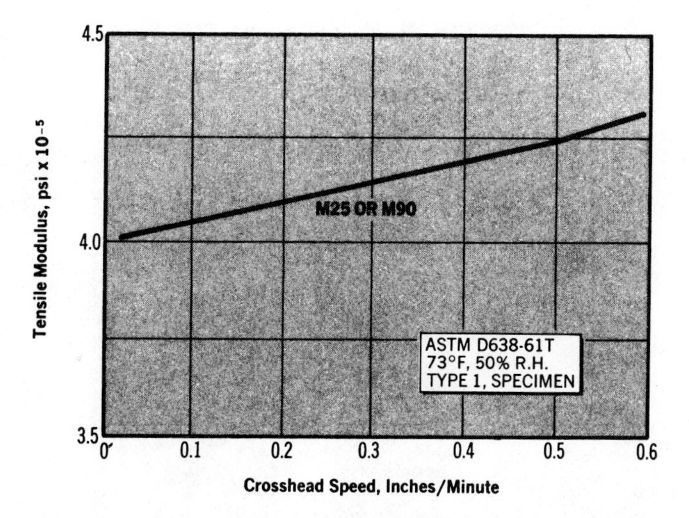

Figure 2-13. The effect of strain rate on modulus. (Courtesy Celanese Plastics Company.)

2.2.4. Factors Affecting the Test Results

A. *Specimen Preparation and Specimen Size*

Molecular orientation has a significant effect on tensile strength values. A load applied parallel to the direction of molecular orientation may yield higher values than the load applied perpendicular to the orientation. The opposite is true for elongation. The process employed to prepare the specimens also has a significant effect. For example, injection molded specimens generally yield higher tensile

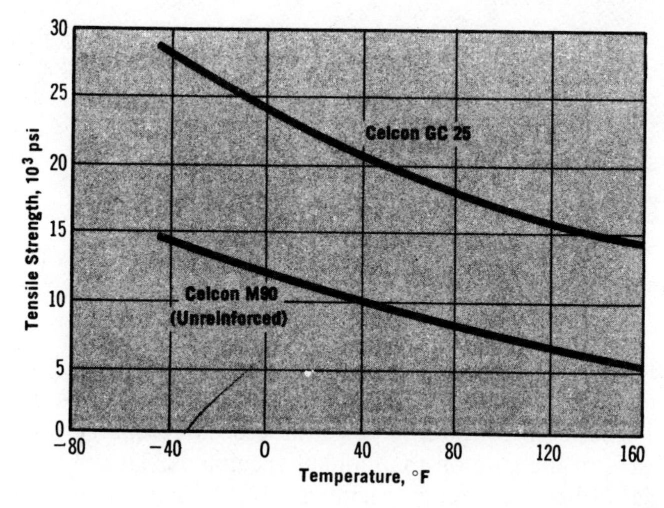

Figure 2-14. The effect of temperature on tensile strength. (Courtesy Celanese Plastics Company.)

Figure 2-15. Environmental test chamber to study tensile properties at different temperatures. (Courtesy Instron Corporation.)

strength values than compression molded specimens. Machining usually lowers the tensile and elongation values because of small irregularities introduced into the machined specimen. Another important factor affecting the test results is the location and size of the gate on the molded specimens. This is especially true in the case of glass-fiber-reinforced specimens (12,13). A large gate located on top of the tensile bar will orient the fibers parallel to the applied load yielding higher tensile strength. A gate located on one side of the tensile bar will disperse the fiber in a random fashion. This effect is shown in Figure 2-12. Tensile properties should only be compared for equivalent sample sizes and geometry.

B. Rate of Straining

As the strain rate is increased, the tensile strength and modulus increases. However, the elongation is inversely proportional to the strain rate. The effect of the crosshead speed on the modulus is shown in Figure 2-13.

C. Temperature

As discussed earlier in this chapter, the tensile properties of some plastics change rapidly with small changes in temperature. Tensile strength and modulus are decreased while elongation is increased as the temperature increases. Figure 2-14 illustrates the effect of temperature on the tensile strength of the materials. Figure 2-15 is a commercially available environmental test chamber to study the effect of temperature on tensile properties.

2.3. FLEXURAL PROPERTIES (ASTM D 790)

The stress–strain behavior of polymers in flexure is of interest to a designer as well as a polymer manufacturer. Flexural strength is the ability of the material to withstand bending forces applied perpendicular to its longitudinal axis. The stresses induced due to the flexural load are a combination of compressive and tensile stresses. This effect is illustrated in Figure 2-16. Flexural properties are reported and calculated in terms of the maximum stress and strain that occur at the outside surface of the test bar. Many polymers do not break under flexure even after a large deflection that makes determination of the ultimate flexural strength impractical for many polymers. In such cases, the common practice is to report flexural yield strength when the maximum strain in the outer fiber of the specimen has reached five percent. For polymeric materials that break easily under flexural load, the specimen is deflected until a rupture occurs in the outer fibers.

Forces involved in bending a simple beam.

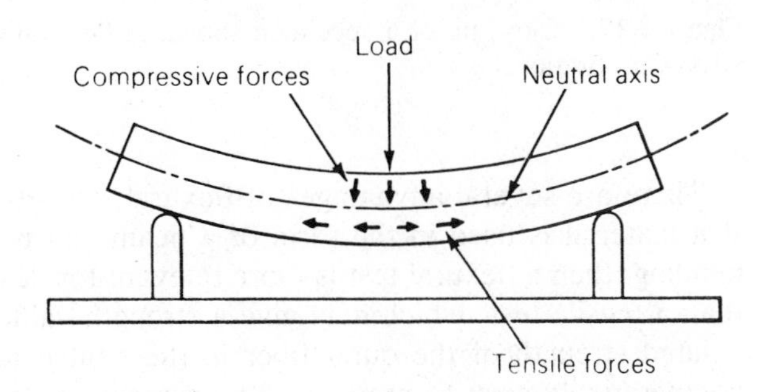

Figure 2-16. Forces involved in bending a simple beam. (Reprinted by permission of McGraw–Hill Book Company.)

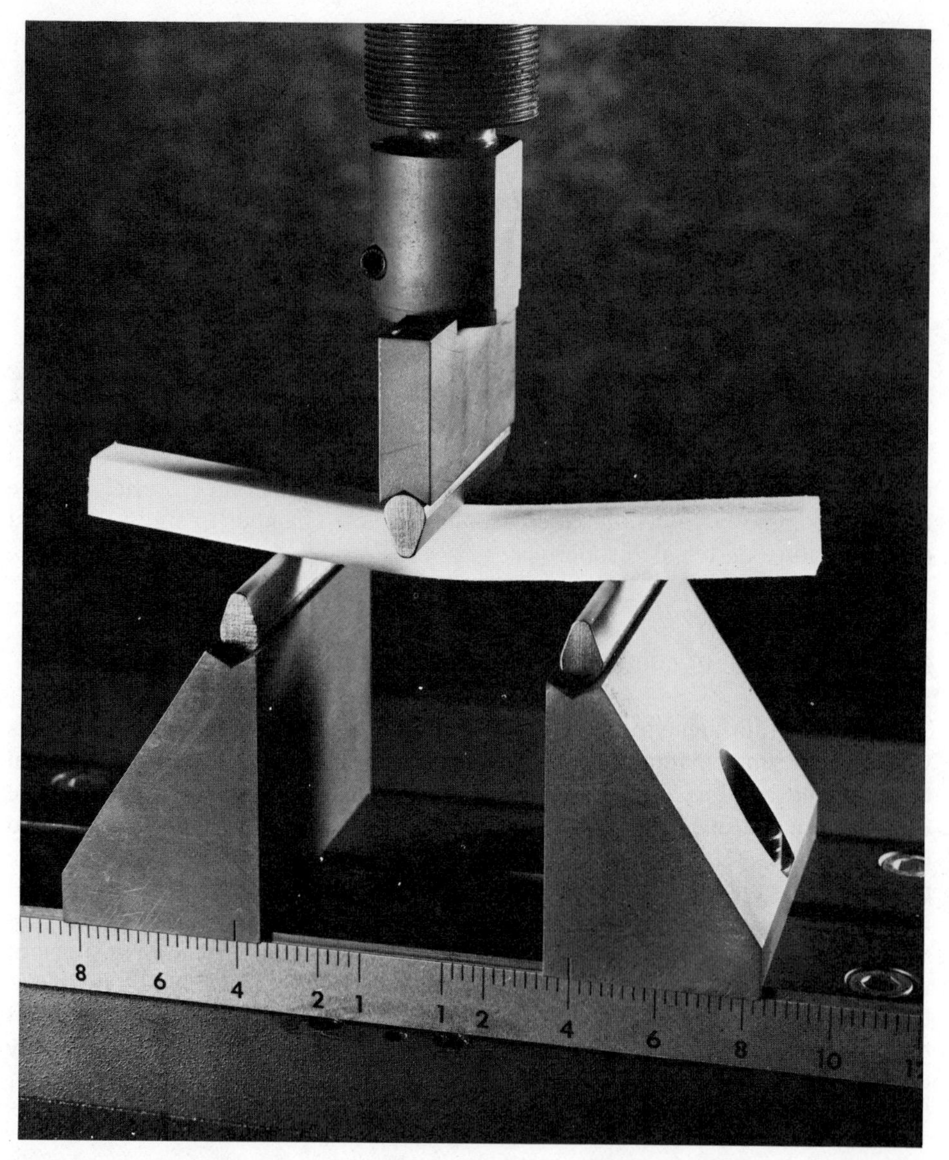

Figure 2-17. Close-up of a specimen shown in flexural testing apparatus. (Courtesy Instron Corporation.)

There are several advantages of flexural strength tests over tensile tests (14). If a material is used in the form of a beam and if the service failure occurs in bending, then a flexural test is more relevant for design or specification purposes than a tensile test, which may give a strength value very different from the calculated strength of the outer fiber in the bent beam. The flexural specimen is comparatively easy to prepare without residual strain. The specimen alignment is also more difficult in tensile tests. Also, the tight clamping of the test specimens creates stress concentration points. One other advantage of the flexural test is that at small strains, the actual deformations are sufficiently large to be measured accurately.

There are two basic methods that cover the determination of flexural properties of plastics. Method 1 is a three-point loading system utilizing center loading on a simple supported beam. A bar of rectangular cross section rests on two supports and is loaded by means of a loading nose midway between the supports. The maximum axial fiber stresses occur on a line under the loading nose. A close-up of a specimen in the testing apparatus is shown in Figure 2-17. This method is especially useful in determining flexural properties for quality control and specification purposes.

Method 2 is a four-point loading system utilizing two load points equally spaced from their adjacent support points, with a distance between load points of one-third of the support span. In this method, the test bar rests on two supports and is loaded at two points (by means of two loading noses), each on equal distance from the adjacent support point. This arrangement is shown schematically in Figure 2-18. Method 2 is very useful in testing materials that do not fail at the point of maximum stress under a three-point loading system. The maximum axial fiber stress occurs over the area between the loading noses.

Either method can be used with the two procedures. Procedure A is designed principally for materials that break at comparatively small deflections. Procedure B is designed particularly for those materials that undergo large deflections during testing. The basic difference between the two procedures is the strain rate, Procedure A being 0.01 in./in./min, and Procedure B being 0.10 in./in./min.

2.3.1. Apparatus

Quite often, the machine used for tensile testing is also used for flexural testing. The upper or lower portion of the movable crosshead can be used for flexural testing. The dual-purpose load cell that indicates the load applied in tension as well as compression facilitates testing of the specimen in either tension or compression. One such universal testing machine is shown in Figure 2-19. The machine used for this purpose should operate at a constant rate of crosshead motion over the entire range and the error in the load measuring system should not exceed ± 1 percent of the maximum load expected to be measured.

The loading nose and support must have cylindrical surfaces. The radius of the nose and the nose support should be at least $\frac{1}{8}$ in. to avoid excessive indentation or failure due to stress concentration directly under the loading nose. A strain-gauge type of mechanism called a deflectometer or compressometer is used to measure deflection in the specimen.

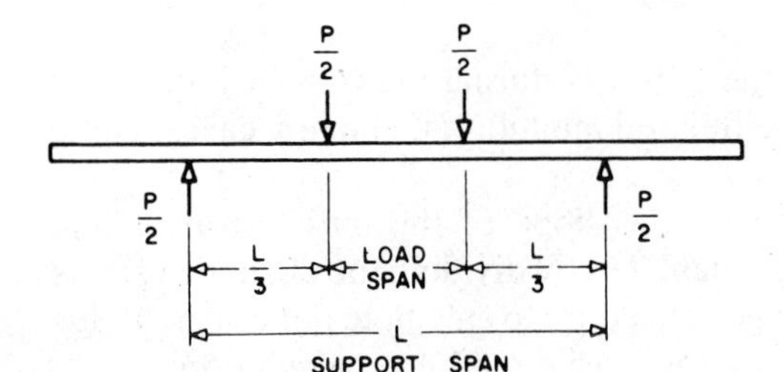

Figure 2-18. Schematic of specimen arrangement for flexural testing (Method II). (Reprinted by permission of ASTM.)

2.3.2. Test Specimens and Conditioning

The specimens used for flexural testing are bars of rectangular cross section and are cut from sheets, plates, or molded shapes. The common practice is to mold the specimens to the desired finished dimensions. The specimens are conditioned in accordance with Procedure A of ASTM methods D 618 as explained in Chapter 11 of this book. The specimens of size $\frac{1}{8} \times \frac{1}{2} \times 4$-in. are the most commonly used.

2.3.3. Test Procedures and Calculations

The test is initiated by applying the load to the specimen at the specified crosshead rate. The deflection is measured either by a gauge under the specimen in contact with it in the center of the support span or by measurement of the motion of the loading nose relative to the supports. A load–deflection curve is plotted if the determination of flexural modulus value is desired.

The maximum fiber stress is related to the load and sample dimensions and is calculated using the following equation:

$$\text{Method 1} \quad S = \frac{3PL}{2bd^2}$$

where S = stress (psi); P = load (lbs); L = length of span (in.); b = width of specimen (in.); d = thickness of specimen (in.).

Flexural strength is equal to the maximum stress in the outer fibers at the moment of break. This value can be calculated by using the above stress equation by letting load value P equal the load at the moment of break.

For materials that do not break at outer fiber strains up to five percent, the flexural yield strength is calculated using the same equation. The load value P in this case is the maximum load at which there is no longer an increase in load with an increase in deflection.

The maximum strain in the outer fibers, which also occurs at midspan, is calculated using the following equation (Method 1).

$$r = 6Dd/L^2$$

where r = strain (in./in.); D = deflection (in); L = length of span (in.); d = thickness of specimen (in.).

The equations for calculating maximum fiber stress and maximum fiber strain are slightly different in Method 2.

2.3.4. Modulus of Elasticity (Flexural Modulus)

The flexural modulus is a measure of the stiffness during the first or initial part of the bending process. This value of the flexural modulus is, in many cases, equal to the tensile modulus.

The flexural modulus is represented by the slope of the initial straight line portion of the stress–strain curve and is calculated by dividing the change in stress by the corresponding change in strain. The procedure to calculate flexural modulus is similar to the one described previously for tensile modulus calculations.

Figure 2-19. Universal testing machine for testing of the specimen in either tension or compression. (Courtesy Tinius Olsen Company.)

2.3.5. Factors Affecting the Test Results

A. *Specimen Preparation*

The molecular orientation in the specimen has a significant effect on the test results. For example, the specimen with a high degree of molecular orientation perpendicular to the applied load will show higher flexural values than the specimen with orientation parallel to the applied load. The injection molded specimen usually shows a higher flexural value than a compression molded specimen.

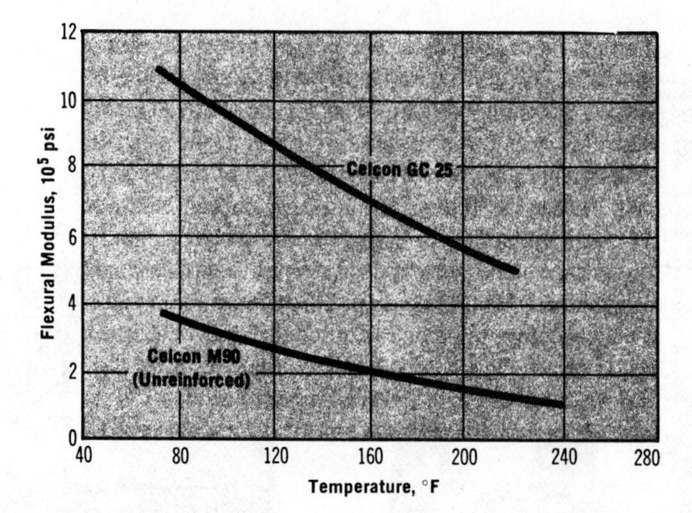

Figure 2-20. The effect of temperature on flexural modulus. (Courtesy Celanese Plastics Company.)

B. Temperature

The flexural strength and modulus values are inversely proportional with temperature. At higher testing temperatures, flexural strength and modulus values are significantly lower. Figure 2-20 shows the effect of temperature on flexural modulus.

C. Test Conditions

The strain rate, which depends upon testing speed, specimen thickness, and the distance between supports (span), can affect the results. At a given span, the flexural strength increases as the specimen thickness is increased. The modulus of a material generally increases with the increasing strain rate (15).

2.4. COMPRESSIVE PROPERTIES (ASTM D 695)

Compressive properties describe the behavior of a material when it is subjected to a compressive load at a relatively low and uniform rate of loading. In spite of numerous applications of plastic products that are subjected to compressive loads, the compressive strength of plastics has limited design value. In practical applications, the compressive loads are not always applied instantaneously. Therefore, the standard test results that fail to take into account the dependence of rigidity and strength of plastics on time cannot be used as a basis for designing a part. The results of impact, creep, and fatigue tests must be considered while designing such a part. Compression tests provide a standard method of obtaining data for research and development, quality control, acceptance or rejection under specifications and special purposes. Compressive properties include modulus of elasticity, yield stress, deformation beyond yield point, compressive strength, compressive strain, and slenderness ratio. However, compressive strength and compressive modulus are the only two values most widely specified in design guides.

In the case of a polymer that fails in compression by a shattering fracture, the compressive strength has a definite value. For those polymers that do not fail by a shattering fracture, the compressive strength is an arbitrary one depending upon the degree of distortion that is regarded as indicating complete failure of material. Some material suppliers report compressive stress at 1 or 10 percent deformation of its original height. Some polymers may also continue to deform in compression

Figure 2-21. A typical test set-up for compression testing. (Courtesy Tinius Olsen Company.)

until a flat disk is produced. The compressive stress continues to rise without any well-defined fracture occurring. Compressive strength has no real meaning in such cases.

2.4.1. Apparatus

The universal testing machine used for tensile and flexural testing can also be used for testing compressive strength of various materials. The machine requirement has been described in detail in the section on tensile testing. A deflectometer or a compressometer is used to measure any change in distance between two fixed points on the test specimen at any time during the test. Figure 2-21 shows a typical set-up for compression testing.

2.4.2. Test Specimens and Conditioning

Recommended specimens for this test are either rectangular blocks measuring $\frac{1}{2} \times \frac{1}{2} \times 1$ in. or cylinders $\frac{1}{2}$ in.-diameter and 1 in. long. Specimens may be prepared by machining or molding. The test specimens are conditioned in accordance with Procedure A of ASTM methods D618 as discussed in Chapter 11.

2.4.3. Procedure

The specimen is placed between the surfaces of the compression tool, making sure that the ends of the specimen are parallel with the surface of the compression tool. The test is commenced by lowering the movable crosshead at a specified speed over the specimen. The maximum load carried by the specimen during the test is recorded. The stress–strain data are also recorded either by recording load at corresponding compressive strain or by plotting a complete load–deformation curve with an automatic recording device. Compressive strength is calculated by dividing the maximum compressive load carried by the specimen during the test by the original minimum cross sectional area of the specimen. The result is expressed in lb/in.^{2} Modulus of elasticity or compressive modulus, like tensile and flexural modulus, is also represented by the slope of the initial straight-line portion of the stress–strain curve and is calculated by dividing the change in stress by the corresponding change in strain. The complete procedure to calculate compressive modulus is described in section 2.2 on tensile properties.

2.5. CREEP PROPERTIES

Today, plastics are used in applications that demand high performance and extreme reliability. Many components, conventionally made in metals, are now made in plastics. The pressure is put on the design engineer to design the plastic products more efficiently. An increasing number of designers have now recognized the importance of thoroughly understanding the behavior of plastics under long-term load and varying temperatures. Such behavior is described in terms of creep properties.

When a plastic material is subjected to a constant load, it deforms quickly to a strain roughly predicted by its stress–strain modulus, and then continues to deform slowly with time indefinitely or until rupture or yielding causes failure (16). This phenomenon of deformation under load with time is called creep. All plastics creep to a certain extent. The degree of creep depends upon several factors, such as type of plastic, amount of load, temperature, and time.

As explained previously in this chapter, the short-term stress–strain data is of little practical value in actual designing the part, since such data does not take into account the effect of long-term loading on plastics. Creep behavior varies considerably among types of plastics; however, under proper stress and temperature conditions, all plastics will exhibit a characteristic type of creep behavior. One such generalized creep curve is shown in Figure 2-22. The total creep curve is divided into four continuous stages. The first stage (OP) represents the instantaneous elastic deformation. This initial strain is the sum of the elastic and plastic strain. The first stage is followed by the second stage (PQ) in which strain occurs rapidly but at a decreasing rate. This stage, where creep rate decreases with time, is sometimes referred to as creep or primary creep. The straight portion of the curve (QR) is characterized by a constant rate of creep. This process is called "cold flow." The final stage (RS) is marked by increase in creep rate until the creep fracture occurs (17–19).

If the applied load is released before the creep rupture occurs, an immediate elastic recovery, substantially equal to elastic deformation followed by a period of slow recovery is observed. The material in most cases does not recover to the

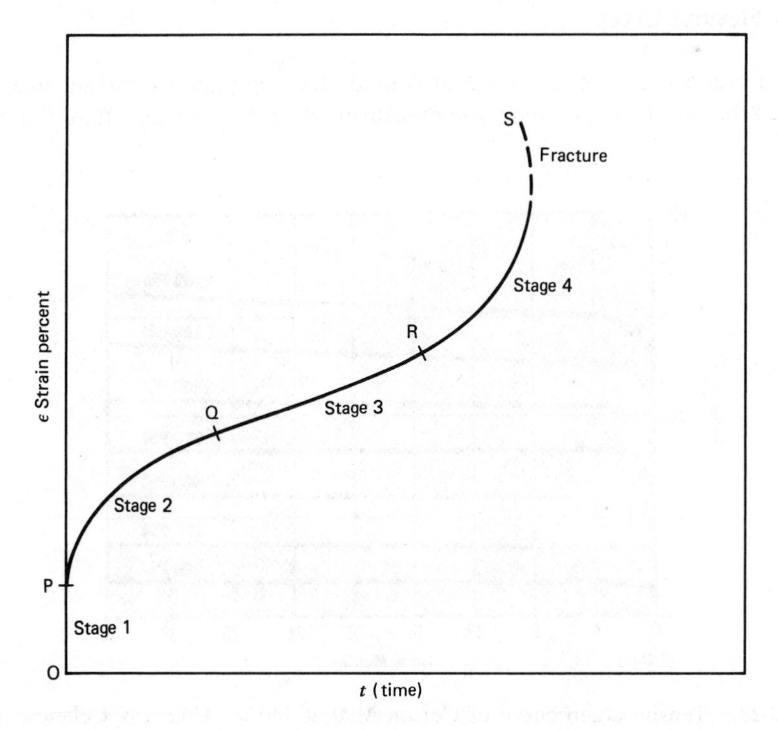

Figure 2-22. Generalized creep curve.

original shape and a permanent set remains. The magnitude of the permanent set depends upon length of time, amount of stress applied, and temperature (20).

The creep values are obtained by applying constant load to the test specimen in tension, compression, or flexure and measuring the deformation as a function of time. The values are most commonly referred to as tensile creep, compressive creep, and flexural creep.

2.5.1. Tensile Creep

Tensile creep measurements are made by applying the constant load to a tensile test specimen and measuring its extension as a function of time. The extension measurement can be carried out several different ways. The simplest way is to make two gauge marks on the tensile specimen and measure the distance between the marks at specified time intervals. The percent creep strain is determined by dividing the extension by initial gauge length and multiplying by 100. The percent creep strain is plotted against time to obtain a tensile creep curve. One such curve is illustrated in Figure 2-23. The tensile stress values are also determined at specified time intervals to facilitate plotting a stress–rupture curve. The more accurate measurements require the use of a strain gauge, which is capable of measuring and amplifying small changes in length with time and directly plotting them on a chart paper. Figure 2-24 illustrates a typical set-up for tensile creep testing. The test is also carried out at different stress levels and temperatures to study their effect on tensile creep properties.

2.5.2. Flexural Creep

Flexural creep measurements are also made by applying a constant load to the standard flexural test specimen and measuring its deflection as a function of time.

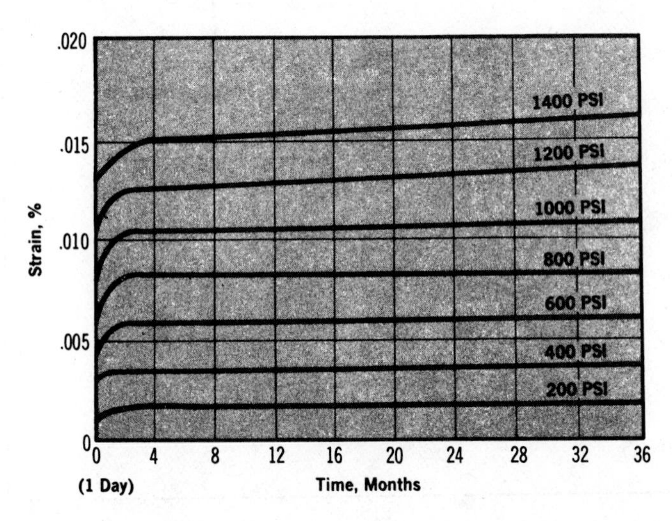

Figure 2-23. Tensile creep curve of Celcon M90 at 140°F. (Courtesy Celanese Plastics Company.)

Figure 2-24. Typical test set-up for tensile creep testing. (Courtesy Instron Corporation.)

A typical test set-up for measuring creep in flexure is shown in Figure 2-25. As illustrated, the deflection of the specimen at mid-span is measured using a dial indicator gauge. The electrical resistance gauges may also be used in place of a dial indicator. The deflections of the specimen are measured at a predetermined time interval. The percent flexural creep strain is calculated using the following formula:

$$r = \frac{6Dd}{L^2} \times 100$$

where r = maximum percent creep strain (in./in.); D = maximum deflection at midspan (in.); d = depth (in.); L = span (in.).

The percent creep strain is plotted against time to obtain a flexural creep curve. An example of one such curve is shown in Figure 2-26. The test is carried out at various stress levels and temperatures and similar flexural creep curves are plotted. If necessary, the maximum fiber stress for each specimen in lb/in.^{2} can also be calculated as follows:

$$S = \frac{3PL}{2bd^2}$$

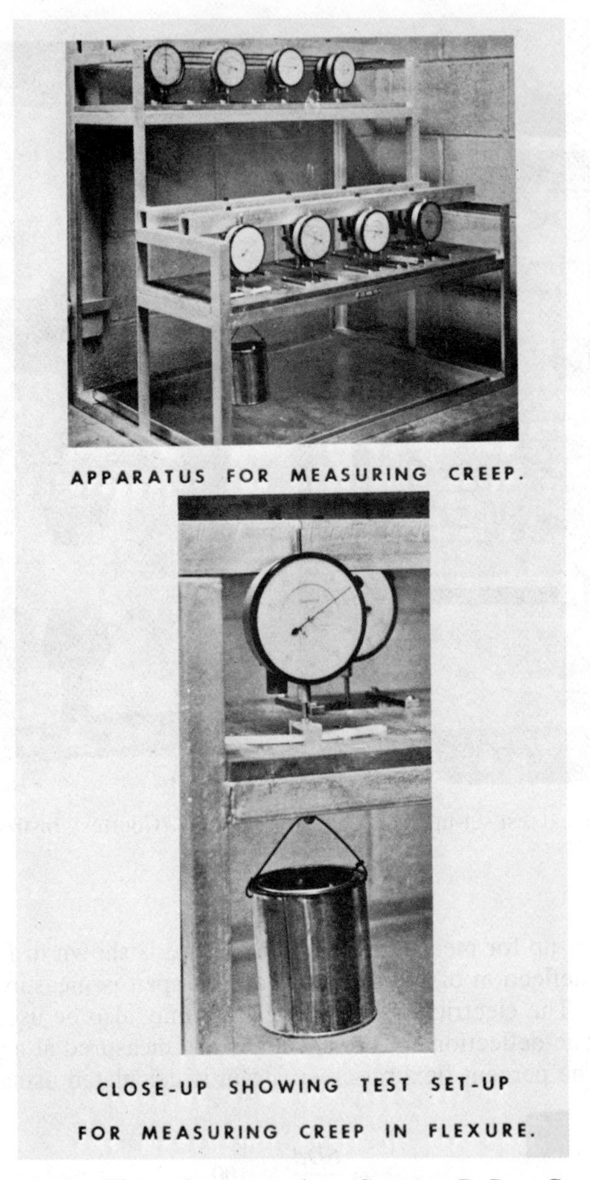

APPARATUS FOR MEASURING CREEP.
CLOSE-UP SHOWING TEST SET-UP
FOR MEASURING CREEP IN FLEXURE.

Figure 2-25. Flexural creep testing. (Courtesy DuPont Company.)

where S = stress (psi); P = load (lbs.); L = span (in.); b = width (in.); d = depth (in.)

2.5.3. Interpretation and Applications of Creep Data

One of the serious limitations of the earlier creep curves such as the one illustrated in Figure 2-26 was the lack of simplicity of the single point stress–strain properties such as tensile modulus and flexural strength, especially when one wanted to measure creep as a function of temperature and stress level. Furthermore, the

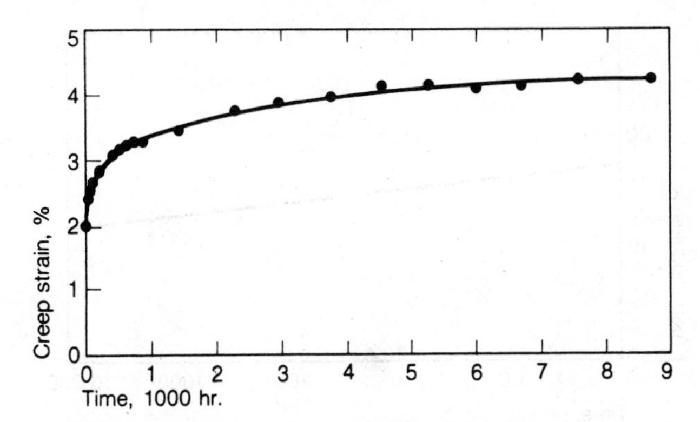

Figure 2-26. Percent creep strain versus time. (Reprinted by permission of *Modern Plastics Encyclopedia* 1980–1981.)

creep data presented in terms of strain were not convenient to use in design or for the purpose of comparing materials. It was obvious that creep curves had to be presented in a more meaningful and convenient way such that they are readily usable. Creep strain curves were easily converted to creep modulus (apparent modulus) by simply dividing the initial applied stress by the creep strain at any time.

$$\text{Creep (Apparent) modulus at time } t = \frac{\text{Initial applied stress}}{\text{Creep strain}}$$

Figure 2-27 shows the same data as shown in Figure 2-26 plotted as creep modulus versus time on Cartesian coordinates. To further simplify the use of the creep curves, the same curve was replotted using logarithmic coordinates as shown in Figure 2-28. These linear plots of creep data are extremely useful in extrapolating the curves to the desired life of the plastic part. Several extrapolation methods are used to extend the range of creep data (21).

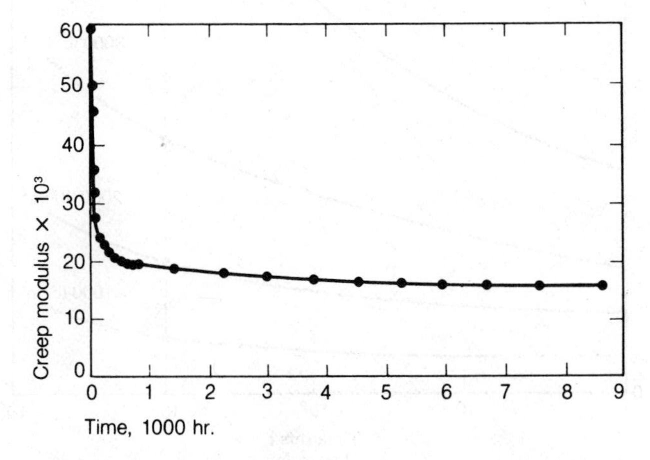

Figure 2-27. Creep modulus versus time on cartesian coordinates. (Reprinted by permission of *Modern Plastics Encyclopedia* 1980–1981.)

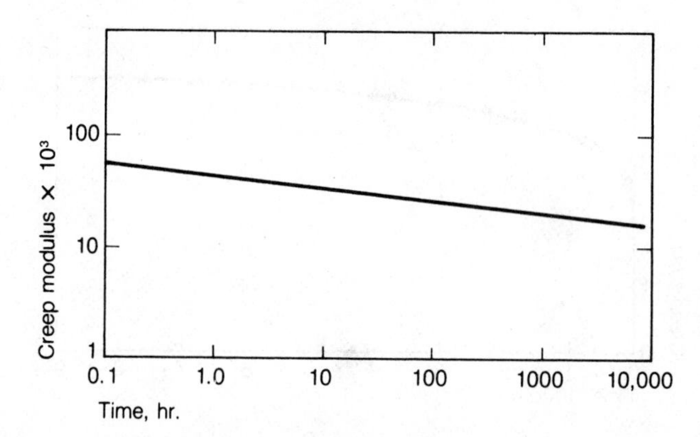

Figure 2-28. Creep modulus versus time on logarithmic coordinates. (Reprinted by permission of *Modern Plastics Encyclopedia* 1980–1981.)

2.5.4. Isochronous Stress–Strain Curves

Quite often, while designing a part, it is necessary to compare various plastic materials. The basic creep curves, such as creep strain versus time or apparent modulus versus time are not completely satisfactory. The comparison is extremely difficult to make, especially when different stress levels are used for different materials.

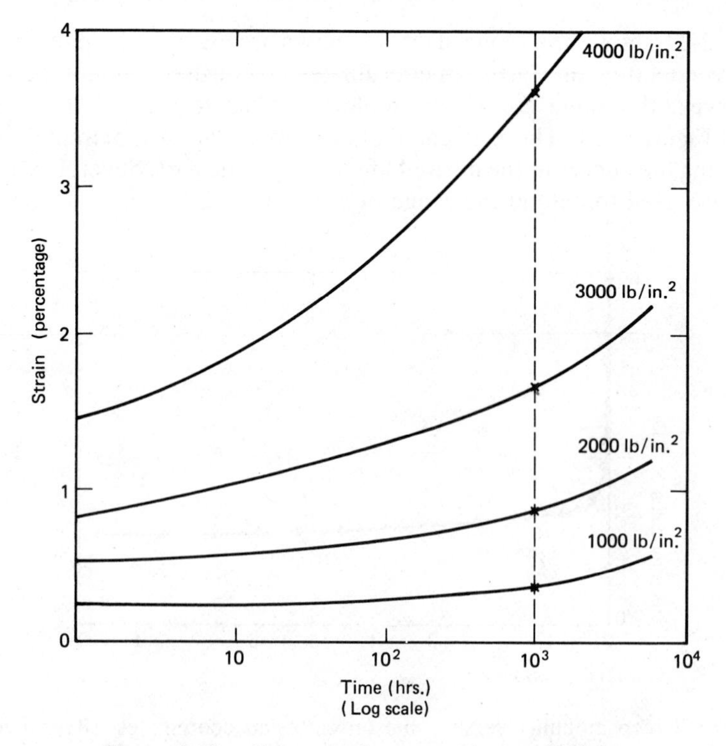

Figure 2-29. Creep strain versus time at 1000 hr.

The method preferred by most design engineers is the use of isochronous (equal time) stress–strain curves. The stress versus corresponding strain is plotted at a specific time of loading pertinent to particular application.

Suppose a designer is asked to design a shelf that is required to withstand a continuous load for 1000 hrs. If after 1000 hrs. of continuous loading, the deformation is not to exceed 2 percent, what should be the maximum allowable stress? To solve this problem, the first thing the designer needs to do is to erect an ordinate on the basic creep strain versus time curve at 1000 hrs, as shown in Figure 2-29. From this curve, now the designer can determine the strain value at different stress levels. These values allow the designer to plot another curve of stress versus strain at 1000 hrs. This isochronous stress–strain curve is shown in Figure 2-30. The maximum allowable stress can be determined by simply erecting an ordinate from 2 percent strain value and reading the corresponding stress, which in this particular case is 3500 psi. A wide variety of materials can easily be compared by studying an isochronous stress–strain curve such as shown in Figure 2-31 (22).

2.5.5. Effects of Stress and Temperature on Creep Modulus

The creep modulus is directly affected with the increase in the level of stress and temperature. With the exception of extremely low strains around one percent or less, the creep modulus decreases as the amount of stress is increased. This effect is illustrated in Figure 2-32. In a very similar manner, as the temperature is increased, the creep modulus significantly decreases. Figure 2-33 shows the creep

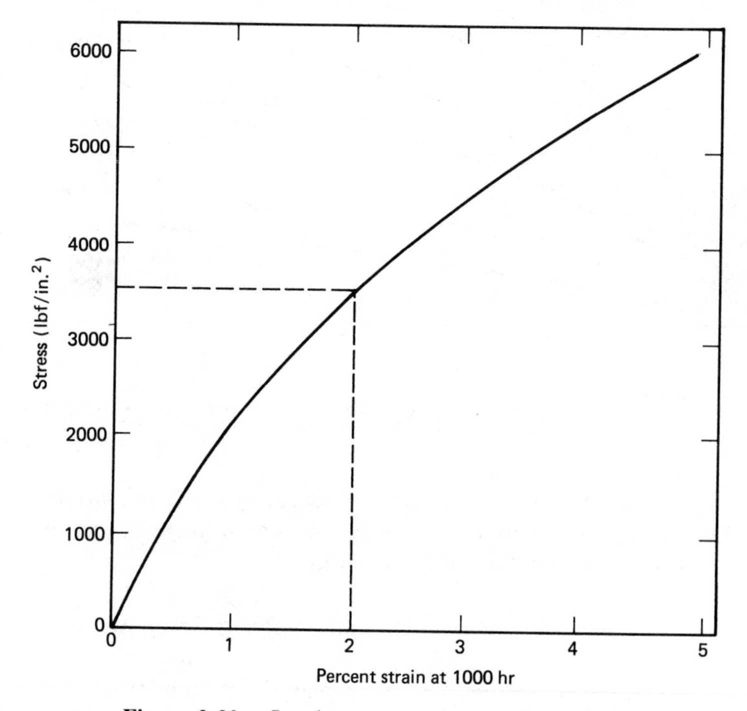

Figure 2-30. Isochronous stress–strain curve.

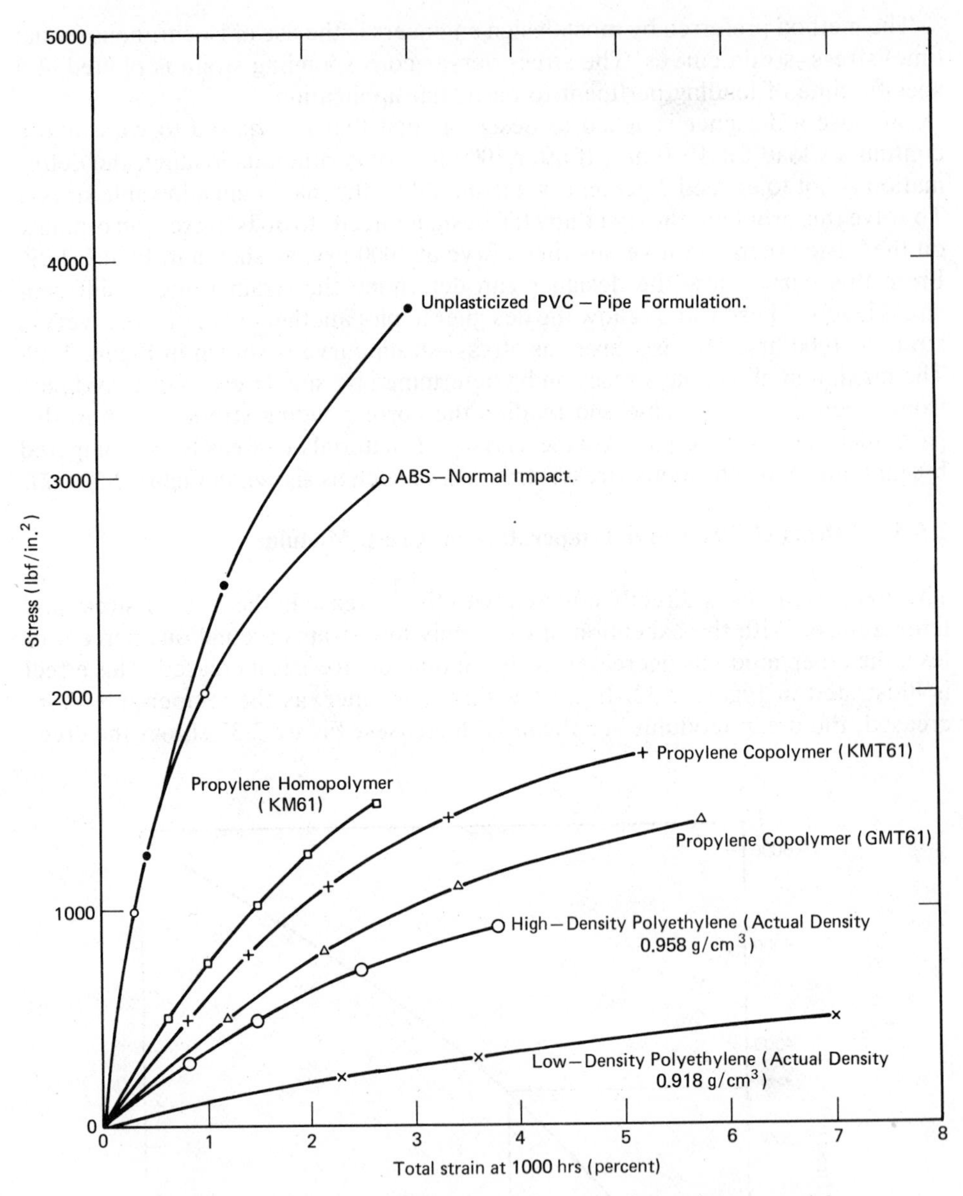

Figure 2-31. Isochronous stress–strain curve for various materials.

modulus versus time plotted at different temperatures. As one would expect, the combined effect of increasing stress level and temperature on creep modulus is much more severe and should not be overlooked.

2.5.6. Basic Procedures for Developing and Applying Creep Modulus Data

In recent years, much emphasis has been placed on standardizing a method to develop and report creep data that would report creep that would be universally

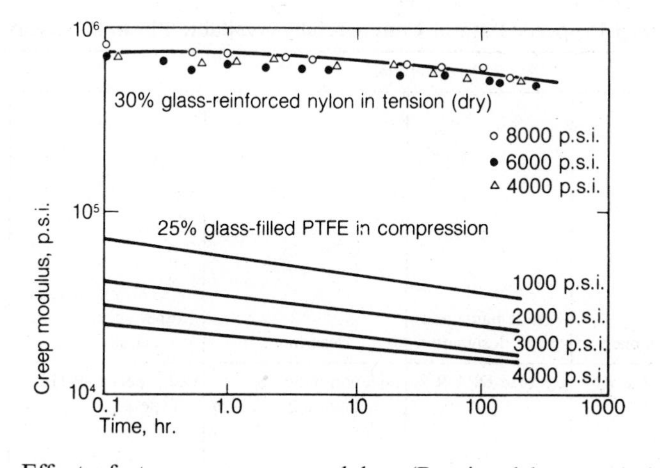

Figure 2-32. Effect of stress on creep modulus. (Reprinted by permission of *Modern Plastic Encyclopedia* 1980–1981.)

acceptable and understood. A systematic accumulation of data on creep properties of commercially available plastics has been developed with the cooperation of major material suppliers and the results have been published in tabular form, such as shown in Table 2-2.

In spite of wide use and acceptance of creep data, it is evident from the table that a standard creep test method agreed upon by everyone in the plastics industry has not yet been developed. All three basic methods—tension, flexure, compression—are used by material suppliers to develop creep data. There are other variables such as sample fabricating methods, type of specimens, and specimen dimension. These variables have a significant effect on creep properties. A designer using the data from such creep tables must be aware of such variables before comparing one material with the other.

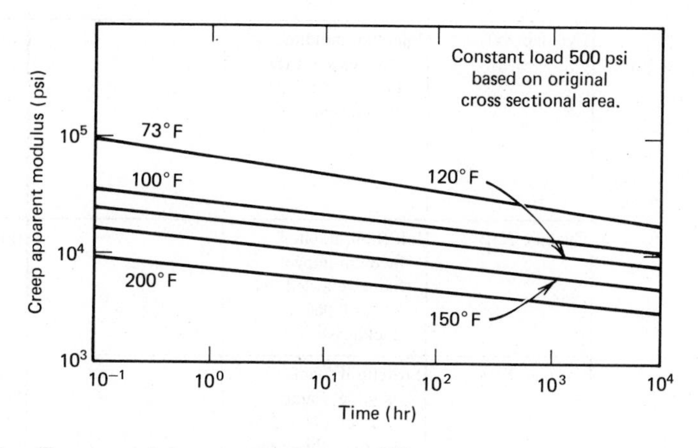

Figure 2-33. Creep modulus versus time at different temperatures. (Reprinted by permission of McGraw–Hill Book Company.)

Table 2-2. Creep Property Data of Commercially Available Plastic Material[a,b]

Material					Test specimen[13]		
Type	Supplier	Trade name and grade designation	Description	ASTM, military, or other specification classification	Molding method	Type or Shape	Dimensions
ABS	Borg-Warner	Cycolac DFA-R	Injection molding, medium-impact, high gloss	Fed. Spec. L-P-1183 Type 2	IM	RB	4
		Cycolac GSE	Extrusion, high-impact, high modulus	Fed. Spec. L-P-1183 Types 2, 5, and 6	IM	RB	4
		Cycolac LS	Extrusion, high-impact, thermoforming	Fed. Spec. L-P-1183	IM	RB	4
		Cycolac DH	Injection molding, heat-resistant		IM	RB	4
		Cycolac X-37	Injection molding, heat-resistant (D TUL = 230°F.), medium-impact		IM	RB	4
		Cycolac KJB	Injection molding, medium-impact, flame-retarded (94-VO at 0.060-in. thickness)		IM	RB	4
		Cycolac FBK	Structural foam molding, flame-retarded (94-VO at 0.240-in. thickness)		IM Solid Sp. Gr. 1.11	RB	4

Table 2-2. (*Continued*)

Creep test conditions[13]					Creep test data[1,2,3]								
					Creep (apparent) modulus,[3] thousand psi								
					Calculated from total creep strain[2] or deflection (before rupture and onset of yielding) at the following test times:								Time at rupture or onset of yielding at initial applied stress in air, hr.
Type of load	Strain measure	Special specimen conditioning	Test temp., °F.	Initial applied stress, psi	1 hr.	10 hr.	30 hr.	100 hr.	300 hr.	1000 hr.	At latest test point	Time at latest test point, hr.	
F2	3		73	1000	345	323	313	278	244	215			
				2000	345	323	308	274	243	211			
			120	1000	303	222	187	154	125	94			
				2000	290	211	167	132	108	90			
			140	1000	137	93	78	63	53	44			
F2	3		73	1000	294	286	278	256	238	208			
				2000	294	286	278	256	233	204			
			120	1000	238	182	159	133	116	100			
				2000	227	172	149	124	106	90			
			140	1000	179	122	100	83	71	60			
			160	1000	105	59	49	41	36	32			
F2	3		73	1000	238	217	204	200	189	159			
				2000	224	211	200	189	179	155			
			120	1000	169	135	118	98	83	71			
				2000	159	121	103	86	74	63			
			140	1000	133	87	72	61	51	43			
F2	3		73	1000	351	333	303	263	227	200			
			100	1000	256	208	185	163	147	132			
			120	1000	238	185	161	139	121	105			
			140	1000	192	132	111	93	81	70			
F2	3		73	1000	364	339	328	299	267	230			
			100	1000	304	256	235	208	187	165			
			120	1000	270	220	200	180	160	143			
			140	1000	238	179	156	135	119	104			
			180	500	167	80	69	59	50	43			
F2	3		73	1000	244	233	222	204	182	156			
			140	1000	93	56	46	36	33	28			
F2	3		73	1000	280	260	230	220	200	160			

Table 2-2. Creep Property Data of Commercially Available Plastic Material[a,b]

Material					Test specimen[13]		
Type	Supplier	Trade name and grade designation	Description	ASTM, military, or other specification classification	Molding method	Type or Shape	Dimensions
					IM foam Sp. Gr. 1.02	RB	4
		Cycovin KAB	ABS/PVC alloy, extrusion, high-impact, flame-retarded (94-VI at 0.058-in. thickness)		IM	RB	4
	Uniroyal	Kralastic W	Pipe extrusion	Fed. Spec. L-P-1183 Types 2, 5, and 6	CM	T1	1
						RS	2
		Kralastic MH	Easy flow, injection molding	Fed. Spec. L-P-1183 Types 1 and 2	CM	T1	1
		Kralastic MV	Sheet extrusion, injection molding	Fed. Spec. L-P-1183 Types 2, 5, and 6	CM	T1	1
		Kralastic K-2938	Heat-resistant, injection molding	Fed. Spec. L-P-1183 Types 1, 2, 3, and 4	CM	T1	1
		Kralastic SRS	Sheet extrusion	Fed. Spec. L-P-1183 Types 1, 2, 5, and 6	CM	T1	1
	Monsanto	Lustran 261	Sheet extrusion	Fed. Spec. L-P-1183 Type 1	ES	T1	1
		Lustran 461	Sheet extrusion	Fed. Spec. L-P-1183 Type 2	ES	T1	1

Table 2-2. (*Continued*)

Creep test conditions[13]					Creep test data[1,2,3]								
					Creep (apparent) modulus,[3] thousand psi								
					Calculated from total creep strain[2] or deflection (before rupture and onset of yielding) at the following test times:								
Type of load	Strain measure	Special specimen conditioning	Test temp., °F.	Initial applied stress, psi	1 hr.	10 hr.	30 hr.	100 hr.	300 hr.	1000 hr.	At latest test point	Time at latest test point, hr.	Time at rupture or onset of yielding at initial applied stress in air, hr.
F2	3		73	1000	230	220	200	190	160	130			
F2	3		73	1000	280	250	235	206	185	160			
			120	1000	133	86	71	62	54	45			
T	1		73	2000	230	220	211	200			194	167	
				2500	223	208	195	180			172	167	
				3000	211	185	165	130			128	111	
T	1		73	1500	230	224	217	211	200	188	140	50,000	
T	1		73	1500	284	278	273	263	254	240			
				2000	284	268	260	245	230	211			
				2500					198	211			
				3000	254	224	203	170	133		125	390	
				3500	233	165					146	17	
				4000	167								
T	1		73	3000	216	196	181	161	140		132	83	
				2500	202	172	152						
				3000	181	136							
			140	500[5]	161	132	109	77	49				
				1000[5]	171	136	114	76	44				
				1500[5]	168	127	99						
T	1		73	2000	330	323	308	290	270	244	167	580	
				3000	328	305	288	265	244	217	188	42	
				4000	308	278	256	230	197				
				5000	279	237	201						
			160	500	263	250	227	200	147	104			
				1000	263	233	189	147	118	84			
				1500	263	203	170	135	105	75			
T	1		73	2000	282	263	247	233	213	192	125	350	
				2500	278	253	236	216	192	156	146	3	
				3000	265	234	214	176	132				
				3500	206								
T	1		73	2000	476	465	465	465	465	465			
T	1		73	2000	571	513	488	417	377	328			

Table 2-2. Creep Property Data of Commercially Available Plastic Material[a,b]

Material					Test specimen[13]		
Type	Supplier	Trade name and grade designation	Description	ASTM, military, or other specification classification	Molding method	Type or Shape	Dimensions
		Lustran 761	Sheet extrusion	Fed. Spec. L-P-1183 Type 5	ES	T1	1
	Schulman	Polyman 511	Alloy, high temperature, injection molding, flame-retarded		IM	RB	4
	LNP	Thermocomp AF-1004	20% glass fiber-reinforced, injection molding		IM	RB	4
		Thermocomp AF 1008	40% glass fiber-reinforced, general-purpose, injection molding		IM	RB	4

[a] Reprinted by permission of *Modern Plastics Encyclopedia 1980–1981*, McGraw–Hill.
[b] Explanation of test specimen and creep test conditions, codes:

Molding methods
CA = Compression molded per ASTM D1928
CB = Comp. molded; annealed 2 hr. 140° C.; quenched 0°C.
CC = Comp. molded; annealed 2 hr. 140° C.; slow cooled
CD = Comp. molded; annealed 2 hr. 140° C.; cooled 40° C./hr.
CE = Comp. molded and free sintered 700° F.; cooled 170° F./hr.
CF = Comp. molded and free sintered 675° F.; cooled 175° F./hr.
CG = Comp. molded and free sintered 720° F.; cooled 180° F./hr.
CH = Comp. molded and free sintered 720° F.; cooled 300° F./hr.
CI = Comp. molded and free sintered 716° F.; cooled 54° F./hr.
CL = Calendered sheet
CM = Compression molded
CS = Cast sheet
DS = Direct blend. screw inj. molded
ES = Extruded sheet
IM = Injection molded
IZ = Inj. molded; mold temp. 265° F.
M = Machined
SF = Structural Foam, injection molded
TM = Transfer molded

Type or shape
CR = Cylindrical rod
DC = Die C per ASTM D412
D1 = Modified Die C with 3.4 in. reduced section
HC = Hollow cylinder—see Note 17
I2 = Tensile bar per ISO R527
RB = Rectangular bar
RC = Rectangular column
RS = Rectangular strip
T = Tensile bar
TB = Tensile bar—see Note 14
T1 = Type 1 tensile bar per ASTM D638
T2 = Type 2 tensile bar per ASTM D638
T3 = Type 3 tensile bar per ASTM D638

Table 2-2. (*Continued*)

Creep test conditions[13]					Creep test data[1,2,3]								
					Creep (apparent) modulus,[3] thousand psi								
					Calculated from total creep strain[2] or deflection (before rupture and onset of yielding) at the following test times:								
Type of load	Strain measure	Special specimen conditioning	Test temp., °F.	Initial applied stress, psi	1 hr.	10 hr.	30 hr.	100 hr.	300 hr.	1000 hr.	At latest test point	Time at latest test point, hr.	Time at rupture or onset of yielding at initial applied stress in air, hr.
T	1		73	2000	328	282	250	222	1290	155			
F4	3		73	1500	360	355	350	340	320	305	283	2200	
				2000	335	330	321	318	315	300	280	2200	
			150	1000	250	230	219	165	138				
F4	3		75	2000		810	800	790	780	780			
				5000		800	790	780	775	770			
F4	3		75	5000[5]		1720	1680	1650	1636	1600			
				10,000[5]		1760	1710	1690	1670	1650			

Dimensions, in. (overall for bars and columns, and reduced section for dog bones)

1 = 1/2 × 1/8
2 = 5 × 1 × 0.090
3 = 4 × 1 × 0.026
4 = 5 × 1/2 × 1/8
5 = 1/4 × 1/16
6 = 1/2 × 1/16
7 = 1/2 diam. × 1 high
8 = 1/4 × 1/8
9 = 1/4 × 1/12
10 = 4 × 1/8 × 0.218
11 = 5 × 1/2 × 1/16
12 = 0.24 × 0.075
13 = 5 × 1/32 × 0.392
14 = 5 × 1/2 × 1/2
15 = 16 × 1 × 1/8 or 1/4
16 = 8 × 1/2 × 1/8
17 = 0.190 × 0.190 × 2 high
18 = 5/8ID × F3/4OD × 2.5 high
19 = 0.024 × 0.035
20 = 3 1/2 × 1/4 × 1/8
21 = 120 × 20 × 6 mm. (4.72 × 0.79 × 0.25 in.)
22 = 0.39 × 0.16
23 = 4 × 0.500 × 0.175
24 = 6.5 × 0.5 × 0.25
25 = 5 × 1/2 × 1/4

Type of load

C = Compression
C1 = Compression—see Note 11
F = Flexure—simple beam bending, load at center
F2 = Flexure—simple beam bending, load at center, 2 in. span
F3 = Flexure—simple beam bending, load at center, 3 in. span
F4 = Flexure—simple beam bending, load at center, 4 in. span.
F6 = Flexure—uniform beam bending moment, 6 in. span
F7 = Flexure—4 point loading, constant outside fiber stress
T = Tension

Strain measurement

1 = Strain in reduced section
1A = Strain in reduced section—see Note 15
1B = Strain in reduced section—see Note 16
1C = Strain in reduced section—corrected for shrinkage
2 = Grip separation
3 = Deflection at center of beam
4 = Reduction in height
5 = Extensometer—see Note 17

The following procedure is recommended for the use of creep modulus data (23).

1. Select the design life of the part.

2. Consult or plot, from data available in the Creep chart, the creep modulus curve of the material of interest for the temperature the part will be used, extrapolating where necessary. Like creep rupture, creep modulus varies greatly with temperature. In addition, it is subject to another variable—stress level. For each material grade, creep modulus data are tabulated in the Creep chart, first by test temperature and second by applied (test) stress. For very rigid plastics, such as thermosets, filled thermoplastics, and amorphous thermoplastics at room temperature and below, the creep modulus curves show minor variation with applied stress. However, for the more flexible and ductile plastics the creep modulus curves will vary significantly and systematically with stress level. The higher the stress, the lower the creep modulus. This is a consequence of viscoelasticity.

To cope with the effect of stress level the designer has two alternatives. If the design problem is such that the stress level is predetermined, then the designer should select the creep modulus curve whose stress level is the closest. If the stress level is not known beforehand, then the designer should choose a creep modulus curve at a conservative stress level and check the choice after calculating a stress level.

3. Read from the selected creep modulus cruve the modulus value corresponding to the design life selected in step 1. above. This is the design modulus.

4. Apply a safety factor to the design modulus to calculate a working modulus. This is rarely necessary in metal design but is in plastics design to correct for any uncertainties arising from extrapolations or other compromises that may have been made. Safety factors of 0.5 to 0.75 are typical.

5. Substitute the calculated working modulus in the part design equation. For example, to calculate the width of a simple rectangular beam required to limit the maximum deflection to specified value when the span and depth of the beam are fixed, the working modulus would be substituted for E in the following design equation.

$$b = \frac{P}{E\Delta} \frac{L^3}{4d^3}$$

where b = width of beam; d = depth of beam; L = span; P = load; Δ = maximum deflection allowed; E = modulus.

2.6. STRESS RELAXATION

Stress relaxation is defined as a gradual decrease in stress with time, under a constant deformation (strain). This characteristic behavior of the polymers is studied by applying a fixed amount of deformation to a specimen and measuring the load required to maintain it as a function of time. This is in contrast to creep measurement, where a fixed amount of load is applied to a specimen and resulting deformation is measured as a function of time. This phenomenon of creep and stress relaxation is further clarified schematically in Figure 2-34.

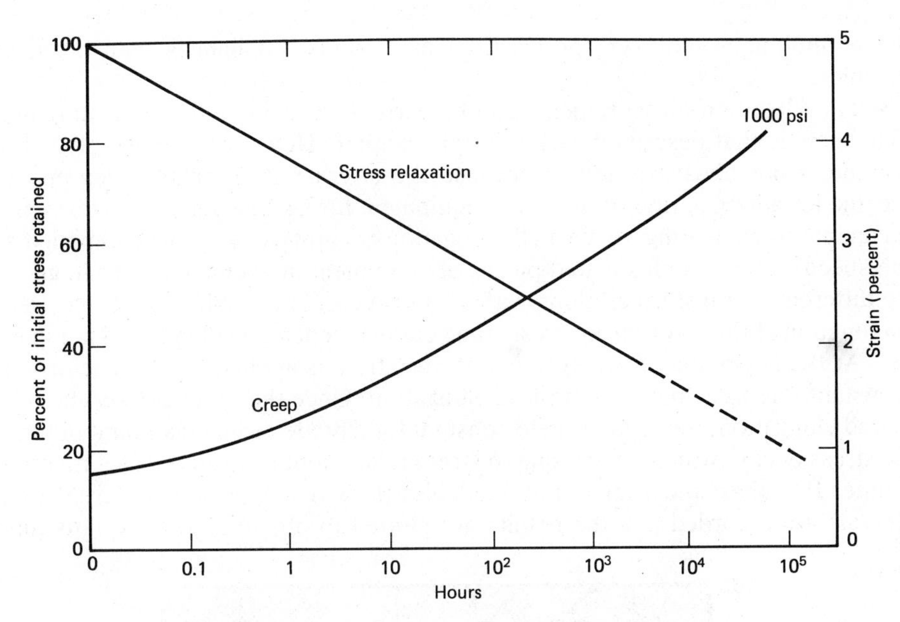

Figure 2-34. Creep and stress relaxation.

Stress relaxation behavior of the polymers has been overlooked by many design engineers and researchers, partly because the creep data is much easier to obtain and is readily available. However, many practical applications dictate the use of stress relaxation data. For example, extremely low stress relaxation is desired in the case of a threaded bottle closure, which may be under constant strain for a long period. If the plastic material used in the closures shows an excessive decrease in stress under this constant deformation, the closures will eventually fail. Similar problems can be encountered with metal inserts in molded plastics, belle-

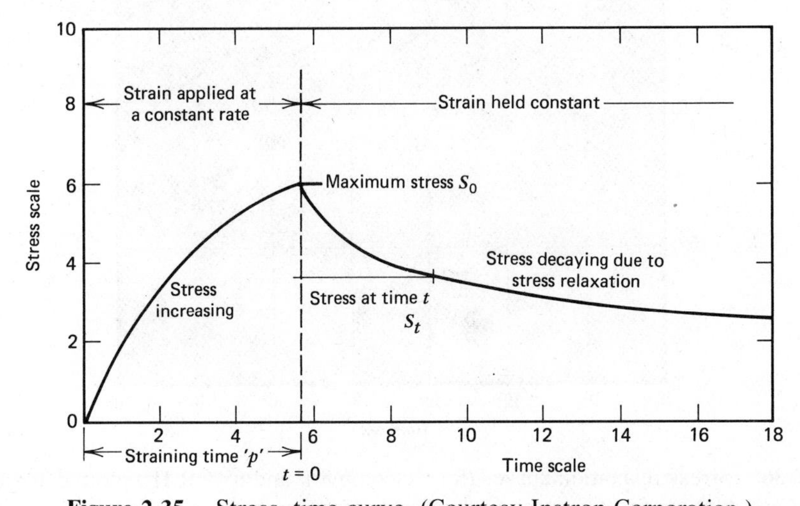

Figure 2-35. Stress–time curve. (Courtesy Instron Corporation.)

ville or multiple cantilever springs used in cameras, appliances, and business machines.

Stress relaxation measurements can be carried out using a tensile testing machine such as that described earlier in this chapter. However, the use of such a machine is not always practical because the stress relaxation test ties up the machine for a long period of time. The equipment for a stress relaxation test must be capable of measuring very small elongation accurately, even when applied at high speeds. Many sophisticated pieces of equipment now employ a strain gauge or a differential transformer along with a chart recorder capable of plotting stress as a function of time. A typical stress–time curve is schematically plotted in Figure 2-35. At the beginning of the experiment, the strain is applied to the specimen at a constant rate to achieve the desired elongation. Once the specimen reaches the desired elongation, the strain is held constant for a predetermined amount of time. The stress decay, which occurs due to stress relaxation, is observed as a function of time. If a chart recorder is not available, the stress values at different time intervals are recorded and the results are plotted to obtain a stress versus time

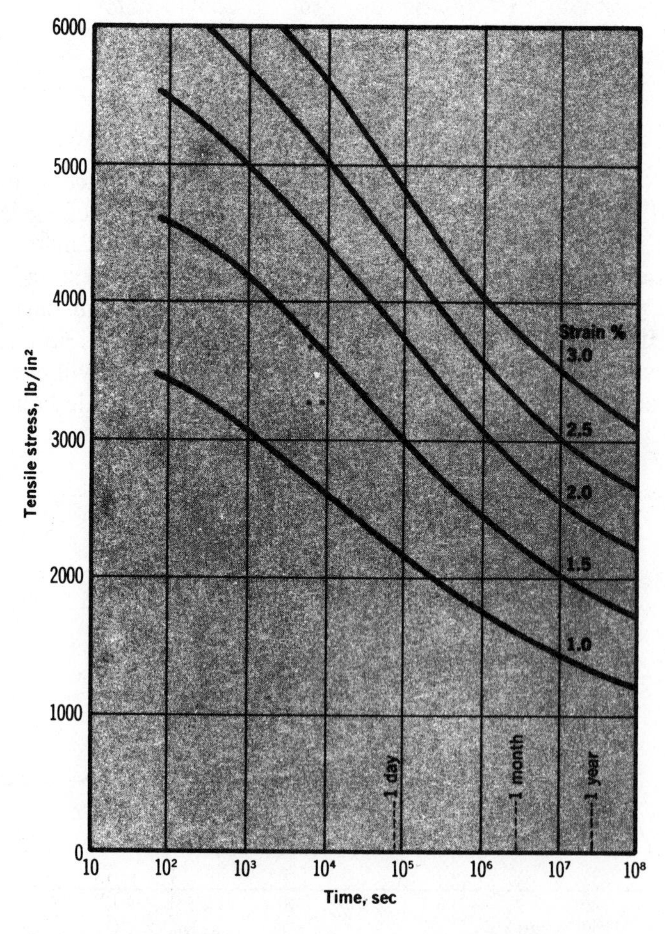

Figure 2-36. Stress relaxation curves (for Celcon, 68°F and 65% R.H.) plotted at various levels of constant strain. (Courtesy Celanese Plastics Company.)

curve. One such stress relaxation curve plotted at various levels of constant strain is shown in Figure 2-36. The stress relaxation experiment is often carried out at various levels of temperature and strain.

The stress data obtained from the stress relaxation experiment can be converted to a more meaningful apparent modulus data by simply dividing stress at a particular time by the applied strain. The curve may be replotted to represent apparent modulus as a function of time. The use of logarithmic coordinates further simplifies the stress relaxation data by allowing us to use standard extrapolation methods such as the one used in creep experiments.

2.7. IMPACT PROPERTIES

2.7.1. Introduction

The impact properties of the polymeric materials are directly related to the overall toughness of the material. Toughness is defined as the ability of the polymer to absorb applied energy. The area under the stress–strain curve is directly proportional to the toughness of a material. Impact strength is a measure of toughness. The higher the impact strength of a material, the higher the toughness and vice versa. Impact resistance is the ability of a material to resist breaking under a shock loading or the ability to resist the fracture under stress applied at high speed.

The theory behind toughness and brittleness of the polymers is very complex and therefore difficult to understand. The molecular flexibility plays an important role in determining the relative brittleness or toughness of the material. For example, in stiff polymers like polystyrene and acrylics, the molecular segments are unable to disentangle and respond to the rapid application of mechanical stress and the impact produces brittle failure. In contrast, flexible polymers such as plasticized vinyls have high-impact strength due to the ability of the large segments of molecules to disentangle and respond rapidly to mechanical stress (24).

Impact properties of the polymers are often modified simply by adding an impact modifier such as butadiene rubber or certain acrylic compounds. The addition of a plasticizer also improves the impact strength at the cost of rigidity. Material such as nylon, which has relatively fair impact strength, can be oriented by aligning the polymer chains to improve the impact strength substantially. Another way to improve the impact strength is to use fibrous fillers that appear to act as stress transfer agents (25).

Most polymers, when subjected to the impact loading, seem to fracture in a characteristic fashion. The crack is initiated on a polymer surface due to the impact loading. The energy to initiate such a crack is called the crack initiation energy. If the load exceeds the crack initiation energy, the crack continues to propagate. A complete failure occurs when the load has exceeded the crack propagation energy. Thus, both crack initiation and crack propagation contribute to the measured impact strength. There are basically four types of failures encountered due to the impact load (26).

BRITTLE FRACTURE: In this type of failure the part fractures extensively without yielding. A catastrophic mechanical failure such as the one in the case of general-purpose polystyrene is observed.

SLIGHT CRACKING: The part shows evidence of slight cracking and yielding without losing its shape or integrity.

YIELDING: The part actually yields showing obvious deformation and stress whitening but no cracking takes place.

DUCTILE FAILURE: This type of failure is characterized by a definite yielding of material along with cracking. Polycarbonate is considered a ductile material.

The distinction between the four types of failures is not very clear and some overlapping is quite possible.

Impact strength is one of the most widely specified mechanical properties of the polymeric materials. However, it is also one of the least understood properties. Predicting the impact resistance of plastics still remains one of the most troublesome areas of product design. One of the problems with some earlier Izod and Charpy impact tests was that the tests were adopted by the plastic industry from metallurgists. The principles of impact mechanisms as applied to metals do not seem to work satisfactorily with plastics because of the plastics' complex structure.

2.7.2. Factors Affecting the Impact Strength

A. *Rate of Loading*

The speed at which the specimen or part is struck with an object has a significant effect on the behavior of the polymer under impact loading. At low rates of impact, relatively stiff materials can still have good impact strength. While at high rates of impact, even rubbery materials may exhibit brittle failure (27). All materials seem to have a critical velocity above which they behave as glassy, brittle materials. Figure 2-37 illustrates some typical velocities encountered during testing and real-life situations. Let us say that we are required to design a football helmet. It is obvious that we cannot use the results from the Izod impact test directly to select our material. The velocity at which the Izod impact test is conducted is approximately 10 times lower than the one encountered in the end use. A more realistic drop impact test at a velocity closer to the actual use condition must be used.

B. *Notch Sensitivity*

A notch in a test specimen or a sharp corner in a fabricated part drastically lowers the impact strength. A notch creates a localized stress concentration and hence the part failure under impact loading. All plastics are notch-sensitive. The rate of sensitivity varies with the type of plastic. Both notch depth and notch radius have an effect on the impact strength of materials. For example, a larger radius of curvature at the base of the notch will have a lower stress concentration and, therefore, a higher impact strength of the base material. It is quite obvious from

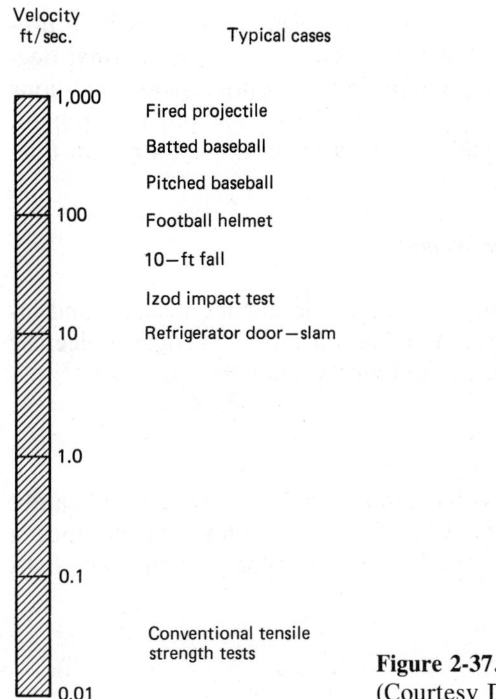

Figure 2-37. Typical velocities of some impact blows. (Courtesy Dow Chemical Company.)

the above discussion that while designing a plastic part, one should avoid notches, sharp corners, and other factors that act as stress concentrators.

C. Temperature

The impact strength of plastic materials is strongly dependent upon the temperature. At lower temperatures the impact strength is reduced drastically. The reduction in impact is even more dramatic near the glass transition temperature. Conversely, at higher test temperatures, the impact strength is significantly improved.

D. Orientation

The manner in which the polymer molecules are oriented in a part will have a major effect on the impact strength of the polymer. Molecular orientation introduced into drawn films and fibers may give extra strength and toughness over the isotropic material (28). However, such directional orientation of polymer molecules can be very fatal in a molded part since the impact stresses are usually multiaxial. The impact strength is always higher in the direction of flow.

E. Processing Conditions and Types

Processing conditions play a key role in determining the impact resistance of a material. Inadequate processing conditions can cause the material to lose its inherent toughness. Voids that act as stress concentrators are created by poor processing conditions. High processing temperatures can also cause thermal degradation and, therefore, reduced impact strength. Improper processing conditions also create a weak weld line that almost always reduces overall impact strength. The compression molded specimen usually shows a lower impact strength than the injection molded specimens.

F. *Degree of Crystallinity, Molecular Weight*

Increasing the percentage of crystallinity decreases the impact strength and increases the probability of brittle failure. A reduction in the average molecular weight tends to reduce the impact strength and vice versa (29).

G. *Method of Loading*

The manner in which the part is struck with the impact loading device significantly affects the impact results. A pendulum type of impact loading will produce a different result from the one produced by falling-weight-type or high-speed ball-impact-type loading.

2.7.3. Types of Impact Tests

In the last two decades, a tremendous amount of time and money have been spent on research and development of various types of impact tests by organizations throughout the world. Attempts have been made to develop different sizes and shapes of specimens as well as impact testers. The specimens have been subjected to a variety of impact loads including tensile, compression, bending, and torsion impacts. Impact load has been applied using everything from a hammer, punches, and pendulums to falling balls and bullets. Unfortunately, very little correlation exists, if any, between the types of tests developed so far. Numerous technical papers and articles have been written on the subject of the advantages of one method over the other. To this date, no industry-wide consensus exists regarding an ideal impact test method. In this chapter, an attempt is made to discuss as many types of impact tests as possible along with the respective advantages and limitations of each test. An impact test can be divided into six major classes and subdivided into many different types having slight variations.

1. Pendulum impact tests.
   - (a) Izod impact test.
   - (b) Charpy impact test.
   - (c) Chip impact test.
   - (d) Tensile impact test.
2. High-rate tension test.
3. Falling weight impact test.
   - (a) Drop weight (tup) impact test.
4. Instrumented impact tests.

5. High-Rate impact testers.
   (a) High-speed ball-impact tester.
   (b) High-speed plunger-impact tester.
6. Miscellaneous impact tests.

A. *Pendulum Impact Tests*

Izod–Charpy Impact Test (ASTM D-265). The objective of the Izod–Charpy impact test is to measure the relative susceptibility of a standard test specimen to the pendulum-type impact load. The results are expressed in terms of kinetic energy consumed by the pendulum in order to break the specimen. The energy required to break a standard specimen is actually the sum of energies needed to deform it, to initiate its fracture, and to propagate the fracture across it, and the energy needed to throw the broken ends of the specimen. This is called the "toss factor". The energy lost through the friction and vibration of the apparatus is minimal for all practical purposes and usually neglected.

The specimen used in both tests is usually notched. The reason for notching the specimen is to provide a stress concentration area that promotes a brittle rather than a ductile failure. A plastic deformation is prevented by such type of notch in the specimen. The impact values are seriously affected because of the notch sensitivity of certain types of plastic materials. This effect was discussed in detail in Section 2.7.2.B.

The Izod test requires a specimen to be clamped vertically as a cantilever beam. The specimen is struck by a swing of a pendulum released from a fixed distance from the specimen clamp. A similar set-up is used for the Charpy test except for the positioning of the specimen. In the Charpy method, the specimen is supported horizontally as a simple beam and fractured by a blow delivered in the middle by the pendulum. The obvious advantage of the Charpy test over the Izod test is that the specimen does not have to be clamped and, therefore, it is free of variations in clamping pressures.

Apparatus and Test Specimens. The testing machine consists of a heavy base with a vise for clamping the specimen in place during the test. In most cases, the vise is designed so that the specimen can be clamped vertically for the Izod test or positioned horizontally for the Charpy test without making any changes. A pendulum-type hammer with an antifriction bearing is used. Additional weights may be attached to the hammer for breaking tougher specimens. The pendulum is connected to a pointer and a dial mechanism that indicates the excess energy remaining in a pendulum after breaking the specimen. The dial is calibrated to read the impact values directly in in.-lbs or ft-lbs. A hardened steel striking nose is attached to the pendulum. The Izod and Charpy tests use different types of striking noses. A detailed list of requirements is discussed in the ASTM Standards Book. Figure 2-38 illustrates a typical pendulum-type impact testing machine. The test specimens can be prepared either by molding or cutting them from a sheet. Izod test specimens are $5 \times \frac{1}{2} \times \frac{1}{8}$ in. size. The most common specimen thickness is $\frac{1}{8}$-in. but $\frac{1}{4}$ in. is preferred since they are less susceptible to bending and crushing. A notch is cut into a specimen very carefully by a milling machine or a lathe. The

Figure 2-38. Pendulum impact tester. (Courtesy Tinius Olsen Company.)

recommended notch depth is 0.100-in. Figure 2-39 illustrates a commercially available notching machine.

TEST PROCEDURES

Izod Test. The test specimen is clamped into position so that the notched end of the specimen is facing the striking edge of the pendulum. A properly positioned test specimen is shown in Figure 2-40. The pendulum hammer is released, allowed

Figure 2-39. Notching machine for impact test bars. (Courtesy Testing Machines, Inc.)

to strike the specimen, and swing through. If the specimen does not break, more weights are attached to the hammer and the test is repeated until failure is observed. The impact values are read directly in in.-lbs or ft-lbs from the scale. The impact strength is calculated by dividing the impact values obtained from the scale by the thickness of the specimen. For example, if a reading of 2 ft-lb is obtained using an $\frac{1}{8}$-in.-thick specimen, the impact value would be 16 ft-lb/in. of notch. The impact values are always calculated on the basis of 1-in.-thick specimens even though much thinner specimens are usually used. The unnotched impact strength is obtained by reversing the position of a notched specimen in the vise. In this

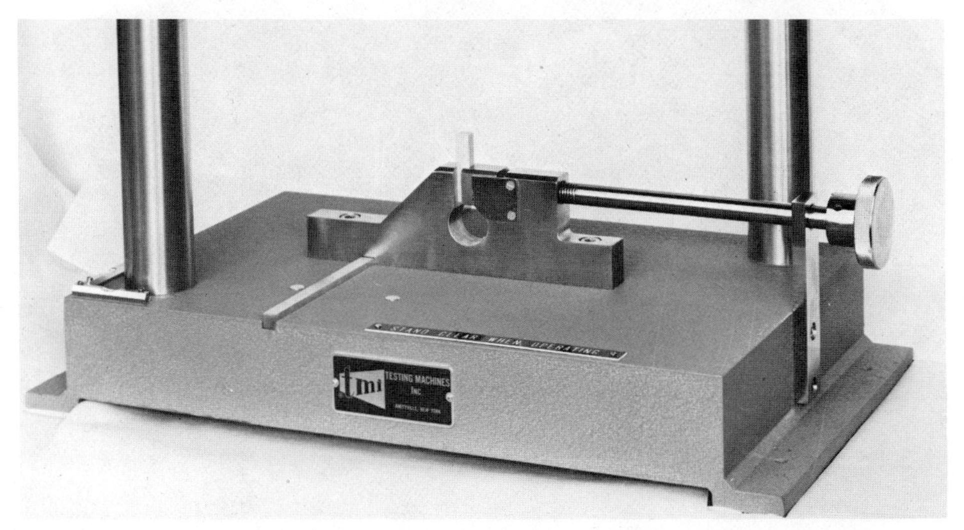

Figure 2-40. Diagram illustrating Izod impact test specimen properly positioned in test fixture. (Courtesy Testing Machines, Inc.)

case, the notch is subjected to compressive rather than tensile stresses during impact. As discussed earlier in this chapter, the energy required to break a specimen is the sum of the energies needed to deform it, initiate and propagate the fracture, and toss the broken end (toss factor).

Notching of the test specimen drastically reduces the energy loss due to the deformation and can generally be neglected. Tough plastic materials that have an Izod impact strength higher than 0.5 ft-lb/in. of notch seem to expend very little energy in tossing the broken end of the specimen. For relatively brittle material, having an Izod impact strength less than 0.5 ft-lb/in. of notch, the energy loss due to the toss factor represents a major portion of the total energy loss and cannot be overlooked. A method to determine such energy loss is devised (30).

Charpy Impact Test. This test is conducted in a very similar manner to the Izod impact strength test. The only difference is the positioning of the specimen. Figure 2-41 illustrates one such set-up. In this test the specimen is mounted horizontally and supported at both ends. Only the specimens that break completely are considered acceptable. The Charpy impact strength is calculated by dividing the indicator reading by the thickness of the specimen. The results are reported in ft-lbs/in. of notch for notched specimens and ft-lbs/in. for unnotched specimens.

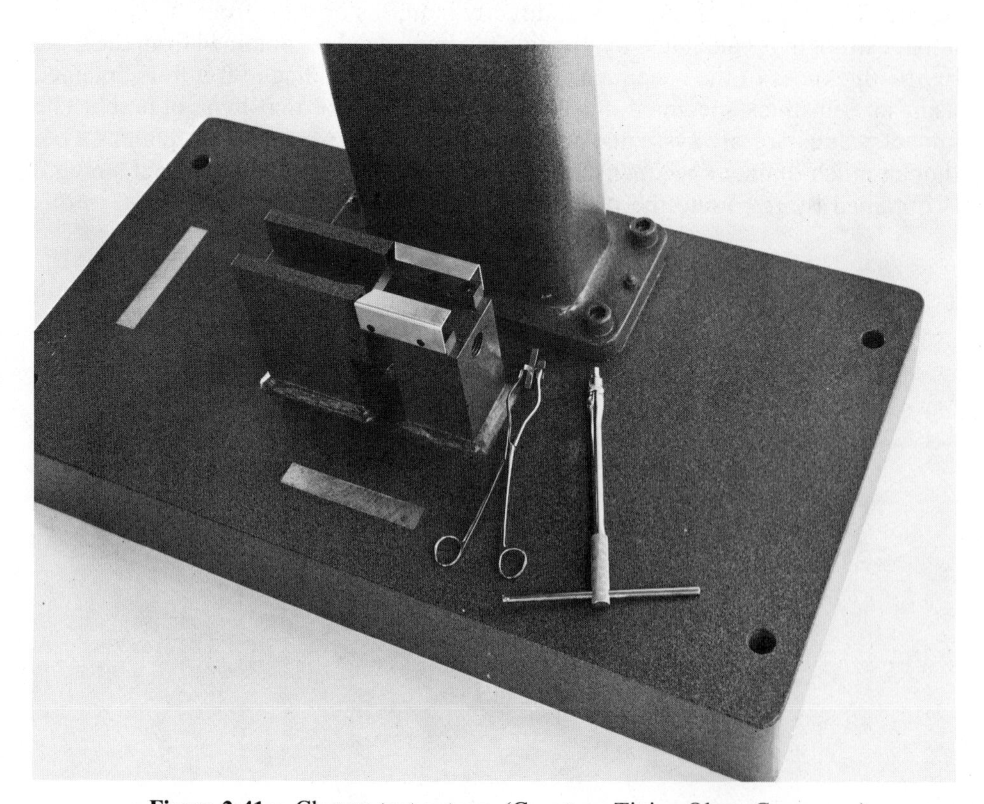

Figure 2-41. Charpy test set-up. (Courtesy Tinius Olsen Company.)

EFFECT OF TEST VARIABLES AND LIMITATIONS

Notch. A slight variation in the radius and depth of a notch affects the impact strength results. Many other variables such as the cutter speed, sharpness of the cutting tooth, feed rate, type of plastic, and quality of the notch cutting equipment, all seem to have a significant effect on the results. Such variations are difficult to control and nonuniformity between the lots is quite common. Certain heat sensitive polymers are also affected by the high cutter speed that seems to contribute to the thermal degradation. The notch in the specimen tends to create a stress concentration area that produces unrealistically low impact values in crystalline plastics.

Specimen Thickness. Although the impact values reported are on the basis of 1-in.-thick specimens, the actual thickness of the specimen used in the test influences the test results.

Specimen Preparation. Injection molded specimens seem to yield higher impact strength values than compression molded specimens. This is due to the molecular orientation caused by the injection molding process. The location of the gate also has a significant effect on the test results, particularly in the case of fiber-reinforced specimens.

Temperature. Impact values increase with the increasing temperature and vice versa.

Fillers and Other Additives. Fillers and reinforcements have a pronounced effect on the test results. For example, unreinforced polycarbonate has an Izod impact value of 15 ft-lb/in. of notch, but reinforcing it with 30 percent glass fibers reduces the impact to a mere 3.7 ft-lb/in. of notch. In contrast, reinforcing polystyrene with 30 percent glass fibers more than doubles the Izod impact strength. Fillers and pigments generally lower the impact strength.

Limitations. The results obtained from the Izod or Charpy test cannot be directly applied to part design since these tests do not measure the true energy required to break the specimen. The notched Izod impact test measures only the notch sensitivity of the different polymers and not the toughness.

Chip Impact Test. The chip impact test was originally developed by Uniroyal* research scientists. The test is considered to be particularly valuable for measuring the effect of surface microcracking caused by the weathering on impact strength retention. The material toughness is measured by this test as opposed to the material's notch sensitivity as measured by the notched Izod impact test. The test also allows a user to determine the orientation, flow effects, and the weld line strength, properties that are difficult to assess with the conventional impact techniques (31).

* Uniroyal Chemical Co. Naugatuck, CT.

The chip impact test is somewhat similar to the Izod impact strength test. The chip impact test requires the use of a pendulum hammer type of device and a specimen holding fixture. The test specimens are usually 1 in.-long × $\frac{1}{2}$ in. wide and 0.065 in. thick. The specimens can be prepared either by injection or compression molding or by simply cutting them from a sheet.

The test is carried out by mounting the test specimen as shown in Figure 2-42. The pendulum hammer is released, allowed to strike the specimen, and swing through. The retained toughness is proportional to the energy absorbed during impact which is measured by the angle of travel of the pendulum after impact. This is schematically shown in Figure 2-43. The value is expressed in in.-lb/in.^{2}

If the test is used exclusively to study the effect of weathering on a polymer, it is required that the sample be struck with a pendulum hammer on the weather exposed side. The chip impact test is also useful in measuring the relative toughness of a rather large and a complex-shaped part that is difficult to hold in a conventional fixture. A small chip can be cut from such a part and subjected to the chip impact test.

Tensile Impact Test (ASTM D 1822). The tensile impact strength test was developed to overcome the deficiencies of flexural (Izod and Charpy) impact tests. The test variables, such as notch sensitivity, toss factor, and specimen thickness, are eliminated in the tensile impact test. Unlike Izod–Charpy-type pendulum impact tests, which are limited to thick specimens only, the tensile impact test allows the user to determine the impact strength of very thin and flexible specimens. Many other characteristics of polymeric materials, such as the anisotropy and the orientation effect, can be studied through the use of the tensile impact test.

Figure 2-42. Chip impact test set-up. (Courtesy Uniroyal Chemical Company.)

The "chip impact test" is a sensitive measure of material toughness retention upon UV exposure. It has been found to correlate very well to part toughness upon outdoor exposure, as measured by drop weight testing. It is an unnotched test and therefore particularly sensitive to surface degradation.

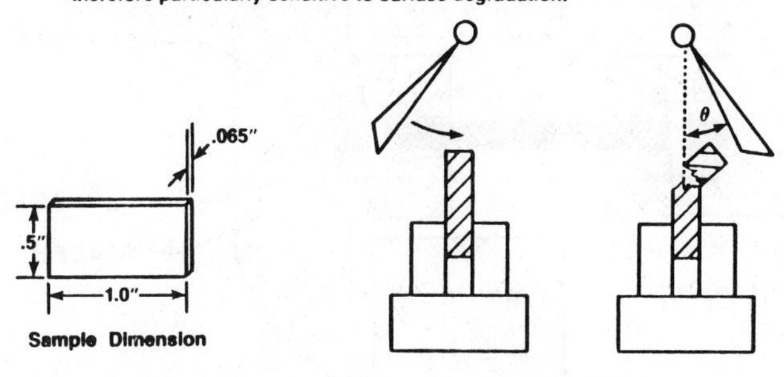

The sample is exposed to accelerated or outdoor aging, clamped in a vise, and struck on the exposed face by a pendulum-type hammer. The retained toughness is proportional to the energy absorbed during impact and is measured by the angle of travel (θ) of the pendulum after impact. The value is expressed as in.-lb./in.^{2}.

Figure 2-43. Diagram illustrating principle of chip impact test. (Courtesy Uniroyal Chemical Company.)

The early development work on tensile impact testers included the specimen-in-base type of set-up (32). This method was excluded by the ASTM Committee D-20 due to the inherent problem of toss factor correction and repeatability. The present acceptable tensile impact test consists of a specimen-in-head type of set-up. In this case, the specimen is mounted in the pendulum and attains full kinetic energy at the point of impact, eliminating the need for a toss factor correction. The energy to break by impact in tension is determined by kinetic energy extracted from the pendulum in the process of breaking the specimen. The test set-up requires mounting one end of the specimen in the pendulum and the other end to be gripped by a crosshead member, which travels with the pendulum until the instant of impact. The pendulum is affected only by the tensile force exerted by the specimen through the pendulum's center of percussion.

As long as the machine base is rigid enough to prevent the vibrational energy losses, the bounce of the crosshead in the opposite direction can be easily calculated. The tensile impact test is far more meaningful than the Izod–Charpy type tests since it introduces strain rate as an important test variable. Many researchers have demonstrated that the tensile impact test results correlate better with the actual field failures and are easier to analyze than the Izod impact test results (33–35).

Apparatus and Test Specimens. The tensile impact testing machine consists of a rigid massive base with a suspending frame. The pendulum is specially designed to hold the dumbbell-shaped specimen so that the specimen is not under stress until the moment of impact. Figure 2-44 schematically illustrates a specimen-in-head tensile impact machine. Illustrated in Figure 2-45 is a commercially available tensile impact tester. The specimens are prepared by molding, machining, or by die-cutting to the desired shape from a sheet. Two basic specimen

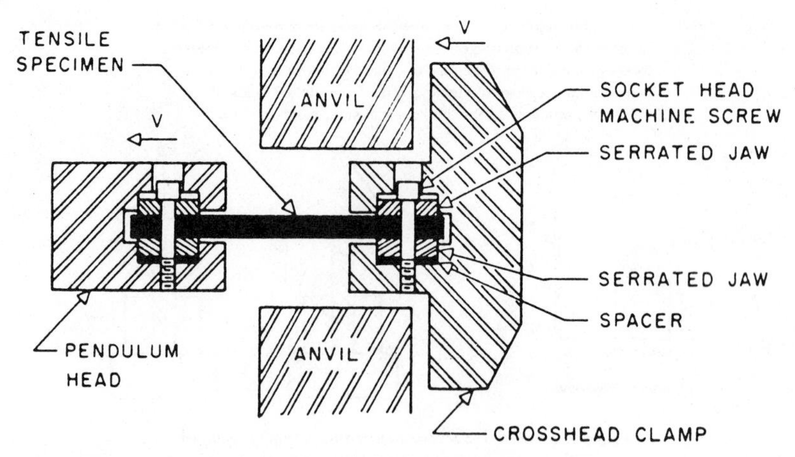

Figure 2-44. Schematic of specimen-in-head tensile-impact machine. (Reprinted by permission of ASTM.)

geometries are used so that the effect of elongation or the rate of extension or both can be observed. Type S (Short) specimens usually exhibit low extension while the Type L (Long) specimen extension is comparatively high. Type S specimens provide a greater occurrence of brittle failures. Type L specimens provide a greater differentiation between materials. Mold dimensions of both types of specimens are shown in Figure 2-46.

Test Procedure. The thickness and width of the test specimen are measured with a micrometer. The specimen is clamped to the crosshead while the crosshead is out of the pendulum. The whole assembly is placed in the elevated pendulum. The other end of the specimen is bolted to the pendulum. The pendulum is released and the crosshead is allowed to strike the anvil. The tensile impact energy is

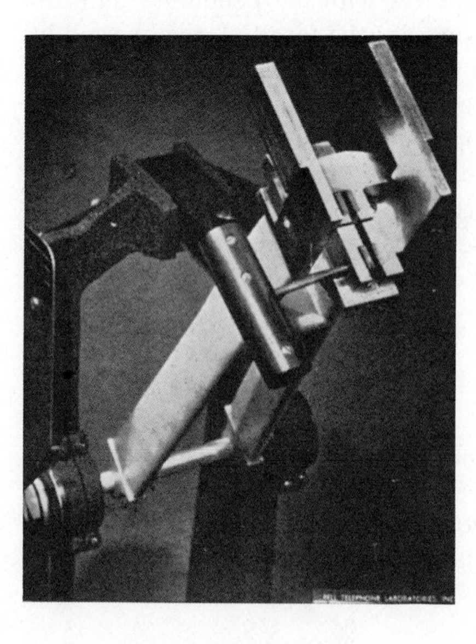

Figure 2-45. Tensile impact tester.

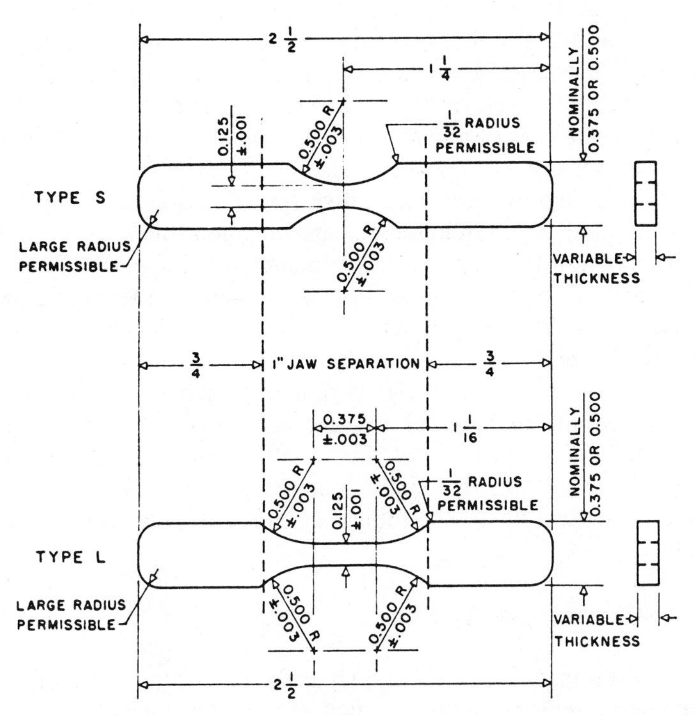

Figure 2-46. Mold dimensions of types S and L tensile-impact specimens. (Reprinted by permission of ASTM.)

measured and recorded from the scale reading. A corrected tensile impact energy to break is calculated as follows:

$$X = E - Y + e$$

where X = corrected impact energy to break (ft-lb);
E = scale reading of energy of break (ft-lb);
Y = friction and windage correction (ft-lb);
e = bounce correction factor (ft-lb).

A bounce correction factor curve for each type of crosshead and pendulum employed must be developed.

B. High-Rate Tension Test

The high-rate tension test was developed to overcome the difficulties involved in the evaluation of impact properties using pendulum-type impact testers. Earlier work was done by using conventional tensile testing equipment. At high speeds, such as 20 in./min., a good correlation with the falling-weight test results was obtained (36). The area under stress–strain curve is proportional to the energy required to break the material. This area is also directly proportional to the impact strength of the material if the curve is generated at a high enough speed. The relationship of tensile strength and elongation with the impact strength of the material is also evaluated by the high-rate tension test. Tensile strength seems to increase with the increase in speed of testing, while elongation decreases with the increasing speed. Therefore, the high-rate tension test must be conducted at a speed suitable to the type of polymer in order to obtain some meaningful results.

Ultra-high-rate tensions testers have also been developed to study the impact behavior of the ductile polymers. These high speed, hydraulically operated testing machines are capable of providing linear displacement rates up to 10,000 in./min (37). High-rate tension tests are usually conducted using a variety of test specimens. Test bars used for the standard tensile test and the tensile impact test are most commonly used. Sample geometrics resembling the actual part are also used to study the mechanical behavior at high speeds.

Many other important impact design parameters such as the yield stress, energy to yield, initial modulus, and the deformation at break can be measured with high-speed tension tests (38). In spite of the capability of high-rate tension tests to provide stress at strain rates that simulate actual service, the tests have not been popular since the delivered stress is uniaxial. Normally, the real-life impact stress is multiaxial.

C. Falling-Weight Impact Test

The falling-weight impact test, also known as the drop impact test or the variable-height impact test, employs a falling weight. This falling weight may be a tup with a conical nose, a ball, or a ball-end dart. The energy required to fail the specimen is measured by dropping a known weight from a known height onto a test specimen. The impact energy is normally expressed in ft-lb and is calculated by multiplying the weight of the projectile by the drop height.

The biggest advantage of the falling-weight impact test over the pendulum impact test or high-rate tension test is its ability to duplicate the multidirectional impact stresses that a part would be subjected to in actual service. The other obvious advantage is the flexibility to use specimens of different sizes and shapes, including an actual part. Unlike the notched Izod impact test which measures the notch sensitivity of the material and not the material toughness, falling-weight impact tests introduce polyaxial stresses into the specimen and measure the toughness. The variations in the test results due to the fillers and reinforcements, clamping pressure, and material orientation are virtually eliminated in the falling-weight impact test. This type of test is also very suitable for determining the impact resistance of plastic films, sheets, and laminated materials.

Three basic ASTM tests are commonly used depending upon the application:

ASTM D 3029	Impact resistance of rigid plastic sheeting or parts by means of a tup (falling weight).
ASTM D 1709	Impact resistance of polyethylene film by the free-falling dart method.
ASTM D 244	Test for impact resistance of thermoplastic pipe and fittings by means of a tup.

DROP IMPACT TEST. This falling-weight impact test is primarily designed for plastic sheeting material or fabricated parts. A free-falling weight or a tup is used to determine the impact strength of the material.

Many different versions of test equipment exist today. They all basically operate on the same principle. Figure 2-47 illustrates one such typical commercially available testing machine. It consists of a cast aluminum base, a slotted vertical guide tube, a round-nose punch and punch holder, an 8-lb weight, a die and die support, and a sample platform. The sample platform is used to position a sheet of desired thickness for impact testing. The die is removable from the base in order that the actual parts of complex shapes can be placed onto the base and impact tested.

The test is carried out by raising the weight to a desired height manually or automatically with the use of a motor-driven mechanism and allowing it to fall freely onto the other side of the round-nosed punch. The punch transfers the

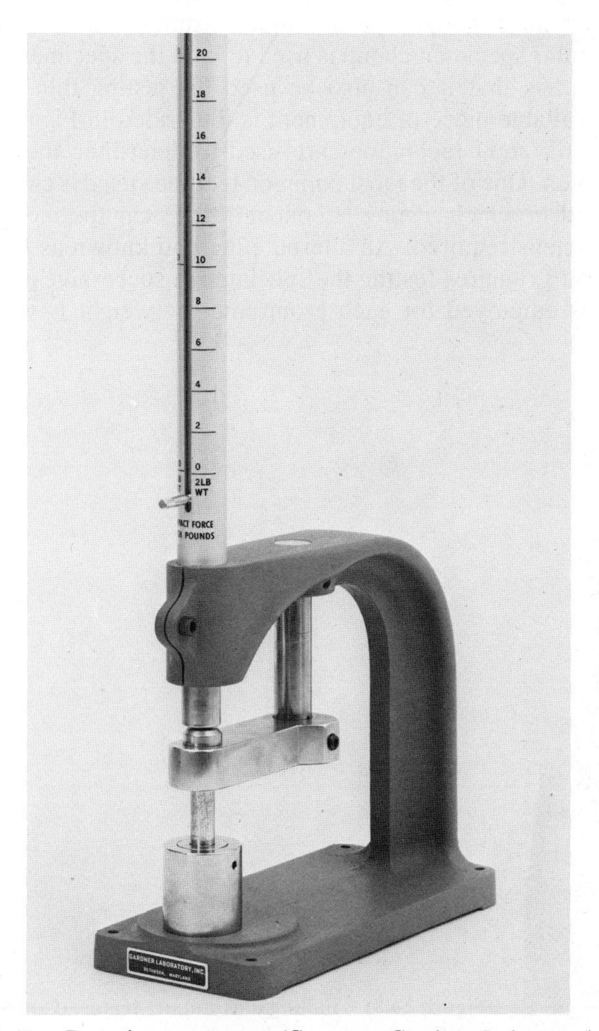

Figure 2-47. Drop impact tester. (Courtesy Gardner Laboratories, Inc.)

impact energy to the flat test specimen positioned on a cylinderical die or a part lying on the base of the machine. The kinetic energy possessed by the falling weight at the instant of impact is equal to the energy used to raise the weight to the height of drop and is the potential energy possessed by the weight as it is released. Since the potential energy is expressed as the product of weight and height, the guide tube can be marked with a linear scale representing the impact range of the instrument in in.-lb. Thus, the toughness or the impact strength of a specimen or a part can be read directly off the calibrated scale in in.-lb. The energy loss due to the friction in the tube or due to the momentary acceleration of the punch is negligible.

An alternate method for achieving the same result utilizes an instrument that employs a free-falling dart dropped from a specified height onto a test specimen. The dart with a hemispherical head is constructed of smooth, polished aluminum or stainless steel. An electromagnetic, air-operated or other mechanical release mechanism with a centering device is used for releasing the dart. The dart is also fitted with a shaft long enough to accommodate removable incremental weights. A two-piece annular specimen clamp is used to hold the specimen. The advantage of this equipment is that it can also be used for testing thin films. One such commercially available piece of equipment is illustrated in Figure 2-48.

A number of different techniques are used to determine the impact strength value of a specimen. One of the most common methods used is called the Bruceton Staircase Method in which testing is concentrated near the mean to reduce the number of specimens required. An alternate method known as the ultimate non-failure level (UNF) requires testing the specimen in successive groups of 10. One missile weight is employed for each group and the weight is varied in uniform

Figure 2-48. Falling dart impact tester. (Courtesy Custom Scientific Instruments, Inc.)

Figure 2-49. Impact tester specifically designed for impact testing pipe and fittings. (Courtesy Custom Scientific Instruments, Inc.)

increments from group to group. These methods are discussed in detail in the ASTM Book of Standards (39).

The impact testing machine, such as the one illustrated in Figure 2-49, is specifically designed for testing plastic pipe and fittings. It employs a much heavier tup with three different sized radii on the tup nose. The drop tube is long enough to provide at least a 10-ft free fall.

TEST VARIABLES AND LIMITATIONS

Specimen Thickness. The drop impact strength of a specimen is directly proportional to the specimen thickness. As the thickness increases, the energy required to fracture the specimen also increases (40). Brittle polymers such as polypropylene show a smaller increase in the impact strength compared to the ductile polymers like ABS and impact polystyrene (41).

Specimen Slippage. If extreme care is not taken in clamping the specimen, some slippage is bound to occur upon impact, causing distorted results. This is particularly true in the case of films which seem to have a greater tendency towards slippage. Unclamped specimens seem to exhibit somewhat greater impact resistance.

Specimen Quality. The surface quality of the test specimen significantly affects the test results. For example, a notch in the test specimen will alter the impact strength values (42).

Stress Concentration. When impact testing a fabricated part, care must be taken not to impact the area of extreme stress concentration such as a bend on a 90° elbow fitting. This usually produces erroneous results because the degree of stress concentration appears to vary with the processing conditions. An area free of stress concentration should be chosen for more reliable results.

Limitations. So far, we have discussed many advantages of falling-weight impact tests over the conventional pendulum impact tests and high-rate tension tests. However, there are serious limitations with the falling-weight tests that cannot be overlooked. One of the biggest disadvantages of this type of test is the large number of samples required to establish an energy level to fail the sample. Since there is no way of determining how many trials will be required to actually fail the sample, a statistical approach must be used. Another serious limitation of a falling-weight test is the problem of isolating the rate of impact. Although a 2-lb tup dropped from 6 ft produces the same 12 ft-lb impact energy as a 6 lb tup dropped from 2 ft, the effect may not be the same. Impact resistance is directly related to the impact and, therefore, a specimen that can survive the impact from a 6-lb tup could fail from the higher velocity impact of the lighter 2-lb tup dropped from a greater height. Many commercially available drop impact testers do not provide flexibility of varying the rate of impact. Furthermore, the velocity impacts produced by such testers are much smaller than the velocity impacts encountered by products in actual use (43). The basic design of the existing drop impact equipment is not suitable for testing large parts such as structural foam cabinets and housings.

D. *Instrumented Impact Testing*

One of the biggest drawbacks of the conventional impact test methods is that it provides only one value—the total impact energy—and nothing else. The conventional tests cannot provide additional information on the ductility, dynamic toughness, fracture, and yield loads or the behavior of the specimen during the entire impact event.

All standard impact testers can be instrumented to provide a complete load and energy history of the specimen (44). Such a system monitors and precisely records the entire impact event, starting from the acceleration (from rest) to the initial impact and plastic bending to fracture initiation and propagation to the complete failure. The instrumentation is done by mounting the strain gauges onto the striking bit in the case of pendulum impact tester or onto the tup in the case of a drop impact tester. During the test, a fiber optic device triggers an oscilloscope just before striking the specimen. The output of the strain gauge is recorded by the oscilloscope depicting the variation of the load applied to the specimen throughout the entire fracturing process. A complete load–time history of the entire specimen is obtained. The apparent total energy absorbed by the specimen can be calculated and plotted against time. Figure 2-50 illustrates one such typical load–energy–time curve. The specimen displacement can be calculated by the double integration of the load–time curve and the load–displacement curve can be plotted (45). With the advent of microprocessor technology, some manufacturers are now capable of offering a unit that automatically calculates the sample

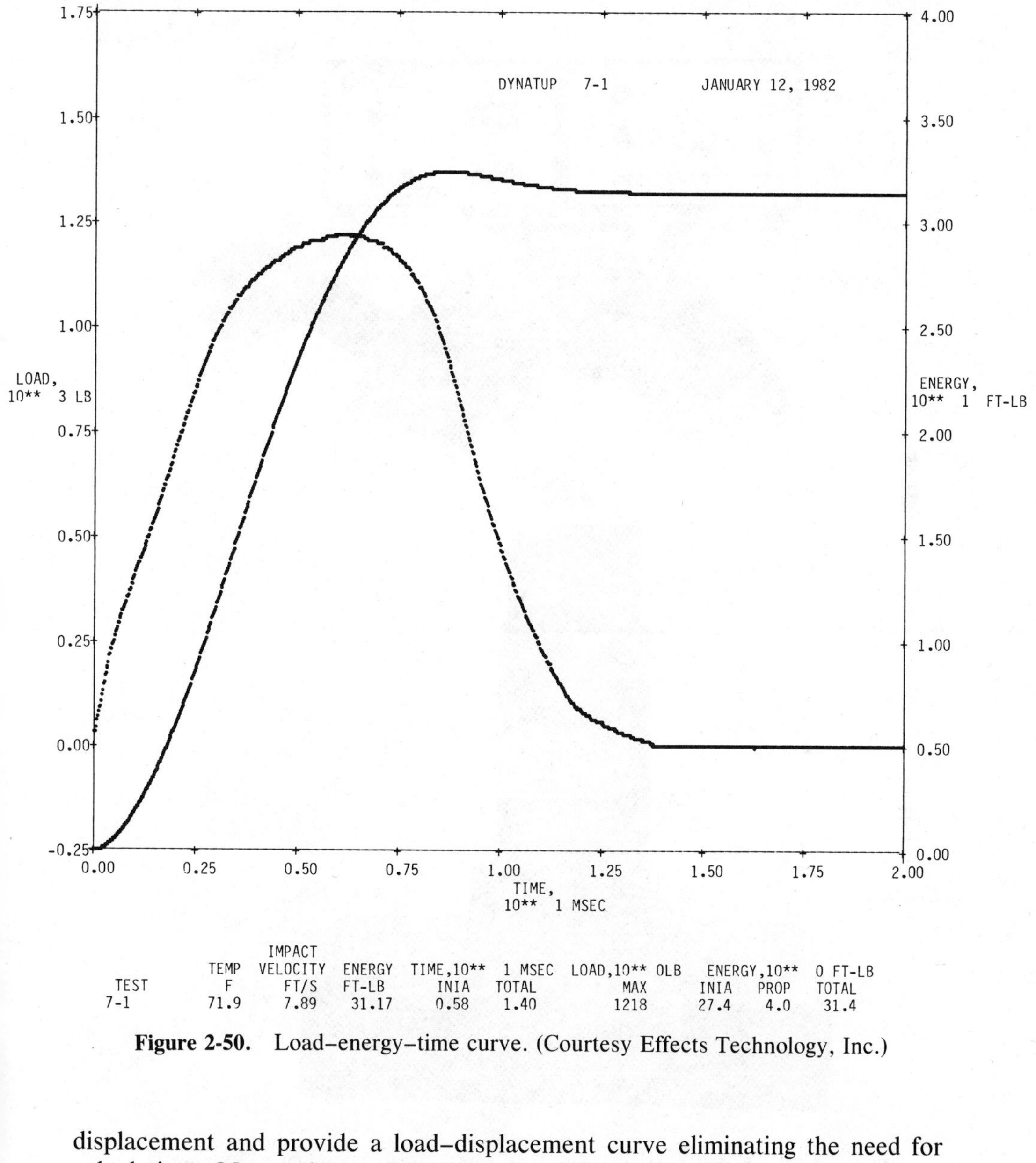

TEST	TEMP F	IMPACT VELOCITY FT/S	ENERGY FT-LB	TIME,10** 1 MSEC INIA	TOTAL	LOAD,10** 0LB MAX	ENERGY,10** 0 FT-LB INIA	PROP	TOTAL
7-1	71.9	7.89	31.17	0.58	1.40	1218	27.4	4.0	31.4

Figure 2-50. Load–energy–time curve. (Courtesy Effects Technology, Inc.)

displacement and provide a load–displacement curve eliminating the need for calculations. Many other useful data such as the impact rate, force and displacement at yield, and break, yield, and failure energies as well as modulus are calculated and printed out. A commercially available instrumented impact tester is shown in Figure 2-51.

E. High-Rate Impact Tests

An ever-increasing demand for engineering plastics and the need for sophisticated and meaningful impact test methods for characterizing these materials has forced

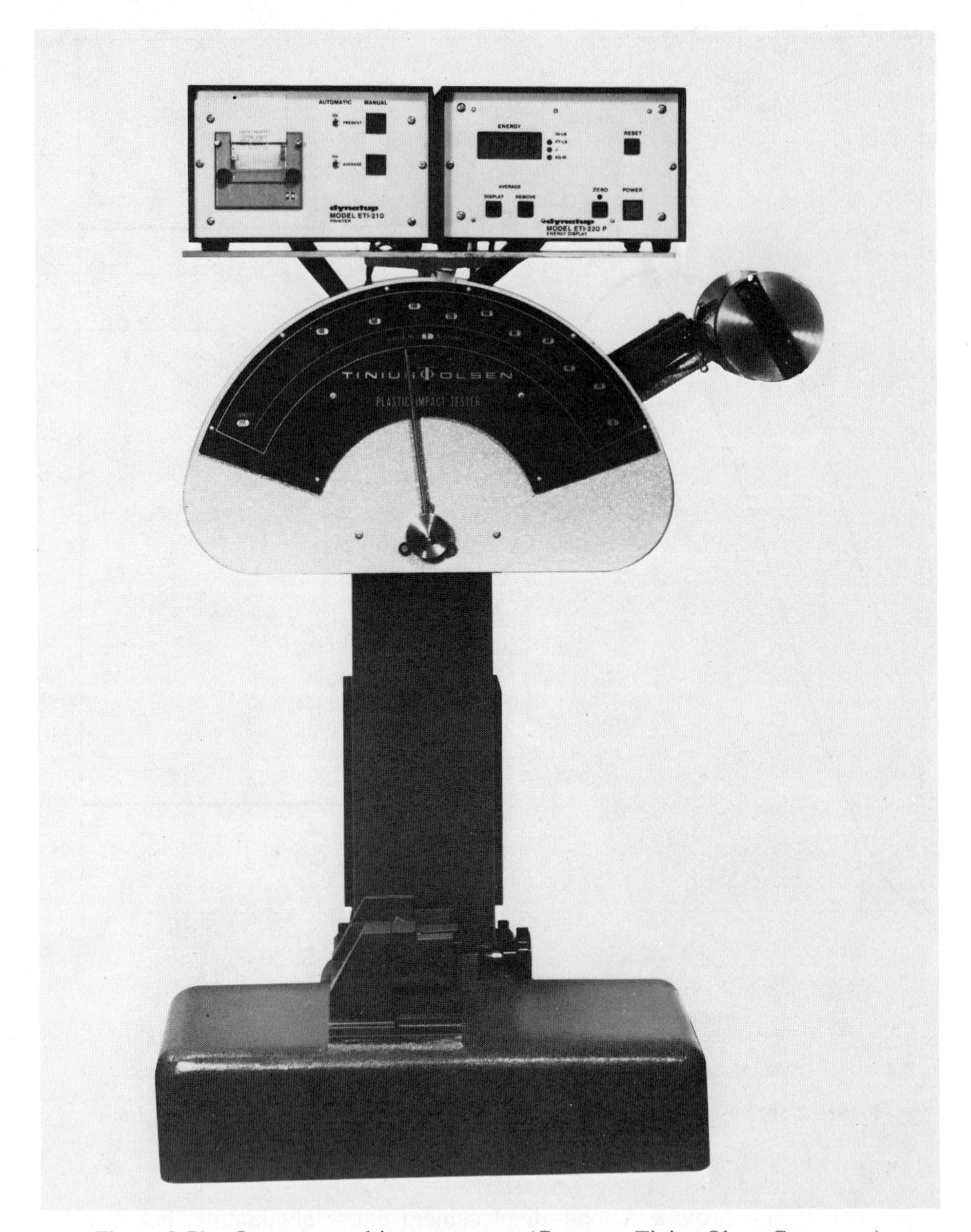

Figure 2-51. Instrumented impact tester. (Courtesy Tinius Olsen Company.)

the industry into developing new high-rate impact tests. These tests not only provide the basic information regarding the toughness of the polymeric materials but provide other important data of interest, such as the load–deflection curve and the total energy absorption. The high-speed impact test overcomes the basic limitations of conventional impact testing methods as discussed previously. The rate of impact can be varied from 30 to 570,000 in./min.

High-rate impact testing has gained considerable popularity in recent years because of its ability to simulate actual impact failures at high speeds. For example, conventional impact testing methods are useless in testing to meet advanced automotive crash standards that require the impact simulation at 28 mph (30,000 in./min). High-speed impact testers are able to meet the challenge. As discussed earlier in this chapter, almost all polymers are strain-rate sensitive. Two polymers, when impact tested at one strain rate, may show similar impact strength values. The same two polymers tested at a high strain rate show a completely different set of values. Two basic types of high speed impact testers are available:

1. High-speed plunger-impact tester.
2. High-speed ball-impact tester.

HIGH-SPEED PLUNGER IMPACT TESTER. This versatile high-speed impact testing machine is capable of testing everything from the thin film which may require as low an impact rate as 30 in./min to the plastic automotive bumper which may require a high impact rate up to 30,000 in./min. The specimen or product can be tested under a controlled environment of temperature and humidity. A commercially available high-rate impact tester is illustrated in Figure 2-52. The equipment basically consists of a horizontal test plunger attached to a hydraulically powered actuator. The hydraulic power is delivered by an accumulator that is charged by a pump. The force is detected with a fast responding quartz load cell mounted directly on the actuator. The velocity can be set digitally from 30–30,000 in./min. The actuator is capable of moving in either direction to make both tensile and impact tests possible. The starting and stopping point of the actuator can be set for the repeated test and for the control of penetration depth in thick samples. The horizontal actuator design permits virtually unlimited variation in sample geometry, size, or orientation. By simply changing an impact probe or a sample clamping fixture, all conventional tests such as Izod, Charpy, and tensile impact, as well as finished part testing, can be done. The tester is equipped with a CRT and x–y plotter that automatically displays load versus displacement data. A built-in microprocessor provides more useful information, such as modulus, yield, and failure energies.

High-rate impact testers have been proven very useful in material evaluation. At low strain rates, some polymers fail in ductile manner. The same polymers appear to show brittle failure at high strain rates. The point at which this ductile to brittle transition takes place is of particular importance. A high-rate impact test can provide such information in a graphical form. Tests can also be carried out at different temperatures to find ductile–brittle transition points at various temperatures (46). Other useful applications of the high-rate impact tester include the process quality control, design evaluation, and assembly evaluation.

F. Miscellaneous Impact Tests

Depending upon the end-use requirement, many different types of impact tests have been devised to stimulate actual impact conditions. Underwriters Labora-

Figure 2-52. High-rate impact tester. (Courtesy Rheometrics, Inc.)

tories requires a fully assembled, working electronic cabinet to withstand 5-ft-lb impact of a swinging ball at the most vulnerable appearing spot to check the damage to the electronics. Small units susceptible to being knocked off a desk must pass a 3 ft drop while the units are plugged in the electrical outlets (47).

Drop impact resistance of a blow molded thermoplastic container is a standard ASTM D 2463 test. The drop impact resistance is determined by dropping conditioned blow molded containers filled with water from a platform onto a prescribed surface. A commercially available bottle drop apparatus is shown in Figure 2-53. One other type of impact test, called the air cannon impact test (ACIT), is used to determine the toughness of rigid plastic exterior building components. The air cannon impact tester consists of a compressed air gun to propel spherical plastic projectiles at a test specimen. Interchangeable barrels allow projectiles of varying sizes and weights to be used. Projectile velocity is controlled by varying

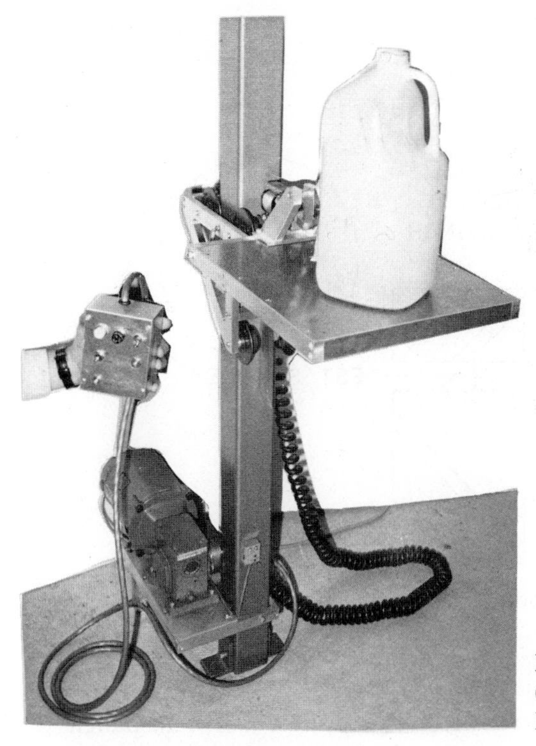

Figure 2-53. Bottle drop impact tester. (Courtesy Custom Scientific Instruments, Inc.)

air pressure. Polypropylene and polyethylene molded balls of different sizes are used as projectiles to simulate the effect of different sized hailstones. The weight and the velocity of the projectile are used to calculate the impact force absorbed by the test specimen. This test more realistically simulates end-use environmental conditions than conventional impact tests and provides important information for evaluating product and establishing product quality (48).

2.8. SHEAR STRENGTH (ASTM D 732)

Shear strength of plastic materials is defined as the ability to withstand the maximum load required to shear the specimen so that the moving portion completely clears the stationary portion. Shear strength test is carried out by forcing a standardized punch at a specified rate through a sheet of plastic until the two portions of the specimen completely separate. The shear strength is determined by dividing the force required to shear the specimen by the area of the sheared edge and is expressed as lb/in.^{2}

2.8.1. Test Specimen and Apparatus

The specimen for the shear strength test can be either molded or cut from a sheet in the form of a 2-in.-diameter disc or 2-in.^{2} plate. The thickness of the specimen may vary from 0.005 to 0.500 in. A hole with 7/16-in. diameter is drilled through the specimen at its center. A universal testing machine with a constant-rate-of-crosshead movement such as the one used for tensile, compressive, and flexural strength testing can be used. The machine requirement has been described in

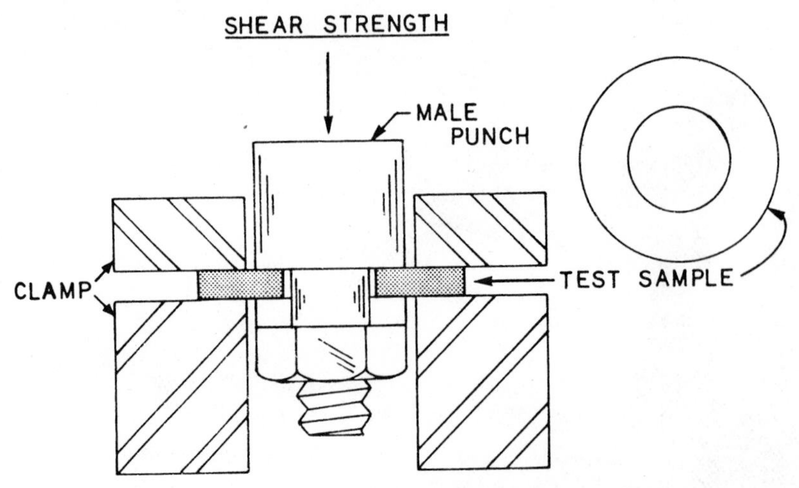

Figure 2-54. Shear strength test set-up. (Reprinted by permission of Van Nostrand–Reinhold Company.)

detail in the section on tensile testing. A specially designed shear tool of the punch type is used for shear testing. One such fixture with a sample already mounted in place is shown in Figure 2-54.

2.8.2. Test Procedures

The test is carried out by properly positioning the sample in the shear fixture and lowering the punch onto the specimen until a complete shear type failure is observed. Shear strength is calculated as follows:

$$\text{Shear strength (psi)} = \frac{\text{Force required to shear the specimen}}{\text{Area of sheared edge}}$$

$$(\text{Area of sheared edge} = \text{Circumference of punch} \times \text{Thickness of specimen})$$

2.8.3. Significance and Limitations

Shear strength data is of great importance to a designer of film and sheet products that tend to be subjected to such shear loads. Most large molded and extruded products are usually not subjected to shear loads. The shear strength data as reported in the material suppliers literature should be used with extreme care when designing a product since these shear strength values can be considerably higher than normal. This method allows a user to test a very thin specimen that may stretch excessively and prevent a true shear type failure, giving high values. For design purposes, shear strength is generally considered to be essentially equal to one-half the tensile strength.

2.9. ABRASION

2.9.1. Introduction

Abrasion resistance of polymeric materials is a complex subject. Many theories have been developed to support the claim that abrasion is closely related to frictional force, load, and true area of contact. An increase in any one of the three generally results in greater wear or abrasion. The hardness of the polymeric materials has a significant effect on abrasion characteristics. For example, a harder material with considerable asperities on the surface will undoubtedly cut through the surface of a softer material to an appreciable depth, creating grooves and scratches. The theory is further complicated by the fact that the abrasion process also creates oxidation on the surface from the build-up of localized high temperatures (49,50). The resistance to abrasion is also affected by other factors, such as the properties of the polymeric material, resiliency, and the type and amount of added fillers or additives.

This all makes abrasion a difficult mechanical property to define as well as to measure adequately. Resistance to abrasion is defined as the ability of a material to withstand mechanical action (such as rubbing, scraping, or erosion) that tends progressively to remove material from its surface.

Resistance to abrasion is significantly affected by factors, such as test conditions, type of abradant, and development and dissipation of heat during the test cycle. Many different types of abrasion-measuring equipment have been developed. However, the correlation between the test results obtained from different machines as well as the correlation between test results and actual abrasion-related wear in real life remains very poor. The tests do, however, provide relative ranking of materials in certain order when performed under specified set of conditions.

2.9.2. Abrasion Resistance Tests

The material's ability to resist abrasion is most often measured by its loss in weight when abraded with an abraser. The most widely accepted abraser in the industry is called the Taber abraser. A variety of wheels with varying degree of abrasiveness is available. The grade of "calibrase" wheel designated CS-17 with 1000-g load seems to produce satisfactory results with almost all plastics. For softer materials less abrasive wheels with smaller load on the wheels may be used. The test specimen is usually a 4-in.-diameter disc or a 4-in.^{2} plate having both surfaces substantially plane and parallel. A $\frac{1}{4}$-in.-diameter hole is drilled in the center. Specimens are conditioned employing standard conditioning practices prior to testing. To commence testing, the test specimen is placed on a revolving turntable. Suitable abrading wheels are placed on the specimen under certain set dead weight loads. The turntable is started and an automatic counter records the number of revolutions. Most tests are carried out to at least 5000 revolutions. The specimens are weighed to the nearest mg.

The test results are reported as weight loss in mg/1000 cycles. The grade of

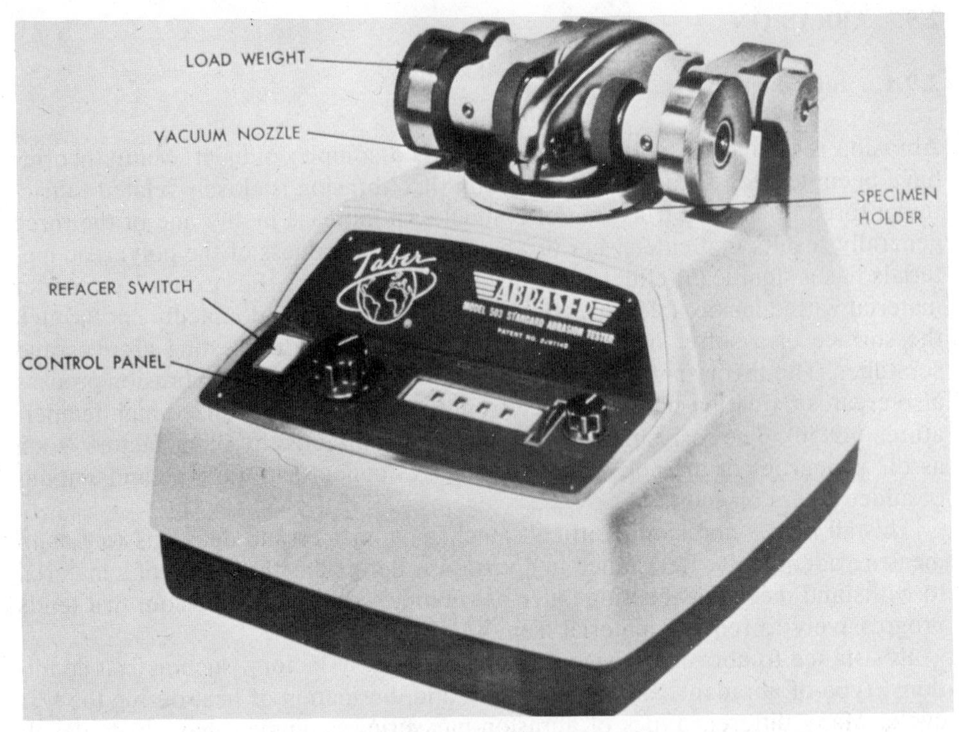

Figure 2-55. Abrasion tester. (Courtesy Gardner Laboratories, Inc.)

abrasive wheel along with amount of load at which the test was carried out is always reported along with results.

Test methods such as ASTM D 1044 (resistance of transparent plastic materials to abrasion) are also developed for estimating the resistance of transparent plastic materials to one kind of abrasion by measurement of its optical affects. The test is carried out in similar manner to that described above, except that 100 cycles with a 500 g. load is normally used. A photoelectric photometer is used to measure the light scattered by abraded track. The percentage of the transmitted light that is diffused by the abraded specimens is reported as a test result.

Another method to study the resistance of plastic material to abrasion is by measuring the volume loss when a flat specimen is subjected to abrasion with loose abrasive or bonded abrasive on cloth or paper. This method is designated as ASTM D 1242. Figure 2-55 illustrates a commercially available abrasion tester.

2.10. FATIGUE RESISTANCE

2.10.1. Introduction

The behavior of materials subjected to repeated cyclic loading in terms of flexing, stretching, compressing, or twisting is generally described as fatigue. Such repeated cyclic loading eventually constitutes a mechanical deterioration and pro-

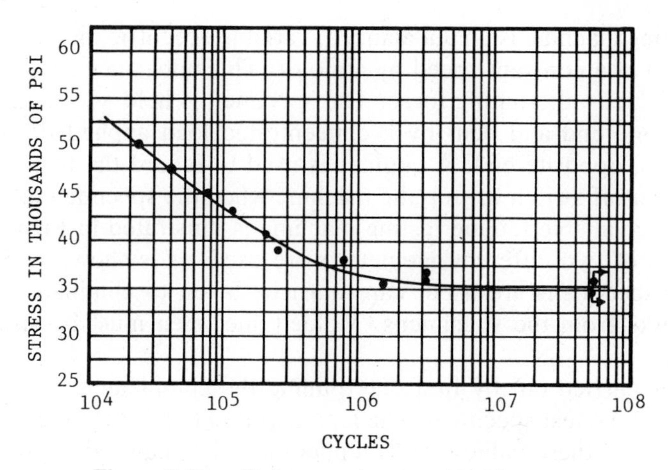

Figure 2-56. Fatigue endurance (S–N) curve.

gressive fracture that leads to complete failure. Fatigue life is defined as the number of cycles of deformation required to bring about the failure of the test specimen under a given set of oscillating conditions (51).

The failures that occur from repeated application of stress or strain are well below the apparent ultimate strength of the material. Fatigue data are generally reported as the number of cycles to fail at a given maximum stress level. The fatigue endurance curve, which represents stress versus number of cycles to failure, also known as S–N curve, is generated by testing a multitude of specimens under cyclic stress, each one at different stress levels. At high stress levels, materials tend to fail at relatively low numbers of cycles. At low stresses, the materials can be stressed cyclically for an indefinite number of times and the failure point is virtually impossible to establish. This limiting stress below which material will never fail is called the fatigue endurance limit. As shown in Figure 2-56, the S–N curve becomes asymptotic to a constant stress line at a certain stress level. The fatigue endurance limit can also be defined as the stress at which the S–N curve becomes asymptotic to the horizontal (constant stress) line. For most polymers, the fatigue endurance limit is between 25–30 percent of the static tensile strength (52). The fatigue resistance data are of practical importance in the design of gears, tubing, hinges, parts on vibrating machinery, and pressure vessels under cyclic pressures.

Two basic types of tests have been developed to study the fatigue behavior of plastic materials:

1. Flexural fatigue test.
2. Tensile fatigue test.

2.10.2. Flexural Fatigue Test (ASTM D 671)

The ability of a material to resist deterioration from cyclic stress is measured in this test by using a fixed cantilever-type testing machine capable of producing a constant-amplitude-of-force on the test specimen each cycle. The main feature of a fatigue testing machine is an unbalanced, variable eccentric, mounted on a shaft that is rotated at a constant speed by a motor. This unbalanced movement of an eccentric produces alternating force. The specimen is held as a cantilever beam in a vice at one end and bent by a concentrated load applied through a yoke fastened to the opposite end. A counter is used to record the number of cycles along with a cutoff switch to stop the machine when the specimen fails. A typical commercially available fatigue testing machine is illustrated in Figure 2-57. The test specimen of two different geometrics are used, as is shown in Figure 2-58. If machined specimens are used, care must be taken to eliminate all scratches and tool marks from the specimens. Molded specimen must be stress-relieved before using.

The test is carried out by first determining the complementary mass and effective mass of the test specimen. The load required to produce the desired stress is calculated from these values. The number of cycles required to produce failure is determined. The test is repeated at varying stress levels. A curve of stress versus cycles-to-failure (S–N diagram) is plotted from the test results.

2.10.3. Tensile Fatigue Test

Unlike the flexural fatigue test which uses the constant deflection (strain) principle, the tensile fatigue test is conducted under constant load (stress) conditions.

Figure 2-57. Flexural fatigue tester. (Courtesy Fatigue Dynamics, Inc.)

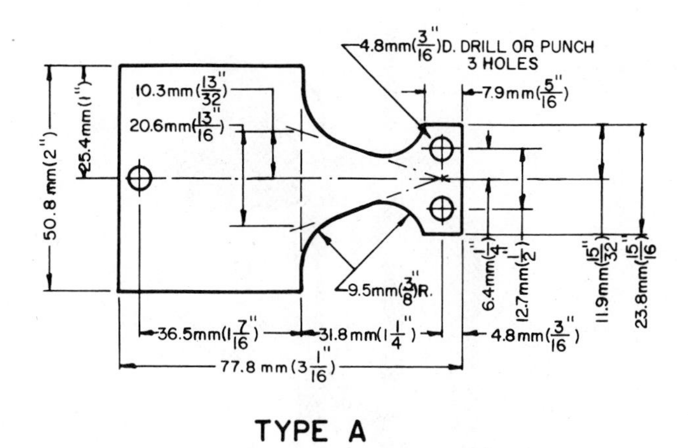

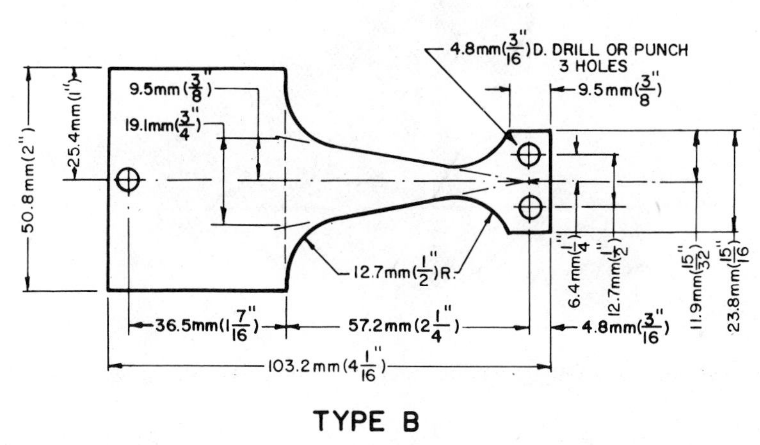

Figure 2-58. Dimensions of constant force fatigue specimen. (Reprinted by permission of ASTM.)

The testing machine most commonly used is shown in Figure 2-59. The specimen is dumbbell-shaped, about 2 in. long with a cylindrical cross section.

The test is conducted by mounting both ends of the dumbbell specimen in the testing machine. The specimen is rotated between two spindles, and stress in the form of tension and compression is applied. The specimen is subjected to the number of cycles of stress specified or until fracture occurs.

2.10.4. Factors Affecting the Test Results and Limitation of Fatigue Tests

1. The data obtained from fatigue tests can be directly applied in designing a specific part only when all variables such as size and shape of the part, method of specimen preparation, loading, ambient and part temperature, and frequency of stressing are identical to those used during testing. In practice, such identical conditions never occur and, therefore, it is very important to conduct the trial on

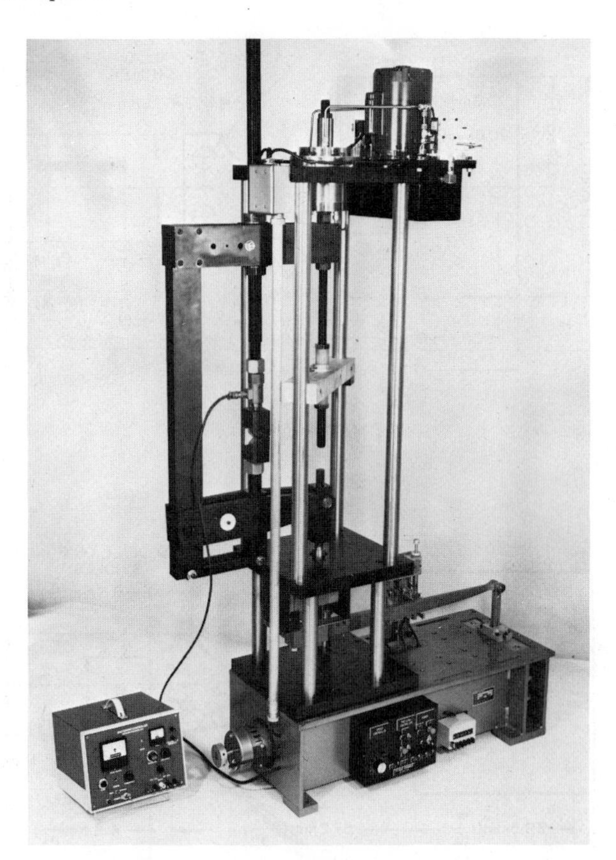

Figure 2-59. Tensile fatigue testing machine. (Courtesy Fatigue Dynamics, Inc.)

an actual part with the end-use conditions simulated as closely as possible. Another reason for not relying very heavily on fatigue test data is the possibility of the presence of a notch, a scratch, or voids in the fabricated part that can cause localized stress concentration and lower the overall fatigue resistance considerably.

2. The correlation of test results obtained from different types of machines is very poor.

3. The thickness of the sample greatly influences the test results.

4. The test results are seriously affected by changes in testing temperature, test frequency, and rate of heat transfer. This is particularly true in the case of plastics having appreciable damping.

5. The fatigue life of a polymer is generally reduced by an increase in temperature (53) although this is not always true for elastoplastic materials.

6. Constant deflection or strain testers have a disadvantage that once a large crack develops, the stress level drops below the fatigue endurance limit and the specimen does not fail for quite some time. In contrast, in the constant stress tester, once the crack develops, the applitude of deformation increases and failure occurs very rapidly (54).

2.11. HARDNESS TESTS

2.11.1. Introduction

Hardness is defined as the resistance of a material to deformation, particularly permanent deformation, indentation, or scratching. Hardness is purely a relative term and should not be confused with wear and abrasion resistance of plastic materials. Polystyrene, for example, has a high Rockwell hardness value but a poor abrasion resistance. Hardness test can differentiate relative hardness of different grades of a particular plastic. However, it is not valid to compare hardness of various types of plastics entirely on the basis of one type of test, since elastic recovery along with hardness is involved. The test is further complicated by a phenomenon such as creep. Many tests have been devised to measure hardness. Since plastic materials vary considerably with respect to hardness, one type of hardness test is not applicable to cover the entire range of hardness properties encountered. Two of the most commonly used hardness tests for plastics are the Rockwell hardness test and the Durometer hardness test. Rockwell hardness is used for relatively hard plastics such as acetals, nylons, acrylics, and polystyrene. For softer materials such as flexible PVC, thermoplastic rubbers, and polyethylene, Durometer hardness is often used. The typical hardness values of some common plastic materials are listed for comparison in Table 2-3. Comparison of various types of hardness scales is also given in the Appendix I.

2.11.2. Rockwell Hardness (ASTM D 785)

The Rockwell hardness test measures the net increase in depth impression as the load on an indentor is increased from a fixed minor load to a major load and then

Table 2-3. Typical Hardness Values of Some Common Plastic Materials

	Hardness		
	Rockwell		Durometer
Plastic material	*M*	*R*	Shore *D*
ABS		75–115	
Acetal	94	120	
Acrylic	85–105		
Cellulosics		30–125	
PPO	80	120	
Nylon		108–120	
Polycarbonate	72	118	
H.D.P.E.			60–70
L.D.P.E.			40–50
Polypropylene			75–85
Polystyrene (G.P.)	68–70		
PVC (rigid)		115	
Polysulfone	70	120	

returned to a minor load. The hardness numbers derived are just numbers without units. Rockwell hardness numbers are always quoted with a scale symbol representing the indentor size, load, and dial scale used. The hardness scales in order of increasing hardness are *R*, *L*, *M*, *E*, and *K* scales. The higher the number in each scale, the harder the material. There is a slight overlap of hardness scales and, therefore, it is quite possible to obtain two different dial readings on different scales for the same material. For a specific type of material, correlation in the overlapping regions is possible. However, due to differences in elasticity, creep, and shear characteristics between different plastics, a general correlation is not possible.

A. *Test Apparatus and Specimen*

Rockwell hardness is determined with an apparatus called the Rockwell hardness tester. Figure 2-60 illustrates a typical Rockwell hardness tester. A standard specimen of $\frac{1}{4}$-in. minimum thickness is used. The specimen can either be molded or cut from a sheet. However, the test specimen must be free from sink marks, burrs, or other protrusions. The specimen also must have parallel, flat surfaces.

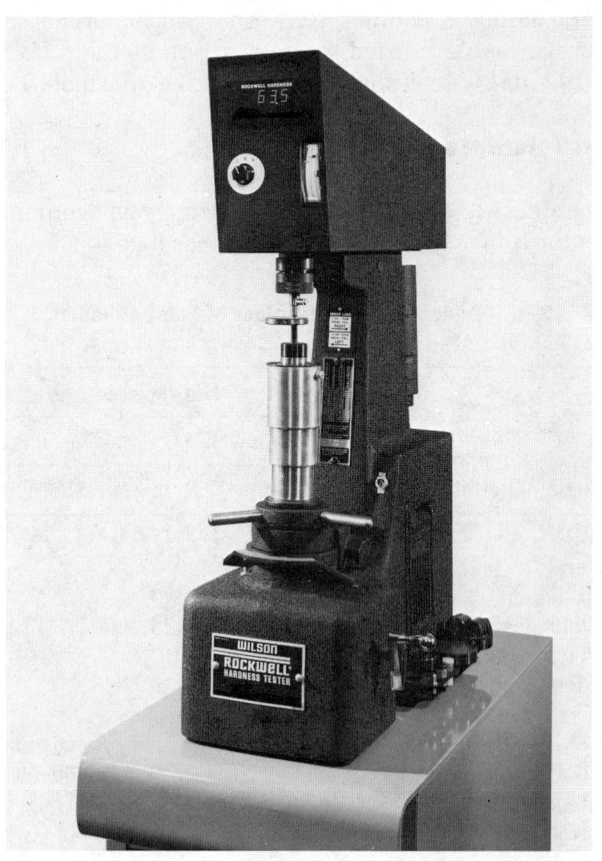

Figure 2-60. Rockwell hardness tester. (Courtesy ACCO Industries, Inc., Wilson Instrument Division.)

B. Test Procedures

The specimen is placed on the anvil of the apparatus and minor load is applied by lowering the steel ball onto the surface of the specimen. The minor load indents the specimen slightly and assures good contact. The dial is adjusted to zero under minor load and the major load is immediately applied by releasing the trip lever. After 15 sec the major load is removed and the specimen is allowed to recover for an additional 15 sec. Rockwell hardness is read directly off the dial with the minor load still applied. Figure 2-61 schematically illustrates the operating principle behind the Rockwell hardness tester.

2.11.3. Durometer Hardness (ASTM D 2240)

The Durometer hardness test is mostly used for measuring the relative hardness of soft materials. The test method is based on the penetration of a specified indentor forced into the material under specified conditions.

The Durometer hardness tester consists of a pressure foot, an indentor, and an indicating device. The indentor is spring loaded and the point of the indentor protrudes through the hole in the base. The test specimens are at least $\frac{1}{4}$ in.-thick and can be either molded or cut from a sheet. Several thin specimens may be piled to form a $\frac{1}{4}$-in.-thick specimen but one piece specimens are preferred. The poor contact between the thin specimens may cause results to vary considerably.

The test is carried out by first placing a specimen on a hard, flat surface. The pressure foot of the instrument is pressed onto the specimen, making sure that it is parallel to the surface of the specimen. The durometer hardness is read within 1 sec after the pressure foot is in firm contact with the specimen.

Two types of durometers are most commonly used—Type A and Type D. The basic difference between the two types is the shape and dimension of the indentor. The hardness numbers derived from either scale are just numbers without any units. Type A durometer is used with relatively soft material while Type D durometer is used with slightly harder material. A commercially available durometer hardness measuring instrument is shown in Figure 2-62.

2.11.4. Barcol Hardness (ASTM D 2583)

The Barcol hardness test was devised mainly for measuring hardness of both reinforced and nonreinforced rigid plastics. The tester is a portable instrument that can be carried around to measure hardness of fabricated parts as well as the test specimens.

Barcol hardness testers consist of an indentor with a sharp, conical tip and indicating device in the form of a dial with 100 divisions. Each division represents a depth of 0.0003-in. penetration. The test specimens are required to be of minimum thickness of $\frac{1}{16}$ in.

The test is carried out by placing the indentor onto the specimen and applying uniform pressure against the instrument. The pressure is applied until the dial indication reaches maximum. The depth of penetration is automatically converted to a hardness reading in absolute Barcol numbers. When measuring the Barcol

Observe the important fact that the depth measurement does not employ the surface of the specimen as the zero reference point and so largely eliminates surface condition as a factor.

Dial now reads B-C plus a constant amount due to the added spring of the machine under major load, but which value disappears from dial reading, when major load is withdrawn.

NOTE—The scale of the dial is reversed so that a deep impression gives a low reading and a shallow impression a high reading; so that a high number means a hard material.

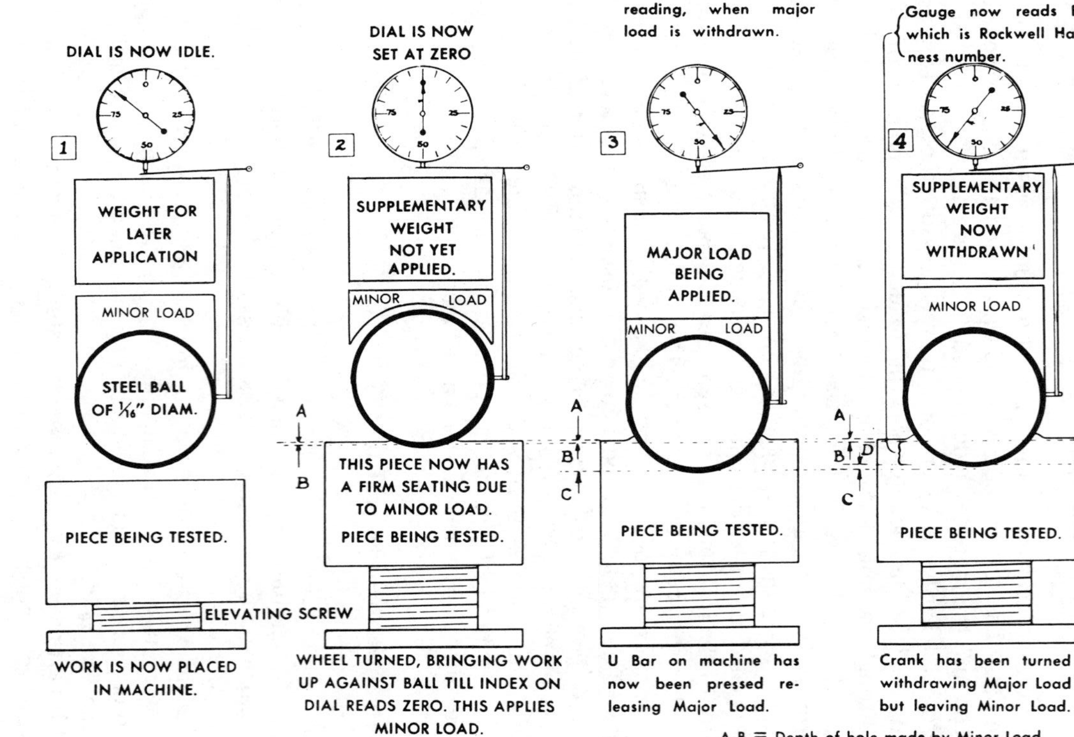

1
DIAL IS NOW IDLE.
WEIGHT FOR LATER APPLICATION
MINOR LOAD
STEEL BALL OF 1/16" DIAM.
PIECE BEING TESTED.
ELEVATING SCREW
WORK IS NOW PLACED IN MACHINE.
2
DIAL IS NOW SET AT ZERO
SUPPLEMENTARY WEIGHT NOT YET APPLIED.
MINOR LOAD
A
B
THIS PIECE NOW HAS A FIRM SEATING DUE TO MINOR LOAD.
PIECE BEING TESTED.
WHEEL TURNED, BRINGING WORK UP AGAINST BALL TILL INDEX ON DIAL READS ZERO. THIS APPLIES MINOR LOAD.
3
MAJOR LOAD BEING APPLIED.
MINOR LOAD
A
B
C
PIECE BEING TESTED.
U Bar on machine has now been pressed releasing Major Load.

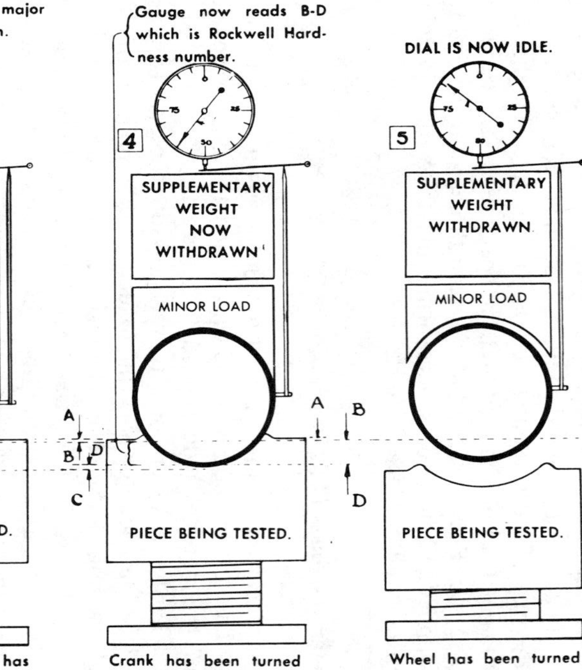

Gauge now reads B-D which is Rockwell Hardness number.
4
SUPPLEMENTARY WEIGHT NOW WITHDRAWN
MINOR LOAD
A
B
D
C
PIECE BEING TESTED.
Crank has been turned withdrawing Major Load but leaving Minor Load.
5
DIAL IS NOW IDLE.
SUPPLEMENTARY WEIGHT WITHDRAWN.
MINOR LOAD
A
B
D
PIECE BEING TESTED.
Wheel has been turned lowering piece.

EXPLANATION—

Diagrammatically the cycle of operation of the Rockwell Direct-Reading Hardness Tester is here shown. To illustrate the principle and show the action of the ball under application and release of minor and major loads, the size of the 1/16" ball has been enormously exaggerated.

A-B = Depth of hole made by Minor Load

A-C = Depth of hole made by Major Load

D-C = Recovery of metal upon reduction of Major to Minor Load. This is an index of the elasticity of metal under test, and does not enter the hardness reading.

B-D = Difference in depth of holes made = Rockwell Hardness number.

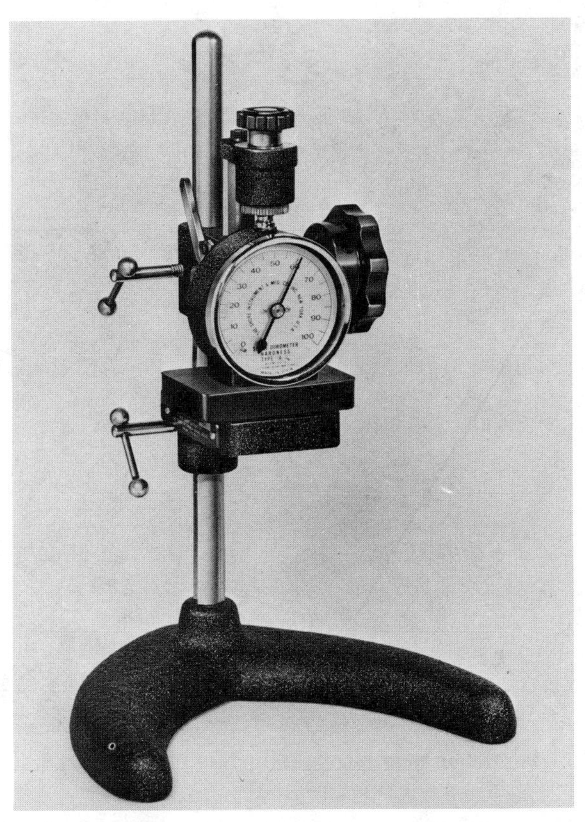

Figure 2-62. Durometer hardness tester. (Courtesy Shore Instrument and Manufacturing Company.)

hardness of the reinforced plastic material, the variation in hardness reading caused by the difference in hardness between resin and filler materials should be taken into account.

Generally, a larger sample size is recommended for reinforced plastic specimen than the sample size used for nonreinforced plastics. Figure 2-63 illustrates a commercially available Barcol hardness tester.

2.11.5. Factors Affecting the Test Results and Limitations

Temperature and Humidity

The hardness of all plastic materials is directly affected by the change in temperature and humidity. Specimens tested at higher than specified temperatures tend to indicate a lower hardness value.

Figure 2-61. Principle of Rockwell hardness tester. (Courtesy ACCO Industries, Inc., Wilson Instrument Division.)

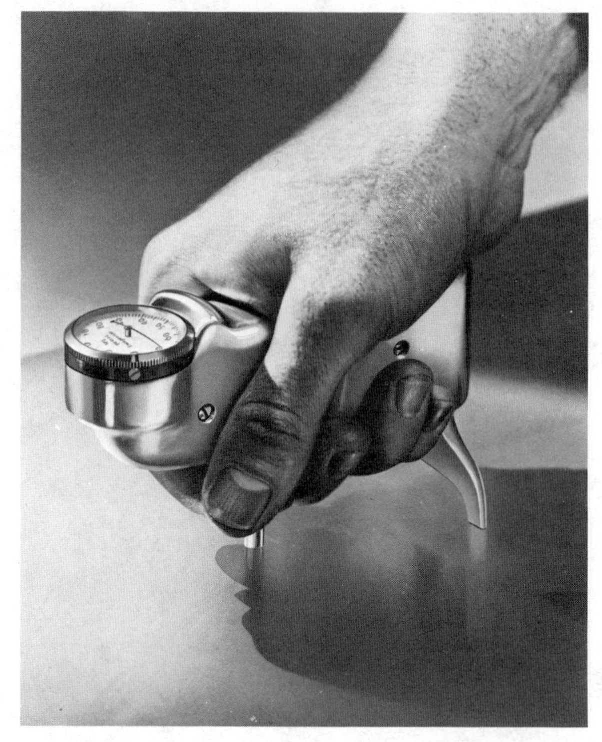

Figure 2-63. Barcol hardness tester. (Courtesy Gardner Laboratories, Inc.)

Surface Conditions of the Specimen

The surface finish has significant effect on hardness test results. A smooth molded surface yields higher value than a machined surface.

Filler

The hardness value may vary depending upon the type and amount of filler present in the resin. The variation in hardness reading may be caused by the difference in hardness between resin and filler material.

Anisotropy

Plastic materials with the anisotropic characteristics may cause indentation hardness to vary with the direction of testing.

REFERENCES

1. Kinney, G. F., *Engineering Properties and Applications,* John Wiley & Sons, New York, 1957, pp. 182–184.
2. *Ibid.,* p. 187.
3. Rodriguez, F., *Principles of Polymer Systems,* McGraw–Hill, New York, 1970, Chapter 8.
4. Baer, E., *Engineering Design for Plastics,* Reinhold, New York, 1964, Chapter 4.
5. Williams, J. G., *Stress Analysis of Polymers,* Longman Group Limited, London, 1973, Chapter 3.

6. Rodriguez, Reference 3, pp. 229–230.
7. Billmeyer, F. W., *Textbook of Polymer Science,* Interscience, New York, 1962, pp. 109–111.
8. Baer, Reference 4, p. 187.
9. Milby, R., *Plastics Technology,* McGraw–Hill, New York, 1973, pp. 490–491.
10. Billmeyer, Reference 7, p. 193.
11. Baer, Reference 4, p. 309.
12. Lubin, G., *Handbook of Fiberglass and Advanced Plastics Composites,* Reinhold, New York, 1969, p. 171.
13. LNP Corporation, *Tech. Bulletin,* "Predict Shrinkage and Warpage of Reinforced and Filled Thermoplastics," Malvern, Pa. (1978).
14. Heap, R. D., and Norman, R. H., *Flexural Testing of Plastics,* The Plastics Institute, London, England, 1969, p. 13.
15. *Ibid.,* p. 13.
16. O'Toole, J. L., "Creep Properties of Plastics," *Modern Plastics Encyclopedia,* McGraw–Hill, New York, 1968–69, p. 48.
17. Kinney, Reference 1, p. 192.
18. Baer, Reference 4, p. 284.
19. Delatycki, O., "Mechanical Performance and Design in Polymers," *Applied Polymer Symposia,* **17**(134), Interscience, New York, 1971.
20. Kinney, Reference 1, p. 192.
21. O'Toole, Reference 16, p. 412.
22. Delatycki, Reference 19, pp. 143–145.
23. "Design Guide," *Modern Plastics Encyclopedia,* McGraw–Hill, New York, 1979–1980, pp. 480–481.
24. Deanin, R. D., *Polymer Structure, Properties and Applications,* Cahners Publishing Co., Boston, Mass., 1972, p. 171.
25. Dubois, J. H. and Levy, S., *Plastics Product Design Engineering Handbook,* Reinhold, New York, 1977, pp. 104–105.
26. Vincent, P. I., *Impact Tests and Service Performance of Thermoplastics,* The Plastics Institute, London, England, 1971, p. 12.
27. Deanin, R. D., Reference 24, p. 171.
28. Vincent, P. I., Reference 26, p. 33.
29. *Ibid.,* p. 44.
30. Spath, W., *Impact Testing of Materials,* Grodon and Breach Science Publishers, New York, 1961, p. 37.
31. Bergen, R. L., "Tests for Selecting Plastics," *Metal Progress* (Nov. 1966), pp. 107–108.
32. Westover, R. F. and Warner, W. C., "Tensile Impact Test for Plastic Materials," *Res. and Stand.,* **1**(11), (1961), pp. 867–871.
33. Bragaw, C. G., "Tensile Impact," *Mod. Plast.,* **33**(10), (1956), p. 199.
34. Westover, R. F., "The Thirty Years of Plastics Impact Testing," *Plastics Technology,* **4,** (1958), pp. 223–228; **4,** (1958), pp. 348–352.
35. Maxwell, B. and Harrington, J. P., "Effect of Velocity on Tensile Impact Properties of PMMA," *Trans. ASME,* **74,** (1952), p. 579.
36. Nielsen, L. E., *Mechanical Properties of Polymers,* Reinhold, New York, 1962, pp. 124–125.
37. Baer, Reference 4, p. 799.
38. *Ibid.,* p. 803.
39. ASTM D 1709 (Part 36); ASTM D 3029 (Part 35), Annual Book of ASTM Standards, Philadelphia, Pa., 1978.
40. Tryson, G. R., Takemori, M. T., and Yee, A. F., "Puncture Testing of Plastics: Effects of Test Geometry," *ANTEC,* **25,** (1979), pp. 638–641.

41. Abolins, V., "Gardner Impact Versus Izod, Which is Better for Plastics?" *Materials Eng.* (Nov. 1973), p. 52.
42. Goldman, T. D. and Lutz, J. T., "Developing Low Temperature Impact Resistance PVC: A New Testing Approach," *ANTEC,* **25,** (1979), pp. 354–357.
43. Starita, J. M., "Impact Testing," *Plast. World* (April 1977), p. 58.
44. Ireland, D. R., "Instrumented Impact Test Method," *Materials Eng.* (May 1976), p. 33.
45. Tardif, H. P. and Marquis, H., "Impact Testing with an Instrumented Machine," *Metal Progress* (Feb. 1964), p. 58.
46. Starita, Reference 43, p. 59.
47. "The Perils of Izod—Part 1," *Plast. Design Forum* (May/June 1980), pp. 13–20.
48. Tanzillo, J. D., "Development of an Impact Test for Evaluation of Toughness of Rigid Plastic Building Components," *ANTEC,* **15,** (1969), pp. 346–349.
49. Neilson, Reference 36, p. 228.
50. Marcucci, M. A., *S.P.E.J.,* **14** 30 (Feb. 1958).
51. Neilson, Reference 36, p. 230.
52. Lazan, B. J. and Yorgiadis, A., "Symposium on Plastics," *ASTM Spec. Tech. Pub.,* No. 59, (Feb. 1944), p. 66.
53. Neilson, Reference 36, p. 230.
54. Neilson, Reference 36, p. 230.

SUGGESTED READING

Dietz, A. G. and Eirich, F. R., *High Speed Testing,* Vol. 1 Interscience, New York, 1960.

Lysaght, V. E., *Indentation Hardness Testing,* Reinhold, New York, 1949.

Alfrey, T., *Mechanical Behavior of High Polymers,* Interscience, New York, 1948.

"Simulated Service Testing in the Plastics Industry," ASTM Spec. Tech. Pub. 375.

Ritchie, P. D., *Physics of Plastics,* Van Nostrand, Princeton, N.J., 1965.

Ward, I. M. and Pinnock, P. R., "The Mechanical Properties of Solid Polymers," *Br. J. Appl. Phys.,* **17**(3) (1966).

Vincent, P. I., "Mechanical Properties of High Polymers: Deformation," in *Physics of High Polymers,* Ritchie, P. D. (Ed.), Iliffe, London, 1965, Chapter 2.

Bueche, F., *Physical Properties of High Polymers,* Wiley, New York, 1960.

Tobolsky, A. V., *Properties and Structure of Polymers,* Wiley, New York, 1960.

Murphy, A., *Properties of Engineering Materials,* 2nd ed., International Textbook Co., Scranton, Pa, 1947, p. 402.

Brown, W. E., *Testing of Polymers,* Interscience, New York, 1969.

Mark, H. F., Gaylord, N. G., and Bikales, M., Eds., *Encyclopedia of Polymer Science and Technology,* Interscience, New York, 1964–70.

Schmitz, J. V., Ed., *Testing of Polymers,* Interscience, New York, 1965.

Kulow, P., *Rubber and Plastics Testing,* Chapman and Hall, London, England, 1963.

Davis, H. E., Troxell, G. E., and Wiskocil, C. T., *The Testing and Inspection of Engineering Materials,* McGraw–Hill, New York, 1955.

Liddicoat, R. T. and Potts, P., *Laboratory Manual of Material Testing,* MacMillan Co., New York, 1952.

Findley, W. N., "Creep Characteristics of Plastics," *ASTM Symposium on Plastics,* Philadelphia, Pa, 1944.

Turner, S., "Creep Studies on Plastics," *J. App. Poly. Sci. Symp.,* **17** (1971).

Gotham, K. V., "A Formalized Experimental Approach to the Fatigue of Thermoplastics," *Plastics and Polymers,* **37**(130), (1969), p. 309.

Larson, F. R. and Miller, J., "A Time–Temperature Relationship for Rupture and Creep Stresses," *Trans. ASME* (July 1952), pp. 765–775.

Constable, I., Williams, J. G. and Burns, D. J., "Fatigue and Cyclic Thermal Softening of Thermoplastics," *J. Mech. Eng. Sci.,* **12**(1), (1970).

Cessna, L. C., Levens, J. A., and Thomson, J. B., "Flexural Fatigue of Glass-Reinforced Thermoplastics," *Poly. Eng. Sci.,* **9**(5), (1969), pp. 339–349.

Horsley, R. A. and Morris, A. C., "Impact Tests—A Guide to Thermoplastics Performance," *Plast.,* **31**(3), (1966), pp. 1551–1553.

Thomas, J. R. and Hagan, R. S., "Meaningful Testing of Plastic Materials for Major Appliances," *S.P.E.J.,* **22** (1966), pp. 51–56.

Hertzberge, R. W. and Manson, J. A., *Fatigue of Engineering Plastics,* Academic, New York, 1980.

CHAPTER 3

Thermal Properties

3.1. INTRODUCTION

Thermal properties of plastic materials are equally as important as mechanical properties. Unlike metals, plastics are extremely sensitive to changes in temperature. The mechanical, electrical, or chemical properties of plastics cannot be looked at without looking at the temperature at which the values were derived. Crystallinity has a number of important effects upon the thermal properties of a polymer. Its most general effects are the introduction of a sharp melting point and the stiffening of thermal mechanical properties. Amorphous plastics, in contrast, have a gradual softening range (1). Molecular orientation also has a significant effect on thermal properties. Orientation tends to decrease dimensional stability at higher temperatures (2). The molecular weight of the polymer affects the low-temperature flexibility and low-temperature brittleness. Many other factors such as intermolecular bonding, cross linking, and copolymerization all have a considerable effect on thermal properties. From the above discussion, it is very clear that the thermal behavior of polymeric materials is rather complex. Therefore, in designing a plastic part or selecting a plastic material from the available thermal property data, one must thoroughly understand the short term as well as the long-term effect of temperature on properties of that plastic material.

3.2. TESTS FOR ELEVATED TEMPERATURE PERFORMANCE

Designers and material selectors of plastic products constantly face the challenge of selecting a suitable plastic for elevated temperature performance. The difficulty arises due to the varying natures and capabilities of various types and grades of plastics at elevated temperatures. Many factors are considered when selecting a plastic for a high-temperature application. The material must be able to support a design load under operating conditions without objectionable creep or distortion. The material must not degrade or lose necessary additives that will cause drastic reduction in the physical properties during the expected service life.

All the properties of plastic materials are not affected in a similar manner by elevating temperature. For example, electrical properties of a particular plastic may show only a moderate change at elevated temperatures, while the mechanical properties may be reduced significantly. Also, since the properties of plastic materials vary with temperature in an irregular fashion, they must be looked at as a function of temperature in order to obtain more meaningful information. From the foregoing, it is quite clear that a single maximum use temperature that will apply to all the important properties in high-temperature applications is simply not possible.

One of the most important considerations while studying the performance of plastics at elevated temperatures is the dependence of key properties such as modulus, strength, chemical resistance, and environmental resistance on time. Therefore, the short-term heat resistance data alone is not adequate for designing and selecting materials that require long-term heat resistance. For the sake of convenience and simplicity, we divide the elevated temperature effects into two categories:

1. Short-term effects.
   (a) Heat deflection temperature.
   (b) Vicat softening temperature.
   (c) Torsion pendulum.
2. Long-term effects.
   (a) Long-term heat resistance test.
   (b) UL temperature index.
   (c) Creep modulus/creep rupture tests.

3.2.1. Short Term Effects

A. *Heat Deflection Temperature (HDT)* (ASTM D 648)

Heat deflection temperature is defined as the temperature at which a standard test bar ($5 \times \frac{1}{2} \times \frac{1}{4}$ in.) deflects 0.010-in. under a stated load of either 66 or 264 psi. The heat deflection temperature test, also referred to as the heat distortion temperature test, is commonly used for quality control and for screening and ranking materials for short-term heat resistance. The data obtained by this method cannot be used to predict the behavior of plastic materials at elevated temperature nor can it be used in designing a part or selecting and specifying material. Heat deflection temperature is a single point measurement and does not indicate long-term heat resistance of plastic materials. Heat distortion temperature, however, does distinguish between those materials that lose their rigidity over a narrow temperature range and those that are able to sustain light loads at high temperatures (3).

APPARATUS AND TEST SPECIMENS. The apparatus for measuring heat deflection temperature consists of an enclosed oil bath fitted with a heating chamber and automatic heating controls that raise the temperature of the heat transfer fluid at a uniform rate. A cooling system is also incorporated to fast cool the heat transfer medium for conducting repeated tests. The specimens are supported on steel supports, 4 in. apart with the load applied on top of the specimen vertically and midway between the supports. The contact edges of the support and of the piece by which pressure is applied is rounded to a radius of $\frac{1}{4}$ in. A suitable deflection measurement device, such as a dial indicator, is normally used. A mercury thermometer is used for measuring temperature. The unit is capable of applying 66 psi or 264 psi fiber stress on specimens by means of a dead weight. A commercially available heat deflection measuring device with a close-up of a specimen holder is illustrated in Figure 3-1.

More recently, automatic heat deflection temperature testers have been developed. These testers typically replace conventional temperature and deflection measuring devices with more sophisticated electronic measuring devices with digital read-out system and a chart recorder that prints out the results. Such an automatic apparatus eliminates the need for the continuous presence of an operator and thereby minimizes operator-related errors.

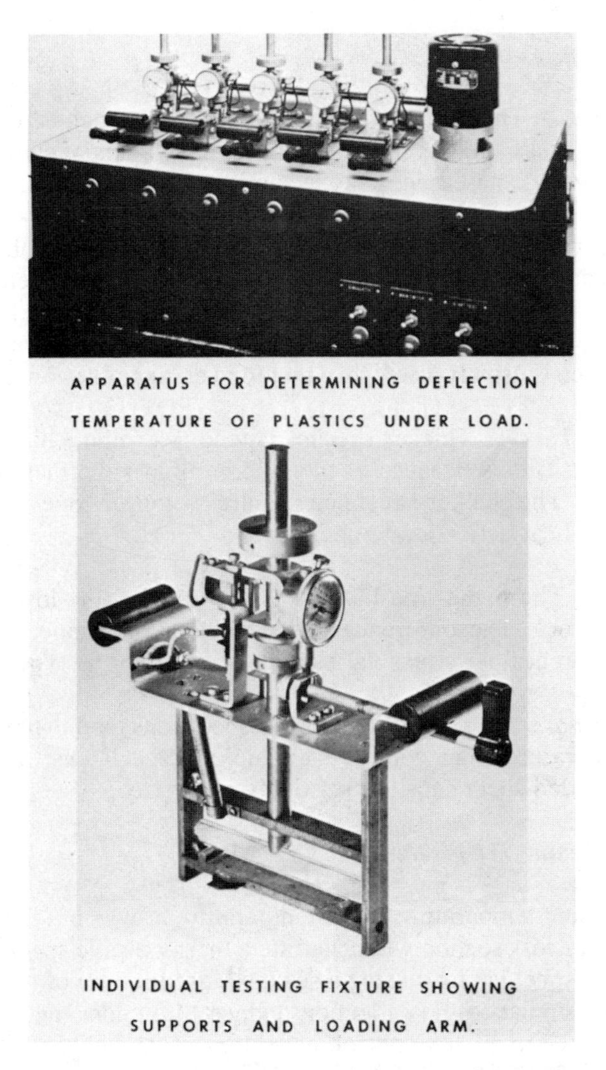

APPARATUS FOR DETERMINING DEFLECTION
TEMPERATURE OF PLASTICS UNDER LOAD.
INDIVIDUAL TESTING FIXTURE SHOWING
SUPPORTS AND LOADING ARM.

Figure 3-1. HDT apparatus and close-up of specimen holder. (Courtesy DuPont Company.)

The test specimens consist of test bars 5 in. in length, $\frac{1}{2}$ in. in depth by any width from $\frac{1}{8}$–$\frac{1}{2}$ in. The test bars may be molded or cut from extruded sheet as long as they have smooth, flat surfaces and are free from excessive sink marks or flash. The specimens are conditioned employing standard conditioning procedures.

TEST PROCEDURE. The specimen is positioned in the apparatus along with the temperature and deflection measuring devices and the entire assembly is submerged into the oil bath kept at room temperature. The load is applied to a desired value (66 psi or 264 psi fiber stress). Five minutes after applying the load, the pointer is adjusted to zero and the oil is heated at the rate of 2 ± 0.2 °C/min. The temperature of the oil at which the bar has deflected 0.010 in. is recorded as the heat deflection temperature at the specified fiber stress.

TEST VARIABLES AND LIMITATIONS

Residual Stress. The specimens consisting of a high degree of "molded-in" residual stresses or a high degree of orientation have a tendency to stress relieve as the temperature is increased. The specimen warpage occurs in a downward direction due to the external loading. The warpage due to stress relaxation and deflection, occurring from the softening of the specimen, combined together, yields a false heat deflection temperature. However, if the specimens are annealed in a controlled oven atmosphere prior to testing, the stresses can be relieved and warpage can be practically eliminated. Heat deflection temperature of the unannealed specimen is usually lower than that for a comparable annealed specimen.

Specimen Thickness. Thicker specimens tend to exhibit a higher heat deflection temperature. This is because of the inherently low thermal conductivity of plastic materials. The thicker specimen requires a longer time to heat through, yielding a higher heat deflection temperature.

Fiber Stress. The higher the fiber stress or loading, the lower the heat deflection temperature. The difference in heat deflection temperature values resulting from different fiber stress varies depending upon the type of polymer.

Specimen Preparation. Injection molded specimens tend to have a lower heat deflection temperature than compression molded specimens. This is because compression molded specimens are relatively stress-free.

B. Vicat Softening Temperature (ASTM D 1525)

The Vicat softening temperature is the temperature at which a flat-ended needle of 1 mm^2 circular cross section will penetrate a thermoplastic specimen to a depth of 1 mm under a specified load using a selected uniform rate of temperature rise. This test is very similar to the deflection temperature under the load test and its usefulness is limited to quality control, development, and characterization of materials. The data obtained from this test is also useful in comparing the heat softening qualities of thermoplastic materials. However, the test is not recommended for flexible PVC or ethyl cellulose or other materials with a wide Vicat softening range.

The test apparatus designed for deflection temperature under load test can be used for Vicat softening temperature test with minor modification. The flat test specimen is molded or cut from a sheet with a minimum thickness and width of 0.12 in. and 0.50 in., respectively.

The test is carried out by first placing the test specimen on a specimen support and lowering the needle rod so that the needle rests on the surface of the specimen. The temperature of the bath is raised at the rate of 50°C/hr or 120°C/hr uniformly. The temperature at which the needle penetrates 1 mm is noted and reported at the Vicat softening temperature. A commercially available test apparatus for measuring the Vicat softening temperature is illustrated in Figure 3-2.

C. *Torsion Pendulum Test* (ASTM D 2236)

The elevated temperature performance of most plastics is dominated by the temperature of occurrence and the temperature range of the glass transition and, in the case of semicrystalline plastics, by the crystalline melting point. The glass transition has a profound negative effect on all mechanical properties except impact, and on certain thermal properties such as the coefficient of expansion. Therefore, a knowledge of the transitional behavior of plastics is necessary in order to understand their elevated temperature capabilities. This knowledge is best derived from a full temperature modulus plot, especially a dynamic modulus, since the elevated temperature behavior of plastics is too complex and too varied to be described by simple tests, such as Vicat softening temperature and deflection temperature under load (4).

Many different experimental techniques exist today to study the effects of elevated temperature on dynamic mechanical properties of plastics. Perhaps, the most meaningful and simplest technique is the torsion pendulum technique. It is through this technique that one can derive a plot of short-time modulus versus temperature up to the beginning of melting and degradation.

The apparatus used for measuring dynamic properties over a wide range of temperatures is quite simple. It consists of a rigidly fixed clamp at one end. The movable clamp is attached to an inertial member, usually a disk or a rod with a known moment of inertia. A differential transformer and strip chart recorder are used to record angular displacement versus time in graphical form as the specimen is oscillating. An insulated chamber is also used, equipped with heating and cooling systems to facilitate testing over a wide range of temperatures. A torsion pendulum test equipment is illustrated in Figure 3-3.

Specimens of rectangular or cylindrical shape and different length and width are used. However, since the thickness of the specimen influences the dynamic results, it is usually specified. To carry out the test, the temperature equilibrium is first established in the chamber. The clamped specimen is put into oscillation and the period and rate of decay of the amplitude of oscillation are measured. The elastic shear modulus is calculated from specimen dimensions, moment of inertia of the movable member, and period of the movement. A quantity known as log decrement is calculated from the rate of amplitude decay. A detailed procedure to calculate these values is given in the literature (5).

Figure 3-2. Vicat softening point apparatus. (Courtesy Tinius Olsen Company.)

The log decrement is observed to be an approximate first derivative of the modulus temperature curve and thus has maxima at the temperatures at which the modulus shows a rapid drop (6). This dramatic drop in modulus indicates the effective maximum load-bearing temperature.

Figure 3-4 illustrates typical torsion pendulum test data graphically. Elastic shear modulas versus temperature graph shows the effects of glass transition temperature. The modulus is very high at the beginning of the curve and decreases very slowly as the temperature increases. Near the glass transition temperature, the modulus decreases very rapidly in a short temperature span. The damping which is expressed as log decrement is calculated from the rate at which the amplitude of the oscillation decreases. The plot of log decrement versus temperature shows the onset of the transition. This temperature is regarded as the maximum usable load-bearing temperature.

Even though torsion pendulum tests run at elevated temperatures provide very useful information regarding the dynamic mechanical properties of plastics, one must not forget that it is, nevertheless, a short-term test. The data obtained from creep modulus and creep rupture tests conducted at elevated temperatures must be relied on for long-term effects.

3.2.2. Long-Term Effects

The long-term effects of elevated temperature on properties of plastics are extremely important, especially when one considers the fact that the majority of

Figure 3-3. Torsion pendulum. Test equipment. (Courtesy Tinius Olsen Company.)

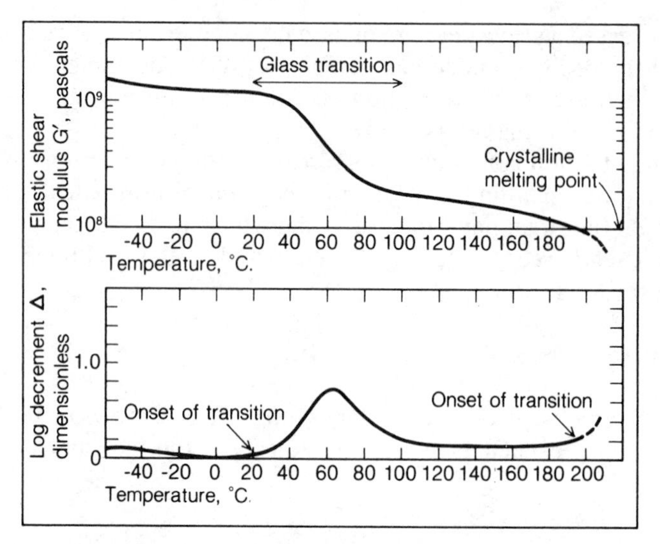

Figure 3-4. Dynamic mechanical properties by torsion pendulum versus temperature of a typical semicrystalline thermoplastic (nylon 6, dry) showing effects of glass transition and crystalline melting transition and method of determining their onsets. (Reprinted by permission of *Modern Plastics Encyclopedia* 1980–1981.)

applications involving high heat are long-term applications. During long-term exposure to heat, plastic materials may encounter many physical and chemical changes. A plastic material that shows little or no effect at elevated temperature for a short time may show a drastic reduction in physical properties, a complete loss of rigidity, and severe thermal degradation when exposed to elevated temperature for a long time. Along with time and temperature, many other factors such as ozone, oxygen, sunlight, and pollution combine to accelerate the attack on plastics. At elevated temperatures, many plastics tend to lose important additives such as plasticizers and stabilizers, causing plastics to become brittle or soft and sticky.

Three basic tests have been developed and accepted by the plastics industry. If the application does not require the product to be exposed to elevated temperature for a long period under continuous load, a simple heat-resistance test is adequate. The applications requiring the product to be under continuous significant load must be looked at from creep modulus and creep rupture strength test data. Yet, another one of the most widely accepted methods of measuring maximum continuous use temperature has been developed by Underwriters Laboratories. The UL temperature index, established for a variety of plastic materials to be used in electrical applications, is the maximum temperature that the material may be subjected to without fear of premature thermal degradation.

A. *Long-Term Heat-Resistance Test* (ASTM D 794)

The long-term heat-resistance test was developed to determine the permanent effect of heat on any property by selection of an appropriate test method and specimen. In ASTM recommended practice, only the procedure for heat exposure is specified and not the test method or specimen.

Any specimen, including sheet, laminate, test bar, or molded part may be used. If a specific property, such as tensile strength loss is to be determined, a standard tensile test bar specimen and procedures must be used for comparison of test results before and after the test. The test requires the use of a mechanical convection oven with a specimen rack of suitable design to allow air circulation around the specimens. The test is carried out by simply placing the specimen in the oven at a desired exposure temperature for a predetermined length of time. The subsequent exposure to temperatures may be increased or decreased in steps of 25°C until a failure is observed. Failure due to heat is defined as a change in appearance, weight, dimension, or other properties that alter plastic material to a degree that it is no longer acceptable for the service in question. Failure may result from blistering, cracking, loss of plasticizer, or other volatile material that may cause embrittlement, shrinkage, or change in desirable electrical or mechanical properties.

There are many factors that affect the reproducibility of the data. The degree of temperature control in the oven, the type of molding, cure, air velocity over the specimen, period of exposure, and humidity of the oven room are some of these factors. The amount and type of volatiles in the molded part or specimen may also affect the reproducibility.

B. UL Temperature Index

The increased use of plastic materials in electrical applications such as appliances, portable electrically operated tools and equipments, and enclosures has created a renewed interest in the ability of plastics to withstand mechanical abuse and high temperatures. A serious personal injury, electric shock, or fire may occur if the product does not perform its intended function (7). Underwirters Laboratories, an independent, nonprofit organization concerned with consumer safety, has developed a temperature index to assist UL engineers in judging the acceptability of individual plastics in specific applications involving long-term exposure to elevated temperatures. The UL temperature index correlates numerically with the temperature rating or maximum temperature in °C above which a material may degrade prematurely and therefore be unsafe (8).

RELATIVE THERMAL INDICES. The relative thermal index of a polymeric material is an indication of the material's ability to retain a particular property (physical, electrical, etc.) when exposed to elevated temperatures for an extended period of time. It is a measure of the material's thermal endurance. For each material, a number of relative thermal indices can be established, each index related to a specific thickness of the material (9).

The relative index of a material is determined by comparing the thermal-aging characteristic of one material of proven field service at a particular temperature level with the thermal-aging characteristics of another material with no field service history. A great deal of consideration is given to the properties that are evaluated to determine relative thermal index. For the relative thermal index to be valid, the properties being stressed in the end-product must be included in the thermal-aging program. If, for any reason, the specific property under stress in the end-product is not part of the long-term aging program, the relative thermal index may not be applicable to the use of the material in that particular application.

RELATIVE THERMAL INDEX BASED UPON HISTORICAL RECORDS. Through experience gained from testing a large volume of complete products and insulating systems over a long period, UL has established relative thermal indices on certain types of plastics. These fundamental temperature indices are applicable to each member of a generic material class. Table 3-1 lists the temperature indices based on past field test performance and chemical structure.

RELATIVE THERMAL INDEX BASED UPON LONG-TERM THERMAL-AGING. The long-term thermal-aging program consists of exposing polymeric materials to heat for a predetermined length of time and observing the effect of thermal degradation. To carry out the testing, an electrically heated mechanical convection oven is preferred, however, with some provisions, a noncirculating static oven may be employed. The specific properties to be evaluated in the thermal-aging program are to be as nearly as possible representative of the properties required in the end-application.

Table 3-1. Relative Thermal Indices Based Upon Past Field Test Performance and Chemical Structure[a]

Material	Generic Thermal Index (°C)
Nylon (Type 6, 11, 6/6, and 6/10)[b]	65
Polycarbonate[b]	65
Molded Phenolic[c,d]	150
Molded Melamine[c,d]	130[e]
Molded Melamine-Phenolic[c,d]	130[e]
Fluorocarbin resins	
(1) Polytetrafluoroethylene	150
(2) Polychlorotrifluoroethylene	150
(3) Fluorinated Ethylene Propylene	150
Silicone Rubber	105
Polyethylene Terephthalate Film	105
Urea Formaldehyde	100
Molded Alkyd[c,d]	130
Molded Epoxy[c,d]	130
Molded Diallyl Phthalate[c,d]	130
Molded Polyester[c,d] (thermosetting)	130

[a] From UL 746 B. Reprinted with permission from Underwriters Laboratories, Inc.

[b] Includes glass fiber-reinforced materials.

[c] Includes simultaneous heat and high-pressure matched metal die molded compounds only. Excludes low-pressure or low-temperature curing processes such as open-mold (hand lay-up, spray-up, contact bag, filament winding), encapsulation, lamination, etc.

[d] Includes materials having filler systems of fibrous (other than synthetic organic) types but excludes fiber reinforcement systems using resins that are applied in liquid form.

[e] Compounds having a specific gravity of 1.55 or greater (including those having cellulosic filler material) are acceptable at temperatures not greater than 150 °C (302 °F).

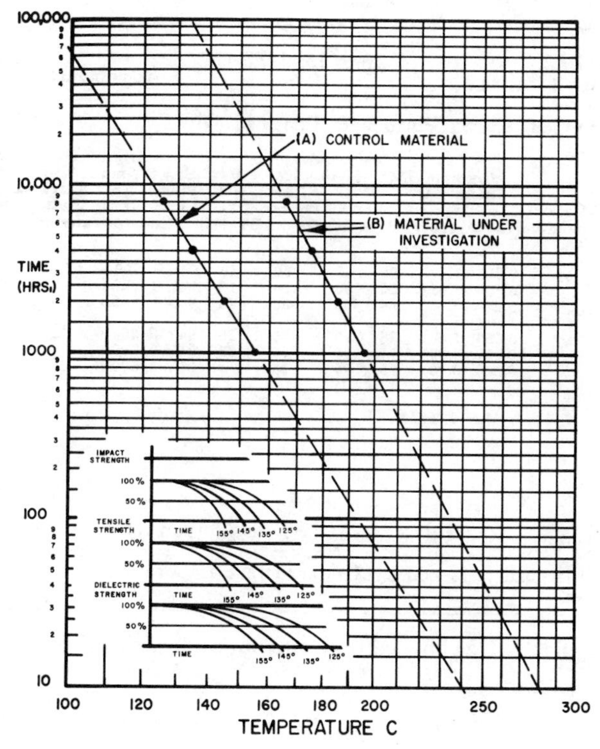

Figure 3-5. Plot of typical time–temperature data. (Courtesy Underwriters' Laboratories, Inc.)

The most common mechanical properties include tensile strength, flexural strength, and Izod impact strength. The electrical properties of concern are dielectric strength, surface or volume resistivity, arc resistance, and arc tracking. The test specimens are standard ASTM test bars, depending upon the type of test. UL publication, "Polymeric Materials—Short-Term Property Evaluations, UL 746A," describes the specimen and test procedures to determine mechanical and electrical properties.

In order to determine the relative thermal index, a control material with a record of good field service at its rated temperature is selected. The control material of the same generic type and the same thickness as the candidate material is preferred. At least four different oven temperatures are selected. The highest temperature is selected so that it will take no more than two months to produce end-of-life of the material. The next two lower temperatures must produce the anticipated end-of-life of 3 and 6 months, respectively. The lowest temperature selected will take 9 to 12 months for the anticipated results.

The end-of-life of a material is based upon the assumption that at least a 2:1 factor of safety exists in the applicable physical and electrical property requirements. The end-of-life of a material is the time at each aging temperature, when a property value has decreased 50 percent of its unaged level. A 50 percent loss of property due to thermal degradation is not expected to result in premature, unsafe failure.

Figure 3-5 is a plot of typical time temperature data obtained from various

Guide QMFZ2. June 13, 1977 [N] E33640D.

Component—Plastics. **(Continued from C card.)**

Mobay Chemical Corp., Plastics & Coatings Div.

Material Designation	Color	Minimum Thickness, In. (mm)	Temp. Index Elec.	Temp. Index Mechanical With Impact	Temp. Index Mechanical Without Impact	Hot Wire Ign.	UL94 Flam. Class	High Amp Arc Ign.	High Volt Track Rate	D-495 Arc Resistance	IEC Track (CTI)
Acrylonitrile-butadiene-styrene/polycarbonate (ABS/PC) blend, designated "Bayblend," furnished in the form of pellets.											
MC-2500††,	Any	0.058 (1.47)	50	50	50	10	94HB	200+	3.3	29	—
-2503††,		0.120 (3.05)	50	50	50	16	94HB	200+	—	87	440
-2505††,		0.240 (6.10)	50	50	50	83	94HB	200+	—	65	—
-2507††											
MD-6500††,	Any	0.058 (1.47)	95	95	95	25	94V-0	200+	22	—	—
-6503††,		0.120 (3.05)	95	95	95	53	94V-0	200+	9	8	270
-6505††,		0.240 (6.10)	95	95	95	—	94V-0	200+	12	—	—
-6507††											

Reports: **May 8, 1977; May 9, 1977.**

This card replaces E33640D dated May 17, 1977.

(Continued on E card.)

PRINTED IN U.S.A. This card is issued by Underwriters Laboratories Inc. ®

Figure 3-6. Typical UL yellow card. (Courtesy Underwriters' Laboratories, Inc.)

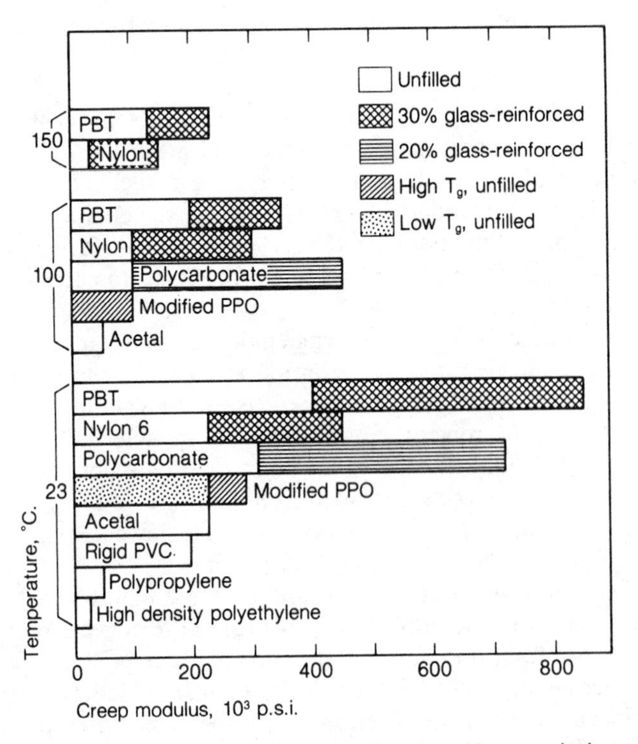

Figure 3-7. Creep modulus versus temperature. (Reprinted by permission of *Modern Plastics Encyclopedia* 1980–1981.)

tests. The insert in Figure 3-5 illustrates the curve obtained as a result of aging a material at four elevated temperatures. The properties of impact strength, tensile strength, and dielectric strength were investigated. At each temperature, the first property to show a reduction of 50 percent of the unaged property value and therefore, failure, is the impact strength. The time-to-failure at each temperature for the impact strength is used to construct an Arrhenius curve B. Curve A represents a plot of a control material having relative thermal index of 100°C. This control material shows a correlation factor of 60,000 hrs in this example. The time–temperature plot of the material under investigation crosses the 60,000 hr line at a temperature of 140°C. Therefore, it can be expected to be safe to use at 140°C in the same manner the known material is useful at 100°C.

A relative thermal index of 140°C in this example is applicable to all applications involving all the properties investigated, including impact strength. In applications where impact strength is not a critical property, a higher relative thermal index may be assigned. To avoid underrating the material's relative temperature index, UL publishes the ratings in three categories: applications involving electrical properties only, applications involving both electrical and mechanical properties, and applications involving both electrical and mechanical properties without impact resistance. UL publishes such data on its widely recognized "yellow card" and in its annual recognized component index. A typical "yellow card" is shown in Figure 3-6. Since the long-term heat-aging resistance of plastics is dependent on the thickness of the part or test specimen, Underwriters Laboratories requires

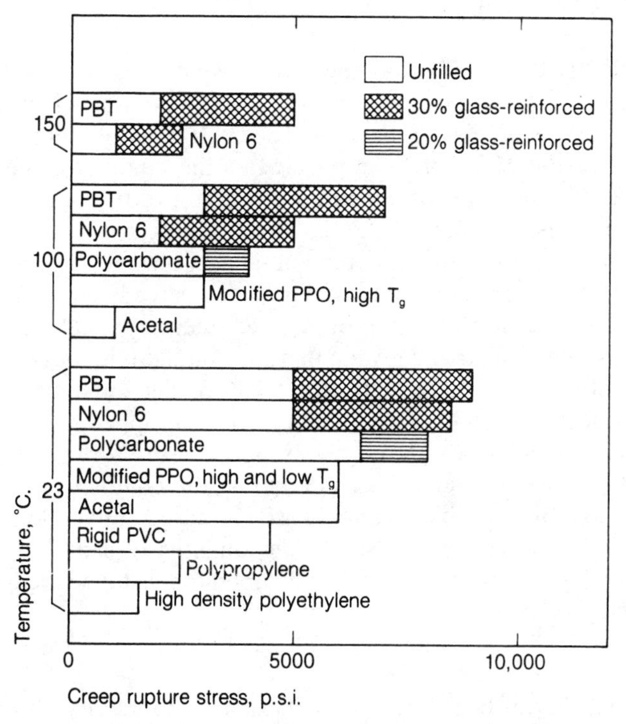

Figure 3-8. Creep rupture strength versus temperature. (Reprinted by permission of *Modern Plastics Encyclopedia* 1980–1981.)

that the thermal index testing be carried out over a wide range of thicknesses. Finally, it is important to understand that the UL temperature index program recognizes that the upper temperature limits of plastics are depndent upon the stresses applied on the end-product in use. Consequently, the temperature index of a particular plastic qualifies it only for those UL applications that UL has specifically approved.

C. Creep Modulus/Creep Rupture Tests

The tests developed to determine creep modulus and creep rupture values of plastic materials have been discussed in the previous chapter. Both creep modulus and creep rupture strength decrease significantly as the temperature is increased. Before selecting a material for a load-bearing application at an elevated temperature, one must carefully evaluate the published creep modulus and creep rupture strength test data. This is accomplished by studying the bar charts such as the ones shown in Figure 3-7 and 3-8. In one chart, creep modulus of various thermoplastics at three different temperatures is compared. A similar comparison is made using creep rupture strength for the same set of thermoplastic materials. A study of the charts reveals that while most of these thermoplastics have substantial rigidity and strength at room temperature, only a few retain enough of these properties at elevated temperatures to bear significant loads (10).

3.3. THERMAL CONDUCTIVITY (ASTM C 177)

Thermal conductivity is defined as the rate at which heat is transferred by conduction through a unit cross sectional area of a material when a temperature gradient exists perpendicular to the area. The coefficient of thermal conductivity, sometimes called the K factor, is expressed as the quantity of heat that passes through a unit cube of the substance in a given unit of time when the difference in temperature of the two faces is 1°C.

One of the major reasons for the tremendous success of plastics in the last two decades is their low thermal conductivity. Plastics have been readily accepted as materials for cooking utensils and automobile steering wheels. In recent years, cellular plastics, which have the lowest thermal conductivity of all materials, have gained popularity in the field of thermal insulation. The outstanding thermal conductivity of cellular plastics is largely due to the entrapped gases and not to the polymeric material which serves merely as an enclosure for entrapment of gases. As the density of the cellular plastic decreases, the conductivity also decreases up to a minimum value and rises again due to increased convection effects caused by a higher proportion of open cells (11). The quantity of heat flow depends upon the thermal conductivity of the material and upon the distance the heat must flow. This relationship is expressed as (12,13):

$$Q \sim K/X$$

where Q = quantity of heat flow; K = thermal conductivity; X = the distance the heat must flow.

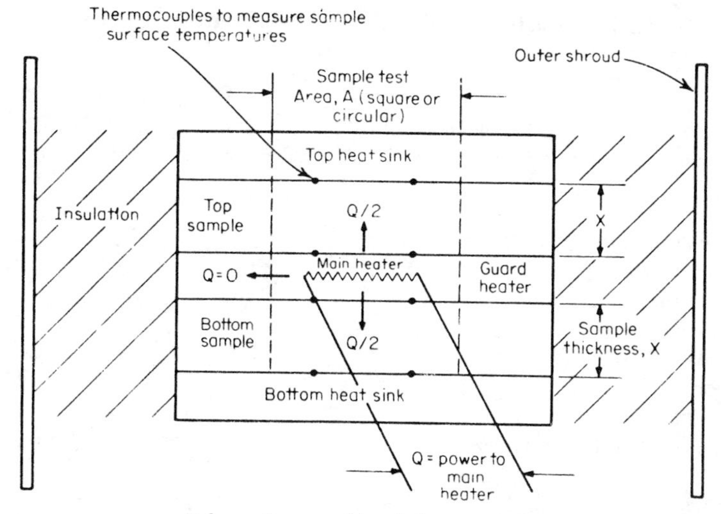

Figure 3-9. Schematic assembly of the guarded hot plate. (Reprinted by permission of McGraw–Hill Book Company.)

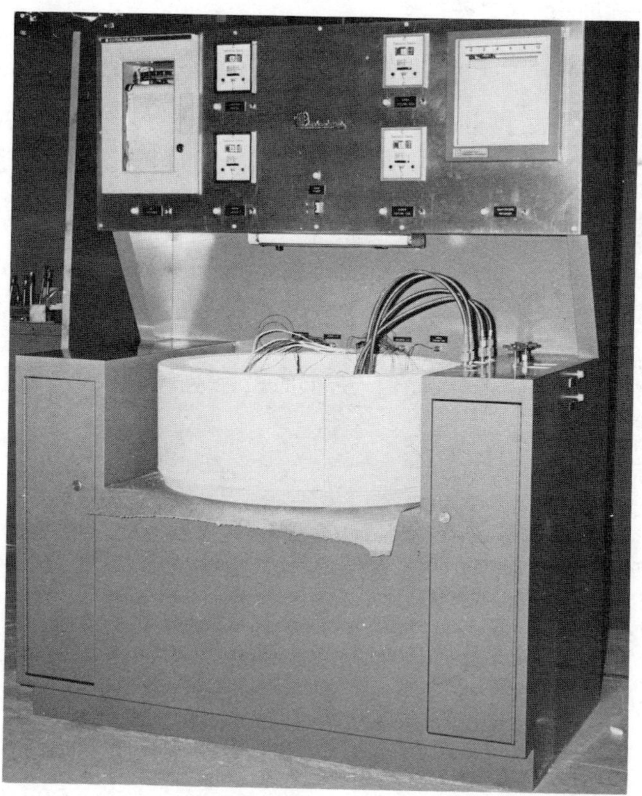

Figure 3-10. Guarded hot plate equipment. (Courtesy Custom Scientific Instruments, Inc.)

Table 3-2. Thermal Conductivity of Different Materials Including Solid and Cellular Plastics[a]

Plastic	Value BTU/(h)(ft^2)(°F/in.)
Solid	
Polystyrenes (medium-impact)	0.29–0.87
Polystyrenes (high-impact)	0.29–0.87
Polystyrenes (heat-resistant)	0.29–0.87
Urethanes elastomers	0.49
Polystyrenes (chemical filler)	0.55–0.87
Polyallomer	0.58–1.20
Polypropylene copolymer	0.58–1.20
Polystyrenes (general purpose)	0.70–0.96
Acrylics	0.7–1.7
Polypropylene (unmodified)	0.81
Styrene acrylonitrile copolymer	0.84
Polyvinylidene fluoride (VF2)	0.87
Polypropylene (impact rubber modified)	0.87–1.20
Chlorinated polyether	0.90
PVC flexible (unfilled)	0.99
PVC flexible (filled)	0.99
Styrene butadiene thermoplastic elastomers	1.00
PVC (rigid)	1.10
Nylons (polyamide) type 12 (glass filler)	1.10
Polyphenylene oxides (30 percent glass filler)	1.10
Ethyl cellulose	1.10–2.00
Methylpentene polymer	1.20
Phenolic (fabric and cord filler)	1.20
Phenolic with butadiene—acrylonitrile copolymer (rag filler)	1.20
Acrylics (impact)	1.20–1.50
Acrylics cast	1.20–1.70
Epoxy (low-density)	1.20–1.70
Phenolic with butadiene—acrylonitrile copolymer (asbestos filler)	1.20–1.70
Melamine (phenolic filler)	1.20–2.00
Cellulose acetate	1.20–2.30
Cellulose propionate	1.20–2.30
Cellulose acetate butyrate	1.20–2.30
Phenolic formaldehyde (wood-flour filler)	1.20–2.40
Phenolic formaldehyde (cotton filler)	1.20–2.40
Epoxy encapsulating (glass filler)	1.20–2.90
Epoxy (glass filler)	1.20–2.90
Epoxy encapsulating (mineral filler)	1.20–2.90
Epoxy (mineral filler)	1.20–8.70
Polyaryl sulfone	1.30
Polycarbonate (unfilled)	1.30
ABS (high-impact)	1.30–2.30
ABS (heat-resistant)	1.30–2.30
ABS (medium-impact)	1.30–2.30
ABS (self-extinguishing)	1.30–2.30

Table 3-2. (*Continued*)

Plastic	Value BTU/(h)(ft^2)(°F/in.)
Polyphenylene oxides (PPO)	1.33
Polycarbonate (10 percent glass filler)	1.40
Polycarbonate (up to 40 percent glass filler)	1.40–1.50
Chlorotrifluoroethylene (CTFE)	1.40–1.50
Acrylic multipolymer	1.50
Nylon (polyamide) type 6/10	1.50
Nylon (polyamide) type 6/10 (20–40 percent glass filler)	1.50
Nylon (polyamide) type 12	1.50
Polyphenylene oxides (modified) (Noryl)	1.50
Phenolic with butadiene—acrylonitrile copolymer (wood-flour and flock filler)	1.50–1.70
Diallyl phthalate (synthetic-fiber filler)	1.50–1.70
Diallyl phthalate (glass filler)	1.50–4.40
Acetal homopolymer	1.60
Acetal copolymer	1.60
Cellulose nitrate	1.60
Nylons (polyamide) type 6 (glass filler)	1.60
Ionomers	1.70
Nylons (polyamide) type 6/6	1.70
Nylons (polyamide) type 6	1.70
Polytetrafluorethylene (TFE)	1.70
Flurocarbon (FEP)	1.70
ABS polycarbonate alloy	1.70–2.60
Phenolic formaldehyde (asbestos filler)	1.70–6.40
Melamine (cellulose filler)	1.90–2.50
Nylons (polyamide) type 11	2.00
Polyaryl ether	2.00
Polyesters (thermoplastic)	2.00
Polyesters (18 percent glass filler)	2.00
Polyphenylene sulfides (unfilled)	2.00
Urea formaldehyde	2.00–2.90
Melamine (alpha cellulose filler)	2.00–2.90
Diallyl phthalate (mineral filler)	2.00–7.30
Silicones (glass filler)	2.20
Polyethylenes (low-density)	2.30
Polyethylenes (medium-density)	2.30–2.90
Phenolic (glass filler)	2.40
Nylons (polyamide) type 11 (glass filler)	2.60
Phenolic (mica filler) mineral	2.90–4.00
Melamine (fabric filler phenolic modified)	2.90
Polyester (premixed chopped glass)	2.90–4.60
Melamic (fabric filler)	3.10
Polyethylenes (high-density)	3.20–3.60
Silicones (mineral filler)	3.20–3.80
Polyester alkyd granular putty type (mineral filler)	3.50–7.30
Polyester alkyd granular putty type (glass filler)	4.40–7.30

Table 3-2. ***(Continued)***

Plastic	Value BTU/(h)(ft²)(°F/in.)
Cellular	
ABS	0.58
Cellulose acetate	0.31
Epoxy	0.13–0.36
Phenolics	0.20–0.28
Polyethylene (low-density)	0.26–0.40
Polyethylene (high-density)	0.80–0.92
Polypropylene	0.27–4.2
Polystyrene	0.17–0.26
Polyurethane	0.11–0.52
Silicone	0.36
SAN	0.29–0.32

[a] From C. Harper, *Handbook of Plastics and Elastomers*. Reprinted by permission of McGraw–Hill Book Company.

Closed cell structures provide the lowest thermal conductivity due to the reduced convection of gas in the cells.

The primary technique for measuring thermal conductivity of insulating materials is the guarded hot plate method. The equipment used for this test is fairly complex and expensive. A schematic assembly of the guarded hot plate is shown in Figure 3-9. The guard heaters are used to prevent the heat flow in all directions except in the axial direction towards the specimens. The test is carried out by placing the specimen in between the main heater and the cooling plate (heat sink). The time required to stabilize the input and the output temperatures is determined. Thermal conductivity is calculated as follows:

$$K = \frac{Qt}{A(T_1 - T_2)}$$

where K = thermal conductivity (BTU/in./h/sq. ft./F); Q = time rate of heat flow (BTU/h); t = thickness of specimen (in.); A = area under test (sq. in.); T_1 = temperature of hot surface (°F); T_2 = temperature of cold surface (°F).

Figure 3-10 illustrates a commercially available guarded hot plate equipment. Table 3-2 lists the thermal conductivity of different materials including solid and cellular plastics.

3.4. THERMAL EXPANSION

The coefficient of thermal expansion is defined as the fractional change in length or volume of a material for a unit change in temperature. The coefficient of thermal expansion values for different plastics are of considerable interest to design engineers. Plastics tend to expand and contract anywhere from six to nine times

Table 3.3. Comparison of Thermal Expansion Values for Some Common Materials[a]

Plastic	Coefficient, 10^{-5} cm/cm/°F
Polyester (low-shrink)	0.366–0.605
Phenolic (glass filler)	0.44–1.14
Phenolic formaldehyde (asbestos filler)	0.44–2.22
Silicones (glass filler)	0.443
Melamine (phenolic filler)	0.555–2.22
Diallyl phthalate (glass filler)	0.555–2.00
Phenolic (fabric and cord filler)	0.555–2.22
Diallyl phthalate (mineral filler)	0.555–2.33
Epoxy (glass filler)	0.610–1.94
Polyester (preformed chopped rovings)	0.66–2.78
Nylons (polyamide) type 6/10 (20–40 percent glass filler)	0.666–1.78
Melamine (glass filler)	0.83–1.10
Nylons (polyamide) type 6/6 (30 percent glass filler)	0.83–1.11
Polyester alkyd granular putty type (glass filler)	0.83–1.39
Polyester woven cloth	0.83–1.66
Phenolic with butadiene acrylonitrile copolymer (wood-flour and flock filler)	0.83–1.94
Phenolic (mica filler)	1.00–1.44
Melamine (fabric filler) (phenolic modified)	1.00–1.55
Styrene acrylonitrile copolymer	1.00–2.50
Polyester (sheet molding compound)	1.11
Polyester (premixed chopped glass filler)	1.11–1.83
Silicones (mineral filler)	1.11–2.22
Melamine (asbestos filler)	1.11–2.50
Polyester alkyd granular putty type (mineral filler)	1.11–2.78
Epoxy (mineral filler)	1.11–2.78
Phenolic with butadiene acrylonitrile copolymer (rag filler)	1.16
Polyphenylene oxides (modified) (20–30 percent glass filler)	1.22
Urea formaldehyde	1.22–2.00
Polyesters (18 percent glass filler)	1.39
Melamine (fabric filler)	1.39–1.55
Styrene acrylonitrile (glass filler)	1.50–2.11
Polypropylene (inert filler)	1.61
ABS (20–40 percent glass filler)	1.61–2.00
Polypropylene (glass filler)	1.61–2.88
Phenolic formaldehyde (wood-flour filler)	1.66–2.50
Epoxy encapsulating (glass filler)	1.66–2.78
Epoxy encapsulating (mineral filler)	1.66–3.33
Polyester alkyd granular putty type (synthetic filler)	1.67–3.06
Nylons (polyamide) type 6 (glass filler)	1.83
Nylons (polyamide) type 11 (glass filler)	1.78
Polystyrenes (medium- and high-impact)	1.88–11.70
Polycarbonate (10 percent glass filler)	1.90
Styrene acrylonitrile copolymer	2.00–2.11
Acetal (20 percent glass filler)	2.00–4.50
Polyphenylene (40 percent glass filler)	2.22
Phenolic with butadiene acrylonitrile copolymer (asbestos filler)	2.22

Table 3.3. ***(Continued)***

Plastic	Coefficient, 10^{-5} cm/cm/°F
Polyamides aromatic	2.22–2.78
Chlorotrifluoroethylene (CTFE)	2.50–3.90
Melamine alpha cellulose	2.22
Melamine (cellulose filler)	2.50
Polyaryl sulfone	2.62
PVC (rigid)	2.77–10.30
Acrylics	2.78–5.00
Polyphenylene oxides (modified) (unfilled)	2.89
Polysulfone	2.89–3.11
Polyphenlene (25 percent asbestos filler)	2.94
Phenoxy	3.00–3.20
Diallyl phthalate (synthetic fiber filler)	3.00–3.33
Polyphenylene sulfides (unfilled)	3.05
Polypropylene (unmodified)	3.22–5.66
Polystyrene (heat and chemical)	3.33
Polyesters (thermoplastic)	3.33
ABS (heat-resistant)	3.33–5.00
Polystyrene (general purpose)	3.33–4.44
Acrylics (impact)	3.33–4.44
Polypropylene (impact rubber modified)	3.33–4.73
Acrylic multipolymer	3.33–5.00
ABS polycarbonate alloy	3.44–4.72
Polyaryl ether	3.61
Polycarbonate (unfilled)	3.64
PVC flexible (unfilled)	3.86–13.90
Nylons (polyamide) type 12 (glass filler)	4.16
Nylons (polyamide) type 6/6	4.44
Chlorinated polyether	4.44
Acetal homopolymer	4.50
ABS (medium-impact)	4.45–5.55
Cellulose nitrate	4.44–6.66
Acetal copolymer	4.72
Polypropylene copolymer	4.44–5.27
Polyallomer	4.60–5.55
Nylons (polyamide) type 6	4.60
Polytetrafluoroethylene (TFE)	4.60–5.80
Cellulose acetate	4.44–10.00
Nylons (polyamide) type 6/10	5.00
ABS (high-impact)	5.27–7.22
Polyethylenes cross-linkable grades	5.55–11.94
Ethyl cellulose	5.55–11.10
Polyethylene (low-density)	5.55–11.10
Urethanes elastomers	5.55–11.10
Fluorocarbon (FEP)	5.60
Nylons (polyamide) type 12	5.78
Polyethylenes (high-density)	6.10–7.22
Cellulose propionate	6.11–9.44

Table 3.3. (*Continued*)

Plastic	Coefficient, 10^{-5} cm/cm/°F
Cellulose acetate butyrate	6.11–9.44
Methylpentene polymer	6.50
Ionomers	6.66
Polyvinylidene fluoride (VF2)	6.70
Styrene butadiene thermoplastic elastomers	7.22–7.60
Polyethylene (medium-density)	7.77–8.89
Polybutylene	8.32
Nylons (polyamide) type 11	8.33
Ethylene vinyl acetate copolymer	8.88–11.10
Ethylene–ethyl acrylate copolymer	8.88–13.80
Polyterephthalate	4.90–13.00

[a] From C. Harper, *Handbook of Plastics and Elastomers*. Reprinted by permission of McGraw–Hill Book Company.

more than materials such as metals. This difference in the coefficient of thermal expansion develops internal stresses and stress concentrations in the polymer, and premature failures occur. Special expansion joints, which generally require the use of rubber gaskets to overcome the expansion of plastics, are commonly used. The use of a filler such as fiberglass lowers the coefficient of thermal expansion considerably and brings the value closer to that of metal and ceramics.

Two basically similar methods have been developed by ASTM to measure the coefficient of linear thermal expansion (ASTM D 696) and the coefficient of cubical thermal expansion of plastics (ASTM D 864), respectively. The values reported in the trade journals are usually the coefficients of linear thermal expansion.

3.4.1. Coefficient of Linear Thermal Expansion

The test method requires the use of a fused quartz-tube dilatometer, a device for measuring changes in length (dial gauge or LVDT) and liquid bath to control temperature. Figure 3-11 illustrates a schematic configuration of a quartz-tube dilatometer.

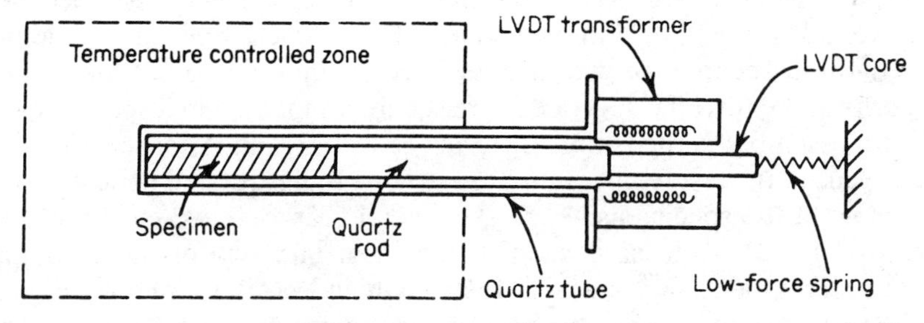

Figure 3-11. Schematic configuration of a quartz-tube dilatometer. (Reprinted by permission of McGraw–Hill Book Company.)

The test is commenced by mounting a preconditioned specimen, usually between 2 and 5 in. long into the dilatometer. The dilatometer, along with the measuring device, is then placed below the liquid level of the bath. The temperature of the bath is varied as specified. The change in length is recorded. The coefficient of linear thermal expansion is calculated as follows:

$$X = \frac{\Delta L}{L_0 \Delta T}$$

where X = coefficient of linear thermal expansion/°C; ΔL = change in the length of the specimen due to heating or cooling; L_0 = length of the specimen at room temperature; ΔT = temperature difference in °C over which the change in the length of the specimen is measured.

Table 3-3 compares the thermal expansion values for some common plastics.

3.5. BRITTLENESS TEMPERATURE (ASTM D 746)

At low temperatures, all plastics tend to become rigid and brittle. This happens mainly because, at lower temperatures, the mobility of polymer chains is greatly reduced. Brittleness temperature is defined as the temperature at which plastics and elastomers exhibit brittle failure under impact conditions. Yet another way to define brittleness temperature is the temperature at which 50 percent of the specimens tested exhibit brittle failure under specified impact conditions.

3.5.1. Test Apparatus and Procedures

The test apparatus consists of a specimen clamp and a striking member. The specimen clamp is designed so that it holds the specimen firmly as a cantilever beam. A typical specimen clamp is shown in Figure 3-12. The test apparatus most commonly used is a motor-driven brittleness temperature tester illustrated in Figure 3-13. The tester has a rotating striking tool that rotates at a constant linear speed of 6.5 ± 0.5 ft/sec. An insulated refrigerant tank with a built-in stirrer to circulate the heat transfer medium is used. The stirrer maintains the temperature equilibrium. Any type of heat transfer medium can be used as long as it remains liquid at the test temperature.

The test specimens are usually die punched from a sheet. The specimen dimensions are 1 in.-long, 0.25 in. wide and 0.075 in. thick. Specimen conditioning is carried out in accordance with the standard conditioning procedures.

To perform the test, the specimens are securely mounted in a specimen clamp. The entire assembly is then submerged in the refrigerant. After immersion for a specified time at the test temperature, the striking tool is rotated to deliver a single impact blow to the specimens. Each specimen is carefully examined for failure. Failure is defined as the division of a specimen into two or more completely separated pieces or as any crack in the specimen that is visible to the unaided eye. The temperature is raised by uniform increments of 2 or 5°C per test and the test is repeated. This procedure is followed until both the no failure and all failure temperatures are determined.

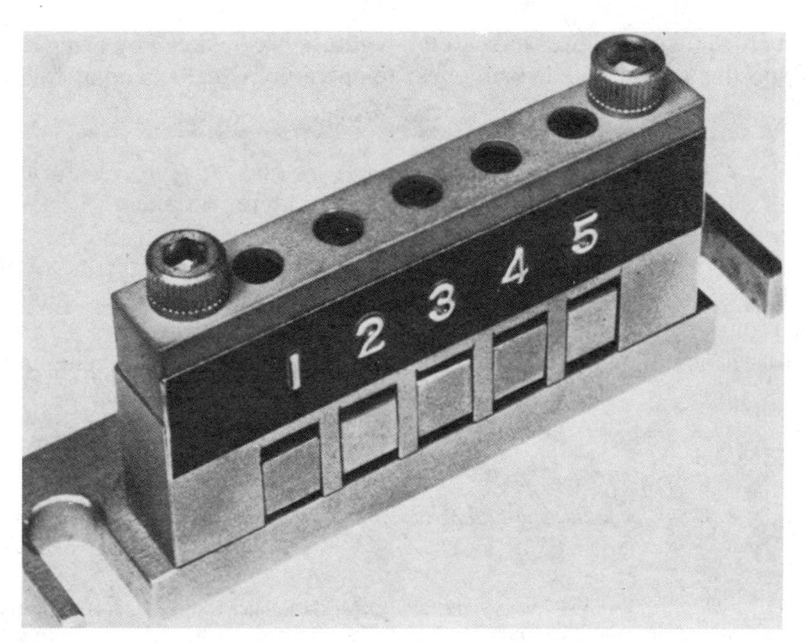

Figure 3-12. Typical specimen clamp for brittleness temperature test. (Reprinted by permission of ASTM.)

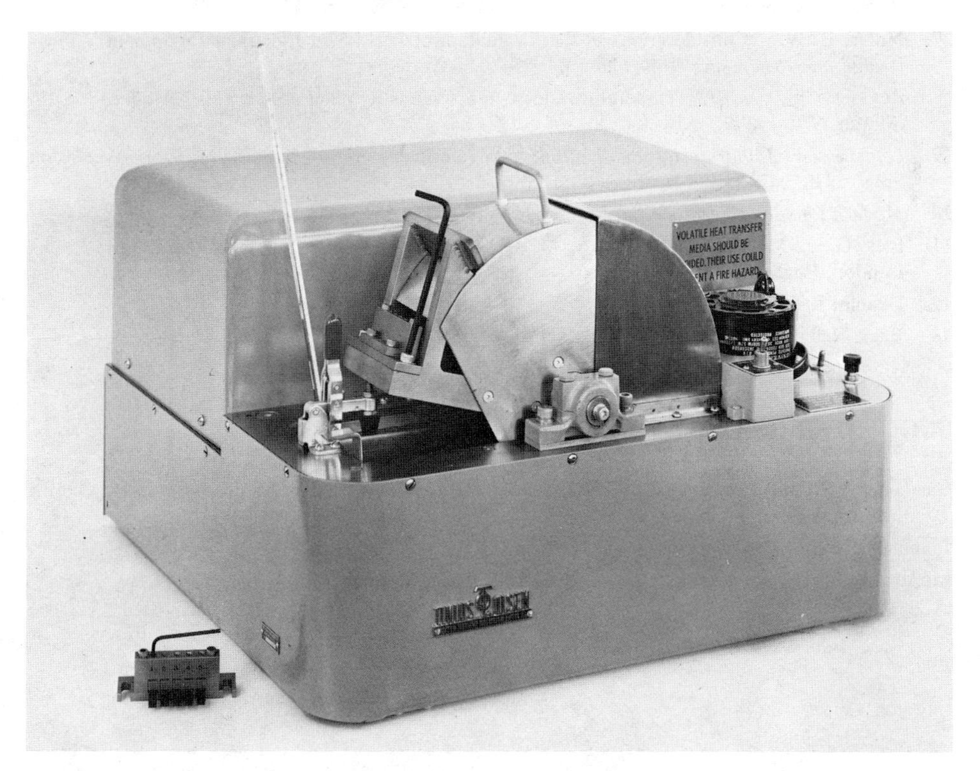

Figure 3-13. Motor-driven brittleness temperature tester. (Courtesy Tinius Olsen Company.)

The percentage of failures at each temperature is calculated by using the number of specimens that failed. Brittleness temperature is calculated as follows:

$$T_b = T_h + \Delta T[(S/100) - (1/2)]$$

where T_b = brittleness temperature (°C); T_h = highest temperature at which failure of all specimens occur (°C); ΔT = temperature increment (°C); S = sum of the percentage of breaks at each temperature.

Brittleness temperature has very little practical value since the data obtained by such a method can only be used in applications in which the conditions of deformation are similar to those specified in the test. The method is, however, useful for specification, quality control, and research and development purposes.

REFERENCES

1. Deanin, R. D., *Polymer Structure Properties and Applications,* Cahners, Boston, Mass., 1972, p. 238.
2. *Ibid,* p. 273.
3. *Symp. on Plastics Testing—Present and Future,* ASTM Publication No. 132, American Society for Testing and Materials, Philadelphia, Pa., 1953, p. 16.
4. "Design Guide," *Modern Plastics Encyclopedia,* McGraw–Hill, New York, 1980–81, p. 490.
5. Nielson, L. E., *Mechanical Properties of Polymers,* Reinhold, New York, 1962, pp. 138–196.
6. Baer, E. *Engineering Design for Plastics,* Reinhold, New York, 1964, p. 185.
7. Miller, R. W., "Considerations in the Evaluation of Plastics in Electrical Equipment," *Plast. Design and Processing* (July 1980), pp. 25–30.
8. Reymers, H., "A New Temperature Index, Who Needs It, What does it Tell," *Modern Plastics* (March 1970), p. 79.
9. *Underwriters Laboratories Publication,* "Polymeric Materials—Long Term Property Evaluations," UL 746B.
10. *Modern Plastics Encyclopedia,* Reference 4, p. 492.
11. Ives, G. C., Mead, J. A., and Riley, M. M., *Handbook of Plastics Test Methods,* Iliffe Books, London, England, 1971, p. 335.
12. Deanin, Reference 1, p. 379.
13. Baer, Reference 6, p. 1025.

SUGGESTED READINGS

Carslaw, H. S., and Jaeger, J. C., *Conduction of Heat in Solids.* Oxford University Press, Oxford, England, 1959.

Jakob, M., *Heat Transfer.* Vols. I and II. Wiley, New York 1949, 1957.

McAdams, W. H., *Heat Transmission,* McGraw–Hill, New York, 1954.

CHAPTER 4

Electrical Properties

4.1. INTRODUCTION

The unbeatable combination of characteristics such as ease of fabrication, low cost, light weight, and excellent insulation properties have made plastics one of the most desirable materials for electrical applications. Although, the majority of applications involving plastics are insulation-related, plastics can be made to conduct electricity by simply modifying the base material with proper additives such as carbon black.

Until recently, plastics were considered a relatively weaker material in terms of load-bearing properties at elevated temperatures. Therefore, the use of plastics in electrical applications was limited to non-load-bearing, general-purpose applications. The advent of new high performance engineering materials has altered the entire picture. Plastics are now specified in a majority of applications requiring resistance to extreme temperatures, chemicals, moisture, and stresses. The primary function of plastics in electrical applications has been that of an insulator. This insulator or dielectric separates two field-carrying conductors. Such a function can be served equally well by air or vacuum. However, neither air nor vacuum can provide any mechanical support to the conductors. Plastics not only act as effective insulators but also provide mechanical support for field-carrying conductors. For this very reason, the mechanical properties of plastic materials used as insulators become very important (1). Typical electrical applications of plastic material include plastic-coated wires, terminals, connectors, industrial and household plugs, switches, and printed circuit boards. The following are the typical requirements of an insulator.

1. An insulator must have a high enough dielectric strength to withstand an electrical field between the conductors.
2. An insulator must possess good arc resistance to prevent damage in case of arcing.
3. An insulator must maintain integrity under a wide variety of environmental hazards such as humidity, temperature, and radiation.
4. Insulating materials must be mechanically strong enough to resist vibration shocks and other mechanical forces.

5. An insulator must have high insulation resistance to prevent leakage of current across the conductors.

The key electrical properties of interest are dielectric strength, dielectric constant, dissipation factor, volume and surface resistivity, and arc resistance.

4.2. DIELECTRIC STRENGTH (ASTM D 149)

The dielectric strength of an insulating material is defined as the maximum voltage required to produce a dielectric breakdown. Dielectric strength is expressed in volts per unit of thickness such as V/mil. All insulators allow a small amount of current to leak through or around themselves. Only a perfect insulator, if there is such an insulator in existence, can be completely free from small current leakage. The small leakage generates heat, providing an easier access to more current. The process slowly accelerates with time and the amount of voltage applied until a failure in terms of dielectric breakdown or what is known as puncture occurs. Obviously, dielectric strength, which indicates electrical strength of a material as an insulator, is a very important characteristic of an insulating material. The higher the dielectric strength, the better the quality of an insulator. Table 4.1 lists some typical dielectric strength values of common plastic materials. Three basic procedures have been developed to determine dielectric strength of an insulator. Figure 4.1 schematically illustrates the basic set-up for a dielectric strength test. A variable transformer and a pair of electrodes are normally employed. Specimens of any desirable thickness prepared from the material to be tested are used. Specimen thickness of $\frac{1}{16}$ in is fairly common. The first procedure is known as the short-times method. In this method, the voltage is increased from zero to breakdown at uniform rate. The rate of rise is generally 100, 500, 1000, or 3000 V/sec until the failure occurs. The failure is made evident by actual rupture or decomposition of the specimen. Sometimes a circuit breaker or other similar devices are employed to signal the voltage breakdown. This is not considered a positive

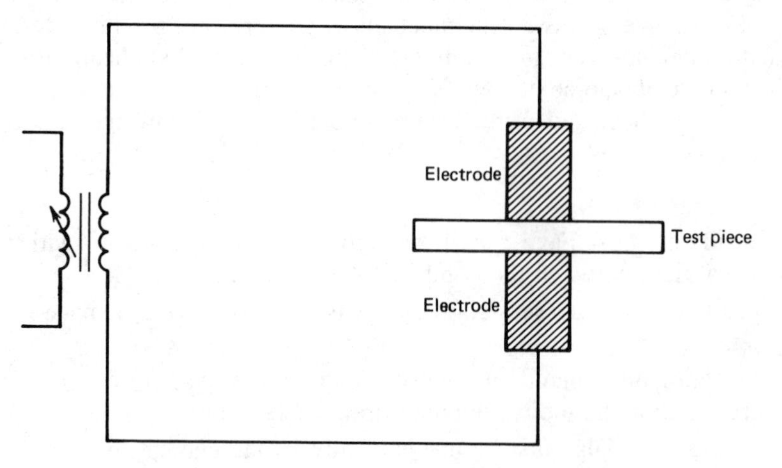

Figure 4-1. Schematic of dielectric strength test.

Table 4-1. Dielectric Strength (Short-Term) of Various Plastics

Plastics	Dielectric strength, V/mil
PFA (fluorocarbon)	2000
CPVC	1200–1500
Rigid PVC	800–1400
Ionomer	1000
Polyester (thermoplastic)	600–750
Polypropylene	650
Polystyrene (high impact)	650
FEP (fluorocarbon)	600
Nylons	350–560
Polstyrene (General Purpose)	500
Acetals	500
PTFE (fluorocarbon)	500
PPO	500
Polyphenylene sulfide	490
Polyethylene	480
Polycarbonate	450
ABS	415
Phenolics	240–340
PVF_2 (fluorocarbon)	260

indication of voltage breakdown since other factors such as flashover, leakage current, corona current, or equipment magnetizing current can influence such indicating devices.

The second method is known as the slow-rate-of-rise method. The test is carried out by applying the initial voltage approximately equal to 50 percent of the breakdown voltage as determined by the short-time test or as specified. Next, the voltage is increased at a uniform rate until the breakdown occurs.

The step-by-step test method requires applying initial voltage equal to 50 percent of the breakdown voltage as determined by the short-time test and then increasing the voltage in equal increments and held for specified time periods until the specimen breaks down. In almost all cases the dielectric strength values obtained by step-by-step method corresponds better with actual use conditions. However, the service failures are generally at voltage below the rated dielectric strength because of the time factor involved (2). The dielectric strength of an insulating material is calculated as follows:

$$\text{Dielectric strength (V/mil)} = \frac{\text{Breakdown Voltage (V)}}{\text{Thickness (mil)}}$$

4.2.1. Factors Affecting the Test Results

A. *Specimen Thickness*

The dielectric strength of an insulator varies inversely with the fractional power (generally 0.4) of the thickness (3). The thicker specimen requires higher voltage to achieve the same voltage gradient. At higher voltage, a reduction in intermolecular bonds is observed resulting from thermal expansion created by heat generation. Thicker sections also have internal voids, flaws, moisture, nonuniformity, and greater current leakage causing early failure of the specimen (4). Thin films with higher dielectric strength value have been used successfully in many critical space-saving applications. Dielectric strength versus thickness curves are readily available from material suppliers.

B. Temperature

Dielectric strength decreases with the increase in the temperature of the specimen. If the design calls for the use of the product at various temperatures, the dielectric strength values at anticipated temperatures should be carefully evaluated. Interestingly enough, the dielectric strength of the material below room temperature is constant and independent of temperature change (5).

C. Humidity

Humidity affects the dielectric strength of the material. Surface moisture as well as moisture absorbed by hygroscopic material affects the results.

D. Electrodes

Dielectric strength of a material is affected by the electrode geometry, the electrode area, as well as the electrode base material composition. Generally the breakdown voltage decreases with increasing electrode area. The effect is more pronounced with thinner test specimens.

E. Time

The rate of voltage application significantly alters the test results. As mentioned earlier in this chapter, dielectric strength values obtained by step-by-step method are lower than those obtained by the short-time test.

F. Mechanical Stress

Mechanical stress tends to reduce the dielectric strength values substantially.

G. Processing

Defects such as poor weld lines, voids, bubbles, and flow lines brought forth by poor processing practices tend to reduce the dielectric strength anywhere from 30–60 percent depending upon the severity of the defect.

4.2.2. Test Limitations and Interpretations

In designing plastic parts for electrical applications, the dielectric strength values must be studied in detail. If the parts are designed based on the values derived from published literature without proper consideration to the effects of thickness, moisture, temperature, mechanical stress, and actual use conditions, the results could prove disastrous. The actual use conditions of the insulating materials are quite different than the conditions of the dielectric strength test and therefore the real emphasis must always be on the performance of the material in actual service.

4.3. DIELECTRIC CONSTANT AND DISSIPATION FACTOR ASTM D150

4.3.1. Dielectric Constant (Permittivity)

Dielectric constant of an insulating material is defined as the ratio of the charge stored in an insulating material placed between two metallic plates to the charge that can be stored when the insulating material is replaced by air (or vacuum). Defined another way, the dielectric constant is the ratio of the capacitance induced by two metallic plates with an insulator placed between them and the capacitance of the same plates with a vacuum between them.

$$\text{Dielectric constant} = \frac{\text{Capacitance, Material as dielectric}}{\text{Capacitance, Air (or Vacuum) as dielectric}}$$

Simply stated, the dielectric constant indicates the ability of an insulator to store electrical energy.

In many applications, insulating materials are required to perform as capacitors. Such applications are best served by plastic materials having a high dielectric constant. Materials with a high dielectric constant have also helped in reducing the physical size of the capacitors. Furthermore, the thinner the insulating material, the higher the capacitance. Because of this fact plastic foils are extensively used in applications requiring high capacitance.

One of the main function of an insulator is to insulate the current-carrying conductors from each other and from the ground. If the insulator is used strictly for this purpose, it is desirable to have the capacitance of the insulating material as small as possible.

For these applications, one is looking for the materials with very low dielectric constant. Table 4.2 lists typical dielectric constant values of various plastics. The dielectric constant of air (or vacuum) is 1 at all frequencies. The dielectric constant of plastics varies from 2 to 20.

The dielectric constant test is fairly simple. The test specimen is placed between the two electrodes, as shown in Figure 4.2, and the capacitance is measured. Next, the test specimen is replaced by air and once again, the capacitance value is measured. The dielectric constant value is determined from the ratio of the two measurements. Dielectric constant values are affected by factors such as frequency, voltage, temperature, humidity, etc.

4.3.2. Dissipation Factor

In all electrical applications, it is desirable to keep the electrical losses to a minimum. Electrical losses indicate the inefficiency of an insulator. The dissipation factor is a measure of such electrical inefficiency of the insulating material. The dissipation factor indicates the amount of energy dissipated by the insulating ma-

Table 4-2. Dielectric Constant[a] of Various Plastics

Plastics	Dielectric constant
Melamines	5.2–7.9
Polyvinylidene fluoride (PVF_2)	7.5
Cellulose acetate	3.0–7.0
Phenolics	4.0–7.0
Nylons (30 percent glass filled)	3.5–5.4
Epoxies	4.3–5.1
Polyesters (thermoset)	4.7
Polystyrene (high impact)	2.0–4.0
Acetal copolymer (25 percent glass filled)	3.9
Nylons	3.5–3.8
Acetals	3.7
Polycarbonate (30 percent glass filled)	3.48
Polysulfone (30 percent glass filled)	3.4
Polyester (thermoplastic)	3.2
ABS	3.2
SAN	3.0
Polystyrene (general purpose)	2.7
Fluorocarbons (PFA, FEP)	2.1

[a] At 10^6 Hz.

terial when the voltage is applied to the circuit (6). The dissipation factor is defined as the ratio of the conductance of a capacitor in which the material is the dielectric to its susceptance or the ratio of its parallel reactance to its parallel resistance. Most plastics have a relatively lower dissipation factor at room temperature. However, at high temperatures, the dissipation factor is quite high, resulting in greater overall inefficiency in electrical system. Loss factor, which is the product of

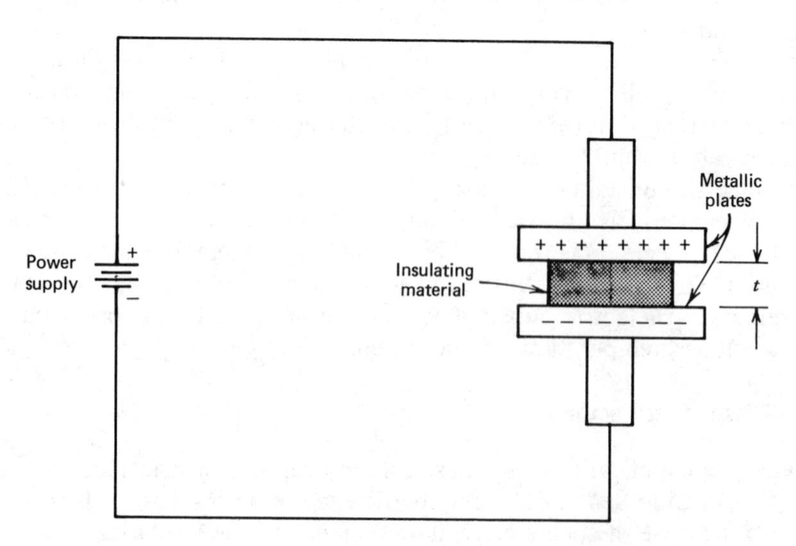

Figure 4-2. Schematic of dielectric constant test.

dielectric constant and the dissipation factor, is a frequently used term, since it relates to the total loss of power occuring in insulating materials.

4.4. ELECTRICAL RESISTANCE TESTS

As was stated earlier, the primary function of an insulator is to insulate current-carrying conductors from each other as well as from ground and to provide mechanical support for components. Naturally, the most desirable characteristic of an insulator is its ability to resist the leakage of the electrical current. The higher the insulation resistance, the better the insulator. Failure to recognize the importance of insulation resistance values while designing products such as appliances and power tools could lead to fire, electrical shock, and personal injury.

Insulation resistance can be subdivided into:

1. Volume resistance.
2. Surface resistance.

Volume resistance is defined as the ratio of the direct voltage applied to two electrodes that are in contact with a specimen to that portion of the current between them that is distributed through the volume of the specimen. Or, simply stated, the volume resistance is the resistance to leakage through the body of the material (7). Volume resistance generally depends upon the material. The term most commonly used by designers is volume resistivity. It is defined as the ratio of the potential gradient parallel to the current in the material to the current density or simply stated, the volume resistivity of a material is the electrical resistance between the opposite faces of a unit cube for a given material and at a given temperature (8).

$$\text{Volume Resistivity } (\rho)\ (\Omega/\text{cm}) = \frac{A}{t}\,(R_v)$$

where A = area; t = thickness of the specimen; R_v = volume resistance (Ω).

High volume resistivity materials are desirable in applications requiring superior insulating characteristics. Table 4.3 lists typical volume resistivity values of some common plastics.

The surface resistance of a material is defined as the ratio of the direct voltage applied to the electrodes to that portion of the current between them that is primarily in a thin layer of moisture or other semiconducting material that may be deposited on the surface. Or simply stated, surface resistance is the resistance to leakage along the surface of an insulator. The surface resistance of a material depends upon the quality and cleanliness of the surface of the product. A product with oil or dirt particles on it gives lower surface resistance values.

The test procedures to determine electrical resistance values are rather complex. ASTM D 257 describes the procedures as well as the complex electrodes and apparatus required to carry out the test in detail.

Temperature and humidity both seem to affect the insulation resistance appreciably. As a rule, the higher the temperature and humidity, the lower the insulation resistance of a material.

Table 4-3. Volume Resistivity of Various Plastics

Plastics	Volume resistivity (Ω/cm)[a]
Melamine (asbestos filler)	1.2×10^{12}
Urea Formaldehyde	10^{12}–10^{13}
Phenolic (asbestos filler)	10^{9}–10^{13}
Acetal copolymer	10^{14}
Acrylics	10^{14}
Epoxy	10^{14}
Polystyrene	10^{16}
SAN	10^{16}
ABS	5×10^{16}
Polycarbonate	2×10^{16}
PVC (flexible)	10^{11}–10^{15}
Nylons, type 6/6	10^{14}–10^{15}
Acetal homopolymer	10^{15}
PVC (rigid)	10^{15}
Polyethylene	10^{16}
Polypropylene	10^{16}
Thermoplastic polyester	3×10^{16}
Polysulfone	5×10^{16}
PPO	10^{17}
PTFE	10^{18}
FEP	2×10^{18}

[a] At 50 percent RH and 73°F.

4.5. ARC RESISTANCE (ASTM D 495)

Arc resistance is the ability of a plastic material to resist the action of a high-voltage electrical arc, usually stated in terms of time required to form material electrically conductive. Failure is characterized by carbonization of the surface, tracking, localized heating to incandescence, or burning. In all applications in which conducting elements are brought into contact, arcing is inevitable. Switches, circuit breakers, and automotive distributor caps, are a few good examples of applications where arcing is known to cause failure. Another term that is generally associated with arcing is tracking. Tracking is defined as a phenomenon where a high voltage source current creates a leakage or fault path across the surface of an insulating material by slowly but steadily forming a carbonized path appearing as a thin, wiry line between the electrodes. Tracking is accelerated by the presence of surface contaminants such as dirt and oil and by the presence of moisture. Resistance to arcing or tracking depends upon the type of plastic materials such as phenolics that tend to carbonize easily and therefore have relatively poor arc resistance. Plastics such as alkyds, melamines, and fluorocarbons are excellent arc-resistant materials. The failures due to arcing are not always because of carbonization or tracking. Many plastics such as acrylics simply do not carbonize. However, they do form ignitable gases that cause the product to fail in a short time. Arc resistance of plastics can be improved substantially by the addition of fillers such as glass, mineral, wood flour, asbestos, and other inorganic fillers. Table 4.4 lists the arc resistance of some common plastic materials.

The determination of the arc resistance of plastics using a standard test method has always been a problem because of the numerous ways the test can be conducted. ASTM has developed four basic test methods to satisfy this concern. ASTM D 495, which is high-voltage, low-current, dry arc resistance of solid electrical insulation, has been the most widely used and accepted. This test is only intended for the preliminary screening of materials, for detecting the effect of changes in formulations, and for quality control testing. Since the test is conducted under clean and dry laboratory conditions that are rarely encountered in practice, the prediction of materials from test results is next to impossible. Figure 4.3 illustrates a typical set-up for an arc resistance test. The voltage is applied intermittantly and severity is increased in steps until the failure occurs. Arc resistance is measured in seconds to failure.

ASTM method D 2132 outlines the procedure for determining dust and fog tracking and erosion resistance of electrical insulating materials. The test is carried out in a fog chamber with a standardized dust applied to the sample surface. Failure is characterized by the erosion of the specimen or tracking. ASTM D 2302 also describes the test for differential wet tracking resistance of electrical insulating materials with controlled water-to-metal discharges. The inclined specimen is partially immersed in a water solution of ammonium chloride and a wetting agent. Failure is characterized by tracking. ASTM D 2303 describes the test for liquid–contaminant inclined plane tracking and erosion of insulating materials. In this test, the specimen is inclined at 45°, and the electrolyte is discharged onto the surface at a controlled rate, increasing the voltage at the same time. The failure is marked by erosion and tracking.

Table 4-4. Arc Resistance of Various Plastics

Plastics	Arc resistance (sec)
Polycarbonate (10–40 percent glass filled)	5–120
Polycarbonate	10–120
Polystyrene (high impact)	20–100
ABS	50–85
Polystyrene (General Purpose)	60–80
Rigid PVC	60–80
Polysulfone	75–190
Urea formaldehyde	80–150
Ionomer	90–140
SAN	100–150
Epoxy	120–150
Acetal (homopolymer)	130
Polyethylene (low-density)	135–160
Polypropylene	135–180
PTFE	>200
Acrylics	No track

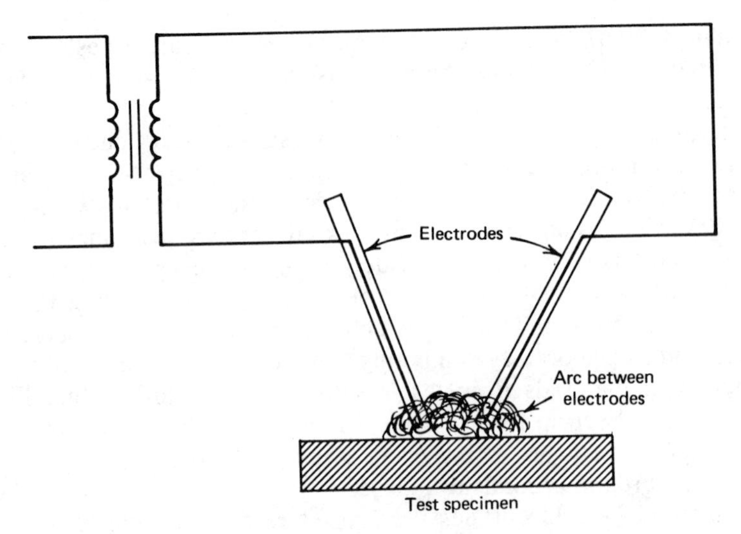

Figure 4-3. Arc-resistance test.

4.6. UL REQUIREMENTS

No discussion on electrical properties and testing can be considered complete without discussing the role of Underwriters Laboratories (UL) in electrical codes, standards, and specifications (9). UL is an independent, nonprofit organization established with the basic goal of testing for public safety and developing standards for safety. A further discussion regarding UL's organizational capabilities is in Chapter 19.

An increasing number of designers are specifying plastics in a variety of applications. Depending upon the application, plastics material may serve as an insulation, a structural member, or a simple enclosure to house uninsulated electrically live parts. Parts must perform, in such cases, their intended function; otherwise, fire, electrical shock, or personal injury may result. From the foregoing, it is clear that a plastic part used in an electrical application must have adequate properties to minimize the risks related to safety. Underwriters Laboratories engineers have developed a set of requirements that one must meet if UL listing is desired.

4.6.1. Material Properties

Properties that are considered by UL in the evaluation of polymeric material for suitability in electrical applications are divided into the two broad areas of short-term properties and long-term properties.

A. *Short-Term Properties UL 746 A*

UL 746 A covers the short-term test procedures. These investigations provide data with respect to the mechanical, electrical flammability, thermal, and other properties of the materials under consideration, and are intended to provide guidance for the material manufacturer, the molder, the end-product manufacturer, and safety engineers. Testing for mechanical properties include tensile testing, Izod impact, flexural properties, shear properties, bond strength properties of adhesives, and durometer hardness. Electrical properties of interest are dielectric strength, DC resistance or conductance of insulating materials, high-voltage, low-current dry arc resistance of solid electrical insulation, and high voltage arc-tracking-rate index of solid insulating materials. Testing for thermal properties include deflection temperature of polymeric materials under load, Vicat softening point, and softening point by ring-and-ball apparatus. Ease of ignition tests are also part of short-term property evaluations. Tests such as hot wire ignitions, high-current arc ignition, high-voltage arc resistance to ignition, and hot gas ignition are briefly described. Three other short-term tests are dimensional change of polymeric parts, resistance of polymeric materials to chemical reagents, and tests for polymer identification. The majority of these test procedures are based on ASTM test methods.

B. Long-Term Properties UL 746 B

UL 746 B covers the procedures for evaluating long-term properties of polymeric materials. Mechanical, electrical, thermal, flammability, and other properties are considered with the intent to provide guidance for the material manufacturer, molder, and safety engineers.

The behavior of plastic materials under long-term exposure to elevated temperatures is of prime importance to UL engineers. A procedure to determine the relative thermal indices of polymeric materials is discussed in detail in this standard. The relative thermal index of a material is an indication of a material's ability to retain a particular property when exposed to elevated temperatures for an extended period of time. It is a measure of a material's thermal endurance. Other long-term property evaluation tests include environmental exposure, creep, and chemical resistance.

4.6.2. Evaluation of Plastic Materials Used in Electrical Equipments UL 746 C

UL 746 C provides guidelines for selecting minimum performance levels based on the application. The use of the UL 746 C standard is best understood through a specific example.

Let us assume that we are designing an enclosure of a stationary equipment that houses uninsulated electrical live parts. In order to determine the material requirements we proceed as follows.

First, from UL standard UL 746 C, we need to find the appropriate table that coincides with our requirements. In this case, for stationary equipment, Table 4.5 (taken from UL 746 C, Table 12.1 and Figure 12.1) is the correct choice. Table 4.5 shows a decision tree and the requirements for enclosure of stationary equipments. The first consideration is the flammability requirements. The letter *D* denotes a 94-5V material is required for this application. Material classed 94-5V should not have any specimens that burn with flaming and/or glowing combustion for more than 60 sec after the fifth flame application, nor should it have any specimens that drip any particles when tested in accordance with vertical burning test. Since the equipment houses uninsulated electrically live parts, path 1 must be followed as shown in Table 4.6. A quick comparison between the different paths indicates that more severe application (path 1) requires more test considerations than the less severe application such as the one in path 4.

Test considerations in Table 4.6 reflect performance levels obtained on standard specimens or end-product tests that stress the material under actual conditions of use and misuse. We need to carefully study each of the enclosure performance requirements that are clearly spelled out in the standard. In many cases, a mere comparison of performance requirements with published material data is sufficient. If the material complies with the minimum performance levels, no additional testing is required. If it does not, the material may be qualified by an appropriate end-product test.

Table 4-5. Decision Tree and Test Requirements for Materials Used in Enclosures for Stationary Equipment[a]

Stationary equipment			
Material is used to enclose uninsulated live parts (enamel-coated wire is to be considered an uninsulated live part)	Material is used to enclose live parts with insulation thickness less than 0.028-in. (0.71 mm)	Material is used to enclose live parts with insulation thickness 0.028 in. (0.71 mm) or greater	Material is used to enclose a metal housing which in turn encloses insulated or uninsulated live parts or material is used as a decorative part or trim of the enclosure
D only Go to path 1 in Table 4-6.	*D* only Go to path 2 in Table 4-6.	*D* only Go to path 3 in Table 4-6.	*D*, *E*, or *F* Go to path 4 in Table 4-6.

[a] Reprinted from *Plastics Design and Processing*, July 1980, p. 29.

[b] *D* Denotes a 94-5V material (as classed by the vertical burning test described in the Standard Tests for Flammability of Plastic Materials for Parts in Devices and Appliances, UL 94) where the temperature and integrity of the material after flame exposure are consistent with the requirements applicable to the product in which a part made of the materials is used. A material is to be considered equivalent if it complies with the flame test on the equipment itself. When flame-retardant paint is applied to the inside of a polymer enclosure for purposes of the flame test, the paint/material interface shall have been found to be acceptable by separate investigation that included consideration of the effects of long-term thermal aging and exposure to environmental conditions such as moisture and high humidity that are likely to be present in the environment associated with the product. *E* Denotes a 95V-0, 94V-1, and 94V-2 material (as classed by the vertical burning test described in the Standard Tests for Flammability of Plastic Materials for Parts in Devices and Appliances, UL 94). *F* Denotes a 94HB material (as classed by the horizontal burning test described in the Standard Tests for Flammability of Plastic Materials for Parts in Devices and Appliances, UL 94).

Table 4-6. Requirements Related to Flammability Symbols Determined in Table 4-5 as Applied to Stationary Equipment Enclosures[a]

	Path 1	Path 2	Path 3	Path 4
1. Ultraviolet radiation resistance	Yes	Yes	Yes	No material tests required other than use of material designated D, E, or F
2. Water and moisture immersion and exposure				
a. Properties	Yes	Yes	Yes	
b. Dimensions	Yes		—	
3. Volume resistivity	Yes		—	
4. Resistance to hot-wire ignition	Yes		—	
5. Distortion under load, either				
a. Heat deflection temperature				
b. Vicat softening temperature				
c. Ball pressure temperature				
6. Impact	Yes	Yes	Yes	
7. Crush resistance	Yes	Yes	—	
8. Mold stress relief distortion	Yes	Yes	Yes	
9. Dielectric voltage-withstand	Yes	Yes	—	
10. Abnormal operation	Yes	Yes	Yes	
11. Resistance to ignition	Yes	Yes	—	
12. Spacings to enclosure	Yes	Yes	—	
13. Severe conditions	Yes	Yes	Yes	
14. Strain relief (if applicable)	Yes	Yes	Yes	
15. Input after mold relief distortion	Yes	Yes	Yes	

[a] Reprinted from *Plastics Design and Processing*, July 1980, p. 29.

Similar tables indicating the decision tree and test requirements for portable equipment and fixed equipment have been developed by Underwriters Laboratories. In each case, the procedure for determining the suitability of a plastic material used in electrical equipment is basically the same. If the plastic parts used in this application are required to go through a special process such as plating, adhesive bonding, flame retardant, or metallic coating etc., some additional testing is required.

4.6.3. Polymeric Materials—Fabricated Parts UL 746 D

This standard covers the requirements for molded or fabricated parts. Material identity control systems intended to provide tracability of material used for polymeric parts through the handling, molding, or fabrication, and shipping operations are described. Guidelines are also provided for the acceptable blending or simple compounding operations that may affect risk of fire, electrical shock, or personal injury.

REFERENCES

1. Levy, S., and Dubois, H., *Plastics Product Design Engineering Handbook,* Reinhold, New York, 1977, p. 276.

2. Kinney, G. F., *Engineering Properties and Applications of Plastics,* John Wiley & Sons, New York, 1957, p. 222.
3. *Ibid.,* p. 223.
4. Milby, R., *Plastics Technology,* McGraw-Hill, New York, 1973, p. 498.
5. Kinney, Reference 2, p. 223.
6. Harper, C. A., "Short Course in Electrical Properties," *Plast. World* (April 1979), p. 73.
7. Kinney, Reference 2, p. 224.
8. Harper, Reference 6, p. 72.
9. *Underwriters Laboratories Standards for Safety,* UL 746 A-B-C-D, U. L. Inc., Melville, L.I., New York.

SUGGESTED READING

Harris, F. K., *Electrical Measurements,* Wiley, New York, 1952.

Birks, J. B., *Modern Dielectric Materials,* Heywood, 1960.

Anderson, J. C., *Dielectrics,* Chapman and Hall, London, England, 1964.

Reddish, W., "Chemical Structure and Dielectric Properties of Polymers," *Pure and Appl. Chem.,* **5**(4) (1962).

Von Hipple, A., *Dielectric Materials and Applications,* Wiley, New York, 1954.

Whitehead, S., *Dielectric Breakdown of Solids,* Oxford University Press, Oxford, England, 1951.

Martin, T. and Hauter, J., "Arcing Tests on Plastics," J. S.P.E., **10**(2), (Feb. 1954), p. 13.

CHAPTER 5

Weathering Properties

5.1. INTRODUCTION

The increased outdoor use of plastics has created a need for a better understanding of the effect of the environment on plastic materials. The environmental factors have significant detrimental effects on appearance and properties. The severity of the damage depends largely on factors such as the nature of the environment, geographic location, type of polymeric material, and duration of exposure. The effect can be anywhere from a mere loss of color or a slight crazing and cracking to a complete breakdown of the polymer structure. Any attempt to design plastic parts without a clear understanding of the degradation mechanisms induced by the environment would result in a premature failure of the product. The major environmental factors that seriously affect plastics are:

1. Solar Radiations—UV, IR, X-rays.
2. Microorganisms, bacteria, fungus, mold.
3. High humidity.
4. Ozone, oxygen.
5. Water: vapor, liquid, or solid.
6. Thermal energy.
7. Pollution: industrial chemicals.

The combined effect of the factors mentioned above may be much more severe than the effect of any single factor, and the degradation processes are accelerated many times. Many test results do not include these synergistic effects that almost always exist in real-life situations.

5.1.1. UV Radiation

All types of solar radiation have some sort of detrimental effect on plastics. Ultraviolet radiation is the most destructive of all radiations. The energy in ultraviolet radiation is sufficiently strong enough to break molecular bonds. This activity in the polymer brings about thermal oxidative degradation which results in embrittlement, discoloration, and an overall reduction in physical and electrical properties. Fluorescent lighting, sun lamps, and other artificial sources also emit a similar type of harmful radiation. Other factors in the environment such as heat, humidity, and oxygen accelerate the UV degradation process.

One of the best methods of protecting plastics against UV radiation is to incorporate UV absorbers or UV stabilizers into the plastic materials. The UV absorbers provide preferential absorption to most of the incident UV light and are able to dissipate the absorbed energy harmlessly. Thus, the polymer is protected from harmful radiation at the cost of UV absorbers which are destroyed in the process with time. Several types of organic and inorganic UV absorbers are developed for this purpose. Almost all inorganic pigments absorb UV radiation to a certain extent and provide some degree of protection. Perhaps the most effective pigments are certain types of carbon black that absorb over the entire range of UV and visible radiation and transform the energy into less harmful radiation (1).

UV stabilizers, unlike UV absorbers, inhibit the bond rupture by chemical means or dissipate the energy to lower levels that do not attack the bonds. The effectiveness of such additives when incorporated with the polymer can be determined by various test methods which will be discussed later in this chapter.

5.1.2. Microorganisms

Polymeric materials are generally not vulnerable to microbial attack under normal conditions. However, low-molecular weight additives such as plasticizers, lubricants, stabilizers, and antioxidants may migrate to the surface of plastic components and encourage the growth of microorganisms. The detrimental effect can be seen readily through the loss of properties, change in aesthetic quality, loss of transmission (optical), and increase in brittleness. The rate of growth depends upon many factors, such as heat, light, and humidity. Preservatives, also known as fungicides or biocides, are added to plastic materials to prevent the growth of microorganisms. These additives are highly toxic to lower organisms, but do not affect the higher organisms.

It is necessary to evaluate the effectiveness of various antimicrobial agents both on a laboratory scale as well as in actual outdoor exposure. Many such methods have been devised to perform these tests. The test methods and their serious limitations are also discussed later in this chapter.

5.1.3. Oxygen, Moisture, Thermal Energy, and Other Environmental Factors

In addition to UV radiation and microorganisms, there are many other relevant factors that can add to polymer degradation. Even though most polymers react very slowly with oxygen, elevated temperatures and UV radiation can greatly promote the oxidation process. Water is considered relatively harmless, however, it can have at least three kinds of effects leading to early polymer degradation: a chemical effect, that is, the hydrolysis of unstable bonds; a physical effect that is, destroying the bond between a polymer and filler resulting in chalking; the third is a photochemical effect. Thermal energy plays an indirect role in polymer degradation by acclerating hydrolyis, oxidation, and photochemical reactions induced by other factors. Many other factors such as ozone, atmospheric contaminants including dirt, soot, smog, sulfur dioxide, and other industrial chemicals, have a significant effect on polymers (2).

5.2. ACCELERATED WEATHERING TESTS

Most data on the aging of plastics are acquired through accelerated tests and actual outdoor exposure. The latter being a time-consuming method, accelerated tests are often used to expedite screening the samples with various combinations of additive levels and ratios. A variety of light sources are used to simulate the natural sunlight. The artificial light sources include carbon arc lamps, xenon arc lamps, fluorescent sun lamps, and mercury lamps. These light sources, except fluorescent, generate a much higher intensity light than natural sunlight. Quite often, a condensation apparatus is used to simulate the deterioration caused by sunlight and water as rain or dew.

There are three major accelerated weathering tests:

1. Exposure to carbon arc lamps.
2. Exposure to xenon arc lamps.
3. Exposure to fluorescent UV lamps.

The xenon arc, when properly filtered, most closely approximates the wavelength distribution of natural sunlight.

5.2.1. Exposure of Plastics to Fluorescent UV Lamps and Condensation (ASTM G 53)

This method simulates the deterioration caused by sunlight and dew by means of artificial ultraviolet light and condensation apparatus (Figures 5-1 and 5-2). Solar radiation ranges from ultraviolet to infrared. Ultraviolet light of wavelengths between 290 and 350 nm is the most efficient portion of terrestrial sunlight that is damaging to plastics. In the natural sunlight spectrum, energy below 400 nm accounts for less than 6 percent of the total radiant energy (3). Since the special fluorescent UV lamps (FS-40 sunlamps are manufactured by Westinghouse Corp.) radiate between 280 and 350 nm, they accelerate the degradation process considerably. Figure 5-3 illustrates the spectral energy distribution of sunlight and the fluorescent sunlamp.

The test apparatus basically consists of a series of UV lamps, a heated water pan, and test specimen racks. The temperature and operating times are independently controlled both for UV and the condensation effect. The test specimens are mounted in specimen racks with the test surfaces facing the lamp. The test conditions are selected based on requirements and programmed into the unit. The specimens are removed for inspection at a predetermined time to examine color loss, crazing, chalking, and cracking.

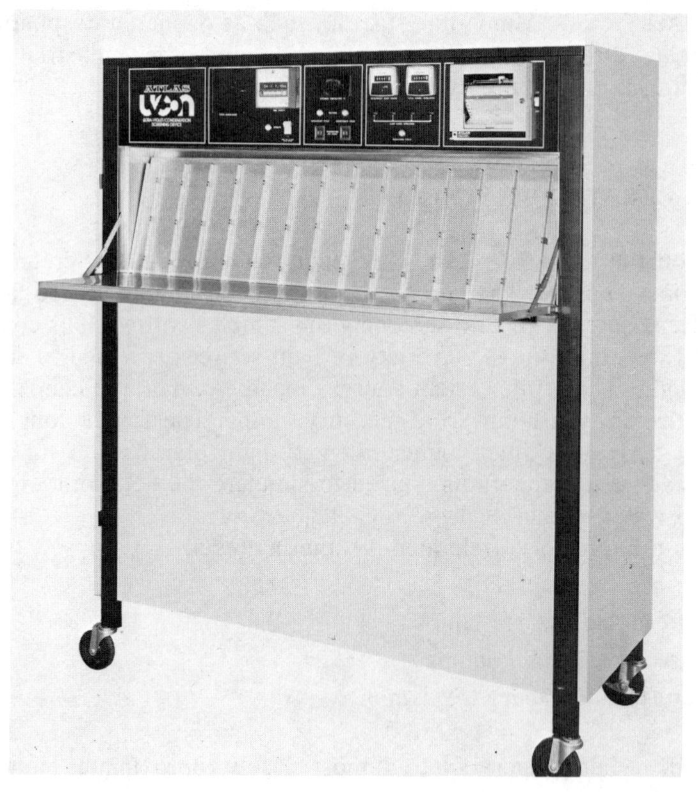

Figure 5-1. UV, light, and condensation apparatus. (Courtesy Atlas Electric Devices Company.)

5.2.2. Exposure of Plastics to Carbon Arc-Type Light and Water (ASTM D 1499)

This method is very useful in determining the resistance of plastic materials when exposed to radiation produced by carbon-arc lamps. There are basically two different types of carbon-arc lamps used as the source of radiation. The first type is enclosed carbon arc lamp. The second type is known as an open-flame sunshine carbon-arc. Figures 5-4 and 5-5 compare both types of light sources with natural sunlight. The enclosed carbon-arc apparatus basically consists of either single or twin enclosed carbon-arc lamps mounted in a chamber. The flame portion of the carbon-arc lamp is surrounded by a bell-shaped borosilicate glass globe. The globe filters out UV radiation below 275 nm not found in direct sunlight and creates a semisealed atmosphere in which the arc burns more efficiently. The globes are normally replaced after 2000 hrs of use. Figure 5-6 illustrates the interior of one such type of twin enclosed carbon-arc apparatus. The open-flame sunshine carbon-arc lamp operates in a free flow of air. This lamp accommodates three upper and three lower electrodes that are consumed during 24–26 hrs of continuous operation. The lamp is surrounded by Corex glass filters that cut off light below 255 nm. The lamp is designed to run with filters in place; however, they can be removed to increase the available UV energy below 300 nm and thereby increase

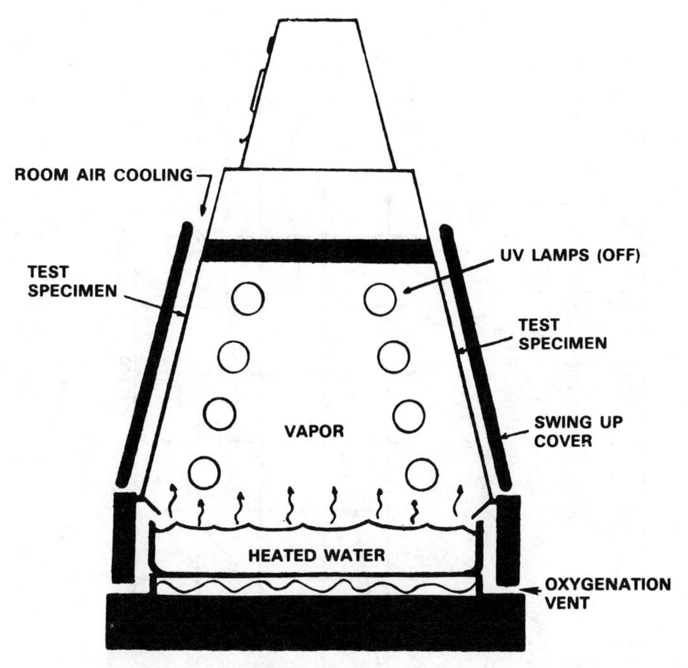

Figure 5-2. Cross section of a UV, light, and condensation apparatus. (Courtesy Q-Panel Company.)

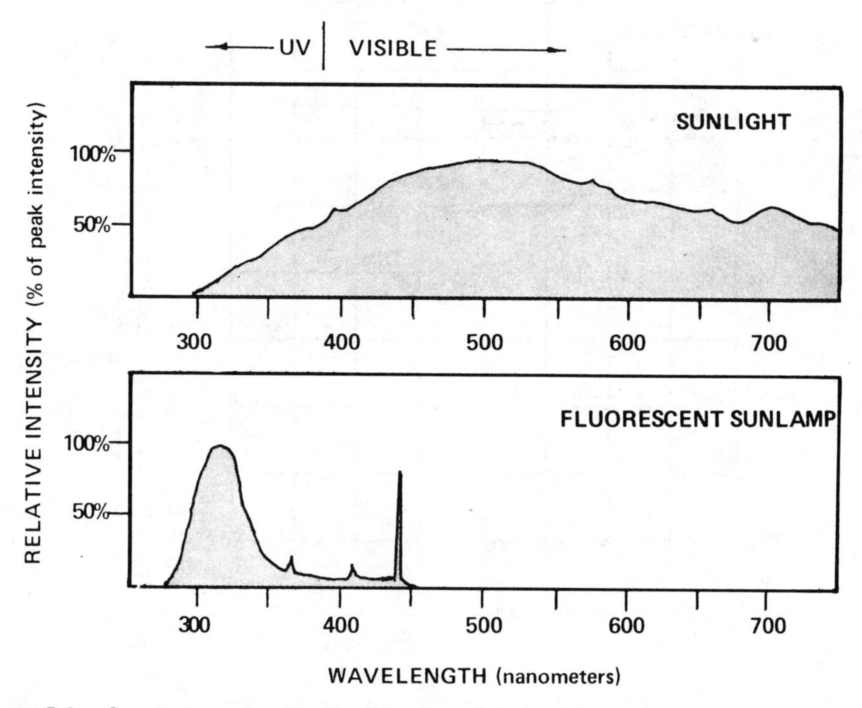

Figure 5-3. Spectral energy distribution of sunlight and fluorescent lamp. (Courtesy Q-Panel Company.)

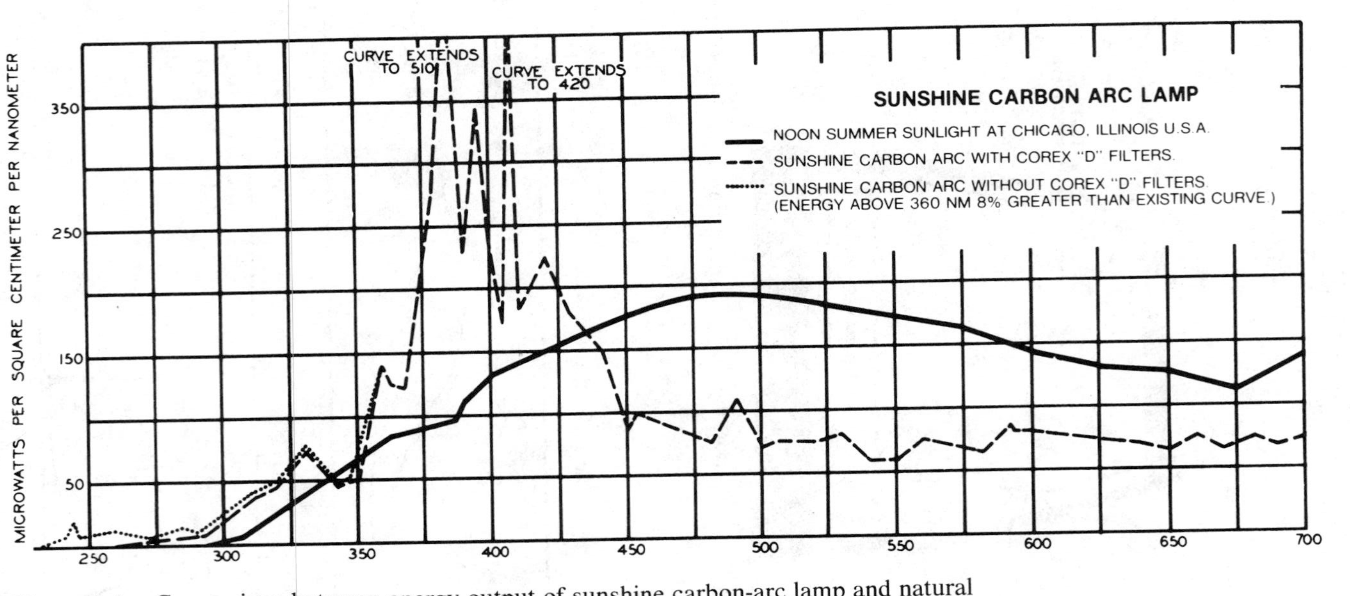

Figure 5-4. Comparison between energy output of sunshine carbon-arc lamp and natural sunlight. (Courtesy Atlas Electric Devices Company.)

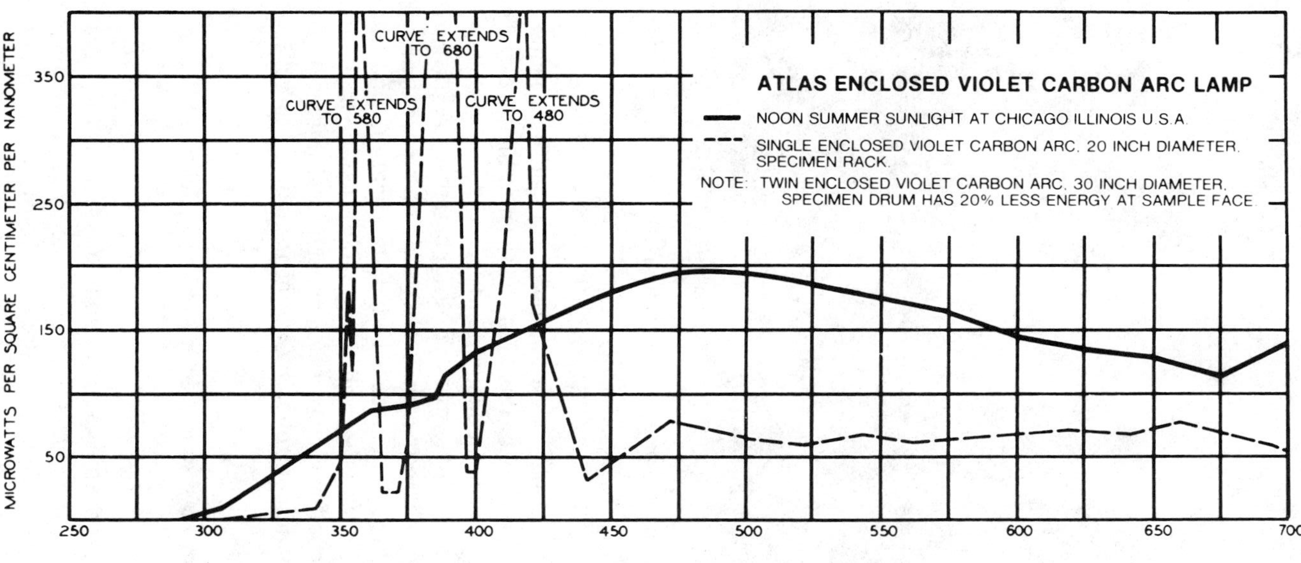

Figure 5-5. Comparison between energy output of enclosed violet carbon-arc lamp and natural sunlight. (Courtesy Atlas Electric Devices Company.)

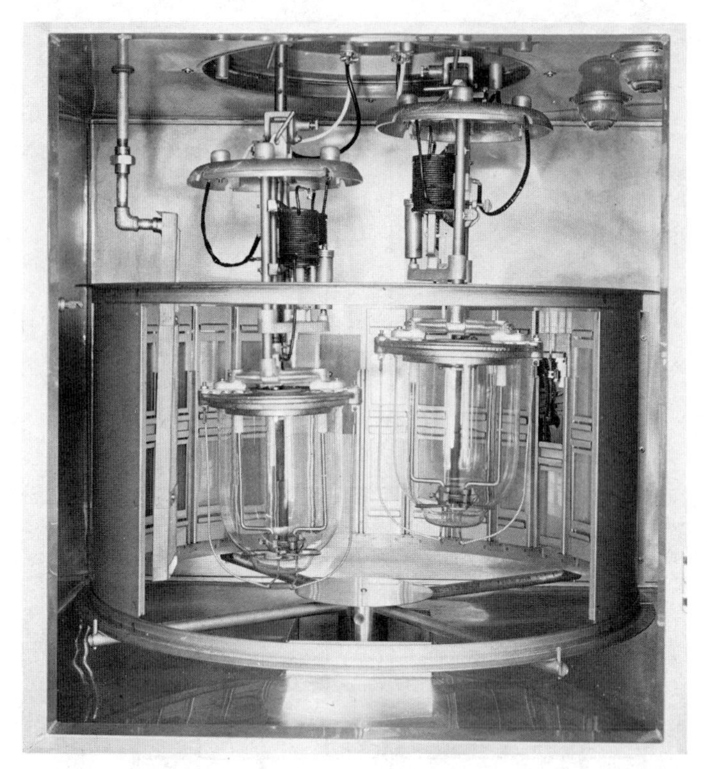

Figure 5-6. Interior of a typical twin enclosed carbon-arc apparatus. (Courtesy Atlas Electric Devices Company.)

significantly the rate of photodegradation of UV-sensitive materials. An interior view of a typical apparatus of this kind is shown in Figure 5-7. The enclosed arc type apparatus uses an open-ended drum-type cylinder, whereas the sunshine arc-type uses a cylindrical framework. Both apparatuses revolve around centrally mounted arc lamps. The provision is also made to expose the specimen to water, which is sprayed through nozzles. The light-on, light-off, and water-spray cycles are independent of each other and the apparatus can be programmed to operate with virtually any combination. A black panel temperature inside the test chamber can be monitored and controlled by a sensor mounted directly on the revolving specimen rack.

5.2.3. Exposure of Plastics to Xenon Arc-Type Light and Water (ASTM D 2565)

A water-cooled xenon arc-type light source is one of the most popular indoor exposure tests since it exhibits a spectral energy distribution of sunlight at the surface of the earth (4). The xenon-arc lamp consists of a burner tube and a light filter system consisting of interchangeable glass filters used in combination to provide a spectral distribution that approximates natural sunlight exposure conditions as shown in Figure 5-8 and Figure 5-9. The apparatus has a built-in recirculating system that recirculates distilled or deionized water through the lamp.

The water cools the xenon burner and filters out long wavelength infrared energy. The interior of one such water-cooled, xenon-arc apparatus is shown in Figure 5-10.

Two basic procedures are recommended. Procedure A is a normal operating procedure for comparative evaluation within a series exposed simultaneously in one instrument. Procedure B is used for comparing results among instruments. Both procedures are described in detail in the ASTM Standards Manual. In either case, it is highly recommended that a controlled irradiance exposure system be used. This is best accomplished through the use of a continuously controlled monitor that can automatically maintain uniform intensity at preselected wavelengths. Figure 5-11 shows a commercially available weathering tester.

5.2.4. Interpretations and Limitations of Accelerated Weathering Test Results

There has been a severe lack of understanding on the part of users regarding the correlation between the controlled laboratory test and the actual outdoor test and application. The questions often asked are: "How many hours of exposure in a controlled laboratory enclosure is equal to one month of outdoor exposure?" "How do the results obtained from one type of weathering device compare to another type?" There is a general agreement among the researchers, manufac-

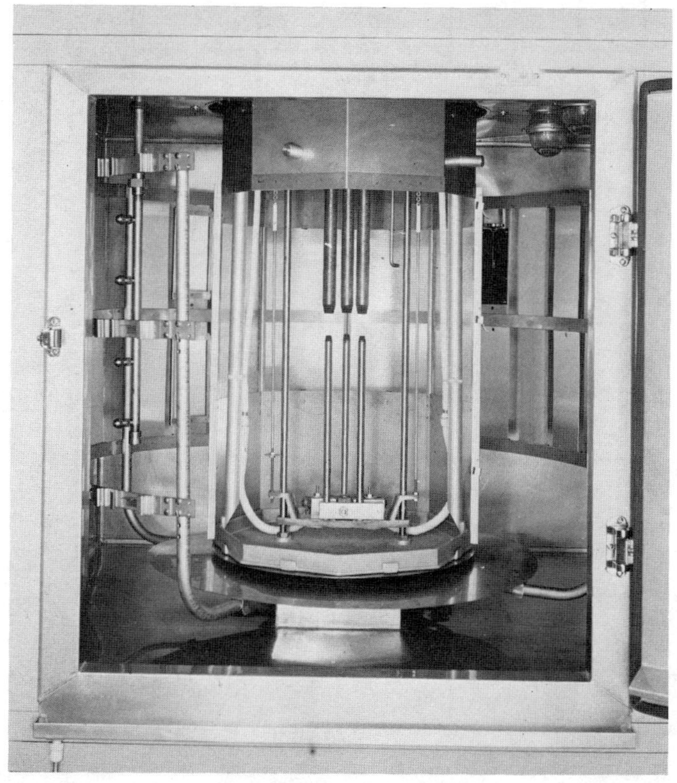

Figure 5-7. Interior of a typical open-flame carbon-arc apparatus. (Courtesy Atlas Electric Devices Company.)

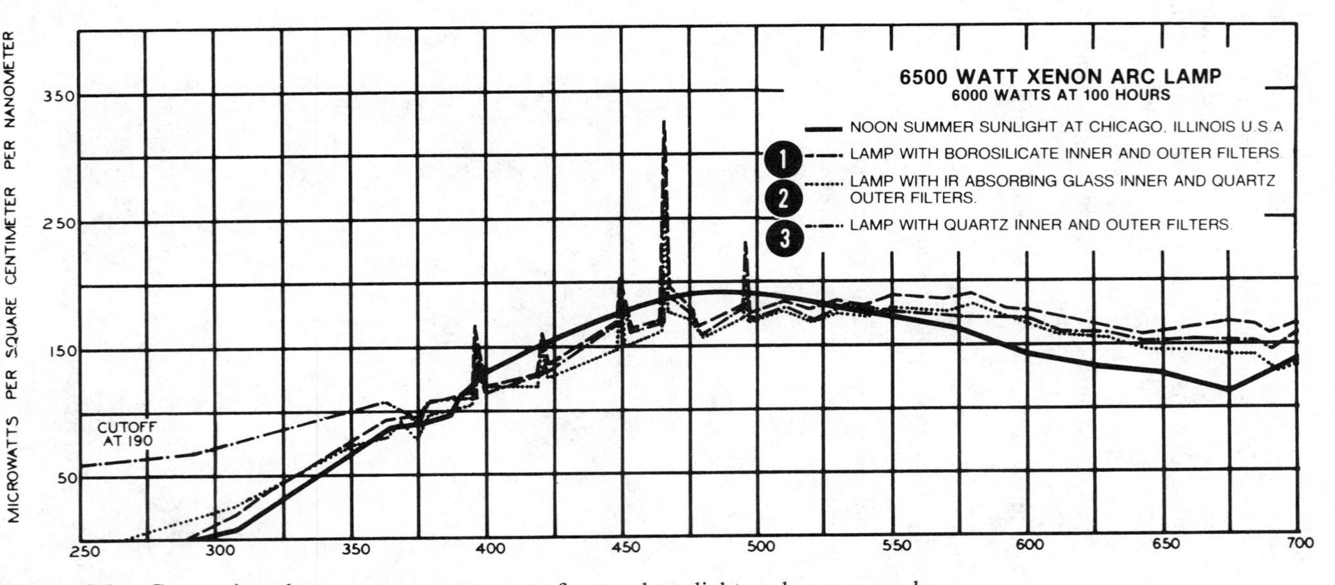

Figure 5-8. Comparison between energy output of natural sunlight and xenon-arc lamp with different types of filters. (Courtesy Atlas Electric Devices Company.)

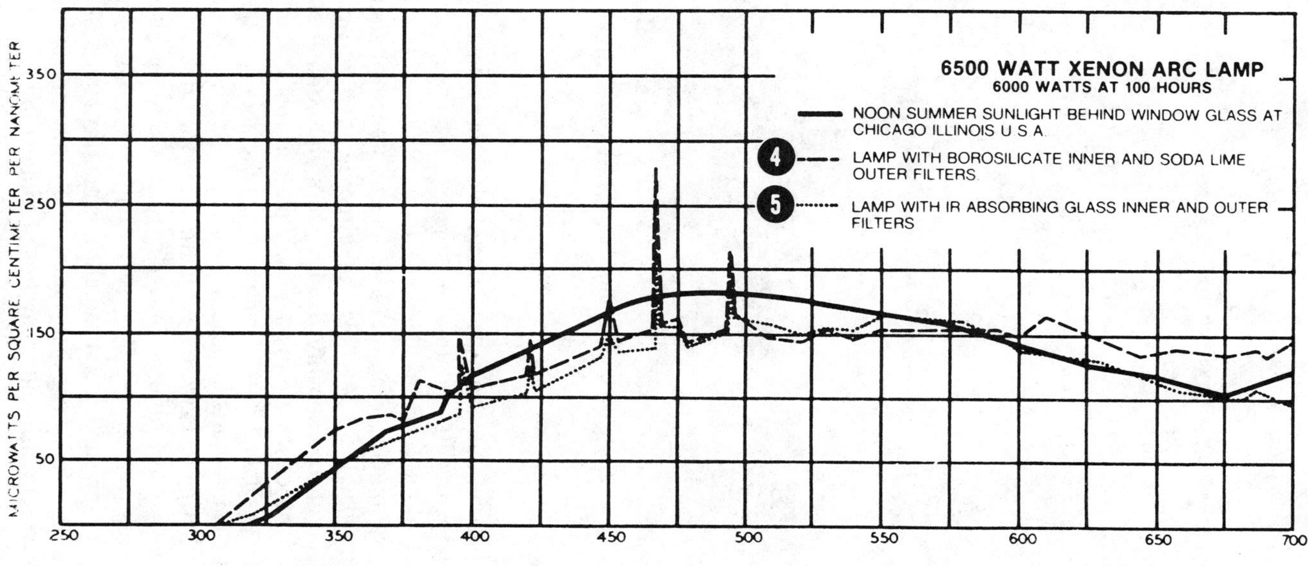

Figure 5-9. Comparison between energy output of natural sunlight and xenon-arc lamp with different types of filters. (Courtesy Atlas Electric Devices Company.)

Figure 5-10. Interior of a typical xenon-arc apparatus. (Courtesy Atlas Electric Devices Company.)

turers, and users that the data from accelerated weathering tests cannot be correlated exactly with the results of natural weathering. However, accurate ranking of the weatherability of most material is possible by using improved test methods and sophisticated equipment. Relative weather resistance of unmodified thermoplastics is shown in Table 5-1. ASTM Committee G3 has proposed the following definition for correlation of exposure results. "For the comparision of the test results to be truly significant, one must satisfy himself that changes in property to be compared are produced by the same or significantly similar chemical reactions throughout the comparison period."

Accelerated weathering tests were devised to study the effect of actual outdoor weather in a relatively short time period. These tests often produce misleading results that are difficult to interpret or correlate with the results of actual outdoor exposure. The reason for such a contradiction is that in many laboratory exposures, the wavelengths of lights are distributed differently than in normal sunlight, possibly producing effects different from those produced by outdoor weathering. All plastics seem to be especially sensitive to wavelengths in the ultraviolet region. If the accelerated device has unusually strong emission at the wavelength of sensitivity of a particular polymer, the degree of acceleration is disproportionately high compared to outdoor exposure. The temperature of the exposure device also

Figure 5-11. Weathering tester. (Courtesy Atlas Electric Devices Company.)

Table 5-1. Relative Weather Resistance of Unmodified Thermoplastics (6)

Polymer	Resistance
Acrylics	High
PTFE and other fluorocarbons	High
Polycarbonate	Medium
Thermoplastic polyester	Medium
Cellulose acetate butyrate	Medium
Nylons	Medium
Polyurethanes	Medium
Polyphenylene-oxide	Medium
Polyphenylene sulfide	Medium
Rigid PVC	Medium
Flexible PVC	Low
ABS	Low
Acetal	Low
Polyethylene	Low
Polysulfone	Low
Polystyrene	Low
Polypropylene	Low
Cellulose acetate	Low

greatly influences the rate of degradation of a polymer. The higher temperature may cause oxidation and the migration of additives which, in turn, affects the rate of degradation. One of the limitations of accelerated weathering devices is their inability to simulate the adverse effect of most industrial environments, and many other factors present in the atmosphere and their synergistic effect on polymers. Some of the newly developed gas exposure cabinets have partially overcome these limitations. These units are capable of generating ozone, sulfur dioxide, and oxides of nitrogen under controlled conditions of temperature and humidity.

Improved ultraviolet sources and more knowledge of how to simulate natural wetness now make it possible to achieve reliable accelerated weathering results if certain procedures are observed:

1. Include a material of known weather resistance in laboratory tests. If such a material is not available, use another similar product that has a history of field experience in a similar use.
2. Measure or estimate the UV exposure, the temperature of the product during UV exposure, and the time of wetness under service conditions of the product. Try to create laboratory cycles that duplicate the natural balance of these factors.
3. Do not use abnormal UV wavelengths to accelerate effects unless you are testing small differences in the same material. Evaluating two different materials by this technique can distort results.
4. Natural weathering is invariably a combination of UV and oxidation from wetness. In laboratory tests, UV or water can be excluded to define the problem area. Do not over look this technique.
5. Do not seek to establish a table of how many hours of accelerated testing equals *x* months of exposure. Natural weathering varies widely, even at one site on one product. Plastic tail light housings do not experience the same weathering effects as a vinyl roof on the same automobile. Reasonable standards of endurance in laboratory tests can be established for specific products, but one tester, one cycle, and one chart for converting laboratory hours to natural months is an unrealistic objective (5).

Before drawing any final conclusions concerning the ability of a polymer to withstand outdoor environment based on accelerated weathering tests, one must conduct actual outdoor exposure tests for a reasonable length of time.

5.2. OUTDOOR WEATHERING OF PLASTICS (ASTM D 1435)

The test is devised to evaluate the stability of plastic materials exposed outdoors to varied influences that comprise weather exposure conditions that are complex and changeable. Important factors are climate, time of year, and the presence of industrial atmosphere. It is recommended that repeated exposure testing at different seasons and over a period of more than one year be conducted to confirm exposure at any one location. Since weathering is a comparative test, control samples are always utilized and retained at standard conditions of temperature

Figure 5-12. Typical aluminum exposure racks. (Courtesy South Florida Test Service, Inc.)

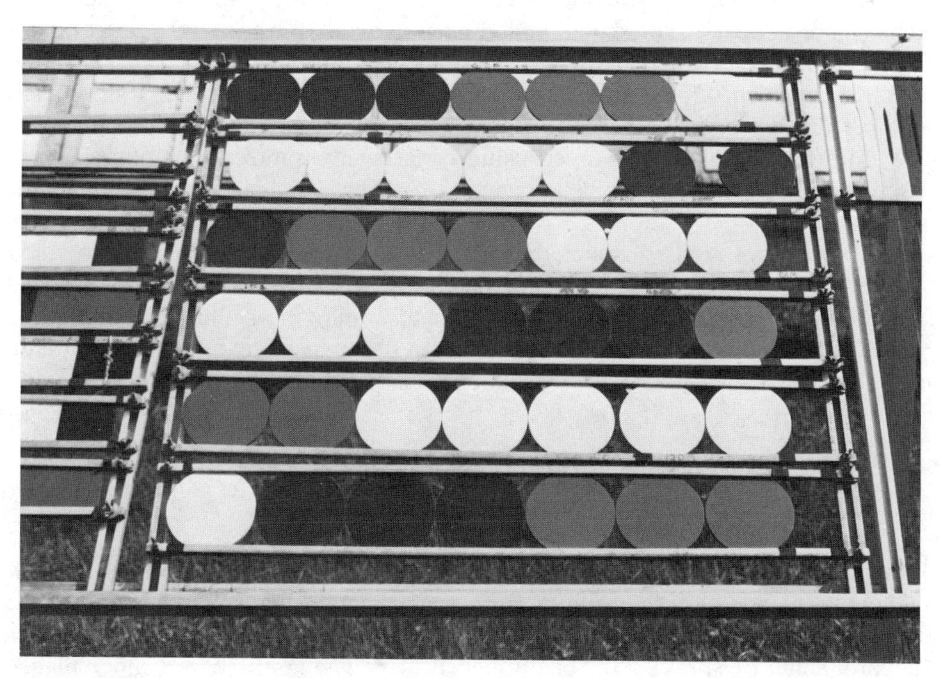

Figure 5-13. Suitably mounted samples. (Courtesy South Florida Test Service, Inc.)

and humidity. The control samples must also be covered with inert wrapping to exclude light exposure during the aging period. However, dark storage does not insure stability (7).

Test sites are selected to represent various conditions under which the plastic product will be used. Arizona is often selected for intense sunlight, wide temperature cycle, and low humidity. Florida, on the other hand, provides high humidity, intense sunlight, and relatively high temperatures.

Exposure test specimens of suitable shape or size are mounted in a holder directly applied to the racks. Racks are positioned at a 45° angle and facing the equator. Many other variations in the position of the racks are also employed, depending upon the requirements.

The specimens are removed from the racks after a specified amount of time and subjected to various tests such as appearance evaluation, electrical tests, and mechanical tests. The results are compared with the test results from testing control specimens. Typical aluminum exposure racks are shown in Figure 5-12 and suitably mounted specimens are shown in Figure 5-13.

5.4. RESISTANCE OF PLASTIC MATERIALS TO FUNGI (ASTM G 21)

This accelerated laboratory test determines the effect of fungi on plastic materials. The effectiveness of antimicrobial additives is also evaluated by such procedures. The test requires preparing a fungus spore suspension from cultures of various fungi that are known to attack polymers. Many other types of organisms can also be used in the test if necessary. The prepared spore suspension is tested for fungal growth without using plastic specimens as a viability control. The plastic specimens can be of any size or shape, including completely fabricated parts, test bars, or pieces from fabricated parts. Optically clear materials are used to study the effect of fungi on optical reflection or transmission. The specimens are placed onto petri dishes or any other suitable glass tray covered with nutrient salts agar. The entire surface is then sprayed using a sterilized atomizer with fungus spore suspension. The inoculated test specimens are covered and placed in the incubator maintained at 28–30°C and 85 percent or more relative humidity for 21 days. The fungal growth is visually inspected after 21 days of incubation. In order to study the effect on physical, optical, or electrical properties, the specimens are washed free of growth and subsequently conditioned, employing standard conditioning procedures. The physical testing is carried out in the usual manner and compared with control.

5.5. RESISTANCE OF PLASTICS MATERIALS TO BACTERIA (ASTM G 22)

This accelerated laboratory test is somewhat similar to the test to determine resistance of plastic material to fungi. The bacterial cell suspension in place of the fungus spore suspension is used to study the effect. The test is conducted in a similar manner by spraying the bacterial cell suspension onto the specimen placed in a petri dish covered with nutrient salts agar. The incubation is carried out for a minimum of 21 days under specified conditions in the incubator. A variation of the standard method is sandwiching the sample between two equal lots of nutrient salts agar and then spraying the bacterial cell suspension on it. This variation provides a more extensive contract between the test bacteria and the specimens.

5.6. LIMITATIONS OF ACCELERATED MICROBIAL GROWTH RESISTANCE TESTING

The data obtained from short-term accelerated testing can be very misleading if the results are not applied correctly and the data is not interpreted properly (8). The limitations of such tests are:

1. Optimum microbial growing conditions dictate the use of high levels of toxic compounds that would "overkill" in actual commercial applications.
2. Exposure normally is limited to a relatively small number of different species of microorganisms.
3. Exposures are limited in time and conditions.
4. Misinterpretation of data by nonexperts.

It is very important to recognize that some chemicals can remain sufficiently toxic for the short duration of 14–21 days in accerated testing but may deteriorate or may cause problems on long-term exterior exposures. For example, a fungicide may volatilize in hot weather, be degraded by UV light, or be bleached out by rain and dew.

5.7. OUTDOOR EXPOSURE TEST FOR STUDYING THE RESISTANCE OF PLASTIC MATERIALS TO FUNGI AND BACTERIA AND ITS LIMITATIONS

The outdoor exposure tests to determine the resistance of plastic materials to microbial attack are carried out many different ways (9). The simplest way is to expose plastic material to an outdoor environment in geographical locations where weather conditions are favorable to microbial growth. The alternate method is called the soil burial method. This method calls for the actual burying of the specimens for four weeks and observing the effects of microorganisms on the specimens.

There are several serious limitations of outdoor exposure testing:

1. The chemical composition of the product.
2. Angle of exposure to weather.
3. Time of year exposures are made.
4. Geographic location of exposures.

The chemical composition of the product being tested influences the degree of microbial attack. Products having good water resistance and weatherability generally have a greater resistance to microbial attack since there is less plasticizer exudation and therefore, less nutrient available for growth. A product that has the ability to allow microbiocides to partially leach out to the surface is advantageous because the microbial attack is on the surface due to exudation of additives.

The angle of exposure to sunlight and weather conditions in general will influence the degree and duration of microbial attack. Some plastics exude plasticizer or other nutrients more rapidly when exposed at 40° south than on vertical exposures. Time of the year and geographic location are also important since microorganisms grow more rapidly under warm, humid climate than cold, dry climate. In conclusion, fungal and bacterial resistance testing must be carried out in two steps: first, a short and accelerated test to screen out unimportant specimens and second, a long-term outdoor exposure test to confirm the results of the laboratory tests.

REFERENCES

1. Mascia, L., *The Role of Additives in Plastics,* Edward Arnold Publishing Company, London, England, 1974, p. 134.
2. Kamal, M. R., "Weatherability of Plastic Materials," *Appl. Poly. Symp.,* No. 4, Interscience Publishers, New York, (1967), p. 2.
3. *Ibid,* p. 64.
4. *Ibid,* p. 62.
5. Dregger, D. R., "How Dependable are Accelerated Weathering Tests for Plastics and Finishes?" *Machine Design* (Nov. 1973), pp. 61–67.
6. Winolow, R. M., Matreyek, W., and Trozzolo, A. M., "Polymers Under Weather," *S.P.E.J.* **28**(7), (July 1972), pp. 19–24.
7. Kinmonth, R. A., Saxon, R., and King, R M., "Sources of Variability in Laboratory Weathering," *Polym. Eng. Sci.* **10**(5), (Sept. 1970), pp. 309–313.
8. Wienert, L. A., and Hillard, M. W., "A Hard Look at Fungal-Resistance Testing," *Plast. Tech.* (Jan. 1977), p. 75.
9. *Ibid,* p. 77.

SUGGESTED READING

Mullen, A., Kinmonth, R. A., and Searle, N. Z., "Spectral Energy Distributions and Aging Characteristics of Fluorescent Sunlamps and Blacklights," *J. Testing and Evaluation* **3**(1) (Jan. 1975), pp. 15–20.

Reinhart, F. W., and Mutchler, M. K., "Fluorescent Sunlamps in Laboratory Aging Tests for Plastics," *ASTM Bull.,* **212**(33), (Feb. 1956), pp. 45–51.

Dunn, J. L., and Heffner, M. H., "Outdoor Weatherability of Rigid PVC," *SPE ANTEC,* **19**(1973), pp. 483–488.

Yaeger, C. C., "The Use of Antimicrobials in Flexible Vinyl Systems," *SPE ANTEC,* **17** (1971), pp. 579.

Weinberg, E. L., *Modern Plastics Encyclopedia.* McGraw–Hill, New York, 1976–1977, p. 217.

Borg–Warner, "Weatherability of Cycolac Brand ABS Resins." *Tech. Bull.,* Parkersburg, W. Va., *PB-120A.*

Atlas Electric Devices Co., "Atlas Sun Spots." *Tech. Bull.,* Chicago, published quarterly.

Metzinger, R., and Kinmonth, R. A., "Laboratory Weathering," *Modern Paint and Coatings* (Dec. 1975).

Int. Symp. on the Weathering of Plastics and Rubber, Reprints, Plastics and Rubber Institute, London, England, 1976.

Rosato, D. V., *Environmental Effects on Polymeric Materials.* Interscience, New York, 1967.

CHAPTER 6

Optical Properties

6.1. INTRODUCTION

Unique properties, such as excellent clarity and transparency, good impact strength, moldability, and low cost have made plastics a number-one choice of many design engineers. Successful applications of transparent and translucent plastics include automotive tail light lenses, safety glasses, window glazing, merchandise display cases, and instrument panels. More recently, plastics have been accepted for more stringent applications such as contact lenses, prisms, low-cost camera lenses, and magnifiers. Plastics are much more resistant to impact than glass and therefore, in applications such as street lamp globes and high school windows, have been replaced by high-impact vandal-resistant plastic materials such as polycarbonate. However, lack of dimensional stability over a wider range of temperatures and poor scratch resistance have prevented further penetration of plastics in the markets for expensive camera lenses, microscopes, and other precision optics where the use of glass is fairly common.

Almost all plastics, below a certain minimum thicknesses, are translucent. Only a few plastics are transparent. Plastic materials' transparency or translucency depends upon its basic polymer structure. Generally, all amorphous plastics are transparent. Crystallinity increases the density of the polymer, which decreases the speed of light passing through it and this increases the refractive index. When crystals are larger than the wavelength of the visible light, the light passing through many successive crystalline and amorphous areas is scattered, and the clarity of the polymer is decreased. A large single crystal scatters light at wide angles and thus causes haze (1). As a rule, crystalline plastics are translucent. However, clarity of crystalline plastics can be improved by quenching or by random copolymerization (2,3).

The primary optical properties are:

Refractive index.
Light transmittance and haze.
Photoelastic properties.
Color.
Gloss.

6.2. REFRACTIVE INDEX (ASTM D 542)

Refractive index is a fundamental property of transparent materials. Refractive index values are very important to a design engineer involved in designing lenses for cameras, microscopes, and other optical equipment. The refractive index, also known as the index of refraction, is defined as the ratio of the velocity of light in a vacuum (or air) to its velocity in a transparent medium.

$$\text{Index of refraction} = \frac{\text{sin of angle of incidence}}{\text{sin of angle of refraction}}$$

Table 6-1 compares the refractive index values of different materials. Note that the refractive index of plastic materials is very close to the refractive index of glass. Two basic methods are most commonly employed to determine the index of refraction. The first method, known as the refractometric method, requires the use of a refractometer. An alternate method calls for the use of a microscope with a magnification power of at least 200 diameters. The refractometric method is generally preferred over the microscopic method since it is much more accurate. The microscopic method, which is dependent upon the operator's ability to focus, is usually less accurate.

6.2.1. Refractometric Method

The Abbé refractometer is the refractometer most widely used to determine the index of refraction (Figure 6-1). The test also requires a source of white light and a contacting liquid that will not attack the surface of the plastic. The contacting liquid must also have a higher refractive index than the plastic being measured.

Table 6-1. Refractive Index Values for Plastics

Plastic	Value
Polytetrafluoroethylene (PTFE)	1.35
Cellulose acetate butyrate	1.47
Cellulose acetate	1.49
Acetal (homopolymer)	1.48
Acrylics	1.49
Polypropylene	1.49
Polybutylene	1.50
Ionomer	1.51
Low-density polyethylene	1.51
PVC (rigid)	1.52
Nylons (type 66)	1.53
Urea formaldehyde	1.54
High-density polyethylene	1.54
SAN	1.56
Polycarbonate	1.58
Polystyrene	1.60
Polysulfone	1.63
Glass	1.60

Figure 6-1. Abbé refractometer. (Courtesy Kernco Instruments Company, Inc.)

A test specimen of any size may be used as long as it conveniently fits on the face of the fixed half of the refractometer prism. The surface of the specimen in contact with the prism must be flat and polished.

The test is carried out by placing a specimen in contact with the prism using a drop of contacting liquid. The polished edge of the specimen is kept towards the light source. The refractive index is determined by moving the index arm of the refractometer so that the field seen through the eyepiece is half dark. The compensator is adjusted to remove all color from the field. Next, the index arm is adjusted using the vernier to coincide the dark and light portion of the field at the intersection of the cross hairs. The value of the index of refraction is read for sodium D lines.

6.2.2. Microscopical Method

In this method, a microscope having a magnifying power of 200 diameters or more is used. A specimen of convenient size, having a fair polish and two parallel surfaces, is used. The test is carried out by alternately focusing the microscope on the top and the bottom surface of the specimen and reading the longitudinal displacement of the lens tube accurately. The difference between the two readings is considered the apparent thickness of the specimen. The index of refraction is determined as follows:

$$\text{Refractive index} = \frac{\text{Actual thickness}}{\text{Apparent thickness}}$$

6.3. LUMINOUS TRANSMITTANCE AND HAZE (ASTM D 1003)

Luminous transmittance is defined as the ratio of transmitted light to the incident light. The value is generally reported in percentage of light transmitted. Polymethyl methacrylate, for example, transmits 92 percent of the normal incident light. There is about 4 percent reflection at each polymer–air interface for normal incident light (4).

Haze is the cloudy appearance of an otherwise transparent specimen caused by light scattered from within the specimen or from its surface. Haze is defined as the percentage of transmitted light which in passing through a specimen deviates from the incident beam by forward scattering. It is generally accepted that if the amount of transmitted light is deviated more than 2.5° from the incident beam, the light flux is considered to be haze. Haze is normally caused by surface imperfections, density changes, or inclusions that produce light scattering. Haze is also reported in percentage.

Light transmittance and haze are extremely important from a practical viewpoint. For example, a window glazing material must have high light transmittance characteristics and must be free from haze. In contrast, the housing material for the light fixture must have maximum diffusion and minimum transparency to conceal the bright light source. The housing material must also have high light transmittance (5).

Two procedures have been developed to measure light transmittance and light scattering properties: Procedure A requires the use of a hazemeter while Procedure B requires the use of a recording spectrophotometer.

6.3.1. Procedure A: Hazemeter

This procedure employs an integrating sphere hazemeter as illustrated schematically in Figure 6-2. The test specimen must be large enough to cover the aperture, but small enough to be tangent to the sphere wall. A disc of 1.375 in. in diameter

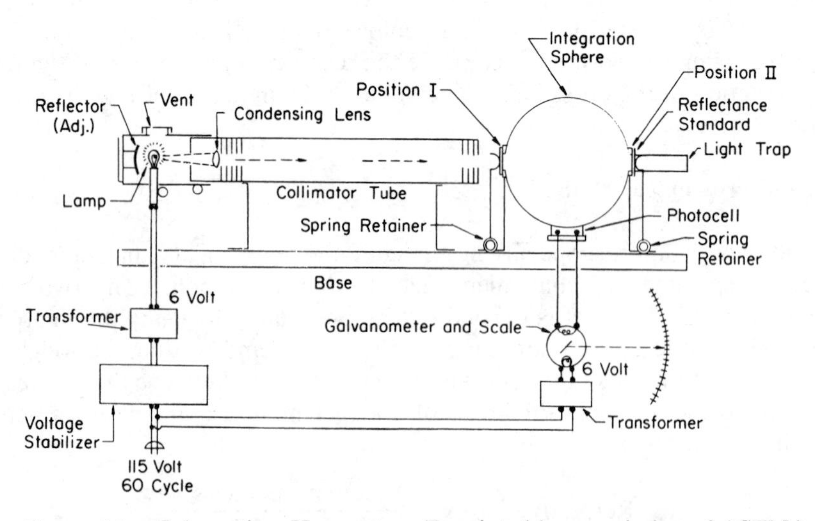

Figure 6-2. Schematic—Hazemeter. (Reprinted by permission of ASTM.)

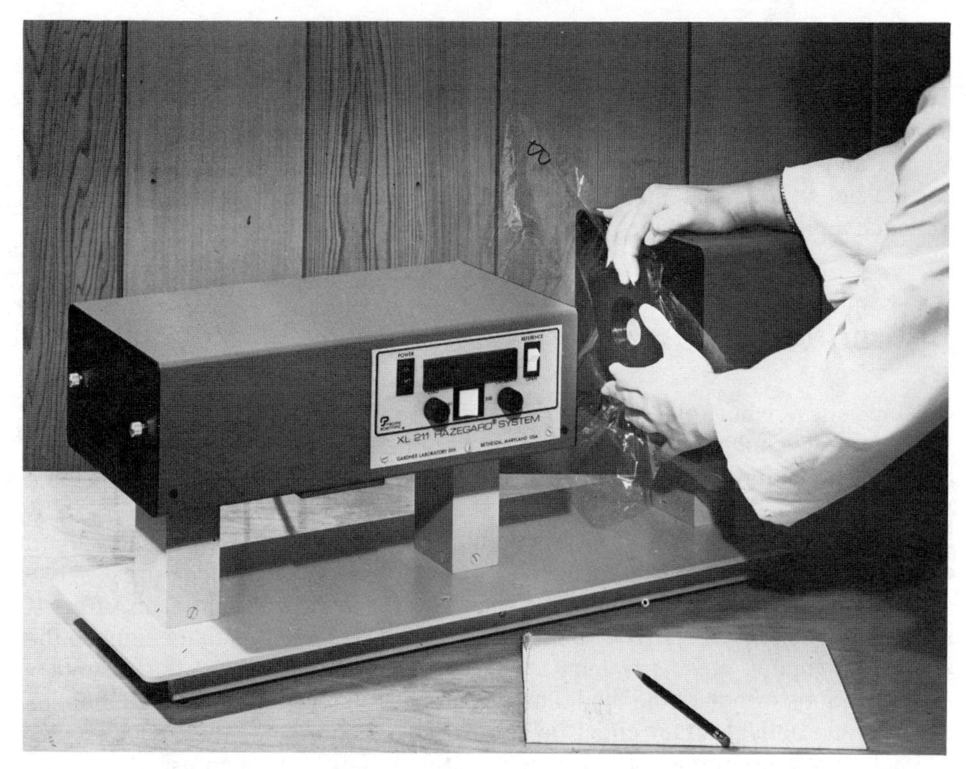

Figure 6-3. Hazemeter. (Courtesy Gardner Laboratories, Inc.)

is most commonly used. A commercially available hazemeter is shown in Figure 6-3. The test is conducted by taking four different consecutive readings and measuring the photocell output as follows:

T_1 = specimen and light trap out of position, reflectance standard in position.
T_2 = specimen and reflectance standard in position, light trap out of position.
T_3 = light trap in position, specimen and reflectance standard out of position.
T_4 = specimen and light trap in position, reflectance standard out of position.

The quantities represented in each reading are incident light, total light transmitted by specimen, light scattered by instrument, and light scattered by instrument and specimen, respectively. Total transmittance T_t and diffuse transmittance T_d is calculated as follows:

$$T_t = \frac{T_2}{T_1}$$

$$T_d = \frac{T_4 - T_3(T_2)}{T_1}$$

Percentage haze is calculated as follows:

$$\text{Haze percent} = \frac{T_d}{T_t} \times 100$$

6.3.2. Procedure B: Recording Spectrophotometer

This procedure is somewhat similar to Procedure A. The recording spectrophotometer is used to generate four different curves. From the recorded curves T_1, T_2, T_3, and T_4 values are computed using automatic integrator. Calculations are carried out in a similar manner to determine total and diffuse luminous transmittance and the percentage haze.

6.4. PHOTOELASTIC PROPERTIES

Most plastics, when placed under internally or externally applied stress and viewed through polarized light, exhibit stress optical properties (6). Light vibrating in one plane within a plastic that is strained travels faster than light vibrating in a plane at right angles. The difference in the two velocities shows up as a birefringence. Stress optical sensitivity is defined as the ability of some materials to exhibit double refraction (birefringence) of light when placed under stress (7).

Polarized lenses are used to plane polarize the incoherent light. This effect is shown in Figure 6-4. Polarized sunglasses and polarizing lenses that eliminate the glare from shining objects by plane polarizing the light are just a few examples of applications of polarizing mediums. When two such cross polarizing mediums are used, an optical rotary effect or birefringence is produced (8).

Photoelastic properties of the transparent materials have been used by design engineers for stress analysis and by process engineers for determining residual stress as well as the degree of orientation in molded parts. Many complicated shapes such as bridges, aircraft wings, and gears can be stress analyzed by building a prototype model out of transparent plastic and viewing them through polarized light. Useful information regarding the location of stress concentration, effects of sharp corners, and changes in cross section can be obtained through such analysis. The individual sections of a model can also be physically stressed to observe the effect. A substantial amount of money can be saved in mold modification and the risk potential of the product can be minimized by making a prototype part in a transparent material and analyzing it for stress concentration area. Such analysis also reveals the effects of gate size and location, weld lines, cored holes, and the

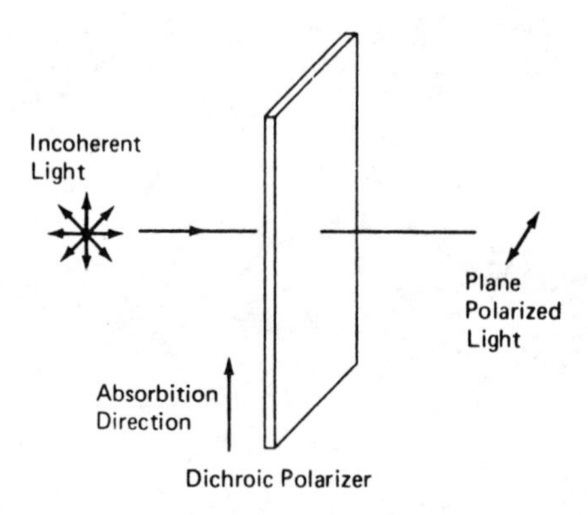

Figure 6-4. Plane polarized light. (Reprinted by permission of Van Nostrand–Reinhold Company.)

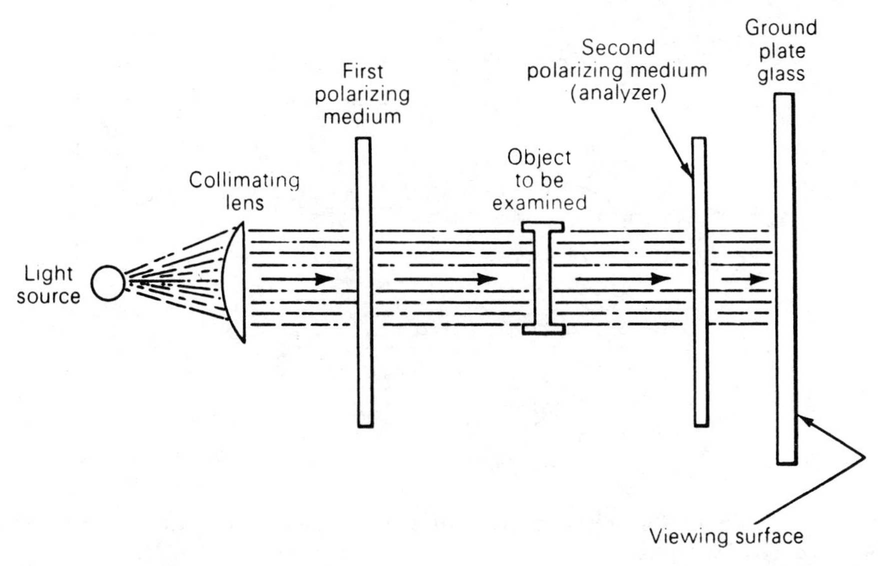

Figure 6-5. Set-up examination of stress–optical sensitivity. (Reprinted by permission of McGraw–Hill Book Company.)

effects of molding, annealing, and machining. In the production of lenses and other transparent parts, this technique can be used as a means of controlling quality. This nondestructive technique can also be used to analyze the field failures by molding a few parts in transparent material and viewing them through a polarized light to detect stress concentrations.

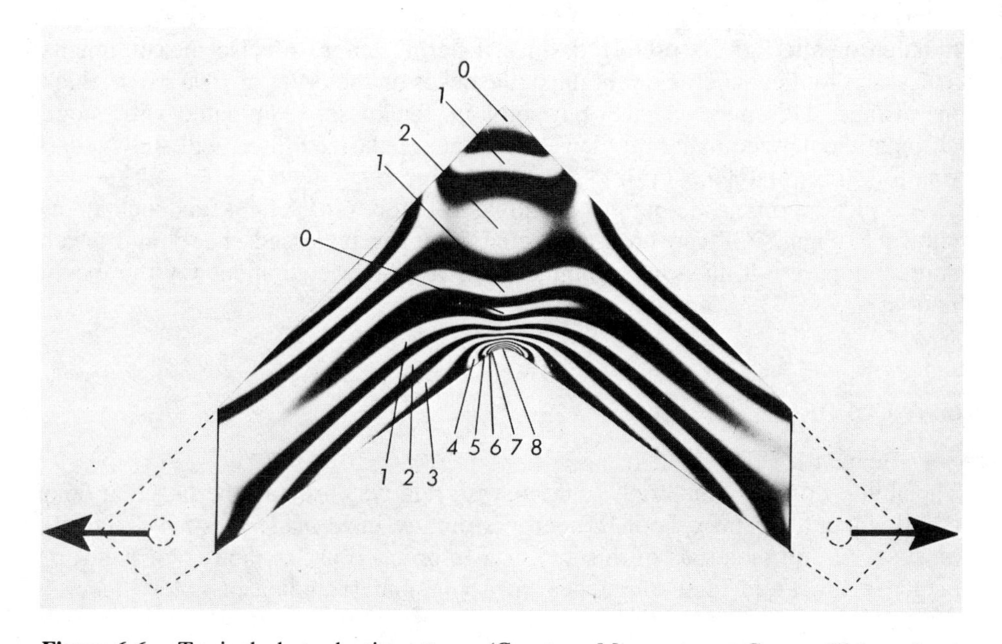

Figure 6-6. Typical photoelastic pattern. (Courtesy Measurement Group, Vishay, Inc.)

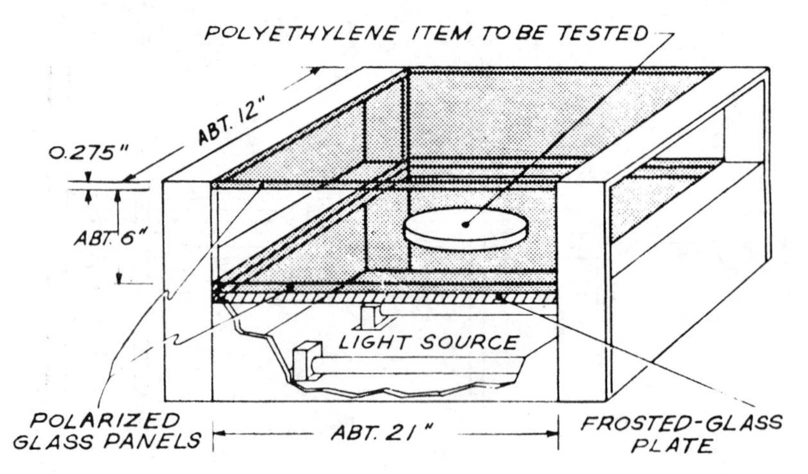

Figure 6-7. Light box for stress–optical sensitivity examination. (Reprinted by permission of *Modern Plastics Encyclopedia*.)

6.4.1. Stress Optical Sensitivity Examination

Transparent plastic parts are examined for stress optical sensitivity by using a relatively simple set-up as shown in Figure 6-5. The object to be examined is placed in the center of the polarizing medium. The object is viewed from the opposite side of the light source. The polarized light indicates the number of fringes or rings, which is related to the amount of stress. If white light is used, interesting colored patterns with all colors of the spectrum are seen, but for stress analysis, monochromatic light is preferred since it permits more precise measurements. The stress–optical coefficient of most plastics is on the order of 1000 psi/in. thickness/fringe. This means that a part of 1 in.-thickness, illuminated with monochromatic polarized light and viewed through a polarized filter, will show a dark ring for stress of 1000 psi (9).

A typical photoelastic pattern is shown in Figure 6-6. A light box such as one shown in Figure 6-7 can be constructed by using polarized sheets and a light source. Figure 6-8 illustrates commercially available equipment for the examination.

6.5. COLOR

The ability of plastic materials to color with relative ease and the fact that color can be integrated throughout the entire structure have made plastics one of the most successful materials of this century. In order to understand color measurement, specification, and tolerances, a proper understanding of color theory is essential.

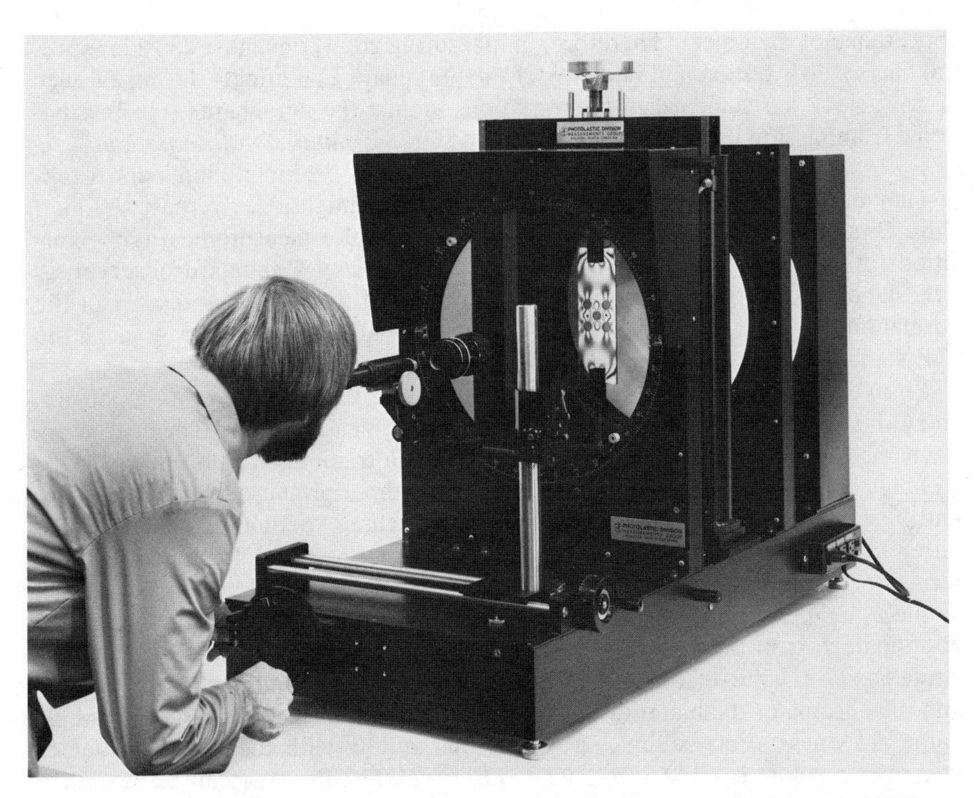

Figure 6-8. Polariscope for two and three-dimensional photoelastic model analysis. (Courtesy of the Photoelastic Division of the Measurements Group.)

Colors range from dark to light—black being darkest, gray being in the middle, and white being the lightest. These are called neutral colors. This aspect of color is termed "value" or "lightness." Color also has another basic difference—red differs from blue, green, or yellow. These distinctions are called "hue." Hue is defined as the attribute of color perception by means of which an object is judged to be red, yellow, green, blue, purple, or intermediate between some adjacent pair of these. One other dimension called "saturation" or "chroma" is defined as the attribute of color perception that expresses the degree of departure from gray of the same lightness. Thus, the entire color spectrum can be described in terms of value, hue, and chroma. This concept is illustrated in the hue, value/chroma chart in Figure 6-9.

The essential difference in the spectrum of light of various hues is the wavelength of the light. The light is dispersed into a spectrum because of the differences in wavelength. The entire visible spectrum ranges from violet light, with the shortest wavelength of about 380 nm, to red light with the longest wavelength of about 760 nm (10).

In order to see any object, the object must be illuminated with an illuminant. The type of illuminant, angle of illumination, and angle of viewing all affect the appearance of the object. Therefore, in measuring color, one must consider spectral energy distribution and intensity of the illuminant as it affects the appearance of the object. To standardize the variations among the illuminants, the International Commission on Illuminants, known as CIE (*Commission Internationale de l'Eclairage*) has established standard illuminants. For example, illuminant A represents an incandescent light; illuminant B, noon sunlight; C, overcast sky daylight. One other factor that must be considered in color measurement is the variations in the color observer. CIE has also established a standard observer. A CIE standard observer is a numerical description of the response to color of the normal human eye (11). CIE spectral tristimulus values are derived by using the CIE standard source, CIE standard observer, and object.

From the above discussion, it is clear that given a CIE standard source, object, and CIE spectral tristimulus values, one can easily measure color. The instrument developed for such color measurement is called a tristimulus colorimeter. A tristimulus colorimeter measures color in terms of three primary colors: red, green and blue, or more properly stated, in terms of three tristimulus values. Many different color scales have been developed to describe the color numerically in terms of lightness and hue. One of the most widely accepted systems is known as the L, a, b tristimulus system. Figure 6-10 illustrates L, a, b color space. The coordinate L is in the vertical direction and corresponds to lightness. A perfect white has an L value of 100 and a perfect black a zero. a_L and b_L identify the hue and the chroma of the material. A plus value of a_L indicates redness and a minus value greenness. For example, a school bus yellow with the following values describes a color: $L = 70.3$; $a_L = 30.3$; $b_L = 23.7$. This color can be described in common terms as fairly light, as indicated by a high L value, and yellowish red as indicated by a_L and b_L values (12).

The filter colorimeter described thus far is based on the theory of standard observer and how the human eye perceives the color. This approach has been described as a "psychophysical" approach to color measurement (13). There are some problems associated with the design of filters and filter material's ability to retain originality with time (14).

In recent years, a new generation of colorimeters based on spectrophotometers have been developed. Spectrophotometric colorimeter does not mimic the human eye. Instead, it makes a spectrophotometric measurement at sixteen 20-nm intervals over the entire range of visible spectrum. The percentage reflectance value obtained by the spectrophotometer is converted into tristimulus values through the use of a microprocessor. One other useful feature added to these spectrophotometric colorimeters is the choice of selecting various types of CIE illuminants. Even though the actual light source remains the same, the microprocessor computes the colors that would be seen if the samples were viewed under various illuminants.

For color development work, a colorimeter that provides the tristimulus value and calculates the color difference is simply not enough. A more sophisticated instrument such as a spectrophotometer is generally required. A spectrophotometer, when used with a chart recorder or CRT, provides a complete spectral reflectance curve covering the entire range of the visible spectrum from 380–700 nm. Additionally, it provides the percentage reflectance value at each 20 nm

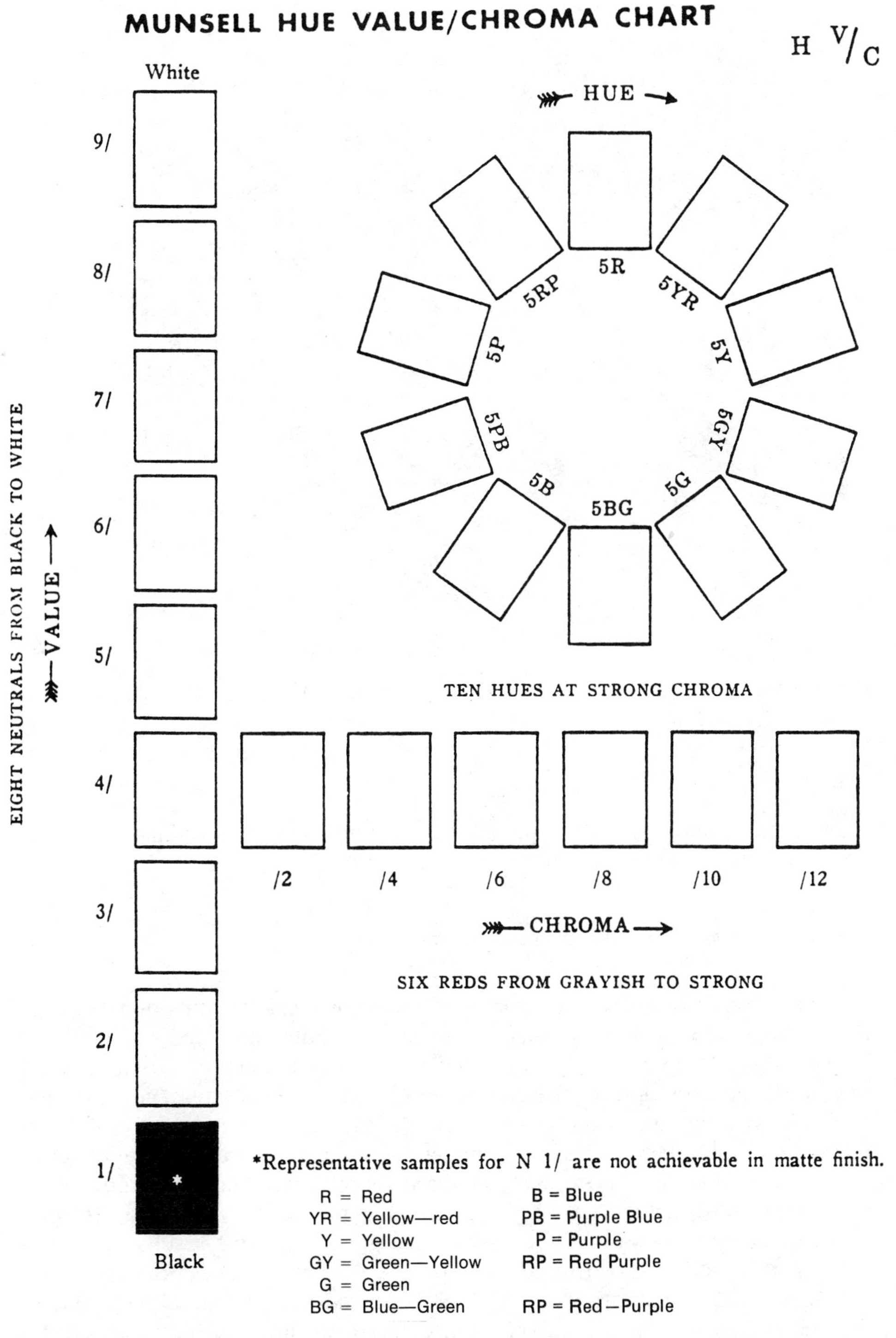

Figure 6-9. Hue, value/chroma chart. (Courtesy Macbeth—Division of Kollmorgen Corporation.)

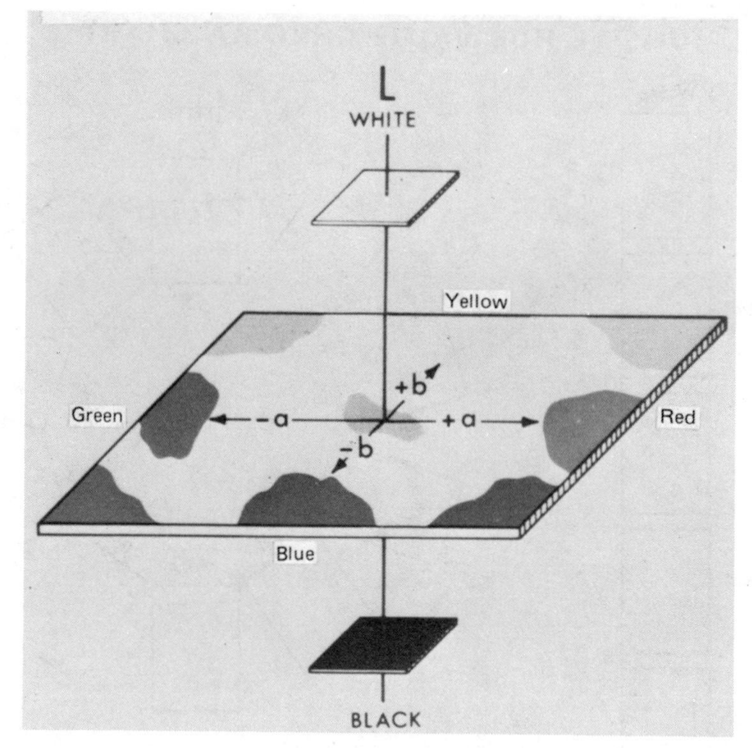

Figure 6-10. *L, a, b* color space. (Courtesy Macbeth—Division of Kollmorgen Corporation.)

interval and computes the tristimulus values, chromaticity coordinates, and the color difference values.

6.5.1. Instrumented Color Measurement

Two basic types of instruments are used for color measurements depending on the requirements. A filter colorimeter or a spectrophotometric colorimeter is employed when the user is only interested in tristimulus values, chromaticity coordinates, and color difference information. Figure 6-11 illustrates a commercially available spectrophotometric colorimeter. Colorimeters are mainly used for production control, quality control, specification, and color matching requirements.

A spectrophotometer is normally required for color formulation and other color development work. Figure 6-12 illustrates a commercially available spectrophotometer equipped with a CRT for displaying and charting the results. The operation of this modern spectrophotometer is very simple. The instrument is first calibrated using a calibration standard. Next, the desired CIE standard illuminant is selected. By placing a relatively flat, colored specimen to be measured in a specimen holder and illuminating a light source, one can instantly obtain percentage spectral reflectance values at 16 nm intervals over the visible spectrum. The microprocessor also computes and displays the CIE *Lab* color space and a spectral reflectance curve as a function of wavelength. The color matching is carried out simply by exposing an unknown specimen and a control specimen to the illuminant and then comparing the two overlapping spectral reflectance curves. A typical printout of spectral reflectance values and spectral reflectance curves for two specimens is illustrated in Figure 6-13.

As an added bonus, most available colorimeters and spectrophotometers calculate the yellowness index (ASTM D 1925) and the whiteness index. ASTM D 2244 describes the standard method for instrumental evaluation of color differences of opaque materials in detail.

6.5.2. Visual Color Evaluation (ASTM D 1729)

A plastic processor requests a color match from a color supplier. The color supplier, through a rigorous color matching process using many sophisticated instruments such as a spectrophotometer or a colorimeter, comes up with a match. When this match is presented to the customer, the customer is totally dissatisfied with the match. A color supplier may have to rematch the color as many as three times before the customer accepts the color match.

The above incident is a common occurrence in plastics and other industries. The reason behind it is very simple. Two objects having the same color, when viewed under one type of illuminant (daylight), appear to match. The same two objects when viewed under different types of illuminants (incandescent) do not match. This phenomenon is known as metamerism. Metamerism is a phenomenon

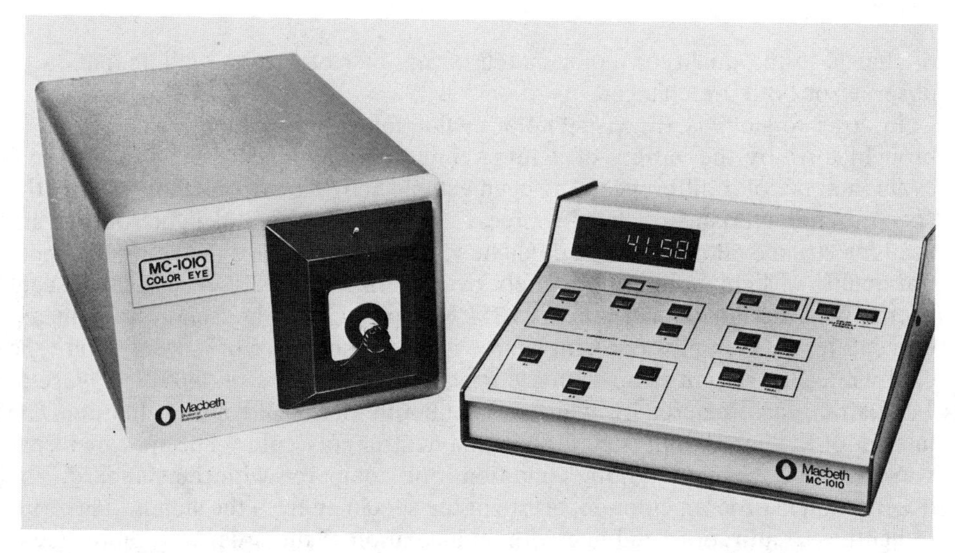

Figure 6-11. Colorimeter. (Courtesy Macbeth—Division of Kollmorgen Corp.)

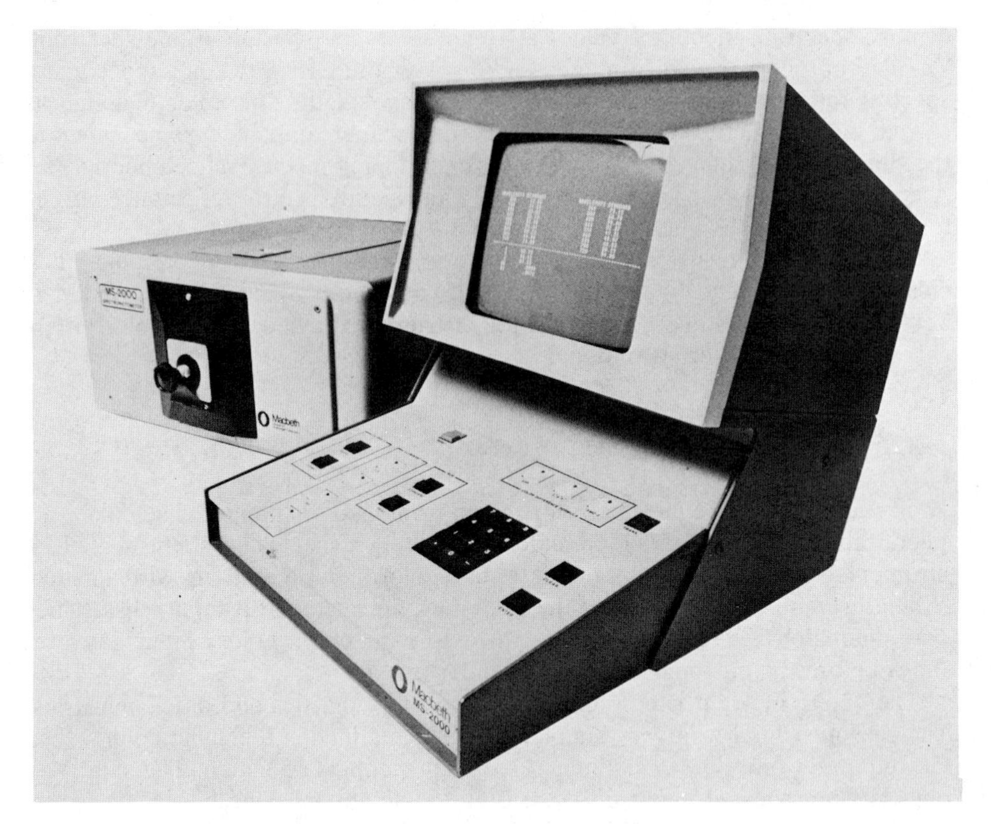

Figure 6-12. Spectrophotometer. (Courtesy Macbeth—Division of Kollmorgen Corporation.)

of change in the quality of color match of any pair of colors, as illumination or observer or both are changed.

In order to simplify the visual color evaluation and to minimize the variations brought forth by the variety of lighting conditions, a standard method of visual evaluation of color differences has been established. The test method defines the spectral characteristics of light sources. Three basic types of light sources are used for visual evaluations. The daylight source approximates the color and spectral quality of light from a moderately overcast northern sky. The incandescent source with a color temperature of 2854°K approximates household-type incandescent light. The minus red light source is a representative of a production-type cool white fluorescent lamp with a correlated temperature of 4400°K. Figure 6-14 illustrates a commercially available light booth for visual color evaluation. The interior of the light booth is painted with a neutral gray color. The specimens are viewed under specified lighting conditions and compared with the standard. The observation in order of color departure of the specimen from the standard in terms of lightness, saturation, and hue with an indication of the order of prominence is made. Many color suppliers provide tolerance chips to the buyer so that the degree of color departure from control can be easily assessed.

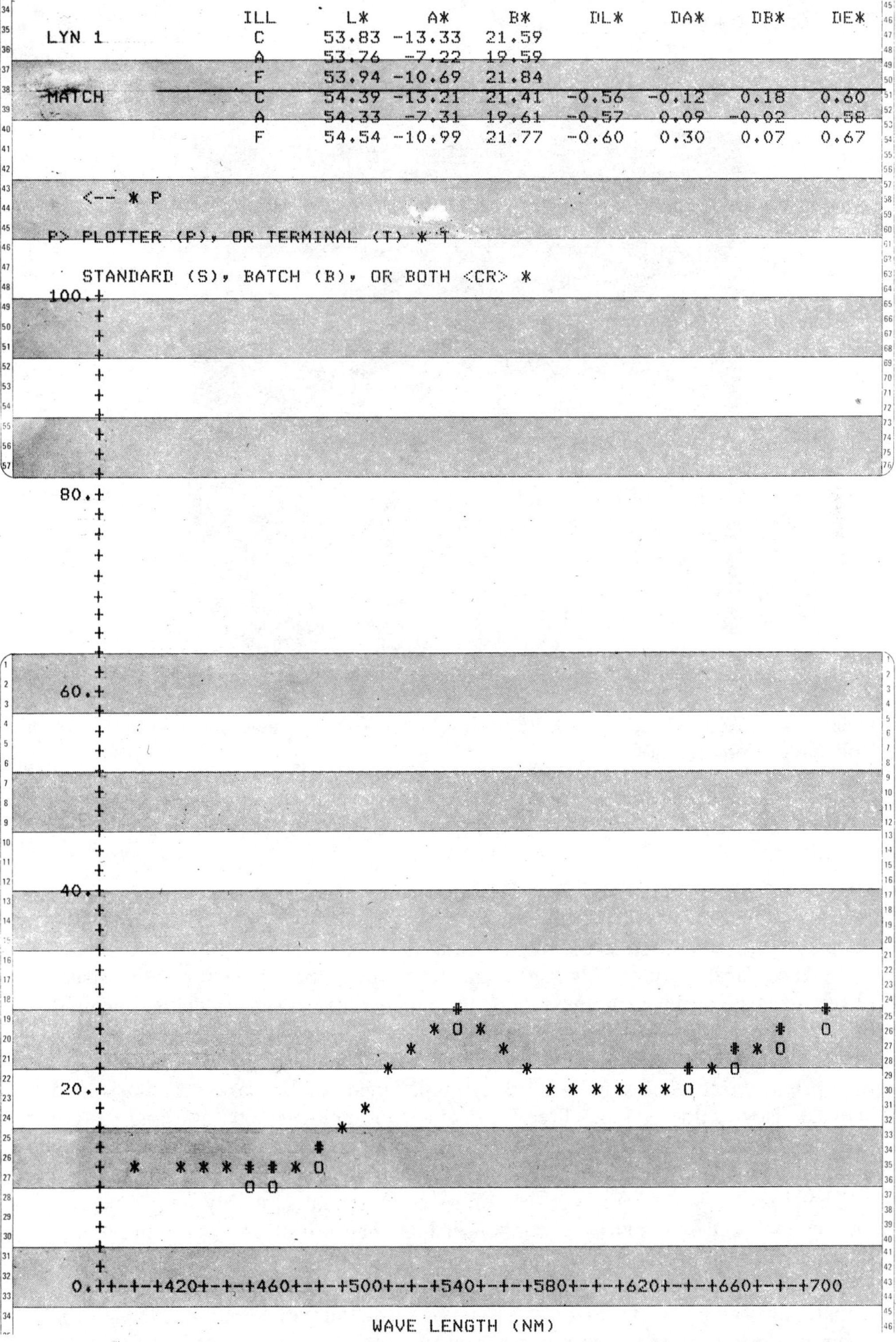

ILL L* A* B* DL* DA* DB* DE*
LYN 1 C 53.83 -13.33 21.59
A 53.76 -7.22 19.59
F 53.94 -10.69 21.84
MATCH C 54.39 -13.21 21.41 -0.56 -0.12 0.18 0.60
A 54.33 -7.31 19.61 -0.57 0.09 -0.02 0.58
F 54.54 -10.99 21.77 -0.60 0.30 0.07 0.67
<-- * P
P> PLOTTER (P), OR TERMINAL (T) * T
STANDARD (S), BATCH (B), OR BOTH <CR> *
100.
80.
60.
40.
20.
0.
420
460
500
540
580
620
660
700
WAVE LENGTH (NM)

Figure 6-13. A typical spectral reflectance curve. (Courtesy Macbeth—Division of Kollmorgen Corporation.)

Figure 6-14. Light booth for visual color evaluation. (Courtesy Macbeth—Division of Kollmorgen Corporation.)

6.6. SPECULAR GLOSS (ASTM D 523)

Specular gloss is defined as the relative luminous reflectance factor of a specimen at the specular direction. This method has been developed to correlate the visual observations of surface shininess made at roughly corresponding angles. The light beam is directed towards the specimen at a specified angle and the light reflected by the specimen is collected and measured. All specular gloss values are based on a primary reference standard—a highly polished black glass with an assigned specular gloss value of 100. Three basic angles of incidence 60°, 20°, and 85° are used for specular gloss measurement of plastic parts. As the angle of incidence increases, the value of gloss of any surface also increases (14).

The glossimeter basically consists of a source optical assembly, which houses an incandescent light source, a condenser lens, and a projection or source lens. An incident beam, created by this assembly, is directed towards the specimen. A sensitive photodetector collects the reflected light and generates an electrical signal that is amplified to energize an analog or digital panel meter to display the value in gloss units. A schematic diagram of a glossimeter is shown in Figure 6-15.

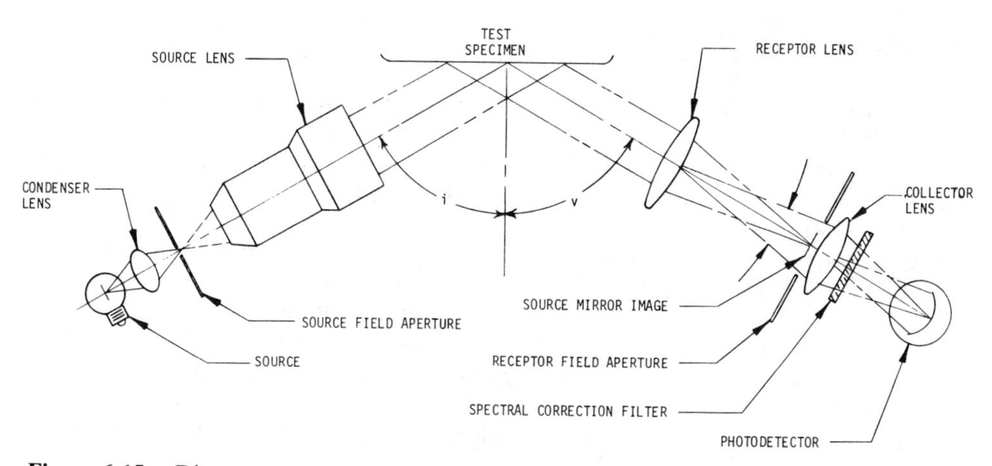

Figure 6-15. Diagram of parallel beam glossmeter showing apertures and source mirror-image position. (Reprinted by permission of ASTM.)

The operation of a glossimeter is very simple. The instrument is turned on and placed on black glass primary standard. The control knob is adjusted so that the meter indicates the value assigned to the primary standard. Next, the sensor is placed on the specimen surface and the gloss value is read directly from the analog or digital display. The linearity of the instrument is routinely checked by placing the sensor on a white secondary standard, which should read within 1.0 gloss unit of the assigned value of that standard. Figure 6-16 illustrates a commercially available glossimeter.

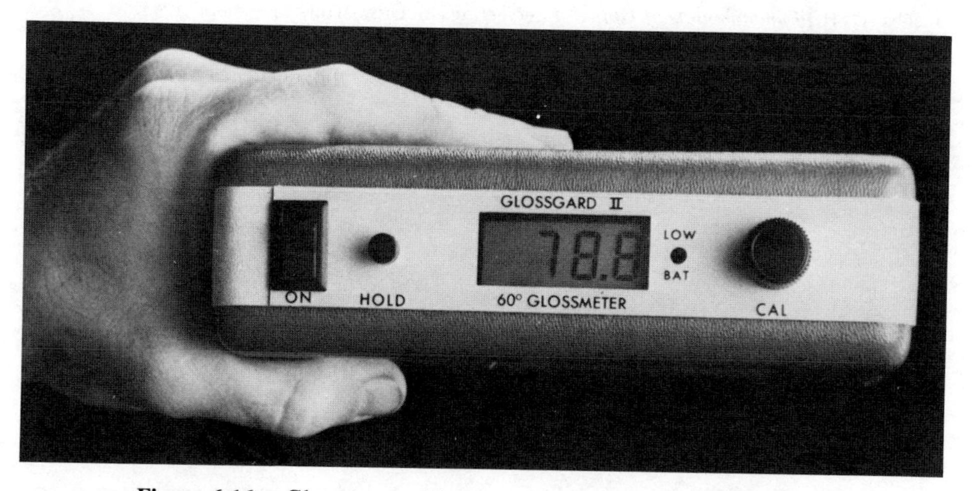

Figure 6-16. Glossmeter. (Courtesy Gardner Laboratories, Inc.)

REFERENCES

1. Deanin, R. D., *Polymer Structure Properties and Applications,* Cahners, Boston, Mass., 1972, p. 248.
2. *Ibid,* p. 248.
3. Byrdson, J. A., *Plastic Materials,* Reinhold, New York, 1970, p. 96.
4. *Ibid,* p. 241.
5. Ives, C. G., Mead, J. A., and Riley, M. M., *Handbook of Plastics Test Methods,* IlLFFE Books, London, England, 1971, p. 432.
6. Baer, E., *Engineering Design for Plastics,* Reinhold, New York, 1964, p. 604.
7. Milby, R. V., *Plastics Technology,* McGraw–Hill, New York, 1973, p. 548.
8. Levy, S. and Dubois, J. H., *Plastics Product Design Engineering Handbook,* Reinhold, New York, 1977, pp. 309–310.
9. Kinney, G. F., *Engineering Properties and Applications of Plastics,* Wiley, New York, 1957, p. 220.
10. McCamy, C. S., *Color Measurement and Specification,* Macbeth Corporation, Newburgh, N.Y., Technical Literature, p. 10.
11. Billmeyer, F. W. and Saltzman, M., *Principles of Color Technology,* Interscience, New York, 1966, p. 38.
12. Gardner Laboratories, *Tech. Bull.: Color and Color Related Properties.* Bull. No. 010, Silver Spring, Md.
13. McCamy, *Reference 10,* p. 9.
14. Gardner Laboratories, *Tech. Bull.: Gloss Measurement.* Bull. No. 073, Silver Spring, Md.

SUGGESTED READING

Webber, T. G. (Ed), *Coloring of Plastics,* Wiley–Interscience, New York, 1979.

Ahmed, M., *Coloring of Plastics Theory and Practice,* Reinhold, New York, 1979.

Ross, L. and Birley, A. W., "Optical Properties of Polymeric Materials and their Measurements," *J. Phys. D: Appl. Phys.* **6,** (1973), p. 795.

Jenkins, F. A., and White, H. E., *Fundamentals of Optics,* McGraw–Hill, New York, 1957.

Jacobs, D. H., *Fundamentals of Optical Engineering,* McGraw–Hill, New York, 1943.

CHAPTER 7

Material Characterization Tests

7.1. INTRODUCTION

The plastics industry has grown phenomenally in the past two decades. Plastic material consumption has quadrupled. Today there are countless numbers of processors consuming in excess of 100 million pounds of material every year. An increasing number of processors are looking into various techniques for characterizing the incoming plastic resin in order to guard themselves against the batch to batch variations. Such variation in the properties and the processibility of the polymer have been very costly.

In order to understand the phenomenon of batch-to-batch variation, one must understand the polymerization process. The polymerization mechanism can often be very complex. Basically, it involves adding a monomer or a mixture of monomers into a reaction kettle along with several other ingredients such as a catalyst, an initiator and water, depending on the type of polymerization process. The contents are agitated at a specific speed. The time, temperature, and pressure are carefully controlled. During this process, the monomer is converted into a polymer. The properties of the final product depend on several factors such as the monomer to water ratio, if water is present, the degree of agitation, the removal of exothermic heat generated during the polymerization, and the ability of the polymer to dissolve in monomer. This is further complicated by the fact that different polymerization techniques are used to form the same chemical type of polymer. Since it is difficult to manufacture a monodispersed polymer (a polymer in which all molecules have the same size) commercially, we are forced to live with variations in the size and weight of molecules in a polymer. Such variations in the size and the weight of the polymer molecules are also extremely difficult to control. The relative proportions of molecules of different weights within a polymer comprise its molecular weight distribution. Depending on the range of distribution (narrow or broad), the processibility and the properties of a polymer vary significantly. This basic nature of the polymerization process creates a need for material characterization tests that can be used as an insurance against variations in the characteristics of the polymer.

There are numerous ways of characterizing a polymer. Some are very basic and simple, others are more sophisticated and complex. The four most common and widely accepted tests are:

1. Melt index (flow rate) test, capillary rheometer test.
2. Viscosity tests.
3. Gel permeation chromatography test.
4. Analytical test (TGA, TMA, DSC).

7.2 MELT INDEX TEST (ASTM D 1238)

7.2.1. Significance

The melt index tests measures the rate of extrusion of a thermoplastic material through an orifice of specific length and diameter under prescribed conditions of temperature and pressure. This test is primarily used as a means of measuring the uniformity of the flow rate of the material. The reported melt index values help to distinguish between the different grades of a polymer. A high molecular-weight material is more resistant to flow than a low molecular-weight material. However, the data obtained from this test does not necessarily correlate with the processability of the polymer. This is due to the fact that plastic materials are seldom manufactured without incorporating additives which affect the processing characteristics of a material such as stability and flowability. The effect of these additives is not readily observed via the melt index test. The rheological characteristics of polymer melts depend on a number of variables. Since the values of these variables may differ substantially from those in large-scale processes, the test results may not correlate directly with processing behavior.

7.2.2. Test Procedures

The melt index apparatus (Figure 7-1) is preheated to a specified temperature. The material is loaded into the cylinder from the top and a specified weight is placed on a piston. The most commonly used test conditions are shown in Table 7-1(a) and Table 7-1(b). The material is allowed to flow through the die. The initial extrudate is discarded since it may contain some air bubbles and contaminants. Depending on the material or its flow rate, cuts for the test are taken at different time intervals. The extrudate is weighed and melt index values are calculated in g/10 min.

An alternate method for making the measurement for materials with a high flow rate involves automatic timing of the piston travel by some electrical or mechanical device. The melt index value is calculated by using the following formula:

$$\text{Flow rate} = (427 \times L \times d)/t$$

where L = length of calibrated piston travel (cm); d = density of resin at test temperature (g/cm^3); t = time of piston travel for length L. (sec).

A commerical melt indexer is shown in Figure 7-2.

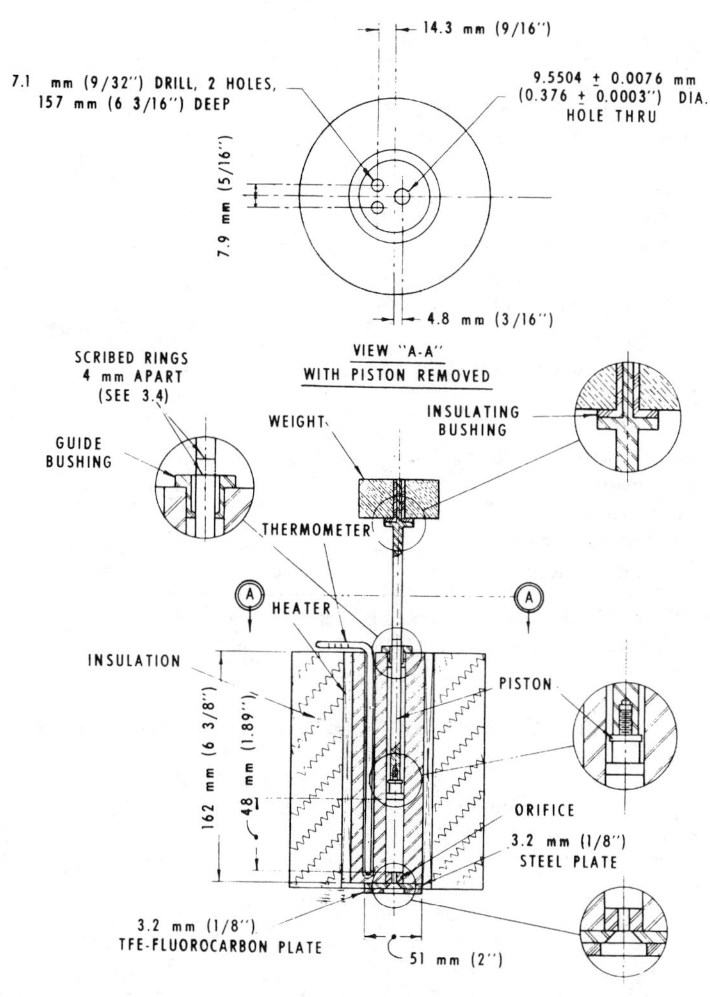

Figure 7-1. Schematic of melt indexer. (Reprinted by permission of ASTM.)

7.2.3. Factors Affecting the Test Results

1. *Preheat Time.* If the cylinder is not preheated for a specified length of time, there is usually some nonuniformity in temperature along the walls of the cylinder even though the temperature indicated on the thermometer is close to the set point. This causes the flow rate to vary considerably.

2. *Moisture.* Moisture in the material, especially a highly pigmented one, causes bubbles to appear in the extrudate which may not be seen with the naked eye. Frequent weighing of short cuts of the extrudate during the experiments reveals the presence of moisture. The weight of the extrudate is significantly influenced by the presence of the moisture bubbles.

3. *Packing.* The sample resin in the cylinder must be packed properly by pushing the rod with substantial force to allow the air entrapped between the resin

Table 7-1(a). Standard Test Conditions[a]

Materials	Test conditions
Acetal	E,M
Acrylics	H,I
ABS	G,I
Cellulose esters	D,E,F
Nylon	K,Q,R,S
Polychlorotrifluorethylene	J
Polyester (thermoplastic)	T
Polyethylene	A,B,D,E,F,N
Ultra-high-density polyethylene	U
Polybutylene	E
Polycarbonate	O
Polyphenylene oxide (PPO)	*[b]
Polypropylene	L
Polystyrene	G,H,I,P
Polyterephthlate	T
Styrene acrylonitrile	I

[a] From ASTM D 1238. Reprinted by permission of ASTM.
[b] Melt Index Test not recommended.

Table 7-1(b). Standard Test Conditions, Temperature, and Load[a]

Condition	Temperature (°C)	Total load, including piston (g)	Approximate pressure	
			kPa	psi
A	125	325	44.8	6.5
B	125	2,160	298.2	43.25
C	150	2,160	298.2	43.25
D	190	325	44.8	6.5
E	190	2,160	298.2	43.25
F	190	21,600	2982.2	432.5
G	200	5,000	689.5	100.0
H	230	1,200	165.4	24.0
I	230	3,800	524.0	76.0
J	265	12,500	1723.7	250.0
K	275	325	44.8	6.5
L	230	2,160	298.2	43.25
M	190	1,050	144.7	21.0
N	190	10,000	1379.0	200.0
O	300	1,200	165.4	24.0
P	190	5,000	689.5	100.0
Q	235	1,000	138.2	20.05
R	235	2,160	298.2	43.25
S	235	5,000	689.5	100.0
T	250	2,160	298.2	43.25
U	310	12,500	1723.7	250.0

[a] From ASTM D 1238. Reprinted by permission of ASTM.

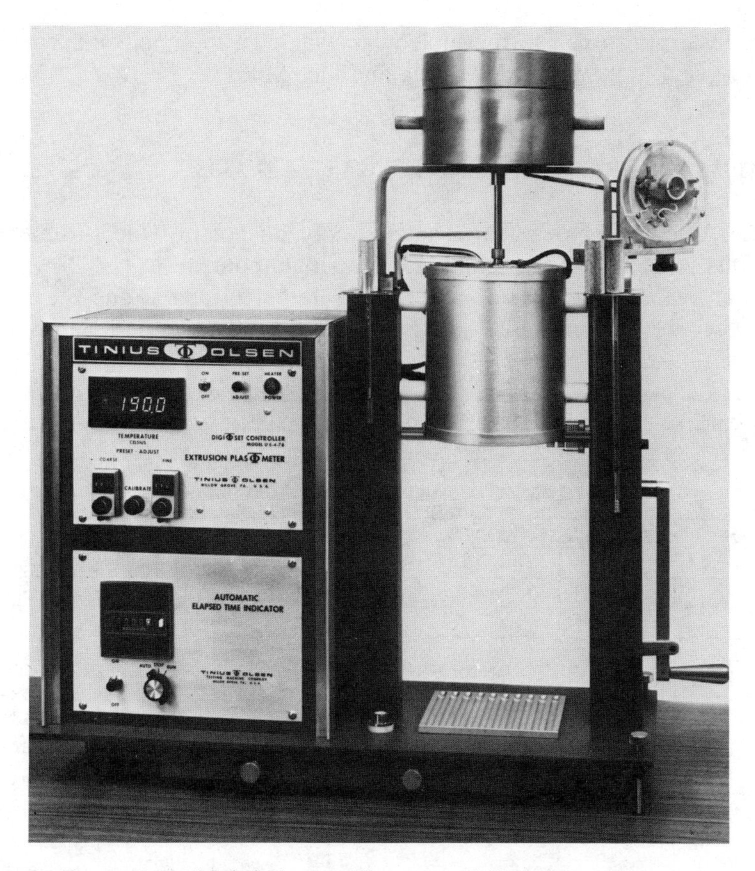

Figure 7-2. Melt indexer. (Courtesy Tinius Olsen Company.)

pellets to escape. Once the piston is lowered, the cylinder is sealed off, and no air can escape. This causes variation in the test results.

4. *Volume of Sample.* In order to achieve the same response curve repeatedly, the volume of the sample in the cylinder must be kept constant. Any change in sample volume causes the heat input from the cylinder to the material to vary significantly.

7.2.4. Interpretation of Test Results

The melt index values obtained from the test can be interpreted in several different ways. First of all, a slight variation (up to 10 percent in the case of polyethylene) in the melt index value should not be interpreted as indicating a suspect material. The material supplier should be consulted to determine the expected reproducibility for a particular grade of plastic material. A significantly different melt index value than the control standard may indicate several different things. The material may be of a different grade with a different flow characteristic. It also means that the average molecular weight or the molecular weight distribution of the material is different than the control standard and may have different properties.

Melt index is an inverse measure of molecular weight. Since flow characteristics are inversely proportional to the molecular weight, a low molecular-weight polymer will have a high melt index value and vice versa.

7.3. CAPILLARY RHEOMETER TEST (ASTM D 1703)

Quite often, the melt index test results simply do not provide sufficient information. This is because the melt index is a single-point test. The flow rates are measured at a single shear stress and shear rate performed at one set of temperatures and geometric conditions. Furthermore, since the melt index measurement

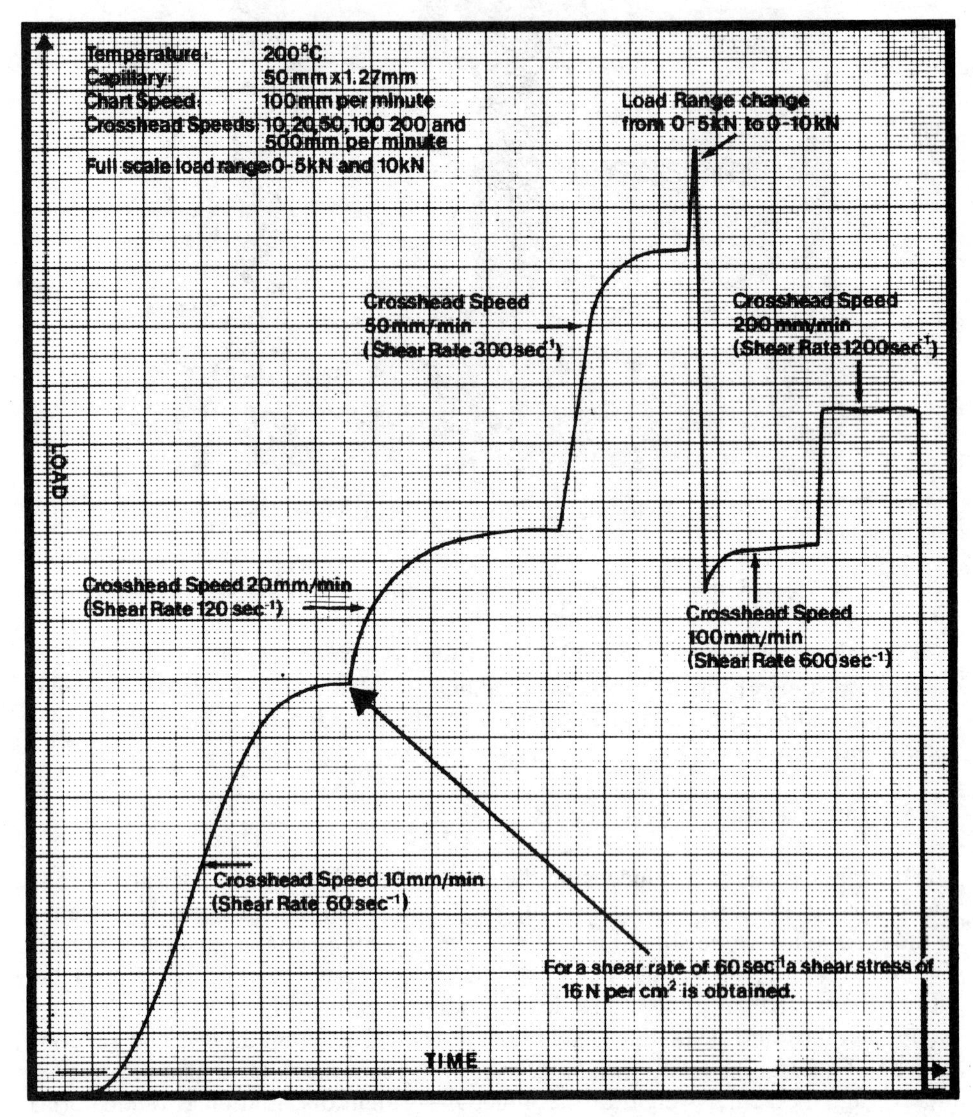

Figure 7-3. A typical load versus time curve as plotted on a Capillary Rheometer chart recorder. (Courtesy Instron Corporation.)

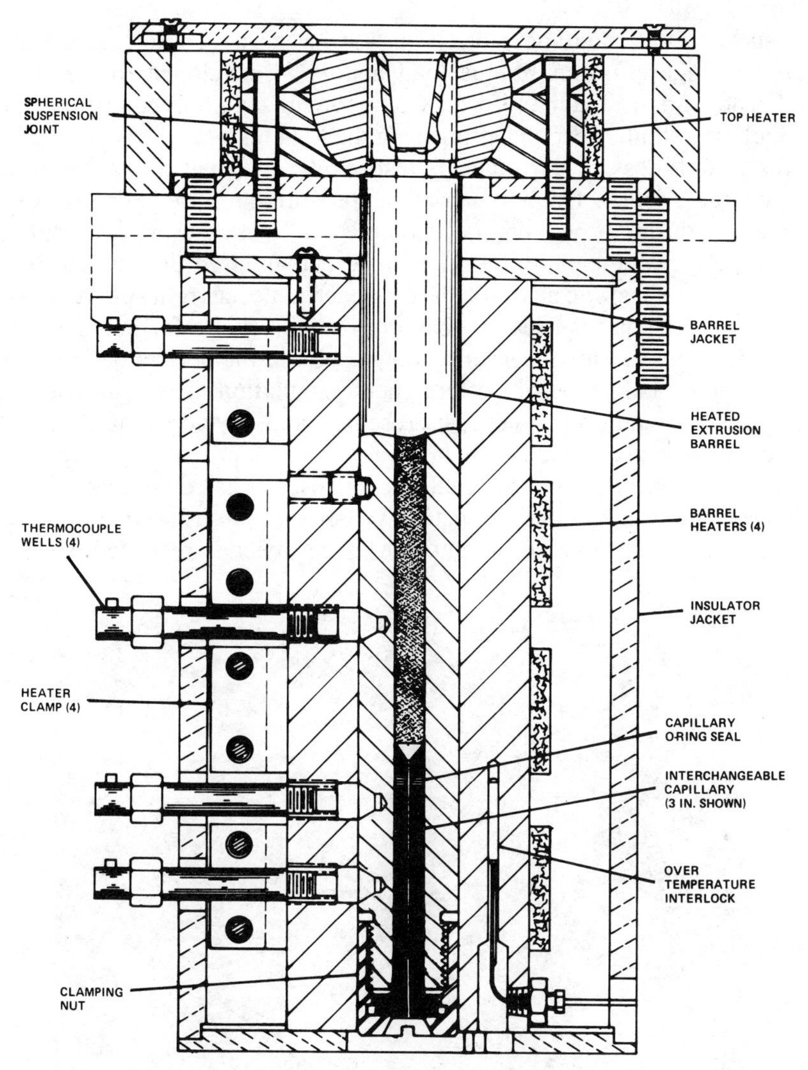

Figure 7-4. Sectional view of rheometer extrusion assembly. (Courtesy Instron Corporation.)

takes account of the behavior of the polymer at only one point, it is quite possible for two materials with the same melt index values to behave completely different at shear stresses that are different from the ones used during the melt index measurements. The capillary rheometer measures apparent viscosity or melt index over an entire range of shear stresses and shear rates encountered in compression molding, calendering, extrusion, injection molding, and other polymer melt processing operations. A rheometer is a precision instrument that provides the accuracy and reproducibility necessary for polymer characterization test.

A quality control test to qualify the incoming material is quick, simple, and accurate. First of all, at least three different load and speed settings are established. Next, the test is carried out and the force required to move the plunger at each speed is detected and plotted on a strip chart recorder against time, as illustrated in Figure 7-3. A unique load history is developed for each type of material. The batch-to-batch uniformity of incoming materials can be checked out by conducting a similar test and comparing the load history. A visual comparison check for smoothness of extrudate is also made to detect gross differences in material characteristics. The polymer characterization for predicting the behavior of the material during processing is accomplished by measuring apparent viscosities at the range of shear rates that may be encountered during processing. A standard or a control curve is established for each material from apparent viscosity and shear rate data and is used as characterization criteria for new materials.

Rheological measurements are also very useful in the determination of molecular, elastic, physical properties of polymers, simulation, determination of large-scale processing conditions, and basic research and development. The capillary rheometer consists of an electrically heated cylinder, a pressure ram, temperature controllers, timers, and interchangeable capillaries (Figure 7-4 and Figure 7-5). The plunger can be moved at a constant velocity which translates to a constant shear rate. The force to move the plunger at this speed is recorded which deter-

Figure 7-5. Capillary Rheometer. (Courtesy Instron Corporation.)

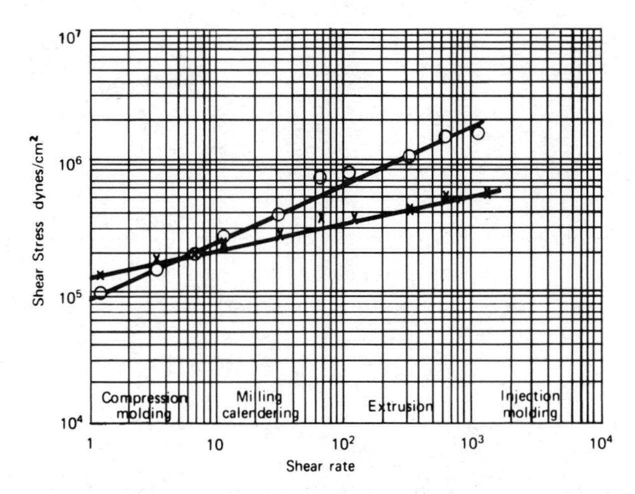

Figure 7-6. Shear stress versus shear rate curve for PVC with two different plasticizers. Note that the two materials exhibit similar characteristics at a shear rate of 5 reciprocal seconds but differ substantially at all others. (Courtesy Instron Corporation.)

mines the shear stress. Alternately, a weight or constant pressure can be applied to the plunger which generates a constant shear stress and the velocity of the plunger can be determined by cutting and weighing the output. Shear rate can be calculated by knowing the melt density.

The sample material is placed in the barrel of the extrusion assembly, brought to temperature, and forced out through a capillary. The force required to move the plunger at each speed is detected by a load cell. Shear stress, shear rate, and apparent melt viscosity are calculated as follows (1):

$$\tau = \frac{F_r}{2\pi R^2 l}$$

where τ = shear stress (psi); F = load on the ram (lbs); r = radius of capillary

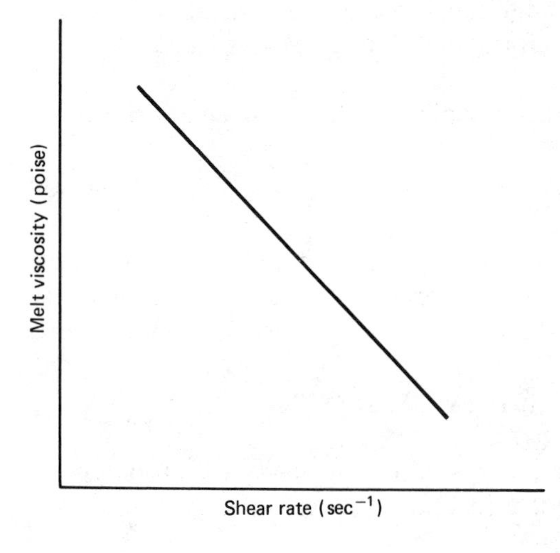

Figure 7-7. Melt viscosity versus shear rate.

orifice (in.); R = radius of the barrel (in.); l = length of capillary orifice (in.).

$$\gamma = \frac{4Q}{\pi r^3}$$

where γ = shear rate (sec^{-1}); Q = flow rate (in^3/sec).

$$\text{Apparent melt viscosity (poise) } \eta = \frac{\text{Shear stress}}{\text{Shear rate}}$$

Shear stress versus shear rate and melt viscosity versus shear rate curves are plotted in Figures 7-6 and 7-7.

7.4. VISCOSITY TESTS

Viscosity is defined as the property of resistance to flow exhibited within the body of a material expressed in terms of a relationship between applied shearing stress and resulting rate of strain in shear. In the case of ideal or Newtonian viscosity, the ratio of shear stress to the shear rate is constant. Plastics typically exhibit non-Newtonian behavior which means that the ratio varies with the shearing stress. There are two different aspects of viscosity. Dynamic or absolute viscosity, best determined in a rotational type of viscometer with a small gap clearance, is independent of the density or specific gravity of the liquid sample and is measured in poises (P) and centipoises (cP). Kinematic viscosity, usually determined in some form of efflux viscometer equipped with a capillary bore or small orifice that drains by gravity, is strongly dependent on density or specific gravity of the liquid, and is measured in stokes (S) and centistokes (cS). The relationship between two types of viscosity is:

$$\text{stokes} \times \text{specific gravity} = \text{poises}$$

The measurement and control of rheological properties are usually performed with simple devices called "viscometers" or "viscosimeters" which do not measure true viscosities of either the dynamic or kinematic type but make relative flow comparisons. Viscosity of plastic materials is measured in three basic ways, employing principles involving liquid deformation due to various forces (2).

1. Downward rate of gravity flow through capillary bores and small orifices.
2. Upward speed of a trapped air bubble.
3. Torque developed by the liquid drag between moving and stationary surfaces.

The first method is basically used for thermoplastic materials. The latter two are more commonly used for thermosetting materials, plastisols, and organosols and are discussed in section 7.7.3 on viscosity tests for thermosets in this chapter.

7.4.1. Dilute Solution Viscosity of Polymers (ASTM D 2857)

This test method is used to determine dilute solution viscosity for all polymers that dissolve completely without chemical reaction or degradation to form solutions that are stable with time at a temperature between ambient and approximately 150°C. The results of the tests are expressed as relative viscosity, inherent viscosity, or intrinsic viscosity. Reduced viscosity and specific viscosity can also be calculated. Table 7-2 shows recommended test conditions for dilute solution viscosity measurements.

The test apparatus consists of volumetric flasks, transfer pipets, a constant temperature bath, a timer, viscometer, and a thermometer. The constant volume device is recommended for use where solution viscosity, reduced viscosity, or inherent viscosity are to be measured at a single concentration. The second type of device is called a dilution viscometer and it does not require constant liquid volume for operation. This type is basically used for measuring intrinsic viscosity. Different types of commercially available viscometers are shown in Figure 7-8.

The test sample is prepared by dissolving a specified amount of polymer in the appropriate solvent. The prepared solution and the viscometer are placed in a constant temperature bath maintained at the test temperature. A suitable amount of solution is transferred into the viscometer using a transfer pipet. Once the temperature equilibrium is attained, the liquid level in the viscometer is brought above the upper graduation mark by means of gentle air pressure applied to the arm opposite the capillary. The timer is started exactly when the meniscus passes the upper graduation mark, and stopped exactly when the meniscus passes the lower graduation mark. The test is repeated at least three times for each solution and for each pure solvent. The time for the liquid to pass through the graduation marks is called efflux time. Relative, specific, reduced, inherent, and intrinsic viscosity values are determined in the following manner (3).

A. *Relative Viscosity (Viscosity Ratio)*

This term refers to the ratio of the time it takes for a specified solution of resin in a pure solvent to flow through an orifice to the flow time for an equal quantity of pure solvent through the same orifice.

$$\text{Relative viscosity} = \eta_{rel} = t/t_0$$

where t = efflux time of solution; t_0 = efflux time of pure solvent.

Table 7-2. Recommended Test Conditions for Dilute Solution Viscosity Measurements

Polymer	Solvent	Solvent dissolving conditions	Test temperature	Solution concentration, (g/ml)
Polyamide	Formic acid or *m*-cresol	30°C 100°C, 2 hrs	30°C	0.0050 ± 0.00002
Polycarbonate	Methylene chloride or *P*-dioxane	30°C 60°C	30°C	0.0040 ± 0.0002
PMMA	Ethylene dichloride	30°C, 24 hrs	30°C	0.0020 ± 0.00002
PVC	Cyclohexanone	85 ± 10°C	30°C	0.0020 ± 0.00002

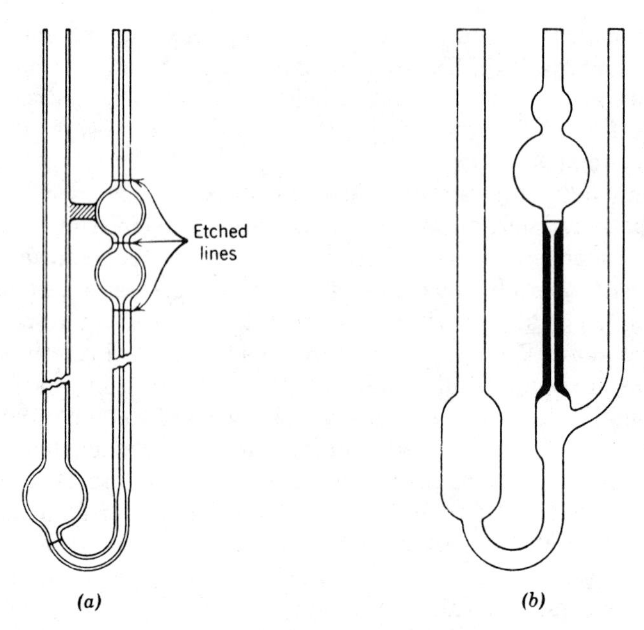

Figure 7-8. Capillary viscometers commonly used for measurement of polymer solution viscosities (*a*) Ostwald–Fenske; (*b*) Ubbelohde. (Reprinted by permission of Wiley—Interscience.)

Or

$$\text{Relative viscosity} = \eta_{rel} = \eta/\eta_0$$

where η = viscosity of solution; η_0 = viscosity of solvent.

B. Specific Viscosity

This value is obtained by subtracting one from the value obtained for relative viscosity:

$$\eta_{SP} = \eta_{rel} - 1$$

where η_{SP} = specific viscosity; η_{rel} = relative viscosity.

C. Reduced Viscosity (Viscosity Number)

This term is used to describe the ratio to the specific viscosity to the concentration.

$$\eta_{red} = \eta_{SP}/c$$

where c = solution concentration in g/ml.

D. Inherent Viscosity (Logarithmic Viscosity Number)

The ratio of the natural logarithm of the relative viscosity to the concentration is called inherent viscosity.

$$\eta_{inh} = \frac{\ln \eta_{rel}}{c}$$

where η_{inh} = inherent viscosity; ln η_{rel} = natural logarithm of relative viscosity; c = solution concentration in g/ml.

E. *Intrinsic Viscosity (Limiting Viscosity Number)*

To determine the intrinsic viscosity of a polymer from dilute solution viscosity data, the reduced and inherent viscosities of solutions of various concentrations are determined at a constant temperature. These values are plotted against the respective concentrations as shown in Figure 7-9. A straight line is drawn through the points and extrapolated to the zero concentration. The intrinsic viscosity is the intercept of the line at zero concentration.

7.4.2. Applications and Limitations of Dilute Solution Viscosity Measurements

Molecular weight of the polymer can be determined by either the absolute or relative method. Absolute methods such as osmotic pressure and light scattering are more accurate but lengthy and complex. In practice, dilute solution viscosity is by far one of the most popular relative methods for characterizing the molecular size of the polymers. Molecular weight from intrinsic viscosity data can be determined as follows (4):

$$\overline{M_n} = \tfrac{1}{2}\,(\eta) \times 10^5$$

where $\overline{M_n}$ = number average molecular weight; η = intrinsic viscosity in cyclohexanone at 30°C.

$$\overline{M_w} = 0.9\,(\eta) \times 10^5$$

where $\overline{M_w}$ = weight average molecular weight; η = intrinsic viscosity in cyclohexanone at 30°C.

The determination of dilute solution viscosity provides one item of information towards the molecular characterization of polymers. When viscosity data is used in conjunction with other molecular parameters, the properties of polymers depending on their molecular structure may be better predicted. Viscosity data alone

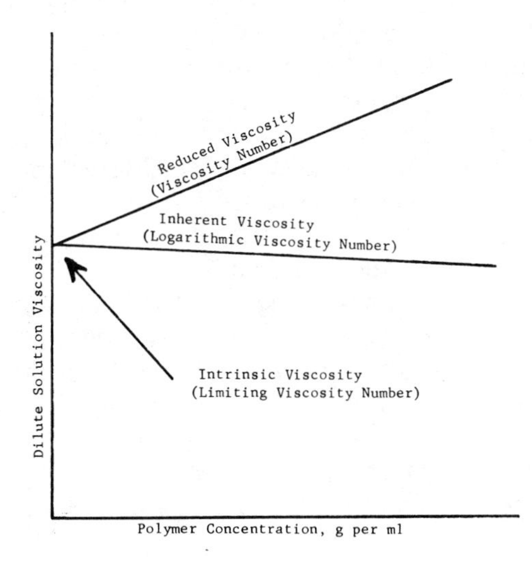

Figure 7-9. Example of plot to determine intrinsic viscosity. (Reprinted by permission of ASTM.)

may be of limited value in predicting the processing behavior of the polymer. However, when used in conjunction with other flow and physical property values, the solution viscosity of polymers may contribute to their characterizations. The viscosity of polymer solutions may be affected drastically by the presence of unknown or known additives in the sample. When dilute solution measurements are used to indicate the absolute physical properties of formulated polymers, many difficulties can arise. Most of these dilute solution measurements are based on one percent or less of a polymer that is in complete solution in a pure solvent. This can hardly be called a normal commercial formulation (5). The solution viscosity of a polymer of sufficiently high molecular weight may depend on the rate of shear in viscometer, and the viscosity of a polyelectrolyte (polymer containing ionizable chemical groupings) will depend on the composition and ionic strength of the solvent. Special precautions are required when measuring such polymers. Dilute solution viscosity data are extremely dependent upon the purity and type of solvent, the type of viscometer, temperature, and time. Even the slightest change in any one or more of the above parameters can seriously affect the result. Satisfactory correlation between solution viscosity and certain other properties is possible from polymers of a single manufacturing process. However, when a polymer is produced by different manufacturing processes, the correlation with other properties of a polymer may be limited.

7.5. GEL PERMEATION CHROMATOGRAPHY (ASTM D 3593)

Quite often traditionally used tests like melt index or viscosity tests do not provide enough information about the processibility of the polymer. Such tests only measure an average value and tell us nothing about the distribution that makes up the average. Take, for instance, average daily temperatures reported for a particular city. These reports can be very misleading since they lack more useful information about extreme high and extreme low temperatures. In a very similar manner, the melt index and viscosity tests relate very well to the average molecular weight of the polymer but fail to provide the necessary information about the molecular weight distribution of the polymer. Two batches of resin may have the same melt index, which simply indicates that their viscosity average molecular weights are similar. Their molecular weight distributions, the number of molecules of various molecular weights that make up their averages, can be significantly different. If an excessively high molecular-weight fraction is present, the material may be hard and brittle. Conversely, if an excessive amount of low molecular-weight fraction is present, the material may be soft or sticky. This is illustrated in Figure 7-10, which shows MWD curves for three different batches of polyethylene resin.

When evaluating the nature of the incoming plastics material is extremely important to a processor, he must look for a more reliable technique that will provide the necessary information, to qualify the material, such as the measurement of molecular weight distribution. The molecular weight distribution is the single most fundamental property of the polymer. The molecular weight distribution not only provides basic information regarding the processability of the polymer but also gives valuable information for predicting its mechanical properties.

Gel permeation chromatography is the method of choice for determining the

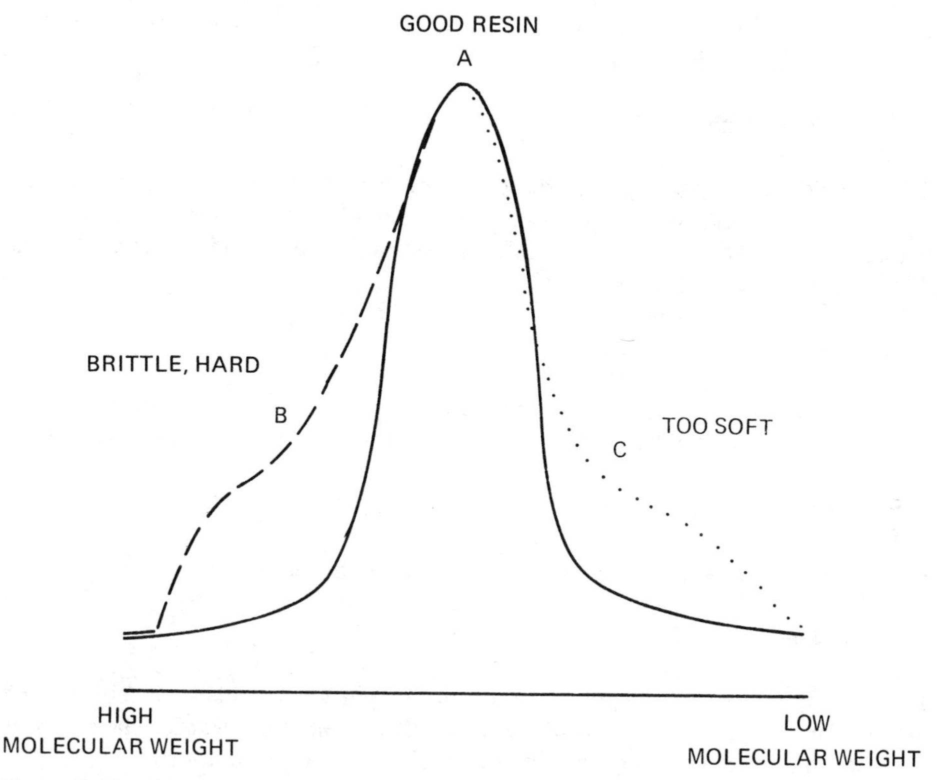

Figure 7-10. Molecular weight distribution curves. Material "A" represents "good" resin. Material "B" with excessively high-molecular weight is brittle and hard. Material "C" with excessively low-molecular weight is soft and sticky. (Courtesy Waters Associates, Inc.)

molecular weight distribution of a polymer. This technique has gained wide acceptance among the plastic material manufacturers and the processors because of its relatively low cost, simplicity, and its ability to provide accurate, reliable information in a very short time. GPC reveals the molecular weight distribution of a polymer compound. It detects not only the resin-based molecules such as polymer, oligomer, and monomer, but most of the additives used in plastic compounds and even low-level impurities. The molecular weight distribution curve is plotted for a well-characterized standard material and the profile of the curve is compared with the test sample. In this manner, batch to batch uniformity can be checked quickly as a means of quality assurance.

The separation of polymer molecules by GPC is based upon the differences in their "effective size" in solution. (Effective size is closely related to molecular weight.) Separation is accomplished by injecting the polymer solution into a continuously flowing stream of solvent that passes through highly porous, tiny, rigid gel particles closely packed together in a tube. The pore size of the gel particles may vary from small to very large. As the solution flows through the gel particles, molecules with small effective size (low molecular weights) will penetrate more pores than molecules with larger effective sizes and therefore, take longer to emerge than the larger molecules. If the gel covers the right range of molecular sizes, the result will be a size separation with the largest molecules exiting the gel-packed tubes (columns) first.

7.5.1. GPC Instrumentation

Gel permeation chromatography basically consists of injecting a polymer in solution into a delivery system that delivers the solution through the columns, detectors, and recorder that records information provided by the detector. This is shown schematically in Figure 7-11.

A. *Solvent Delivery System*

The delivery pumps produce constant flow rates independent of viscosity differences. The system rapidly pumps the polymer in solution through the system.

B. *Injector*

An injector accurately places the polymer solution in the flowing solvent stream.

C. *Columns*

The columns consist of highly porous, tiny, rigid gel particles closely packed together in a tube. The pore size varies from very small to very large. The columns separate the sample and provide the molecular weight distribution.

D. *Detectors*

Once the polymer is separated into various molecular weight species, they are detected by two basic types of detectors. A universal detector, called a differential refractometer, is used to monitor polymer molecular weight distribution. The UV detector, which provides higher sensitivity than the refractometer, is used for determining low molecular-weight additives and impurities.

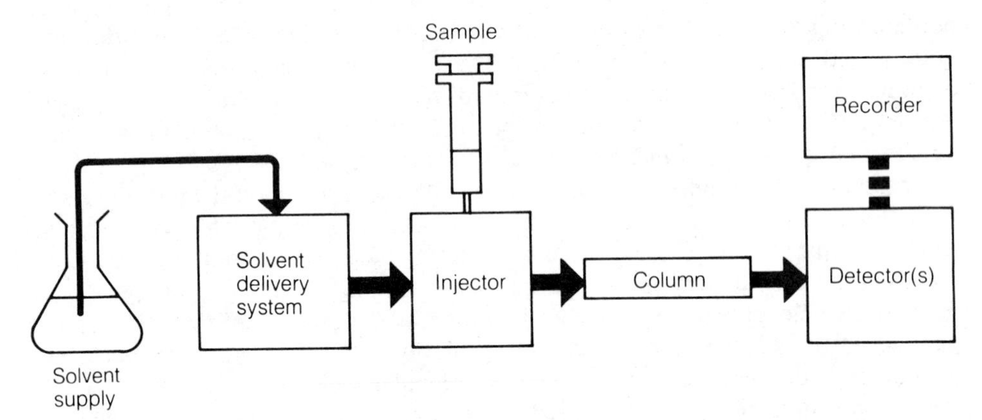

Figure 7-11. Schematic of a GPC system. (Courtesy Waters Associates, Inc.)

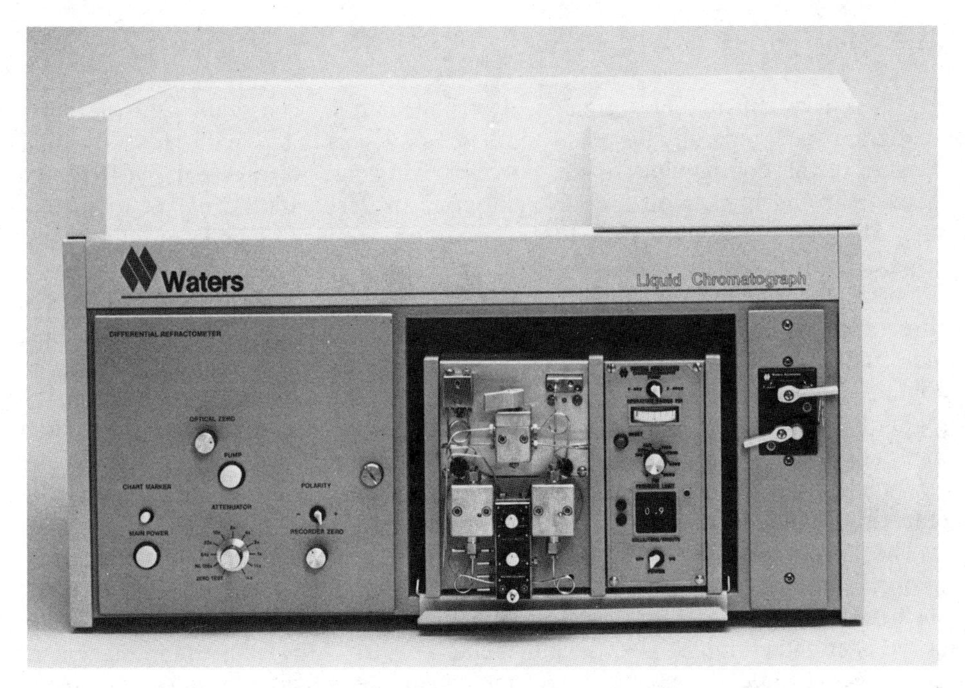

Figure 7-12. GPC equipment. (Courtesy Waters Associates, Inc.)

E. Recorder

The separation is recorded immediately to provide a continuous chromatogram.

7.5.2. Test Procedure

A test sample is prepared by dissolving a small amount of polymer in the solvent and filtering the solution to remove the undissolved impurities. The next step is to select the proper size columns, connect them, set the sensitivity setting on the detectors, and allow the instrument to equilibrate. A trial analysis is done by injecting the polymer solution into the instrument. The chromatogram is carefully analyzed. If the chromatogram shows all the desired information, the final analysis is carried out. If not, the operating parameters, such as column size, flow rate, and number of columns are optimized. The trial step is repeated before proceeding to the final analysis.

During the final analysis, as the sample flows through the column, the molecules are separated according to size by a simple mechanical effect. Because of their smaller size, the smaller molecules enter into the gel pores more readily, and therefore take extra time to reach the bottom of the column. The various molecular weight species are separated by the difference in travel time through the column and pass through a detector in descending order of size. The detector measures the concentration of each molecular size and plots the molecular weight distribution of the sample on a strip chart. Commerically available GPC equipment is shown in Figure 7-12.

7.5.3. Interpreting the GPC Curve

A GPC curve can reveal a great deal of information. The regions of the chromatogram can generally be easily and quickly correlated with the molecular weights of the components of the polymer. Molecules always emerge from the instrument and appear on the chromatogram in descending order of molecular size. The polymer is always the first to emerge since its molecules are the largest in size. Typically, it is shown as a broad curve in region A of the chromatogram (Figure 7-13). The intermediate or low molecular-weight material appears next as sharper peaks in region B of the chromatogram. Last to appear are the smallest molecular weight components—unreacted monomer and low molecular weight contaminants such as moisture. These species are represented by sharp peaks in region C of the chromatogram. In addition, the relative proportions of high and low ends of the polymer can be quickly determined because the height of the curve at any given point is proportional to the concentration of the component emerging at that point.

The broad curve in region A of the chromatogram provides the most useful information regarding the characteristic of the polymer. The distance from the start of the chromatogram to the midpoint of the molecular weight distribution curve is an indication of average molecular weight. This distance, the placement of the curve, and the shape of the molecular weight distribution, lead to the prediction of the physical and processing characteristics of the polymer. Any changes in the placement or shape of the curve reflects a change in the behavior of the polymer. An excessive amount of high molecular-weight fraction present in the polymer is indicated by the skewing of the curve towards the high end. Conversely, an excessive amount of low molecular-weight fraction is indicated by the skewing of the curve towards the low end. A discrete peak at the high or the low end of the distribution curve shows the change in behavior of the polymer.

The most direct—and sometimes the most informative—use of the GPC curve is in the comparison of different materials by overlaying their chromatograms on a light box. Differences in molecular weight distributions, peak shapes, shifts, and tailing are readily observable. Comparisons of additives and other lower molecular weight species are straightforward.

Frequently, a master chromatogram representing the acceptable range of GPC profiles is established with samples from "good" batches. All subsequent batches are then chromatographed and compared with the master curve as a rapid quality

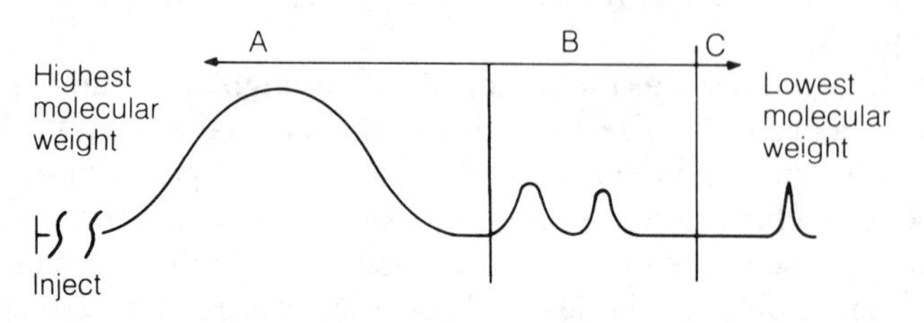

Figure 7-13. A typical chromogram showing emergence of molecules from the point of injection of polymer into a GPC equipment. (Courtesy Waters Associates, Inc.)

Figure 7-14. A sample material GPC curve is compared with control material GPC curve using a light box. (Courtesy Waters Associates, Inc.)

control method. Figure 7-14 illustrates one such use of the light box. A number of correlations between the GPC curve and the physical and processing behaviors of a polymer have been developed.

7.6. THERMAL ANALYSIS TECHNIQUES

Thermal analysis instrumentation has found wide acceptance in the plastics industry for quality control and basic material characterization. Thermal analysis consists of three primary techniques that may be used individually or in combination.

1. Differential scanning calorimetry (DSC).
2. Thermogravimetric analysis (TGA).
3. Thermomechanical analysis (TMA).

The maximum benefit of thermal analysis can be gained by using combination of all three techniques to characterize a polymer. The schematic representation of the three analyzers together with some of their most common uses is shown in Figure 7-15.

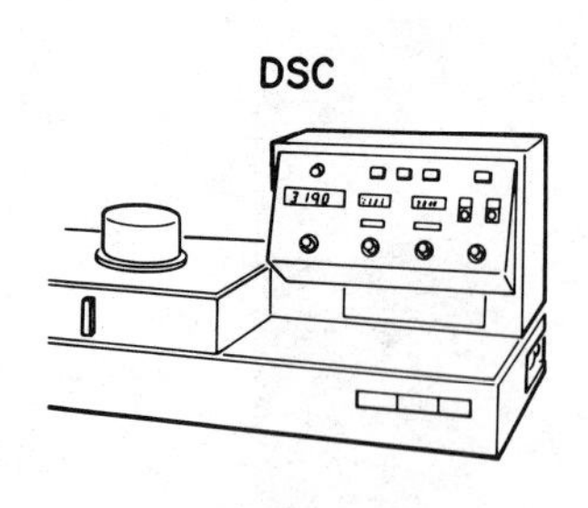

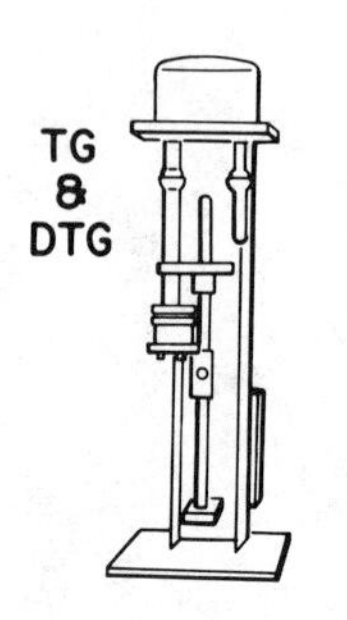

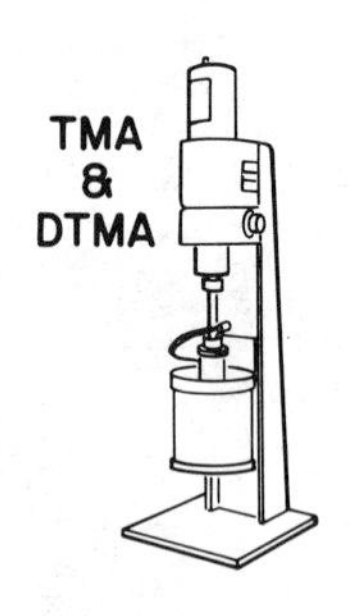

MEASURE HEAT FLOW

- MELTING PROFILE OR T_g
- PROCESSING ENERGY, C_P
- EXPLOSIVE HAZARD
- % CRYSTALINITY
- CURING PROFILE
- BLEND, COPOLYMER ANALYSIS
- ADDITIVE ANALYSIS (MOLD-RELEASE, ANTISTAT, IMPACT AND VISCOSITY MODIFIERS)

MEASURES WEIGHT CHANGES

- % VOLATILES - H_2O, SOLVENT, MONOMER
- % PLASTICIZER, OIL-EXTENDER
- % CARBON BLACK
- % CARBONATE
- % INERT MATERIAL
- DEGRADATION PROFILES
- THERMAL, OXIDATIVE STABILITY

MEASURES DIMENSIONAL CHANGES

- THERMAL EXPANSION
- SOFTENING POINT
- HEAT DEFLECTION TEMP.
- MODULUS, CREEP
- THERMAL REORIENTATION
- SHRINKAGE FROM MOLD

Figure 7-15. Thermal analysis of plastics. (Courtesy Perkin–Elmer Corporation.)

7.6.1. Differential Scanning Calorimetry (ASTM D 3417, ASTM D 3418)

DSC is a technique for calorimetric measurements either by recording and integrating the temperature difference or recording the energy necessary to establish zero temperature difference between a substance and a reference material against either time or temperature as the two specimens are subjected to identical temperature regimes in an environment heated or cooled at a controlled rate. DSC is a thermal analysis technique that measures the quantity of energy absorbed or evolved (given off) by a sample in calories as its temperature is changed. This is accomplished by heating a sample and an inert reference and measuring the difference in energy required to heat the two at a programmed rate. Initially, constant energy input is required to heat both the sample and the reference at a constant rate. At the transition point, the sample requires either more or less energy than the reference, depending on whether the change is endothermic or exothermic. For example, when a polymer reaches a melting point, it requires more energy than reference. This change is called endothermic. Conversely, when the curing takes place, more energy is released than reference and the change is considered exothermic. A typical DSC curve showing both endothermic and exothermic changes in a polymer is shown in Figure 7-16.

The test procedure is very simple. A small quantity of sample, usually 5–10 mg, is weighed out. The sample and reference are placed in a sample holder and heated at a predetermined constant rate in an inert atmosphere of nitrogen. The changes are recorded graphically for measurements of either temperature or energy differential against temperature or time. A commercial DSC apparatus is shown in Figure 7-17.

A polymer characterization with DSC technique is carried out by studying the

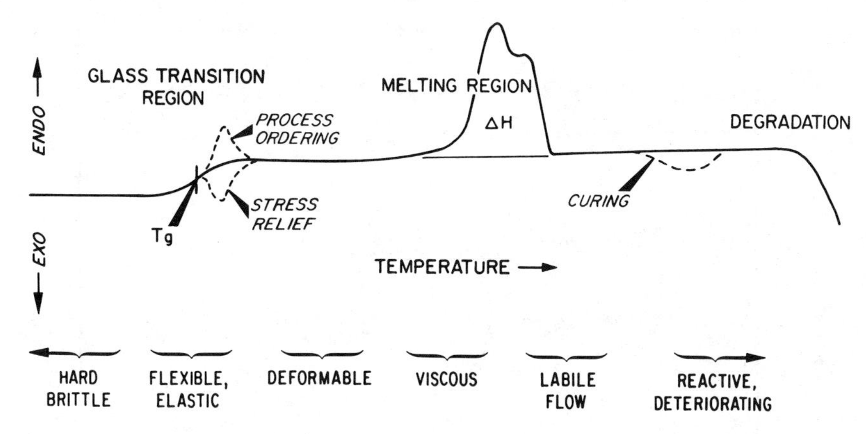

Figure 7-16. A typical DSC thermogram. (Courtesy Perkin–Elmer Corporation.)

melting behavior, crystalline melting point, degree of cure, percentage of additives in a polymer, oxidative stability, and degree of cross linking. The use of DSC in foam formulations (6,7) and epoxy-curing studies (8,9) is discussed in the literature.

7.6.2. Thermogravimetric Analysis (TGA)

Thermogravimetric analysis is a test procedure in which changes in weight of the specimens are recorded as the specimen is progressively heated. The sample weight is continuously monitored as the temperature is increased at a constant rate and components of a polymer that volatilize or decompose at different temperatures are quantitatively measured. A typical apparatus consists of an analytical balance supporting a platinum crucible for the specimen, the crucible situated in an electric furnace, and means for plotting the percentage weight change as a function of temperature. A typical TGA thermogram is illustrated in Figure 7-18 for the determination of mineral filler in polypropylene. The sample is heated in an air atmosphere to completely decompose the polypropylene such that the remainder is mineral filler. The amount of filler is read directly off the *y* axis which

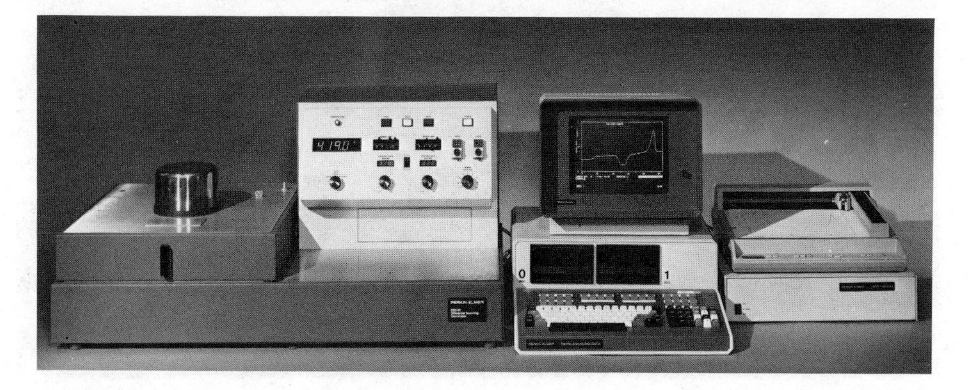

Figure 7-17. Differential scanning calorimeter. (Courtesy Perkin–Elmer Corporation.)

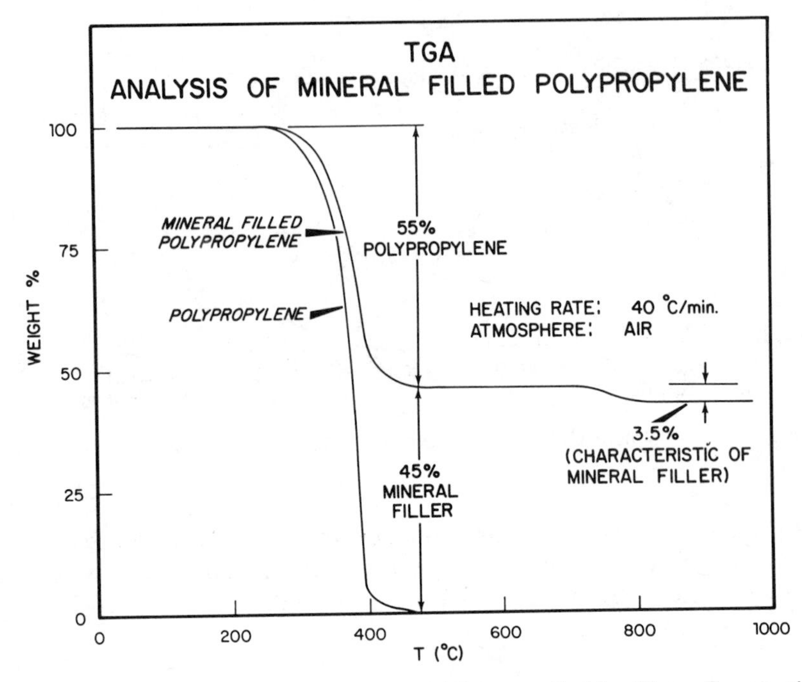

Figure 7-18. A typical TGA thermogram. (Courtesy Perkin–Elmer Corporation.)

is 45 percent in this case. The small weight loss (3.5 percent) occurring over the temperature range of 750°–800°C is characteristic of the particular mineral filler used in this sample. An unfilled polypropylene thermogram is also run to compare with the filled formulation.

TGA is very useful in characterizing polymers containing different levels of

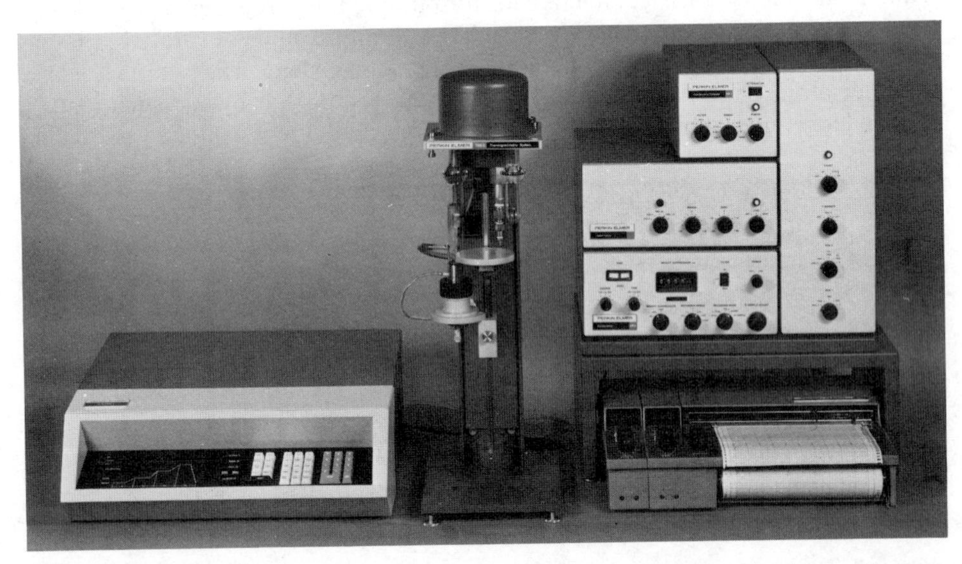

Figure 7-19. Thermogravimetric analysis equipment. (Courtesy Perkin–Elmer Corporation.)

additives by measuring the degree of weight loss. TGA can also be used to identify the ingredients of blended compounds according to the relative stabilities of individual components. Yet another way of characterizing polymers through the use of TGA is by studying the degradation profiles of various polymers. The use of TGA in determining the track resistance of epoxy compounds is described (10). A typical commercially available TGA instrument is shown in Figure 7-19.

7.6.3. Thermomechanical Analysis (TMA)

Thermomechanical analysis consists of measuring physical expansion or contraction of a material, or changes in its modulus or viscosity as a function of temperature. TMA equipment consists of (Figure 7-20) a probe connected mechanically to the core of a linear variable differential transformer (LVDT) (11). Samples to be examined are placed in a sample holder tube that is surrounded

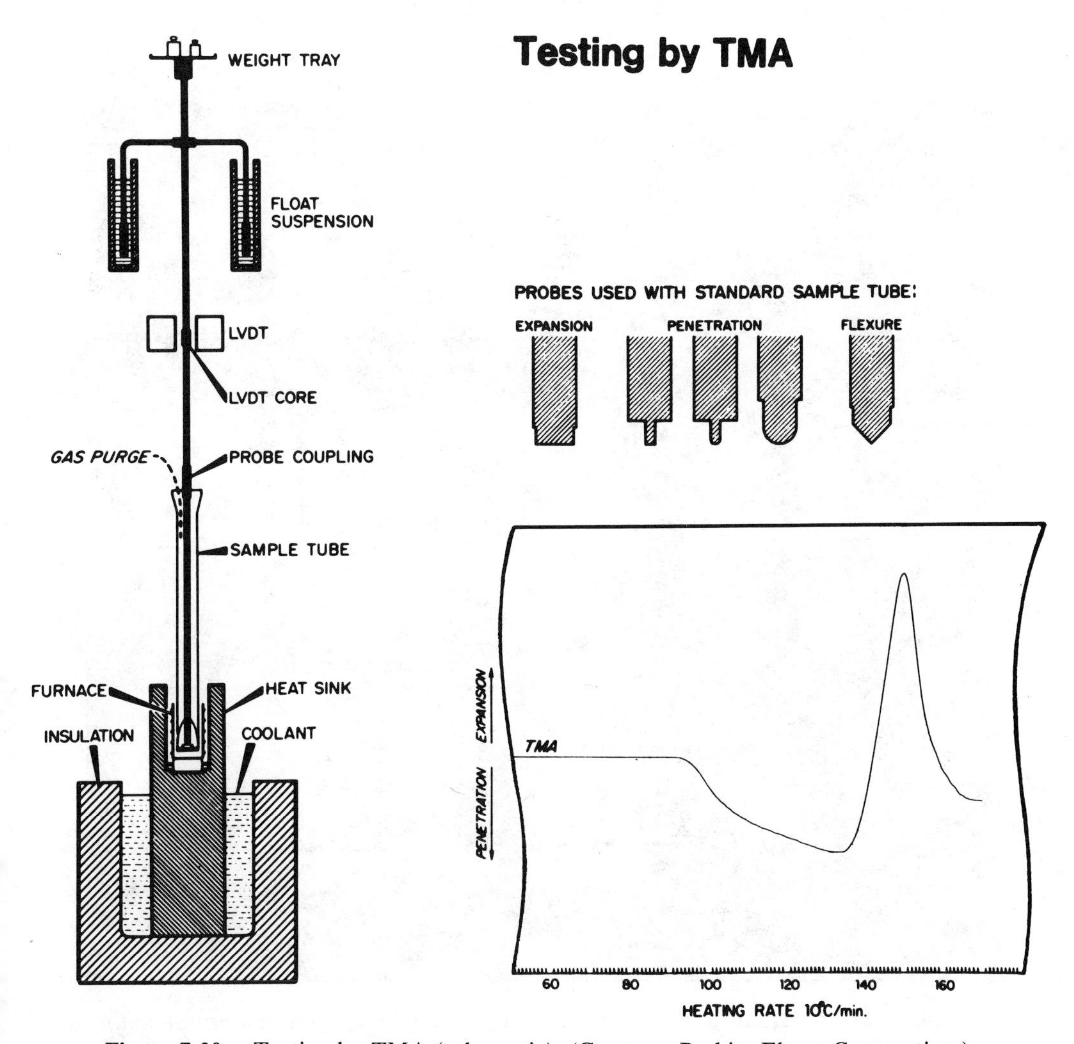

Figure 7-20. Testing by TMA (schematic). (Courtesy Perkin–Elmer Corporation.)

by a heating furnace and thermocouple. A fused silica probe is used as a compression probe. The probe is placed onto the sample and thermal expansion or contraction is transferred, amplified, and recorded on the chart recorder as a function of temperature. Commercial TMA equipment is illustrated in Figure 7-21. A typical TMA thermogram is shown in Figure 7-22. The polymer used in the test is modified with a chemical blowing agent. Over the first 60°C, the TMA probe is slightly pushed up by the expansion of solid. As the melting point is reached, the lightly loaded probe penetrates the sample at a rate inversely related to the degree of crosslinking. The slope "s" in the figure corresponds to the crosslink density. This penetration is later reversed by the expansion of the foam as the blowing agent decomposes. The polymer melting temperature, crosslink density, and foaming temperature can be read directly from the TMA trace.

Polymer characterization through the use of TMA is quite common. This is accomplished by determining the glass transition temperature, the coefficient of expansion, and the elastic modulus. TMA data also correlates with the Vicat softening point and the heat distortion temperature.

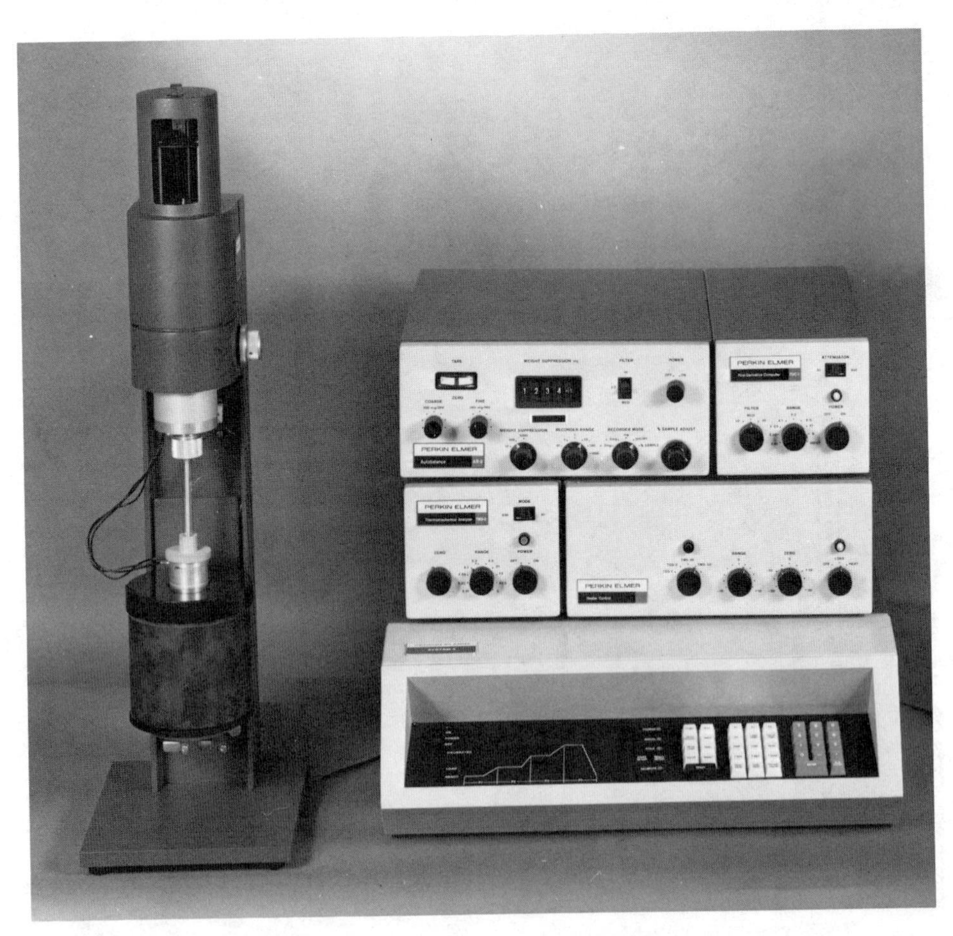

Figure 7-21. Thermomechanical analyzer. (Courtesy Perkin–Elmer Corporation.)

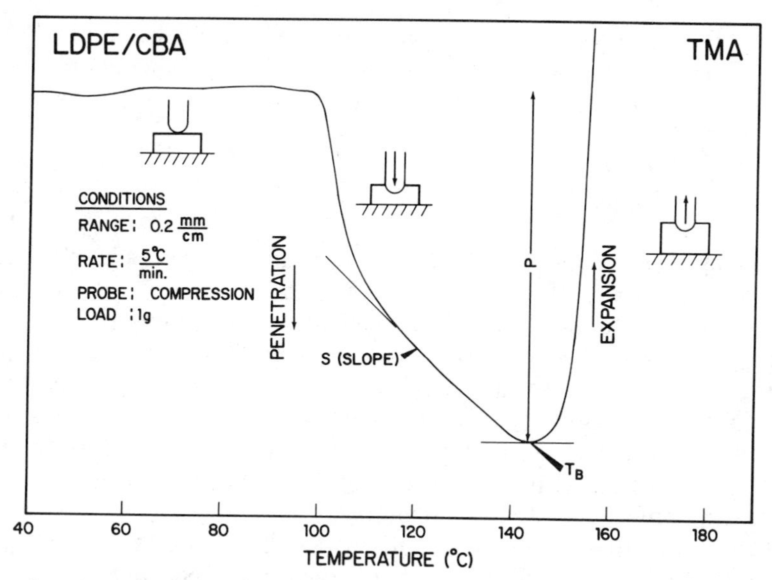

Figure 7-22. TMA analysis of a fully formulated foam shows that the compression probe penetrates the sample during melting, and then is pushed back up by the expanding foam when CBA decomposes. Note that this one test provides a check on resin melting temperature, cross link density, and foaming temperature. (Courtesy Perkin–Elmer Corporation.)

7.7. MATERIAL CHARACTERIZATION TESTS FOR THERMOSETS

Thermoplastic materials are usually purchased in a ready to process form. Most thermplastic materials do not require further treatment in terms of blending and compounding except for the addition of color. The material supplier is generally expected to provide quality resin and only a few characterization tests are conducted to determine the uniformity of the resin as discussed earlier in this chapter.

Thermoset materials, unlike thermoplastics, are purchased in an unpolymerized intermediate chemical stage. The polymerization takes place during processing. In order to ascertain the quality of the raw materials, material characterization tests are routinely conducted by thermoset processors.

7.7.1. Apparent (Bulk) Density, Bulk Factor, and Pourability of Plastic Materials (ASTM D 1895)

The information regarding the bulk factor of the material is very important to a molder prior to molding for several reasons. The bulk factor of the material directly affects the preform size, the loading space in the cavities, and the design of the loading trays. Any change in the bulk factor can cause serious molding problems such as porous or incomplete moldings since the molding charge is measured rather than weighed. The bulk factor is defined as the ratio of the volume of any given quantity of the loose plastic material to the volume of the same quantity of the material after molding or forming. The bulk factor is also equal to the ratio of density after molding or forming to the apparent density of material as received.

Bulk factor is a measure of volume change that may be expected in fabrication. Apparent density is defined as the weight per unit volume of a material including voids inherent in the material as tested. Apparent density is a measure of the fluffiness of a material.

A. *Determination of Apparent Density*

This is a simple test requiring a measuring cup and a funnel of a specified size. The funnel is closed from the small end with the hand or a suitable flat strip. The material is poured into the funnel. After the material is poured, the bottom of the funnel is quickly opened and the material is allowed to fall into a cylindrical measuring cup. The excess material laying on top of the cup is immediately scraped off with a straight edge without shaking the cup. The material in the cup is weighed to the nearest tenth of a gram. Apparent density values are reported as g/cm^3 or lbs/ft^3. This method is used primarily to determine the apparent density of free-flowing powders and fine granules.

An alternate method requires a funnel with a larger diameter opening at the bottom so that a coarse, granular material in the form of pellets or dice can easily pass through. The procedure and calculations are carried out in a similar manner.

Yet another variation of the same test is used for measuring the apparent density of coarse flakes, chips, cut fibers, or strands. Such materials cannot be poured through the funnels described in methods A and B. Also, since they ordinarily are very bulky when loosely poured and since they usually are compressible to a lesser bulk, even by hand, a measure of their density under a small load is appropriate. An apparatus for all three apparent density tests are shown in Figure 7-23.

B. *Determining Bulk Factor*

The density of molded or formed plastic material is determined as described in Chapter 10, Section 10.3. The calculation is carried out as follows:

$$\text{Bulk factor} = \frac{\text{Average density of molded part}}{\text{Average apparent density of material prior to molding}}$$

C. *Pourability*

Pourability is defined as a measure of the time required for a standard quantity of material to flow through a funnel of specified dimensions. Pourability characterizes the handling properties of finely divided plastic material. The procedure simply calls for pouring the material through a specified size funnel and measuring the time required for material to completely pass through the funnel.

7.7.2. Flow Tests

The ability of the material to flow is measured by filling a mold with the plastic material under a specified condition of applied temperature and pressure with a controlled charge mass. The flow tests are used as a quality control test and as an acceptance criterion for incoming raw materials.

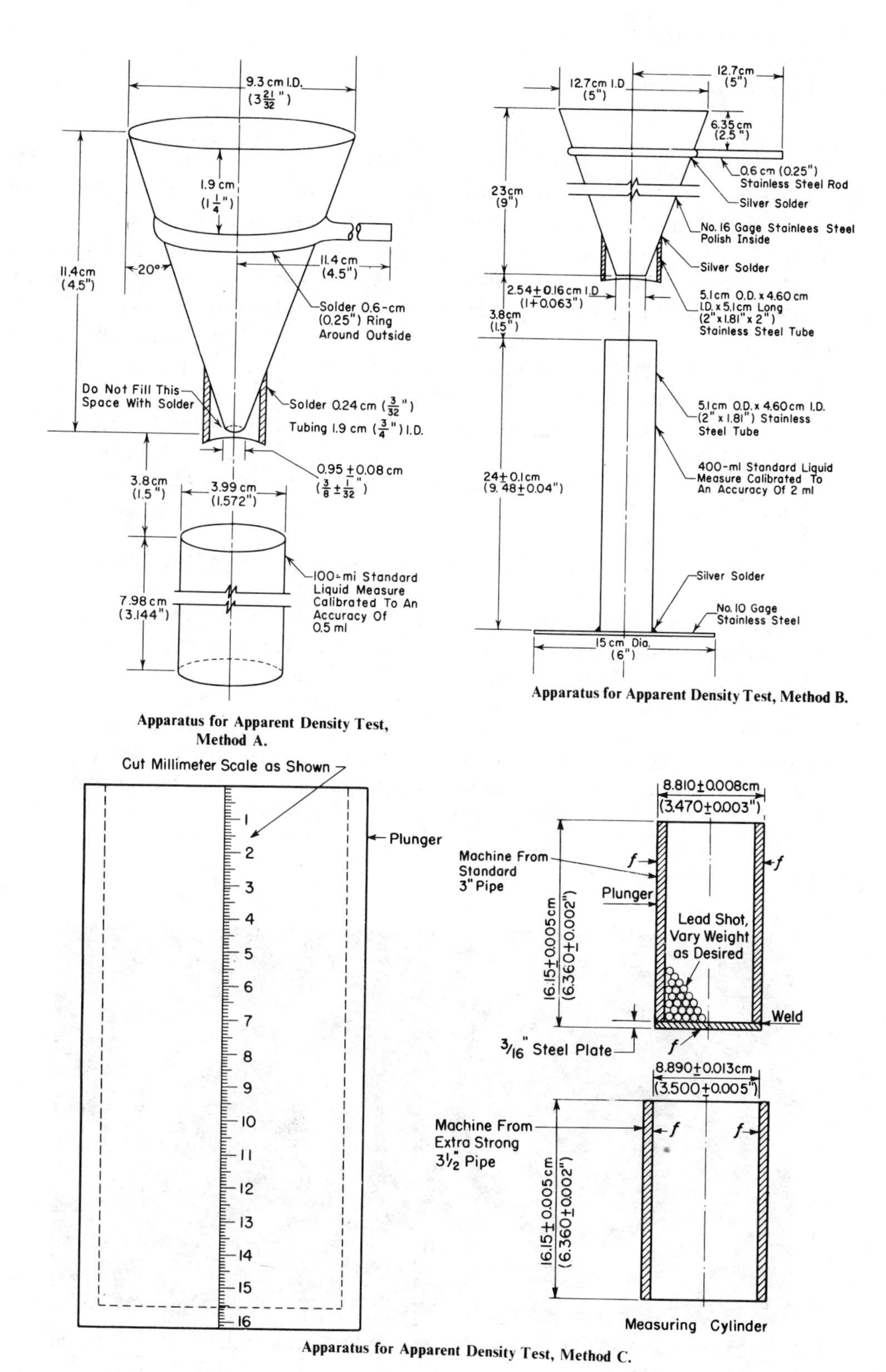

Apparatus for Apparent Density Test, Method A.

Apparatus for Apparent Density Test, Method B.

Apparatus for Apparent Density Test, Method C.

Figure 7-23. Apparatus for apparent density test. (Reprinted by permission of ASTM.)

A. *Factors Affecting Flow*

RESIN TYPES: All resins flow differently because of basic differences in the structure of the polymers. For example, melamine formaldehyde exhibits longer flow than urea formaldehyde. Phenolics, because of the variety of resin types, enable the molder to select the flow best suited for a particular design.

TYPE OF FILLERS: The small particle size of wood flour, mica, and minerals creates less turbulence and less frictional drag during mold filling. The size of the glass fibers, short or long, can adversely affect the flow.

DEGREE OF RESIN ADVANCEMENT: The degree of advancement is generally controlled by the resin manufacturers. Molders can advance resin polymerization with oven or radiant heat or electronic preheating.

STORAGE TIME: All resins have a natural tendency to polymerize in storage, causing partial precure which reduces flow. An exception might be a polyester in which catalyst decomposition slows or prevents curing which increases flow duration. (12)

B. *Spiral Flow of Low-Pressure Thermosetting Compounds (ASTM D 3123)*

The spiral flow of a thermosetting molding compound is a measure of the combined characteristics of fusion under pressure, melt viscosity, and gelation rate under specific conditions. The test requires a transfer molding press, a standard spiral flow mold, and a thermosetting molding compound. The molding temperature, transfer pressure, charge mass, press cure time, and transfer plunger speed are preselected as specified. The preconditioned compound is forced through a sprue into a spiral flow mold. Once the curing is complete, the part is removed and the spiral flow length is read directly from the molded specimen. Compounds are classified as low (1–10), medium (11–22), and high (23–40) plasticity. A typical moded specimen is shown in Figure 7-24.

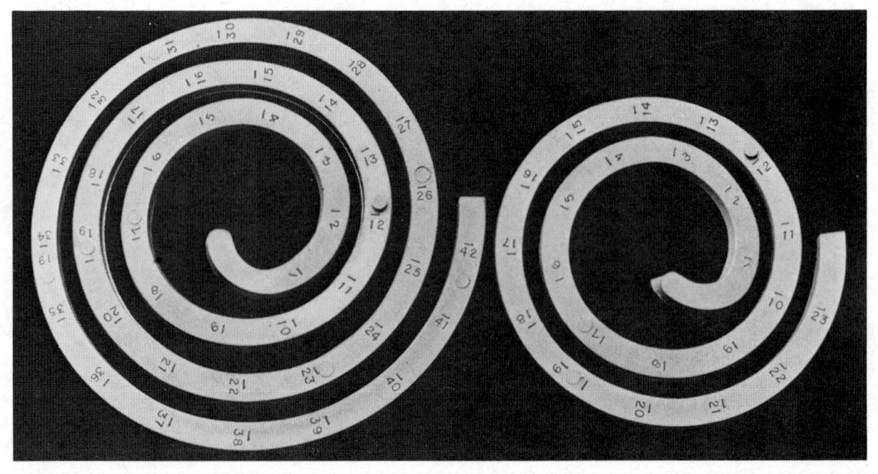

Figure 7-24. Spiral flow test specimens. (Courtesy B. F. Goodrich Chemical Company.)

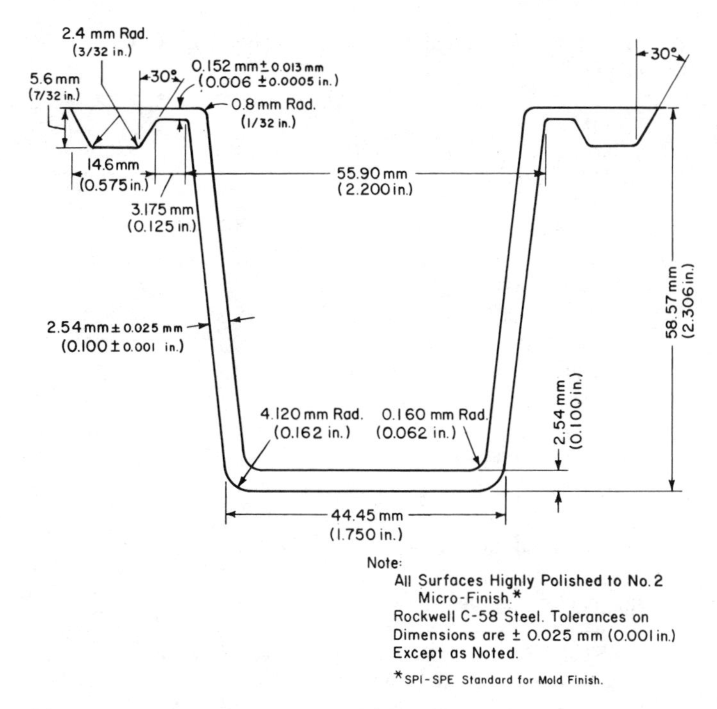

Figure 7-25. Cup mold. (Reprinted by permission of ASTM.)

C. Cup Flow Test (ASTM D 731)

MOLDING INDEX OF THERMOSETTING MOLDING POWDER. This test is primarily useful for determining the minimum pressure required to mold a standard cup and the time required to close the mold fully. The preconditioned and preweighed material is loaded into the mold. The mold is closed using sufficient pressure to form a required cup. The pressure is reduced step by step until the mold cannot close. The next higher pressure and time to close the mold is reported as the molding index of the material. Figure 7-25 illustrates a cup mold to produce a molded cup.

7.7.3. Viscosity Tests for Thermosets

Viscosity tests are very important in assessing the quality of incoming thermosetting materials. Variation in the viscosity can occur because of chemical changes due to storage conditions, manufacturing variations, and the presence of impurities. Viscosity limits are of particular interest in achieving uniform impregnation and wetting of fiber/fabric reinforcements, and flow during molding and curing operations. Viscosity measurements also serve as an indicator of working life in the case of catalyzed resin systems. The increase in viscosity indicates advancing stages of polymerization or solvent evaporation.

Table 7-3. Recommended Spindles for Brookfield RVF Viscometer[a]

Range (cP)[b]	Spindle	Speed (rpm)[c]	Factor
100–400	1	20	5
400–800	1	10	10
800–1,600	2	20	20
1,600–3,200	2	10	40
3,200–4,000	3	20	50
4,000–8,000	4	20	100
8,000–16,000	4	10	200
16,000–20,000	3	4	250
20,000–40,000	4	4	500
40,000–80,000	4	2	1,000
80,000–160,000	5	2	2,000
160,000–200,000	6	4	2,500
200,000–400,000	6	2	5,000
400,000–800,000	7	4	10,000
800,000–2,000,000	7	2	20,000

[a] From ASTM D 2393. Reprinted by permission of ASTM.
[b] To obtain the viscosity in centipoises, multiply the reading on the "100" scale by the factor for the given spindle and speed.
[c] If the scale reading is below 20 or above 80, move to the spindle and speed recommended for the next lower or higher viscosity range.

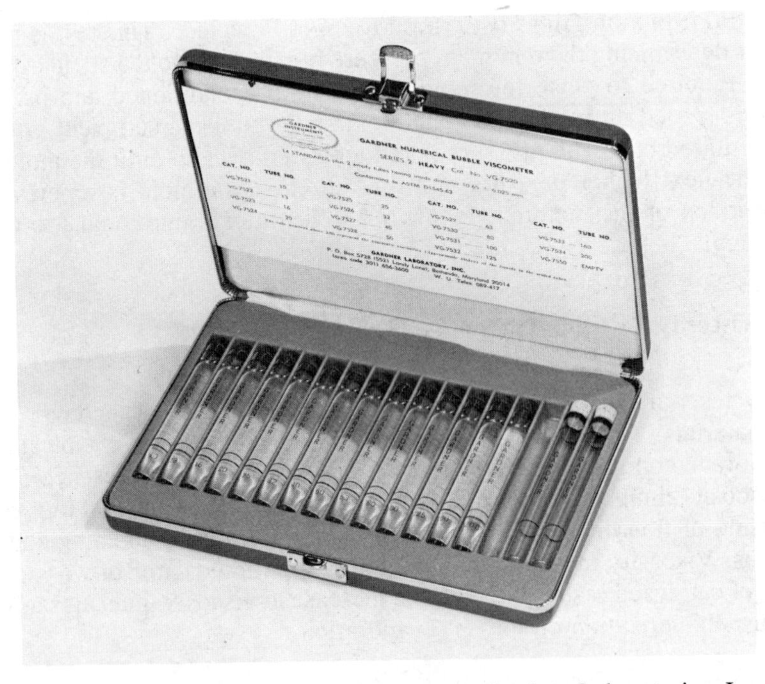

Figure 7-26. Bubble viscometers. (Courtesy Gardner Laboratories, Inc.)

A. *Bubble Viscometer*

In a bubble viscometer, a liquid streams downward in the ring-shaped zone between the glass wall of a sealed tube and a rising air bubble. The rate at which the bubble rises is a direct measure of the kinematic viscosity. The rate of bubble rise is compared with a set of calibrated bubble tubes containing liquids of known viscosities. Bubble viscometers are shown in Figure 7-26.

B. *Brookfield Viscometers (ASTM D 2393: ASTM D 1824)*

Brookfield viscometers are used for measuring viscosity based on the principles of rotational rheology. Brookfield model RVF (Figure 7-27) which was designed for materials in the medium viscosity region is the most commonly used viscometer for plastic materials. The sample is preconditioned by placing it in a constant temperature bath at the specified test temperature. The proper sized spindle is allowed to rotate in the sample for 30 sec. The instrument is stopped through the use of a clutch and the reading is taken from the dial. The test is repeated until a constant reading is obtained. The reading obtained from the dial is converted in centipose using a conversion table as shown in Table 7-3.

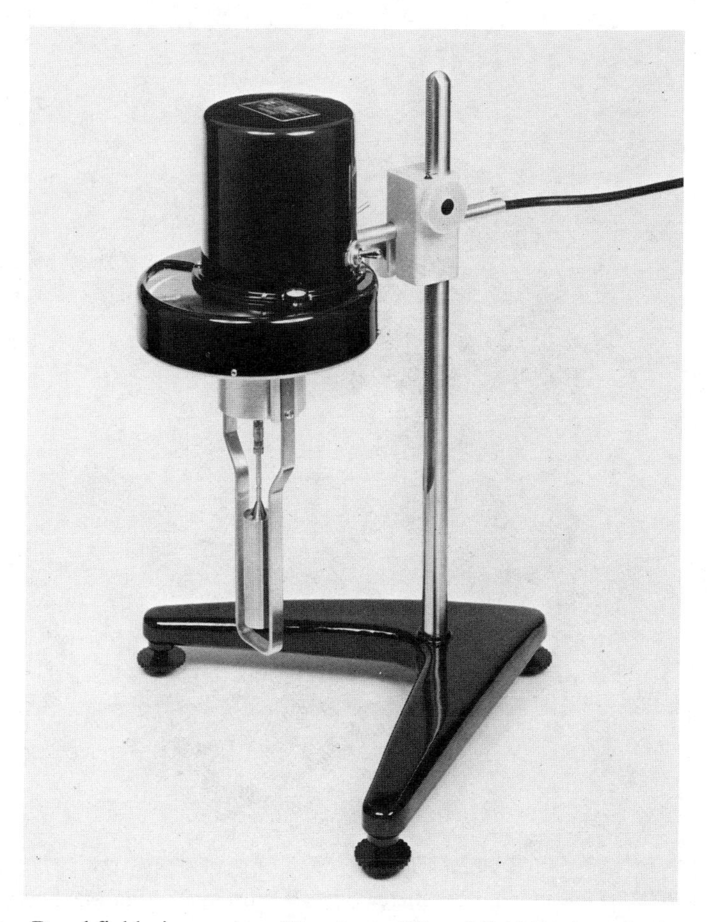

Figure 7-27. Brookfield viscometer. (Courtesy of Brookfield Engineering Laboratories.)

7.7.4. Gel Time and Peak Exothermic Temperature of Thermosetting Resins (ASTM D 2471)

Gel time and peak exothermic temperature are two of the most important parameters for a thermosetting material processor. Gel time is the interval of time between introduction of the catalyst and the formation of gel. Such information regarding viscosity change with time of resin–catalyst mixture helps to determine working-life characteristics of the material. The maximum temperature reached by reacting thermosetting plastic composition is called peak exothermic temperature. Resin producing high exothermic heat is more susceptible to cure shrinkage and craze cracking. This is due to the thermal expansion that occurs as the heat is generated and shrinkage that takes place when the thermosetting three-dimensional network forms. In the laminating process, excessive heat build-up during cure tends to contain weak interlaminar bonds and hence, relatively poor physical properties. To insure against storage changes and batch variations, the processor should measure gel characteristics before using the materials.

The test method requires sample containers, a wooden probe, a constant temperature bath, a stop watch, and a temperature-measuring device. All the components are placed in the temperature-controlled bath for a specified amount of

Figure 7-28. Gel time meter, stirrer type. (Courtesy Testing Machines, Inc.)

time and temperature. When all the components have reached the test temperature, they are combined in the recommended ratio. The start of mixing is recorded as starting time. Components are agitated with a stirrer. Simultaneously, the temperature of the reactants are monitored and recorded. Every 15 sec, the center surface of the reacting mass is probed with an applicator stick. When the reacting material no longer adheres to the end of a clean probe, the "gel time" is recorded as the elapsed time from the start of mixing. The time and temperature recording is continued. The highest temperature is recorded as "peak exothermic temperature." "Peak exothermic time" is recorded as the elapsed time from the start of mixing.

Commercially available mechanical gel time meters or gel timers can be used in place of manual stirring to determine gel time. There are various types of gel time meters available in the market. One such version measures a liquid's gel point by dropping a definite weight into a sample once per minute. Once the gelation point is reached, the weight does not sink during the downstroke. At this point, the equipment automatically turns off and records the elapsed time from the start of the experiment. Yet another version of a gel timer (Figure 7-28) consists of a motor-driven stirrer. As the gelation commences, drag soon exceeds torque, and the motor stalls. Gel time can then be directly read off the counter to the nearest 0.1 min. Gel time of resins at elevated temperature can also be determined by using a thermostatically controlled heated pot in place of a regular pot.

REFERENCES

1. Slysh, R. and Guyler, K. E., "Prediction of Diallyl Phthalate Molding Performance from Laboratory Tests," *SPE ANTEC*, **23** (1977), p. 4.
2. Gardner Laboratories, *Tech. Bull.: Rheology*, Silver Spring, Md., Sept 1976.
3. Parks, R. A., "Re-evaluation of Dilute Solution Viscosity Test Methods Proves Boon to PVC Quality Control," *Plast. Design and Processing* (Aug. 1974), pp. 24–25.
4. McKinney, P. V., *J. Polym. Sci.*, **9** (1965) pp. 583–587.
5. Parks, Reference 3, p. 26.
6. May, W. P., "New Test Methods for Plastisol Foams," *Plast. Tech.* (June 1977), p. 97.
7. Breakey, D. and Cassel B., "What Foam Processors Should Know about Thermal Analysis Techniques," *Plast. Tech.* (Nov. 1979), p. 75.
8. Abofalia, O. R., "Application of DSC to Epoxy Curing Studies," *SPE ANTEC* **15** (1969), p. 610.
9. Prime, R. B., "DSC of Epoxy Cure Reaction," *SPE ANTEC* **19** (1973), p. 205.
10. Jaegers, G. and Gedmer, T. J., "The Use of TGA in Determining Track Resistance of Epoxy Compounds," *SPE ANTEC* **16** (1970), p. 450.
11. Millder, G. W., and Cassey, D. L., "A Versatile Thermomechanical Analyzer," *SPE ANTEC* **15** (1969), p. 475.
12. Milby, R. V., *Plastics Technology*, McGraw–Hill, New York, 1973, pp. 457–474.

GENERAL REFERENCES

Waters Associates, *Tech. Bull.: Gel Permeation Chromatography.*

Instron, *Tech. Bull.: Rheological Equipments.*

Perkin–Elmer, *Tech. Bull.: Thermal Analysis.*

SUGGESTED READING

Sward, G. G. (Ed.), *Paint Testing Manual*, American Society for Testing and Materials, Philadelphia, Pa., 1970.

Shida, M. and Cancino, L. V., "Prediction of High Density Polyethylene Resin Processibility from Rheological Measurements," *SPE ANTEC*, **16** (1970), pp. 620–624.

Macosko, C., and Starita, J. M., "Polymer Characterization with a New Rheometer," *SPE ANTEC*, **17** (1971), p. 595–600.

Terry, B. W., and Yang, K., "A New Method for Determining Melt Density as a Function of Pressure and Temperature," *S.P.E. J.* **20**(6), p. 37 (June 1964).

Merz, E. H. and Colwell, R. E., "A high Shear Rate Capillary Rheometer for Polymer Melts," *ASTM Bull.* No. 232, (Sept. 1958).

Mendelson, R. A., "Melt Viscosity," *Encyclopedia of Polymer Science and Technology*, Vol. 8, John Wiley & Sons, New York, 1970, pp. 587–619.

Bernhardt, E. C., *Processing of Thermoplastic Materials*, Reinhold, New York, 1959.

McKelvey, J. M., *Polymer Processing*. John Wiley & Sons, New York, 1960.

Van Wurzer, J. R., Lyons, J. W., Kim, K. Y., and Colwell, R. E., *Viscosity and Flow Measurement, A Laboratory Handbook of Rheology*. John Wiley & Sons, New York, 1963.

Hertel, D. L., and Oliver, C. K., "Versatile Capillary Rheometer," *Rubber Age* (May 1975).

Miller, B., "Why Good Resin Makes Bad Parts—and What you Can Do about It," *Plast. World* (Feb. 1976), p. 44.

Ekmanis, J., and Church, S., "Simple Test of Incoming Resins Rates Batch to Batch Quality Level," *Plast. Design and Processing* (March 1977), pp. 30–34.

Willard, P. E., "Determination of Cure of DAP Using DSC," *SPE ANTEC*, **17** (1971), pp. 464–468.

Perkin–Elmer, Thermal Analysis Literature, *TA Application Studies Bulletin*: TAAS-19, "Characterization of Thermosets"; TAAS-20, "Polymer Testing by TMA"; TAAS-22, "Characterization and Quality Control of Engineering Thermoplastics by Thermal Analysis"; TAAS-25, "Applications of TA in the Electrical and Electronic Industries"; TAAS-26, "Applications of TA in the Automotive Industries"; TAAS-29, "Use of Thermal Analysis Method in Foam Research and Development".

Cassel, B., and Gray, A. P., "Thermal Analysis Simplifies Accelerated Life Testing of Plastics," *Plast. Eng.* (May 1977), pp. 56–58.

Cassel, B., and Breakey, D., "What Foam Processors Should Know About Thermal Analysis Techniques," *Plast. Tech.* (Nov. 1979), pp. 75–78.

Slade, P. E., and Jenkins, L. T., (Eds.), *Techniques and Methods of Polymer Evaluation*, Vol 1, *Thermal Analysis*, 1966; Vol. 2, *Thermal Characterization Techniques*, 1970, Marcel Dekker, New York.

CHAPTER 8

Flammability

8.1. INTRODUCTION

Because of its increased use in homes, buildings, appliances, automobiles, aircraft and many other sectors of our lives, plastic materials have been under considerable pressure to perform satisfactorily in situations involving fire. A good deal of time and money has already been spent on research and in-depth study of the behavior of polymeric materials exposed to fire. Many new tests for preliminary screening of materials and simulating actual fire have been developed. Before getting into a detailed discussion on tests and testing procedures, it is necessary to understand polymers as they relate to flammability.

When a polymeric material is subjected to combustion, it undergoes decomposition which produces volatile polymer fragments at the polymer surface. The fuel produced in this process diffuses to the flame front, where it is oxidized, producing more heat. This, in turn, causes more material decomposition. A cyclic process is established, solid material is decomposed, producing fuel which burns, giving off more heat, which results in more material decomposition (1). This process is illustrated in Figure 8-1.

To reduce the flammability of a material, this cycle must be attacked in either the vapor phase or at the solid material surface. In the vapor phase, the cycle can be inhibited by adding certain additives to the polymer that disrupt the flame chemistry when vaporized. Bromocompounds and chlorocompounds with antimony oxide operate in this manner and are commonly used in polystyrene or ABS structural foams. Solid phase inhibition may be achieved by including additives in the polymer that promote the retention of fuel as carbonaceous char as well as providing a protective insulating layer. This layer prevents further fuel evolution. Such an approach is effective in polycarbonate and polyphenylene oxide-based structural foams. Other solid phase approaches involve the use of heat sinks, such as hydrated alumina, which absorb heat and release water of hydration when heated, or alter the decomposition chemistry to consume additional heat in the decomposition process (2).

Polymer's inherent flammability can be divided into basic classes as listed in Table 8-1. The first group consists of inherently flame retardant structures containing either halogen or aromatic groups that confer high thermal stability as well as the ability to form char on burning. The second group of materials is relatively

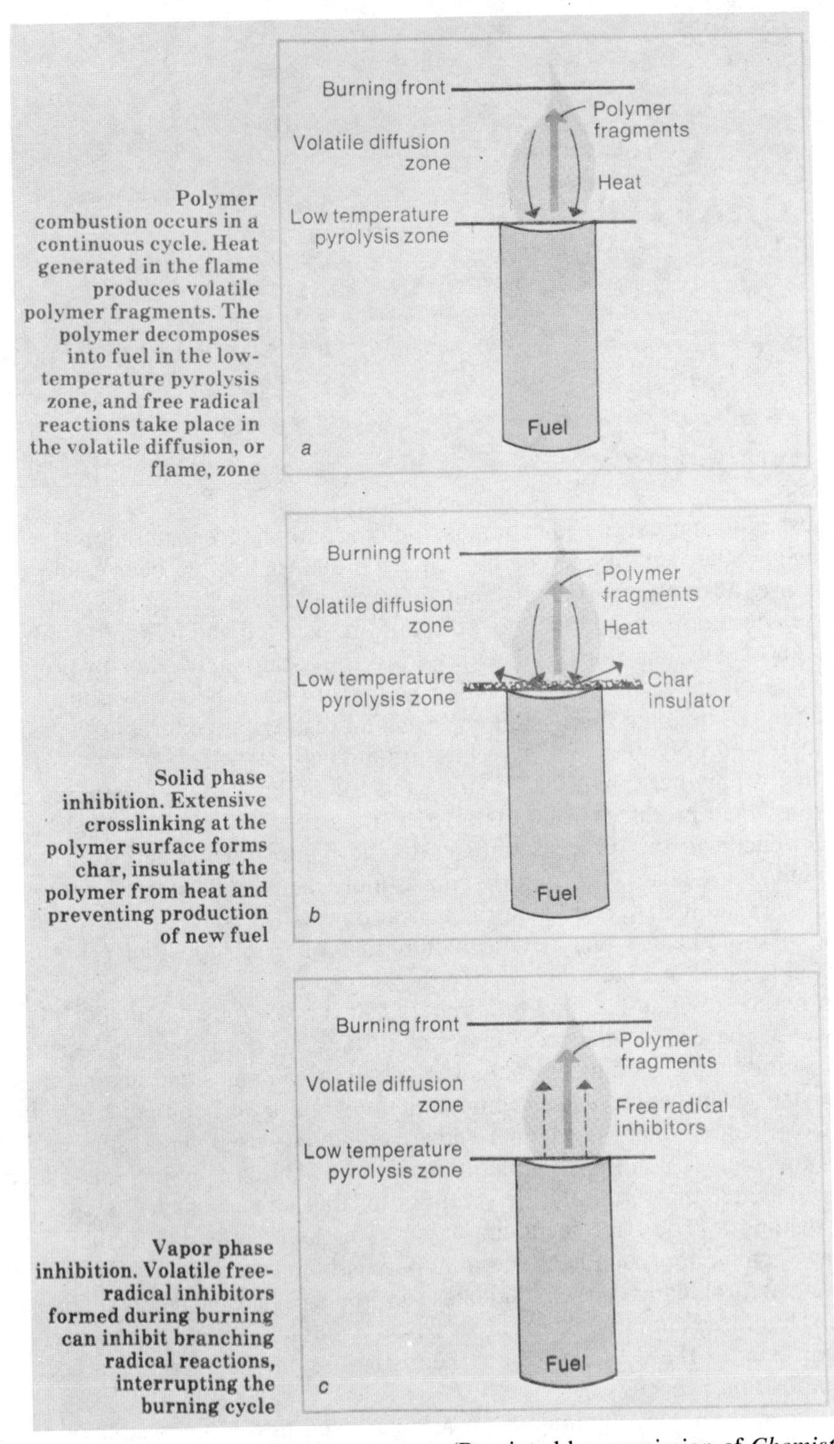
Polymer combustion occurs in a continuous cycle. Heat generated in the flame produces volatile polymer fragments. The polymer decomposes into fuel in the low-temperature pyrolysis zone, and free radical reactions take place in the volatile diffusion, or flame, zone
Burning front
Polymer fragments
Volatile diffusion zone
Heat
Low temperature pyrolysis zone
Fuel
a
Solid phase inhibition. Extensive crosslinking at the polymer surface forms char, insulating the polymer from heat and preventing production of new fuel
Burning front
Polymer fragments
Volatile diffusion zone
Heat
Low temperature pyrolysis zone
Char insulator
Fuel
b
Vapor phase inhibition. Volatile free-radical inhibitors formed during burning can inhibit branching radical reactions, interrupting the burning cycle
Burning front
Polymer fragments
Volatile diffusion zone
Free radical inhibitors
Low temperature pyrolysis zone
Fuel
c

Figure 8-1. Polymer combustion process. (Reprinted by permission of *Chemistry*.)

Table 8–1. Polymers and Flammability[a]

Inherently Flame Retardant
Polytetrafluoroethylene
Aromatic polyethersulfone
Aromatic polyamides
Aromatic polyimides
Aromatic polyesters
Aromatic polyethers
Polyvinylidene dichloride
Less Flame Retardant
Silicones
Polycarbonates
Polysulfone
Quite Flammable
Polystyrene
Polyacetal
Acrylics
Polyethyleneterpthalate
Polypropylene
Polyethylene
Cellulose
Polyurethane

[a] Reprinted from *Chemistry*, June 1978, p. 23.

less flame retardant. Many of the second group can be made more flame retardant by the use of appropriate flame retardants. The third group consists of quite flammable polymers that are difficult to make flame retardant because they decompose readily, forming large quantities of fuel (3).

No discussion on polymer's flammability can be considered complete without discussing the formation of smoke and the generation of toxic gases. Smoke impairs the ability of occupants to escape from a burning structure as well as the ability of fire fighters to carry out rescue operations. All polymers, to some extent, produce smoke and toxic gases, although some polymers inherently generate more smoke and toxic gases than others. Many tests have been developed to measure smoke density and toxicity.

The material's ability to burn depends upon fire conditions as well as polymer composition. Actual fire conditions are difficult to simulate and therefore we are forced to rely upon small and large-scale laboratory tests to predict combustibility, smoke density and toxicity. The flammability of materials is influenced by several factors. (4)

Ease of ignition—how rapidly a material ignites.

Flame spread—how rapidly fire spreads across a polymer surface.

Fire endurance—how rapidly fire penetrates a wall or barrier.

Rate of heat release—how much heat is released and how quickly.

Ease of extinction—how rapidly the flame chemistry leads to extinction.

Smoke evolution.

Toxic gas generation.

Table 8-2. Flammability Tests

Sponsoring organizatin	Name of test	Procedure identification number	Specimen size (in.)	Number of specimens	Angle of specimen	Ignition source	Properties measured
ASTM	Rate of burning flexible plastics 0.050-in. and under	D-568	1×18	3	Vertical	Bunsen burner	Burning rate, average time and extent of burning
ASTM	Rate of burning rigid plastics over 0.050-in.	D-635	$\frac{1}{2} \times 5 \times$ thickness	10	Horizontal	Bunsen burner	Burning rate, and average time and extent of burning
ASTM	Incandescence resistance of rigid plastics	D-757	$\frac{1}{2}$(W) $\times$ 4.75(L) $\times$ $\frac{1}{4}$(T)	3	90° to source	Silicone carbide incandescent rod	Burning rate, and burning time and extent of burning
ASTM	Ignition properties of plastics Procedure B	D-1929	3 g wt. sheet size $\frac{3}{4} \times \frac{3}{4} \times t$	14	Horizontal	Hot air ignition furnace	Flash and self-ignition temperatures Visual characteristics
ASTM	Oxygen index flammability test	D-2863	$\frac{1}{4} \times \frac{1}{8} \times 5$	10	Vertical	Propane flame in oxygen and nitrogen atmosphere	Oxygen index

ASTM	Surface burning characteristics of building material	E-84	20 × 300	1	Horizontal	Gas/air mixture	Flame spread index Smoke density
ASTM	Flame height, time of burning, and loss of weight Cellular plastics Vertical position	D-3014	10 × $\frac{3}{4}$ × $\frac{3}{4}$	6	Vertical	Gas burner	Loss of weight and time to extinguishment
ASTM	Rate of burning cellular plastics horizontal position	D-1692	. . .	. . .	. . .	. . .	Discontinued
ASTM	Rate, extent, and time of burning of flexible thin plastic sheeting	D-1433	3 × 9	10	45°	Butane burner	Burning rate, extent and time of burning, visual characteristics
Federal Test Method Standard	Flame resistance of difficult to ignite plastics	FTMS 406 Method 2023	5 × $\frac{1}{2}$ × $\frac{1}{2}$	5	Vertical	Electric coil and spark plugs	Ignition time, burning time, flame travel
UL	Flammability of plastic materials	UL 94V-0	$\frac{1}{2}$ × 5 × thickness	5	Vertical	Bunsen burner	Rate of burning
UL	Flammability of plastic materials	UL 94 HB	$\frac{1}{2}$ × 5 × thickness	3	Horizontal	Bunsen burner	Rate of burning

Table 8-3. Smoke Evolution Tests

Sponsoring organization	Name of test	Procedure identification number	Specimen size (in.)	Number of specimens	Angle of specimen	Ignition source	Properties measured
ASTM	Smoke density from burning of plastics	D-2843	$1 \times 1 \times \frac{1}{4}$	3	Horizontal	Propane burner	Percent light absorption, smoke density
NBS	NBS smoke test	NFPA 258	3×3	6	Vertical	Electric furnace	Smoke density
Arapahoe Chemical Company	Arapahoe smoke test	–	$1\frac{1}{2} \times \frac{1}{2} \times \frac{1}{8}$	6	Horizontal	Micro-bunsen burner	Percent smoke
ASTM	Radiant panel test	E-162	$6 \times 18 \times$ thickness	4 minimum	60°	—	—
OSU	OSU release rate test	—			Horizontal or Vertical	Gas fired radiant panel or electric globar	Maximum smoke release rate

Table 8-2 lists existing flammability tests and other pertinent information. Table 8-3 lists smoke evolution tests and related information.

8.2. FLAMMABILITY TEST (FLEXIBLE PLASTICS) (ASTM D 568)

This test was developed basically for flexible thin sheets or films of plastics with thickness of 0.050-in. and under. The method covers a small-scale laboratory screening procedure for comparing the relative rate of burning and/or extent and time of burning of plastics tested in the vertical position. The test is primarily useful for screening, comparing, quality control, and meeting specification requirements in terms of flammability of plastics.

The test requires the use of a Bunsen burner, a clamp for holding the test specimen, a safety shield, and a timing device such as stop watch. The test specimen of size $\frac{1}{2} \times 5 \times$ thickness of the specimen as prepared is used. A gauge mark is drawn across the specimens so that the burning rate can be measured over the gauge length. After appropriate conditioning, the specimen is clamped vertically in a holding clamp. The Bunsen burner flame is ignited and adjusted to a specified height. The tip of the flame is applied to the end of the specimen until it is ignited, but not longer than 15-sec. If the specimen melts or shrinks away from the flame as far as the gauge mark, the test is considered inapplicable. The time for the burning edge of the specimen to reach the gauge mark is determined. If the flame is extinguished before reaching the gauge mark, the length from the gauge mark remaining unburned is measured. Many visual observations are made, such as cause of extinguishment, melting, and extent of dripping of the specimen.

Calculations are carried out to determine the burning rate, the average time of burning (ATB), and the average extent of burning. Burning rate is plotted as a function of thickness to study the correlation. A typical test set-up is shown in Figure 8-2.

The application of the test is limited to screening, development work, production control, specification testing, and quality control. The test results cannot be used as a criterion for fire hazard. Many papers have been presented (5,6) to illustrate the poor correlation between such small-scale tests and large-scale flame tests that are closer to predicting real fire behavior.

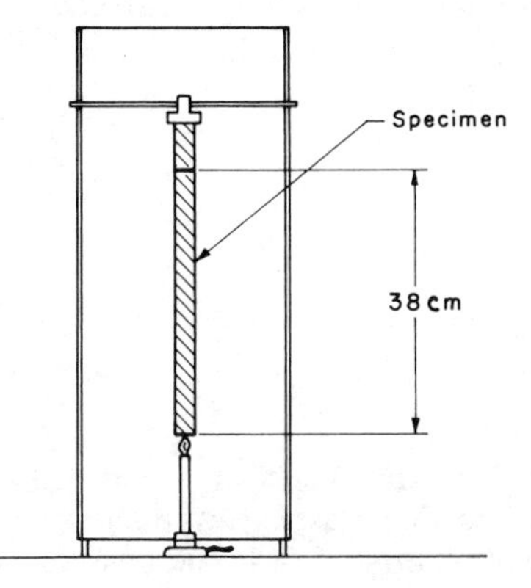

Apparatus for Burning Test.

Figure 8-2. Flammability test set-up for flexible plastics in vertical position. (Reprinted by permission of ASTM.)

8.3. FLAMMABILITY TEST (SELF-SUPPORTING PLASTICS) (ASTM D 635)

This method covers the procedures for determining relative rate of burning and/or extent and time of burning of self-supporting plastics, generally with thickness over 0.050-in. Self-supporting plastics are defined as those plastics which, when mounted with the clamped end of the specimen 0.4-in. above the horizontal screen, do not sag initially so that the free end of the specimen touches the screen.

This test is very similar to the rate-of-burning test for flexible plastics except for the specimen support. Unlike the previous test, the specimen is supported horizontally at one end. The free end is exposed to specified gas flame for 30-sec. Time and extent of burning are measured and reported along with the visual observations.

The test is mainly useful for quality control, production control, comparison, screening, and specification testing. A poor correlation between test results obtained by this method and flammability under actual use condition exists and therefore it cannot be used as a criterion for fire hazard. Figure 8-3 shows a typical set-up for the rate-of-burning test for rigid plastics.

8.4. INCANDESCENCE RESISTANCE TEST (ASTM D 757)

The incandescence resistance test provides laboratory comparisons of the resistance of rigid plastics over 0.050 in. thickness to an incandescent surface at 950 ± 10°C. The test requires the specimen to be clamped in a horizontal position. The data obtained from this test is useful for research and development, quality control, and verification testing. This method, like the previous test methods, is not intended to be a criterion for fire hazard.

The test requires a specially designed apparatus consisting of parts as shown in Figure 8-4. A silicone carbide incandescent rod is used as the igniting source. A variable autotransformer or rheostate provides necessary current to heat the silicone carbide rod to 950 ± 10°C. The temperature of the incandescent rod is measured optically by an optical pyrometer.

The design of a specimen holder is such that the specimen is clamped horizontally and at a right angle to the axis of the silicone carbide rod. A dash pot slows down the travel of the specimen before contacting the silicone carbide rod. A flexible cable is used as a remote control device to control the movement of the specimen. Figure 8-5 illustrates a commercially available incandescent resistance tester.

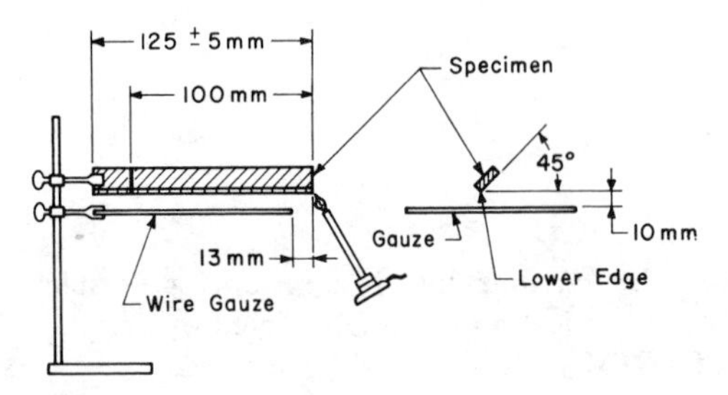

Figure 8-3. Flammability test set-up for self-supporting plastics in horizontal position. (Reprinted by permission of ASTM.)

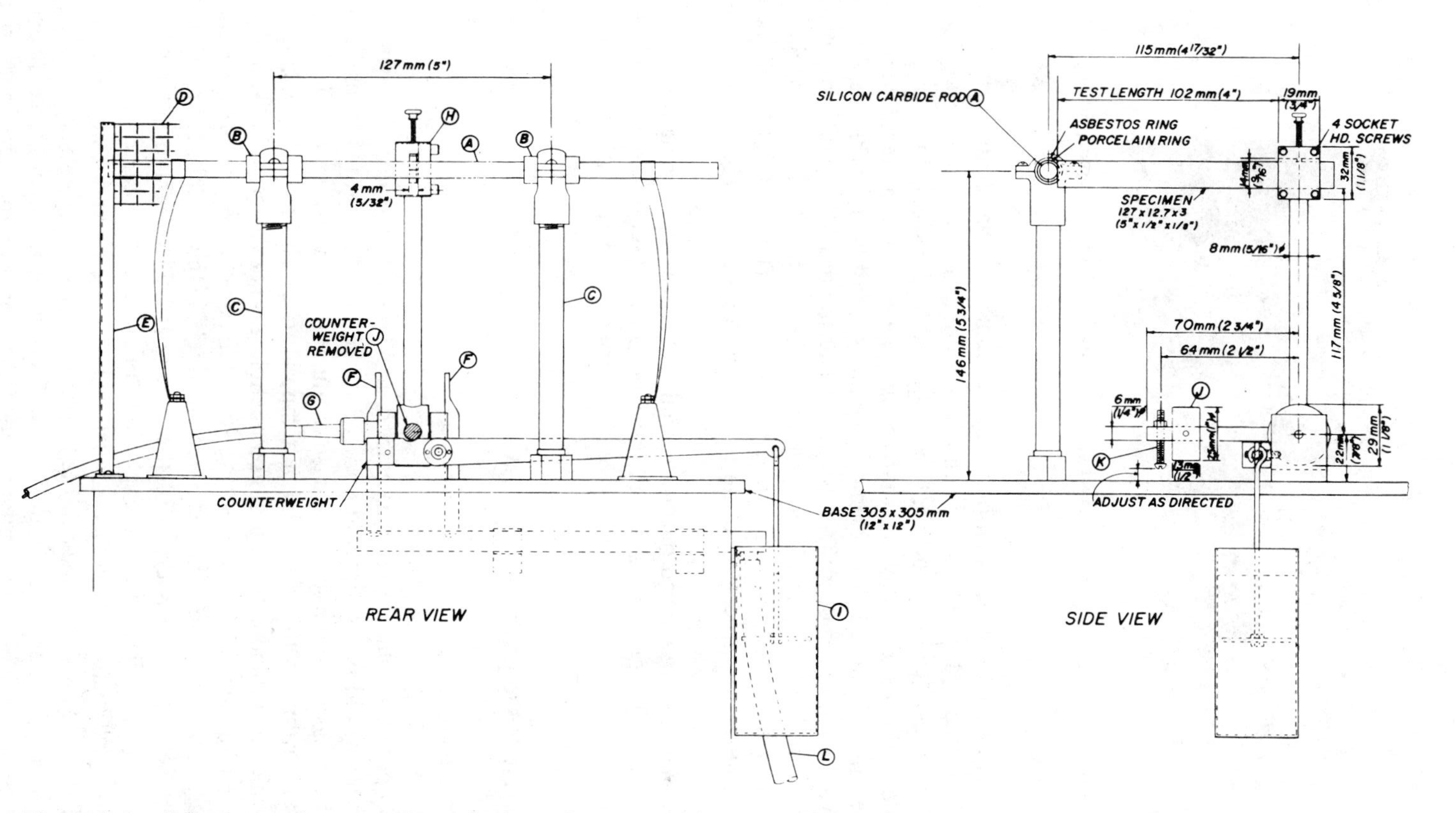

Figure 8-4. Schematic of test apparatus for incandescence resistance test. (Reprinted by permission of ASTM.)

Figure 8-5. Incandescent test apparatus. (Courtesy Custom Scientific Instruments, Inc.)

The test is carried out by positioning the specimen in the clamp so that the length to be tested is 4-in. The specimen is maintained in a vertical position until the start of the test. The incandescent rod is heated to the specified temperature and the specimen is slowly lowered onto the igniting rod with the remote control cable. The timer is started. The specimen changes length as it burns and finally slides down the igniting rod. If the specimen ceases to burn, the timer is stopped, and time is recorded. If the specimen continues to burn, after 3 min. the specimen is moved away from the igniting rod and extinguished. The length of the burned specimen is measured precisely to the point at which no charring or melting is visible.

The calculations are carried out to determine the average burning rate in in./min, the average time of burning, and the average extent of burning for a specimen that does not burn 3 min. No rating is given if the specimen melts without burning.

8.5. IGNITION PROPERTIES OF PLASTICS

8.5.1. Introduction

Ignition properties of plastics are very important when one considers that without ignition there would be no fire and no need for fire containment. Ignitability is defined as the ease of ignition or the ease of initiating combustion. (7) The flash-ignition temperature is the lowest initial temperature of air circulating the specimen, at which a sufficient amount of combustible gas is evolved to be ignited by a small external pilot flame. The self-ignition temperature is the lowest initial temperature of air circulating the specimen, at which, in the absence of an ignition source, the self-heating properties of the specimen lead to ignition or ignition occurs of itself, as indicated by an explosion, a flame, or a sustained glow.

Two basic test methods have been developed to measure ignitability. The first test (ASTM D 1929) determines the temperatures that are necessary to cause sufficient decomposition to generate volatile fuel that will either spontaneously ignite or that can be ignited by a pilot flame. The second approach for measuring ignitability involves the application of a Bunsen burner flame to polymer samples for short periods of time and determining whether sustained ignition occurs after the removal of the flame.

8.5.2. Ignition Temperature Determination (ASTM D 1929)

This method covers the laboratory determination of the self-ignition and flash-ignition temperatures of plastics using a hot-air ignition furnace. The equipment used is known as Setchkin apparatus as illustrated in Figure 8-6. A schematic cross section view of a hot-air ignition furnace assembly is shown in Figure 8-7. The apparatus consists of a furnace tube and an inner ceramic tube placed inside the furnace tube that is capable of withstanding 750°C temperature. The provision is made to introduce clean, metered outside air, tangentially near the top of the annular space between the ceramic tubes. Air is heated with an electrical heating unit, circulated in the space between the two tubes, and allowed to enter the inner furnace tube at the bottom. Copper tubing attached to a gas supply and placed horizontally above the top surface of the divided disc provides a pilot flame. A specimen pan is used to hold the specimen in place inside the furnace. Two thermocouples, one located near the specimen and the other located below the specimen holder, are used to measure specimen temperature and air temperature, respectively. Thermoplastic materials are tested in pellet form normally supplied for molding. Sheet specimens of size $\frac{3}{4} \times \frac{3}{4}$ in. are also used. The specimens are conditioned using standard conditioning procedures prior to testing.

Two procedures are used to determine ignition temperatures. Procedures A is quite lengthy and therefore less commonly used. Procedure B accomplishes similar results in a considerably smaller period of time. To commence the test, the air flow rate is set at 5 ft/min at 400°C. Once the constant air temperature is

Figure 8-6. Setchkin apparatus. (Courtesy Custom Scientific Instruments, Inc.)

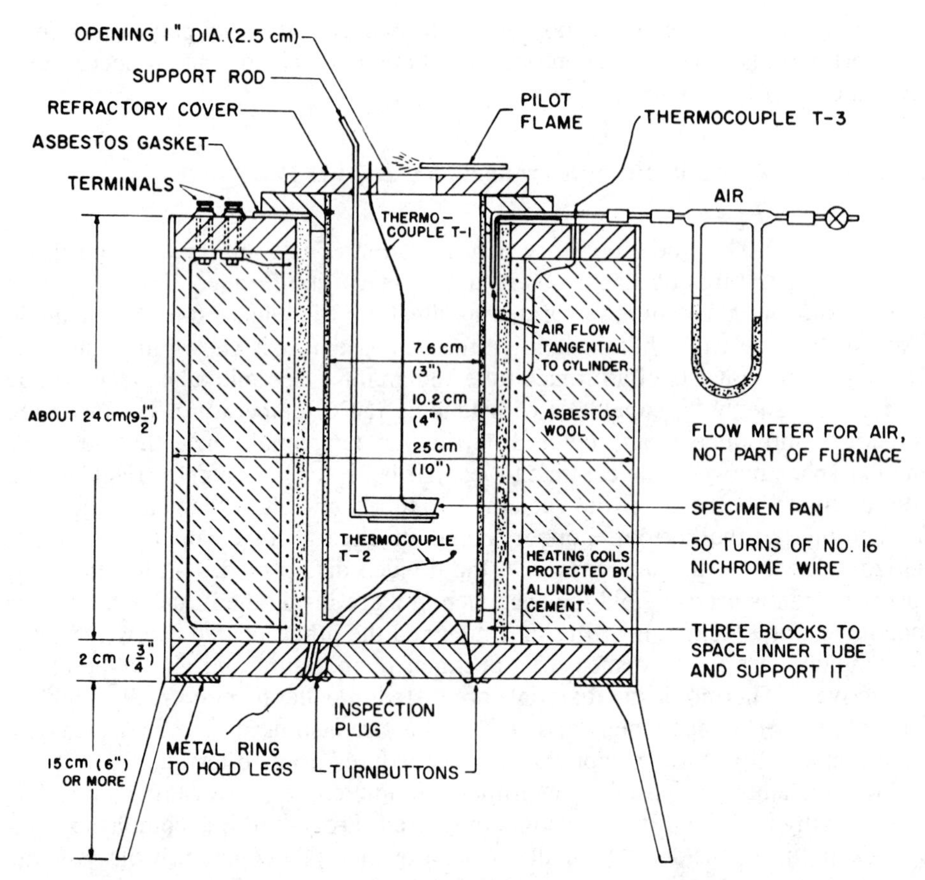

Figure 8-7. Cross section of hot-air ignition furnace assembly. (Reprinted by permission of ASTM.)

attained, the specimen holder is lowered into the furnace. Immediately after that, the timer is started and the pilot flame ignited. A mild explosion of combustible gases or flash indicates flash ignition. This is usually followed by continuous burning of the specimen. If the specimen ignites before the end of a 5-min period, the test is repeated at a lower temperature setting using a fresh specimen. If ignition has not occurred at the end of 5-min, the temperature is raised, and the test is repeated. The lowest air temperature at which a flash is observed is recorded as the minimum flash-ignition temperature. The self-ignition temperature is determined in a similar manner without the gas pilot flame. The lowest air temperature at which a specimen burns is recorded as the minimum self-ignition temperature. The visual observations such as melting, bubbling, and smoking are also recorded.

This test is useful in comparing the relative ignition characteristics of different materials. The test cannot be regarded as the sole criterion for fire hazard. An alternate method for screening and comparing materials for ignitability has been reported (8).

8.6. OXYGEN INDEX TEST (ASTM D 2863)

8.6.1. Introduction

Oxygen index is defined as the minimum concentration of oxygen, expressed as volume percent, in a mixture of oxygen and nitrogen that will just support flaming combustion of a material initially at room temperature under specified conditions. The oxygen index test is considered one of the most useful flammability tests since it allows one to precisely rate the materials on a numerical basis and simplifies the selection of plastics in terms of flammability. The oxygen index test overcomes the serious drawbacks of conventional flammability tests. These drawbacks are variation in sample ignition techniques, variation in the description of the end-point from test to test, and operation of tests under nonequilibrium conditions (9). Table 8-4 compares the oxygen index values of a variety of materials. Note that red oak wood has a higher oxygen index (lower flammability) than polystyrene, but a much lower oxygen index compared to polycarbonate or PVC.

8.6.2. Test Procedures

The test determines the minimum concentration of oxygen in a mixture of oxygen and nitrogen flowing upward in a test column that will just support combustion. This process is carried out under equilibrium conditions of candlelike burning. It is necessary to establish equilibrium between the heat removed by the gases flowing past the specimen and the heat generated from the combustion. The equilibrium can only be established if the specimen is well-ignited and given a chance to reach equilibrium when the percent oxygen in the mixture is near limiting or critical value. (10)

The equipment used for measuring the oxygen index consists of a heat-resistant glass tube with a brass base. The bottom of the column is filled with glass beads that allows the entering gas mixture to mix and distribute more evenly. A specimen-holding device to support the specimen and hold it vertically in the center

Table 8-4. Oxygen Index Rating of Some Materials

Material	Oxygen-Index, percent
Red oak	24.6
Acetal	16.2
Polyethylene	17.4
Polypropylene	17.4
Polystyrene	18.3
Polycarbonate	27.0
Nylon 6-6	28.0
40 percent glass-filled polycarbonate	30.5
PVC	47.0
Polyvinylidene chloride	60.0
PTFE	95.0

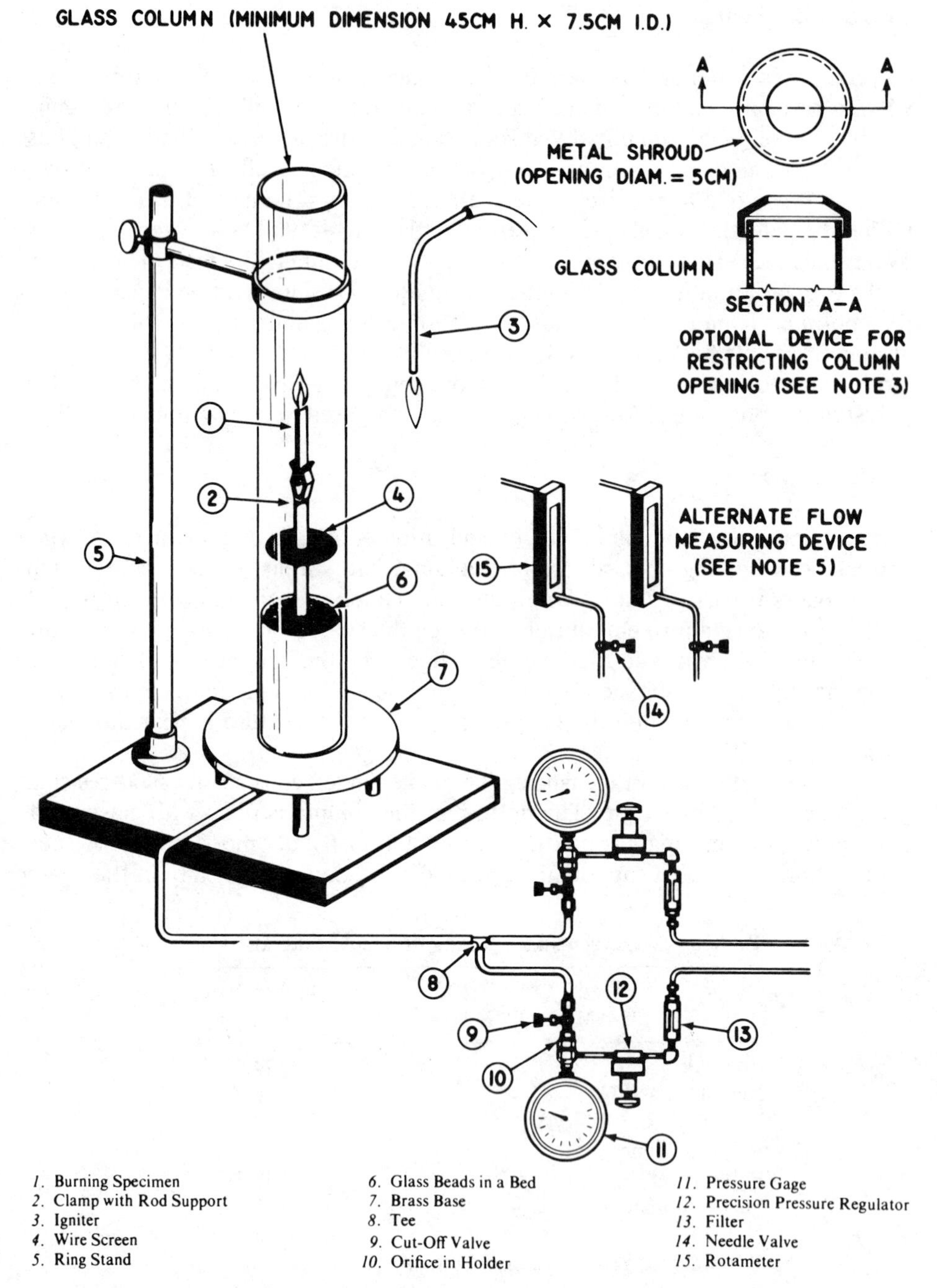

Figure 8-8. Typical equipment layout for Oxygen index test. (Reprinted by permission of ASTM.)

of the column is used. A tube with a small orifice having propane, hydrogen, or other gas flame, suitable for inserting into the open end of the column to ignite the specimen, is used as an ignition source. A timer, flow measurement, and control device are also used. Figure 8-8 illustrates a typical equipment layout. A commerically available oxygen index tester is shown in Figure 8-9.

The test specimen used in the experiment must be dry since the moisture content of some materials alters the oxygen index. Four different types of specimens are specified. They are physically self-supporting plastics, flexible plastics, cellular plastics, and plastic film or thin sheet. The dimension of the specimen varies according to the type.

The specimen is clamped vertically in the center of the column. The flow valves are set to introduce the desired concentration of oxygen in the column. The entire top of the specimen is ignited with an ignition flame so that specimen is well-lighted. The specimen is required to burn in accordance with set criteria, which spell out the time of burning or the length of specimen burned. The concentration of oxygen is adjusted to meet the criteria. The test is repeated until the critical concentration of oxygen which is the lowest oxygen concentration that will meet the specified criteria is determined. The oxygen index is calculated as follows:

$$\text{Oxygen index percent} = (100 \times O_2)/(O_2 + N_2)$$

where O_2 = volumetric flow of oxygen at the concentration determined; N_2 = volumetric flow of nitrogen.

8.6.3. Factors Affecting the Test Results

1. *Thickness of Specimen.* As the specimen thickness increases, the oxygen index also increases steadily.

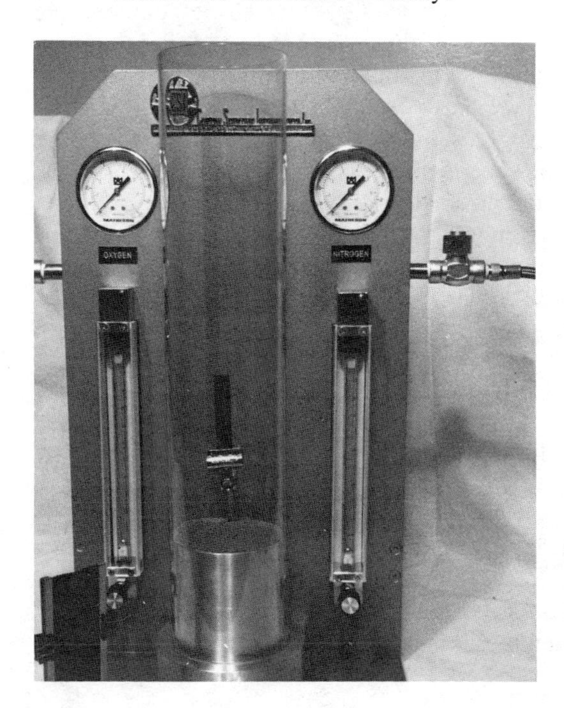

Figure 8-9. Oxygen index tester. (Courtesy Custom Scientific Instruments, Inc.)

2. Fillers. Fillers such as glass fibers tend to increase the oxygen index up to a certain percentage loading. In the case of polycarbonate, the oxygen index peaks at about 25 percent loading. Higher loading beyond this point, subsequently decreases the oxygen index.
3. *Flame Retardants.* Flame retardants increase the oxygen index, making polymers more suitable for applications requiring improved flammability.

8.7. SURFACE BURNING CHARACTERISTICS OF MATERIALS

Two major tests have been developed to study the surface burning characteristics of materials.

8.7.1. Surface Flammability of Materials Using a Radiant Heat Energy Source (ASTM E 162)

This test, also known as the radiant panel test, is one of the most widely used laboratory-scale flame-spread tests. Although not recommended for use as a basis of ratings for building code, the test does provide a basis for measuring and comparing the surface flammability of materials when exposed to a prescribed level of radiant heat energy.

The test apparatus consists of a radiant panel with an air and gas supply, a specimen holder, a pilot burner, a stack, thermocouples, a hood with an exhaust blower, a radiation pyrometer, a timer, and an automatic potentiometer recorder. Figure 8-10 is a schematic diagram of a radiant panel test apparatus.

The test specimen of size 6 × 18 in. is employed. A specimen thickness of 1 in. or less is desirable. Prior to the test, the specimens must be dried for 24 hrs at 140°F and then conditioned to equilibrium at 73°F and 50 percent relative humidity. The specimen is mounted in the specimen holder so that it is inclined at 30° to a 12 × 18 in. vertical radiant panel maintained at 1238°F. A small pilot flame is positioned at the top of the specimen to ignite the developing volatile gases. Once started, the flame front progresses downward. The time for the flame to reach each of the 3-in. marks on the specimen holder is recorded. Observations such as dripping of the burning specimen are also recorded.

The flame spread index of a specimen is determined by measuring the rate of burning down of the specimen and the heat rise associated with the burning in the stack above the burning specimen.

$$I_s = F_s \times Q$$

where I_s = flame spread index; F_s = flame spread factor; Q = heat evoluation factor.

The reproducibility of this test within laboratory and between laboratories is very poor for thermoplastics.

8.7.2. Surface-Burning Characteristics of Building Materials (ASTM E 84)

This method for determining surface-burning characteristics of building materials, commonly known as the "E-84 tunnel test," is applicable to any type of building

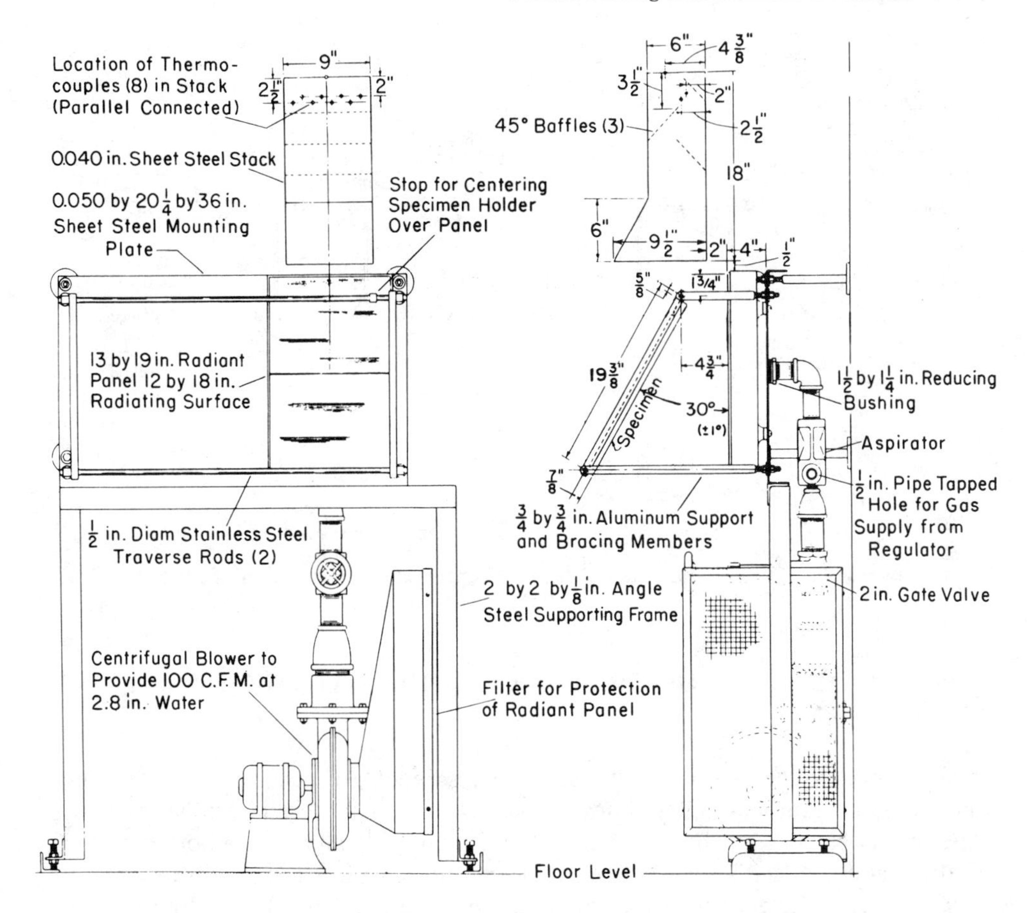

Figure 8-10. Schematic radiant panel test. (Reprinted by permission of ASTM.)

material capable of supporting itself or being supported in a test furnace at a thickness comparable to its recommended use. The purpose of this test is to determine the comparative burning characteristics of the material and smoke density.

The test is carried out by placing a 2-in.-wide and 24-ft-long specimen in a fire test chamber. The specimen may be continuous or sections joined end to end. Figure 8-11 shows a schematic diagram of a typical test furnace. One end of the test chamber is designed as the "fire end," where two gas burners capable of delivering the flames upward against the surface of the test specimen are positoned. The other end if designated as the "vent end" and is fitted with an induced draft system.

The flame spread rating is determined by exposing the specimen to a fire at one end of a 25-ft-long tunnel for 10 min. The flame spread rating is stated as a comparative measurement of the progress of flame over the surface of the tested material on the 0–100 scale where cement asbestos board is rated at 0 and red oak is rated at 100. A photoelectric cell is used to measure the smoke density.

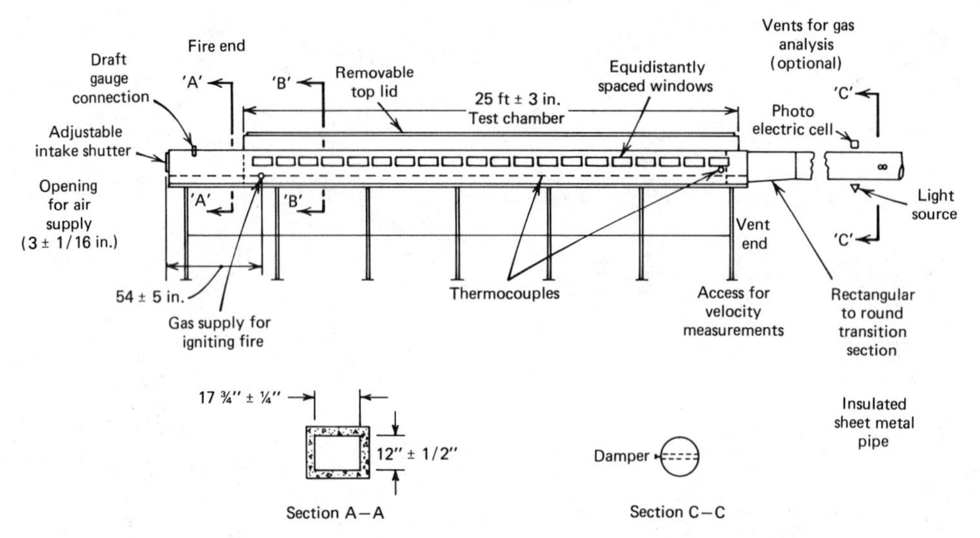

Figure 8-11. Schematic "tunnel test." (Reprinted by permission of ASTM.)

8.8. FLAMMABILITY OF CELLULAR PLASTICS—VERTICAL POSITION (ASTM D 3014)

This method covers a small-scale laboratory screening procedure for comparing the relative extent and time of burning and loss of weight of rigid cellular plastics. The test is carried out by mounting a specimen vertically on support pins in a vertical chimney with a glass front and igniting it with a Bunsen burner for 10 sec. The flame height, time of burning, and weight percent retained by the specimen are determined. A commercially available test apparatus is shown in Figure 8-12. This test is useful only for comparing relative flammability of cellular plastics and does not give any information regarding the behavior of cellular plastics in actual fire conditions.

8.9. FLAMMABILITY OF CELLULAR PLASTICS—HORIZONTAL POSITION (ASTM D 1692)

This test which covers a small-scale laboratory screening procedure for measuring the rate of burning and/or extent and time of burning of rigid or flexible cellular plastics, was discontinued in 1978 and has not been replaced.

8.10. FLAME RESISTANCE OF DIFFICULT-TO-IGNITE PLASTICS (Federal STD. No 406 Method 203)

This federal test method was developed primarily to measure ignition time, burning time, and flame travel of plastics that are difficult to ignite. An appartus called an ignition tester, illustrated in Figure 8-13, is employed. It consists of an enclosure

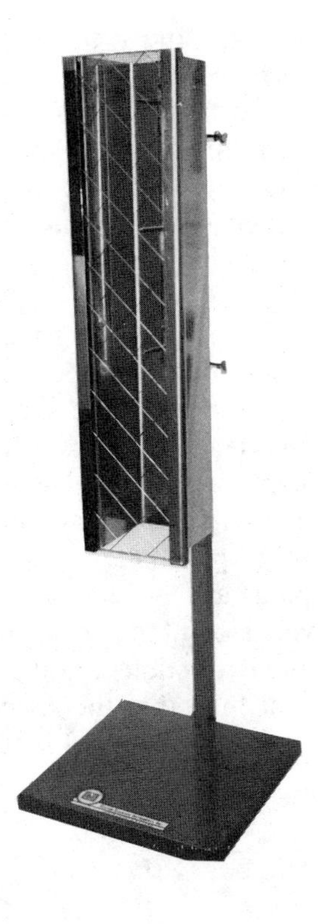

Figure 8-12. Apparatus to determine the flammability of rigid cellular plastics. (Courtesy Custom Scientific Instruments, Inc.)

Figure 8-13. Ignition tester. (Courtesy Scientific Instruments, Inc.)

that houses heating coils, spark plugs, specimen supports, and a flame travel gauge. The specimen supports hold the specimen in vertical position. Two spark plugs with extended electrodes are spaced $\frac{1}{8}$ in. from the surface of the specimen. The function of these spark plugs is to ignite the gases emitted from the specimen preheated by the heating coil. A suitable electric circuit is provided to maintain continuous sparking at the electrodes during the specified time. A shatterproof glass window allows clear viewing of the interior and protects the operator.

The test is started by positioning the specimen ($5 \times \frac{1}{2} \times \frac{1}{2}$ in.) in the specimen support and applying a constant current of 55 amps to the heating coil. The spark plugs are energized at the same time and the timer is started. As soon as the ignition starts, the timer is stopped, and the time required to ignite the specimen is recorded. Ignition is considered as occuring when the flame transfers from the escaping gases to the surface of the specimen and continues there. Heating is discontinued 30-sec after ignition occurs. If ignition does not take place in 600-sec, the test is discontinued. The electrical supply to the spark plugs.is also discontinued immediately after ignition occurs and the plugs are moved away from the flame. The maximum distance that the flame travels along the surface of the specimen, measured from the top of the heater coil before extinction, is considered the flame travel distance. The burning time is the total time that the specimen continues to burn until the cessation of all flaming after the heater coil is turned off.

8.11. SMOKE GENERATION TESTS

There is a growing sense of urgency regarding the need to study and develop realistic smoke generation tests. The smoke generation from burning plastics is of as much concern as the flames from burning structures, since the smoke obscures the visibility and seriously impairs the ability of a person to escape the fire hazard. Many small-scale laboratory tests as well as large-scale tests have been developed. However, there is a considerable controversy over the practicality, usefulness, and reliability of these test methods. Some tests require a large number of specimens while others are very cumbersome and time-consuming. Some are too expensive, some are misleading, and others simply lack correlatability. An attempt is made in this section to briefly describe many of the current tests, test equipments, procedures, advantages, and limitations. Table 8-3 summarizes smoke generation tests.

8.11.1. Smoke Density Test (ASTM D 2843)

This test measures the loss of light transmission through a collected volume of smoke produced under controlled, standardized conditions. The test employs a $12 \times 12 \times 31$ in. aluminum test chamber with a heat-resistant glass observation door. The chamber is completely sealed except for 1×9 in. openings on four sides of the bottom of the chamber. A specimen holder holds the specimen in a horizontal position. A photoelectric cell and a light source is used to measure light absorption. A test specimen of size $1 \times 1 \times \frac{1}{4}$ in. is placed in the specimen holder and exposed to a propane–air flame so that the flame is directly under the specimen. The percent light absorbed by the photoelectric cell is measured and recorded at 15 sec. intervals for 4 min. A visual comparison of smoke density is made by observing the illuminated "EXIT" sign. The light absorption data (light absorption in percent) is plotted versus time on a graph recorder. The total smoke produced is determined by measuring the area under the curve. This area under the curve in percent is the smoke density rating. The maximum smoke density is the highest point on the curve. Figure 8-14 illustrates a commercially available smoke density chamber.

8.11.2. Surface Flammability Tests (ASTM E 84)

This test has been discussed in detail in Section 8.7.

8.11.3. NBS Smoke Test

This test, developed by the National Bureau of Standards (NBS), measures the density of smoke accumulated in a chamber by exposing the specimen to a flame

Figure 8-14. Smoke density chamber. (Courtesy United States Testing Company, Inc.)

or radiant panel. The smoke chamber consists of a 36 × 24 × 36 in. enclosure with a specimen holder capable of holding the specimen vertically. The specimen of size 3-sq. in. is exposed to heat under flaming or smoldering conditions. The heat source used in this test is an electric furnace capable of generating heat flux of 2.5 W/cm^2 at the specimen surface. The generated smoke is measured by a photoelectric cell and a light source positioned at top and bottom of the chamber. This vertical positioning of the photometer and light source minimizes measurement differences due to smoke stratification that could occur with a horizontal photometer path at a fixed height (11). Figure 8-15 illustrates a commercially available NBS smoke test apparatus.

8.11.4. Arapahoe Smoke Test

The Arapahoe smoke test, developed by Arapahoe Chemicals, is a gravimetric test as opposed to an optical test. In this test, the smoke evolution is measured gravimetrically by the weight of the smoke particulate produced rather than op-

Figure 8-15. NBS smoke chamber. (Courtesy Superpressure, Inc.)

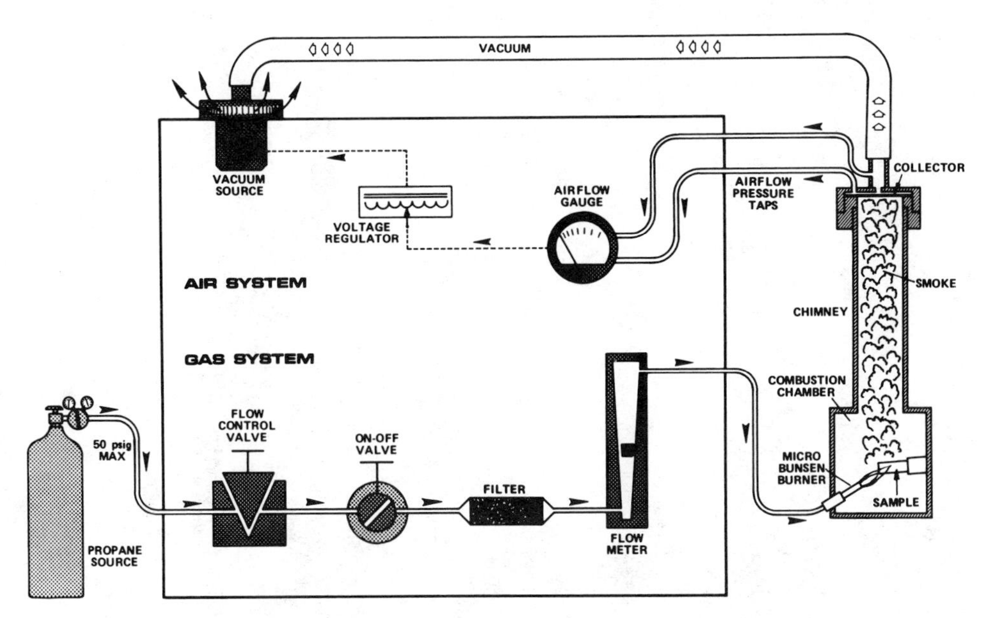

Figure 8-16. Schematic of Arapahoe test apparatus. (Courtesy Arapahoe Chemicals, Inc.)

tically by the light obscuration caused by the particulates (12). The test was introduced with the idea of providing the industry with a quick, simple, and repeatable technique for measuring smoke generated from burning plastics.

The test equipment, shown schematically in Figure 8-16, consists of a chimney extending from a cylindrical combustion chamber; a filter assembly, positioned at the top of the chimney, is connected to a high capacity vacuum source; a Micro-bunsen burner, located at the bottom of the combustion chamber, is mounted at an angle of 10° from horizontal; a specimen holder holds $1\frac{1}{2} \times \frac{1}{2} \times \frac{1}{8}$ in. specimen horizontally. The specimen is exposed to the flame for 30 sec. At the end of 30 sec, the burner is shut off, and the specimen is extinguished. The smoke generated by the burning specimen is drawn up the chimney by the draft caused by vacuum suction and is collected on the surface of filter paper. The weight of smoke deposited is determined by weighing the filter paper. The value reported is percent smoke produced from burning plastics. Figure 8-17 shows a commerically available Arapahoe smoke test apparatus.

8.11.5. Radiant Panel Test (ASTM E 162)

A radiant panel apparatus consists of a radiating heat source maintained at 1238°F. A vertically mounted, porous refractory panel acts as a radiating heat source. A specimen of size 6 × 18 in. is supported in front of the panel. The specimen is ignited from the top by a pilot flame so that the flame travels downward along the underside exposed to the radiant panel. A glass fiber filter paper is positioned at the top of the stack. The smoke particles are collected on the surface of the filter by drawing air through the filter. The filter paper is weighed before and after the test and the difference in weight is reported as smoke deposit in mg.

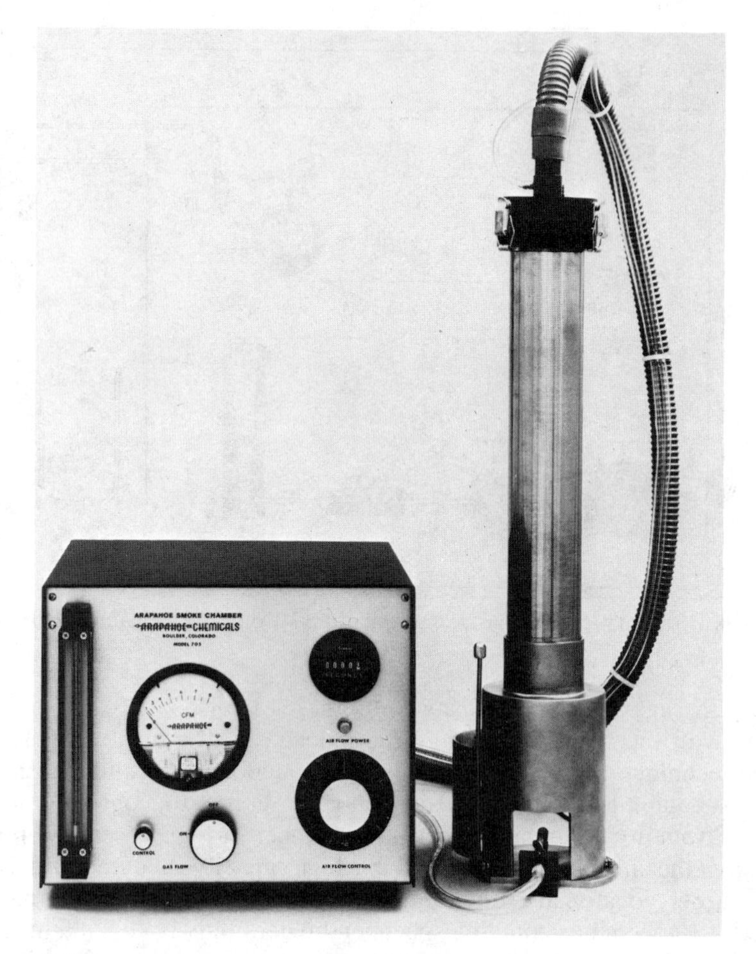

Figure 8-17. Arapahoe test apparatus. (Courtesy Arapahoe Chemicals, Inc.)

8.11.6. OSU Release Rate Test

The Ohio State University (OSU) release rate apparatus consists of a chamber 35 × 16 × 8 in., with a top section shaped like a pyramid connected to an outlet. The chamber contains an electrically heated radiant panel, a gas-fired radiant panel, or electrically heated elements. This test offers the flexibility of orienting the specimen either vertically or horizontally to simulate wall and floor applications (13). A photocell and a light source positioned above the outlet measures the light absorption.

8.12. UL 94 FLAMMABILITY TESTING

UL 94, developed by Underwriters Laboratories, is one of the most widely used and most frequently cited sets of flammability tests for plastic materials. These tests are for the flammability of plastic materials used for parts in devices and appliances. They are intended to serve as a preliminary indication of their suitability with respect to flammability for a particular application. The UL flammability tests include a standard burning test applied to vertical and horizontal test bars, from which a general flammability rating is derived.

The UL 94 standard for tests for flammability of plastic materials consists of four basic tests for classifying materials in different categories. These four basic tests are:

1. Horizontal burning test for classifying materials (94 HB).
2. Vertical burning test for classifying materials (94V-0, 94V-1).
3. Vertical burning test for classifying materials (94-5V).
4. Vertical burning test for classifying materials (94 VTM-0, 94 VTM-1, or 94 VTM-2).

8.12.1. Horizontal Burning Test for Classifying Materials, 94HB

Materials are classified 94HB if they burn over a 3 in. span in a horizontal bar test at a rate not more than 1.5 in./min. for specimens 0.120 to 0.500 in. thick and not more than 3 in./min. for specimens less than 0.120 in. thick, or if the specimens cease to burn before the flame reaches the 4.0 in. reference mark.

The apparatus employed for the test consists of a test chamber, an enclosure or laboratory hood, a laboratory burner, wire gauze, technical grade methane gas, a ring stand, and a stopwatch. The test is conducted in a humidity and temperature-controlled room. Test specimens are usually $\frac{1}{2} \times 5$ in. and should have smooth edges. Before conducting the test, the specimens are marked across the width with two lines, 1.0 and 4.0 in. from one end of the specimen. The specimen is clamped in the ring stand, as shown in Figure 8-18. The burner is ignited to produce a 1 in. high blue flame. The flame is applied so that the front edge of the specimen, to a depth of approximately $\frac{1}{4}$ inch, is subjected to the test flame for 30 sec. without changing the position of the burner, and is then removed from the burner. If the specimen burns to the 1.0 in. mark before 30 sec., the flame is withdrawn. If the specimen continues to burn after removal of the flame, the time for the flame front to travel from the mark 1.0 in. from the free end to the mark 4.0 in. from the free end is determined and rate of burning is calculated.

8.12.2. Vertical Burning Test for Classifying Materials 94V-0, 94V-1, 94V-2

In this test, the specimens are clamped vertically. The materials are classified 94V-2, the least stringent classification or 94V-1 or 94V-0, which is the most stringent or the highest classification for this test. Table 8-5 summarizes the requirements for each classification.

The apparatus employed for vertical burning tests are similar to the ones employed in the horizontal burning test, except for a few additional items such as a desiccator, a conditioning oven, and dry absorbent surgical cotton. The test is conducted on a $\frac{1}{2} \times 5$-in. specimen. A small $\frac{3}{4}$-in.-high blue flame is applied to the bottom of the specimen for 10 sec., withdrawn, then reapplied for an additional 10 sec.; the duration of flaming and glowing is noted as soon as the specimen has

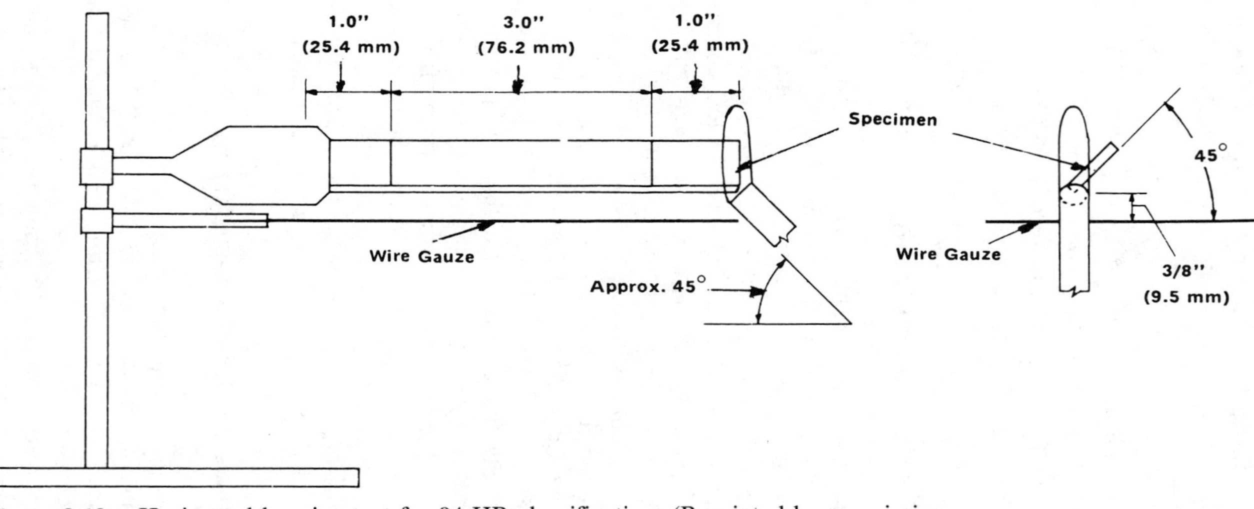

Figure 8-18. Horizontal burning test for 94 HB classification. (Reprinted by permission of Underwriters' Laboratories.)

Table 8-5. Summary of UL 94 Vertical Burning Test for Classifying Materials, 94V-0, 94V-1, 94V-2.[a]

Criteria	Classification		
	V-2	V-1	V-0
Number of specimens	5	5	5
Number of ignitions	2	2	2
Maximum flaming time per specimen per flame application, sec	30	30	10
Total flaming time, five specimens, 2 ignitions, sec	250	250	50
Flaming drips ignite cotton	yes	no	no
Maximum afterflow time, per specimen, sec	60	60	30
Burn to holding clamp	no	no	no

[a] Reprinted from *Plastics Compounding,* May/June 1978.

extinguished. A layer of cotton is placed beneath the specimen to determine whether dripping material will ignite it during the test period.

8.12.3. Vertical Burning Test for Classifying Materials, 94-5V

For any material to achieve this somewhat stringent classification, the test specimens must not burn with flaming and/or glowing combustion for more than 60 sec. after the fifth flame. Also, the test specimens must not drip.

The test apparatus consists of a test chamber, a laboratory burner, an adjustable ring stand for vertical positioning of the specimens, a gas supply, a mounting block capable of positioning the burner at an angle of 20° from the vertical, a stop watch, a desiccator, and a conditioning oven. The test specimens are in two forms—$\frac{1}{2} \times 5$-in.-long bars or 6×6 in. plaques.

Method A for the testing of specimens $\frac{1}{2} \times 5$ in. involves positioning the test specimen vertically on the ring stand and supporting the burner on the inclined plane of a mounting block so that the burner tube may be positioned 20° from the vertical. This arrangement is shown in Figure 8-19. The burner is ignited at a remote location from the specimen and adjusted so that the overall flame height is 5 in. and height of the inner blue cone is $1\frac{1}{2}$ in. The flame is then applied to one of the lower corners of the specimen at a 20° angle from the vertical so that the tip of the blue cone touches the specimen. The flame is applied for 5 sec and removed for 5 sec. This process is repeated four additional times. Duration of flaming plus glowing, the distance the specimen burned, dripping, and deformation of specimen are observed and recorded.

Method B, which involves use of test plaques, is similar to Method A, the only exception being the positioning of the test plaques. The five different positions include plaque vertical with the flame applied to lower corner of the plaque, plaque vertical with the flame applied to the lower edge, plaque vertical with the flame applied to the center of one side of the plaque, plaque horizontal with the flame applied to the center of the bottom surface of the plaque, and plaque horizontal with the flame directed downward to the top surface of the plaque.

This test is also conducted using an actual molded part. The flame is applied to the most vulnerable areas of the part (14).

8.12.4. Vertical Burning Test for Classifying Materials 94VTM-0, 94 VTM-1, or 94VTM-2

This particular test is designed for very thin materials which may distort, shrink, or get consumed up to the holding clamp if tested using previously described test conditions. Such materials must possess physical properties that will allow an 8-in.-long by 2-in.-wide specimen to be wrapped longitudinally around a 0.5-in.-diameter mandrel. The requirements for three different classifications are summarized in Table 8-6.

The specimens are cut from the sheet to the specified size and conditioned prior to the test. A 5 in. mark is made across the specimen width. The specimen is wrapped around a $\frac{1}{2}$-in.-diameter mandrel. The test is carried out by subjecting the specimens to a $\frac{3}{4}$-in.-high blue flame for 3 sec. and withdrawn until flaming of the specimen ceases. The second application is made immediately and repeated. Observations such as the duration of the flame after the first flame application, the duration of flaming after the second application, the duration of flaming plus

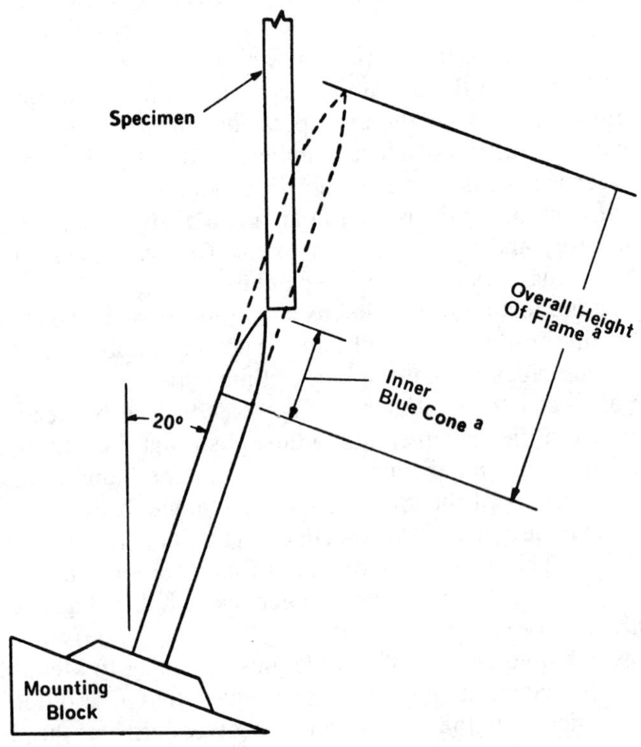

Figure 8-19. Vertical burning test for 94-5V classification. (Reprinted by permission of Underwriters' Laboratories.)

Table 8-6. Summary of UL 94 Vertical Burning Test for Classifying Materials, 94VTM-0, 94VTM-1, 94VTM-2

	Classification		
Criteria	VTM-0	VTM-1	VTM-2
Number of test specimens	5	5	5
Number of ignitions	2	2	2
Maximum flaming time per specimen per flame application, sec	10	30	30
Total flaming time, two flame applications, five specimens, sec	50	250	250
Flaming drips ignite cotton	no	no	yes
Maximum afterglow time, per specimen, sec	30	60	60
Burn to 5-in. bench mark	no	no	no

[a] Reprinted from *Plastics Compounding,* May/June 1978.

glowing after the second application, whether or not specimens burn up to a 5 in. mark, and whether or not specimens drip flaming particles which ignite the cotton, are recorded.

8.12.5. Factors Affecting UL 94 Flammability Testing

The previously described UL 94 tests are considered to be very subjective tests. Identical specimens tested by different operators using the same testing equipment can give different flammability ratings. Such variations are attributed to the differences in interpretation of end points, differences in observation techniques, and differences in operators.

The application of the flame to the test specimen is also very critical. If proper care in observing the procedure is not taken, the specimen may preheat, overheat, underheat, or unevenly heat, giving inconsistent results. Other factors affecting the test results are molding of the test specimens, variations in calibration procedures, equipment variations, test burners, and testing environments (15).

8.13. MEETING FLAMMABILITY REQUIREMENTS

In this day and age of rules and regulations, no design engineer can afford to overlook the flammability requirements put out by government agencies, private consumer protection institutions, and insurance underwriters. Today, no matter what one is designing, whether it is a TV cabinet, an appliance housing, furniture, business machine components, or a building product, it is more than likely that the plastic materials used in these applications will be required to meet specific criteria regarding their ability to withstand ignition and burning. Meeting flammability requirements is simply not enough, one must also consider smoke formation and generation of toxic gases and how it affects overall flammability picture.

8.13.1. Agencies Regulating Flammability Standards

A. *Government Agencies*

National Bureau of Standards (NBS). NBS develops and issues a variety of standards and testing techniques. The NBS smoke test is one of the most widely accepted smoke density tests.

The Department of Housing and Urban Development (HUD). This agency sets the standards for products used in the building and construction industry.

Consumer Product Safety Commission. This agency issues standards requiring voluntary as well as mandatory compliance in areas relating to consumer safety.

Department of Transportation (DOT). DOT's main concern is in the area of flammability as it relates to public safety in transportation. Many tests have been developed by DOT to study burning of simulated mass transit interiors.

Federal Aviation Administration (FAA). FAA regulates the standards and specifications concerning flammability of materials used in the interior of aircraft. One of the main concerns of FAA is the formation of smoke and toxic gases from a burning object.

B. *Industry Associations*

Society of Plastics Industry (SPI). This association representing the plastic industry is actively involved in developing realistic flammability tests and standards. SPI works closely with many government and private agencies to develop new standards.

Society of Automotive Engineers (SAE), Manufactured Housing Institute, and *Manufacturing Chemists Association.* These are also actively involved in developing flammability standards.

C. *Private Institutions*

Underwriters Laboratories (UL). UL is an independent nonprofit organization that develops standards and test methods, operates laboratories, and issues certification, all in the interest of public safety.

American Society for Testing and Materials (ASTM). ASTM is a scientific and technical organization whose prime function is to develop standards on characteristics and performance of materials, products, systems and services. Numerous test methods concerning flammability of plastics have been developed by ASTM.

D. *Insurance Underwriters*

An increasing number of insurance underwriters, such as Factory Mutual, have become interested in the flammability of plastic materials. Many realistic tests have been developed by Factory Mutual Research Corporation.

E. *Other Agencies*

National Fire Protection Association (NFPA), Southern Furniture Manufacturers Association, Building Officials and Code Administrators (BOCA), National Electric Manufacturers Association (NEMA), Southern Building Code Congress (SBCC).

8.13.2. Steps in Meeting Flammability Requirements

1. The first step in meeting flammability requirements is to carefully define the application in detail. This will help narrow the list of agencies you may have to deal with.

2. Determine the appropriate agency that deals with your application. For example, if the application has something to do with the building industry, you may want to contact one of the building and construction organizations. If the application is a plastic cabinet that houses electrical components, UL is the organization to contact. One good source of general information is the Society of Plastics Industry.

3. Once the application is defined and the governing agency is narrowed down, you may proceed with the designing and material selection. An important thing to remember at this stage is to specify the material to your design and not vice versa. The material selection process can be expedited by consulting published sources, such as the UL-recognized component directory that lists plastics according to performance as tested by UL 94. *Modern Plastics Encyclopedia's* flammability chart is another source of information for preliminary screening of the materials.

4. Once the preliminary decisions have been made on the type of material that will meet the requirement of the application, a resin supplier or a custom compounder can be consulted for specific grade of material.

5. If the code or standard requires testing, an independent laboratory should be consulted.

REFERENCES

1. Nelson, G. L., "Flame Tests for Structural Foam Parts," *Plast. Tech.* (Nov. 1977), p. 88.
2. *Ibid*, p. 89.
3. Nelson, G. L., "Fire and Polymers," *Chem.* **51** (June 1978), p. 23.
4. *Ibid*, p. 26.
5. Hill, B. J., "How Predictive Are Small-Scale Flame Tests?" *SPE ANTEC* **24** (1978), p. 587.
6. Fang, J. B., NBS Technical Note 879, U.S. Department of Commerce, National Bureau of Standards, Washington, D.C., June 1976.
7. Hilado, C. J. and Murphy, R. M., "Screening Materials for Ignitability," *Mod. Plast.* (Oct. 1978), p. 52.
8. *Ibid*, p. 52.
9. Goldblum, K. B., "Oxygen Index: Key to Precise Flammability Rating," *S.P.E. J.*, **25**(2) (Feb. 1969), p. 50.
10. *Ibid*, p. 51.

11. Hilado, C. J., Cumming, H. J., and Machdo, A. M., "Screening Materials for Smoke Evolution," *Mod. Plas.* (July 1978), p. 62.
12. Kracklauer, J., Sparkes, C., and Legg, R., "New Smoke Test—Fast, Simple, Repeatable," *Plast. Tech.* (Mar. 1976), pp. 46–49.
13. Hilado, C. J., Reference 7, p. 62.
14. Nelson, G. L., Reference 2, p. 89.
15. Howard, J. M., "Factors Affecting UL 94 Flammability Testing," *SPE ANTEC*, **25** (1979), p. 942.

SUGGESTED READING

Hilado, C. J., *Flammability Test Methods Handbook*, Technomic, West Port, 1973.

Hilado, C. J., *Flammability Handbook for Plastics*, Technomic, West Port, 1969.

Zabetakis, "Flammability Characteristics of Combustible Gases and Vapors," *U.S. Bureau of Mines Bull.*, **627** (1965).

Bradley, J. N., *Flame and Combustion Phenomena*, Methuen and Co., London, England 1969.

Lyons, J. W., *The Chemistry and Uses of Fire Retardants*, John Wiley, New York, 1970.

Hilado, C. J., *Fire and Flammability Series*, Technomic, West Port, 1973. Vol. 1, "Flammability of Cellulosic Materials"; Vol. 2, "Smoke and Products of Combustion"; Vol. 3, "Flammability of Consumer Products"; Vol. 4, "Oxygen Index of Materials"; Vol. 5, "Surface Flame Spread"; Vol. 6, "Flame Retardants."

CHAPTER 9

Chemical Properties

9.1. INTRODUCTION

Chemical resistance of plastics is a complex subject. The test results are often misinterpreted by engineers and designers. Material selection is made without a proper understanding of the tests' limitations and how the results are derived. Extremely strong and tough plastic like polycarbonate has limited applications because of its poor chemical resistance. Polypropylene, on the other hand, has poor physical properties but is impervious to most chemicals and solvents. Plastic's resistance to chemicals is best understood through the study of its basic polymer structure. The type of polymer bonds, the degree of crystallinity, branching, the distance between the bonds, and the energy required to break the bonds are the most important factors to consider while studying the chemical resistance of plastic materials (1). For example, highly crystalline structure, lack of branching, and presence of very strong covalent bonds between carbon and fluorine atoms in the main chain makes polytetrafluoroethylene resistant to almost all chemicals and solvents. Similarly, in the case of polyamides (nylons), the regular symmetrical structure and the molecular flexibility that produces high crystallinity and the presence of greater intermolecular forces help the polymer to be rigid, strong, and resistant to chemicals (2). Polycarbonate is easily attacked by most common solvents due to its intermediate polarity and lack of major intermolecular attraction. (3) Excessive molecular inflexibility and low intermolecular attraction combine to make polystyrene rigid but unable to withstand surface attack by surfactants and solvents. (4) Another important consideration when studying the chemical resistance of plastics is the effect of additives such as plasticizers, fillers, stablizers, and colorants.

Chemical resistance tests are conducted four different ways:

1. Immersion Test.
2. Stain-Resistance Test.
3. Solvent Stress-Cracking Resistance.
4. Environmental Stress-Cracking Resistance.

Plastic materials should not be selected solely on the basis of published chemical resistance data. The type of test conducted, test temperature, media concentration, duration of exposure, type of loading, and additives used in the base polymer must be considered, since each of the above-mentioned factors can have a significant effect on the chemical resistance of plastics. The risk potential of premature failure can be minimized by conducting the test under anticipated end-use conditions and media (5).

9.2. IMMERSION TEST (ASTM D 543)

The method of measuring the resistance of plastics to chemical reagents by simple immersion of processed plastic specimens is a standard procedure used throughout the plastics industry. The method can only be used to compare the relative resistance of various plastics to typical chemical reagents. The test results do not provide a direct indication of suitability of a particular plastic for end-use application in certain chemical environments. The limitations of results such as duration of immersion, temperature of the test, and concentration of reagents should be considered when studying the test data. For applications involving continuous immersion, the data obtained in short-time tests are useful only in screening out the most unsuitable materials.

The test equipment consists of a precision chemical balance, micrometers, immersion containers, an oven, or a constant temperature bath, and a testing device for measuring physical properties. The dimensions and type of test specimens are dependent upon the form of the material and tests to be performed. At least three test specimens are used for each material being tested and each reagent involved. For studying the weight and dimension change, each specimen is weighed and thickness is measured. The specimens are totally immersed in a container for seven days in a standard laboratory atmosphere, in such a way that no contact is made with the wall or the bottom of the container. After seven days, the specimens are removed from the container and weighed. The dimensions are remeasured. The procedure remains unchanged for studying the mechanical property changes after immersion of the test bars in reagents. The mechanical properties of nonimmersed and immersed specimens are determined in accordance with standard methods for tests prescribed in the specifications and comparison is made. Observations such as loss of gloss, swelling, clouding, tackiness, crazing, and bubbling are also reported in the test results.

9.3. STAIN-RESISTANCE OF PLASTICS (ASTM D 2290)

Plastics have deeply penetrated the household products market in the last two decades. Determination of stain-resistance of plastic materials has become increasingly important since such household products come in contact with many types of chemicals and staining reagents everyday. The test developed for determining stain-resistance applies only to the incidental contact of plastic materials with miscellaneous staining reagents. Any long-term intimate contact of the reagent with plastics must be dealt with in a different manner. Certain types of additives in plastic materials seem to contribute substantially to the staining process.

The test requires an oven, an applicator, and closed glass containers for low-viscosity liquids. A wide variety of staining reagents are used. The most common ones are found among food, cosmetics, solvents, detergents, pharmaceuticals, beverages, and cleansing agents. Jelly, tea, blood, coffee, bleach, shoe polish, crayons, lipstick, and nail-polish remover, are some examples of staining reagents.

A test specimen of any size may be used as long as it has a flat, smooth surface and is large enough to permit the test and visual examination. It is recommended that all thermosetting decorative laminates be wet-rubbed with a grade FF or equivalent grade of pumice to remove the surface gloss and then washed with mild soap. The staining reagent is applied onto the specimen with an applicator, forming a thin coat. In the case of low-viscosity liquids, the specimen is immersed in a liquid staining reagent kept in a glass container. The container is then closed and the specimens are placed in an oven at 50 ± 2°C for 16 hrs.

Excess staining material is removed from the surface after exposure and the specimen is visually observed for residual staining. Depending upon the specific requirement, the residual staining may or may not be acceptable. The color of plastic has a significant bearing on the noticeability of stains and therefore one must consider testing end-use color specimens.

9.3.1. Resistance of Plastics to Sulfide Staining (ASTM D 1712)

Many plastic compositions contain salts of lead, copper, and antimony in the form of pigments, stabilizers, fillers, and other additives. When these materials come in contact with external materials containing sulfide such as hydrogen sulfide they stain easily. For example, if a lead-stabilized PVC compound is mixed with a tin-stabilized PVC compound that contains sulfide the staining is quite evident. Industrial fumes and rubber are other two major sulfide-containing external agents.

The test to determine the resistance of plastics to sulfide staining is very simple and requires only a freshly prepared solution of hydrogen sulfide and a test specimen of any size or shape. The specimen is partially immersed in a saturated hydrogen sulfide solution for 15 min along with a control specimen with a known tendency to sulfide stain. After 15-min, the specimens are removed and examined for staining. The comparision between control, unexposed, and exposed specimens is made to determine the relative degree of staining.

9.4. SOLVENT STRESS-CRACKING RESISTANCE

One of the most difficult challenges a design engineer faces is selecting the right plastic for the right application. The chemical resistance of plastics is a prime consideration in selecting the proper material. The chemical resistance data published by material suppliers is the most convenient source of information. Such published data is usually derived from a simple immersion test, such as described earlier in this chapter. Most polymers will undergo stress cracking when exposed to certain chemical environments under high stress for a given period of time. Such cracking will occur even though some chemicals have no effect on unstressed parts, and therefore simple immersion of test specimens is an inadequate measure of chemical resistance of polymers (6). At this point it is important to understand how solvent stress-cracking occurs in a polymer. Initially, the polymer-to-polymer bond is replaced by a polymer–solvent bond by lowering the cohesive bond energies of the surface layers of the affected materials. These new polymer–solvent bonds cannot contribute to the overall strength of the material. If the stresses present exceed the cohesive strength of the weakened polymer, rupture occurs. The type and number of such fractures depend upon the stress pattern present in the material. The solvent penetrates deeper and cracks becomes more extensive with time (7).

The solvent stress-cracking phenomenon occurs in all plastics at varying degrees. However, the presence of stress, internal or external, is essential. The internal or molded-in stresses pose the biggest problem since complete removal of such stresses is practically impossible. The internal stresses can be minimized through proper design, optimizing processing conditions, and annealing the parts after fabrication. When a polymeric material is exposed simultaneously to a chemical and a stress, it can be characterized as exhibiting "critical stress," below which chemical media has no apparent effect. Critical stress is defined as the stress at which the first sign of crazing is observed when a specimen is exposed to chemical environment. Two different tests have been developed to determine critical stress. One test, often referred to as the calibrated solvent test, employs a tensile testing machine along with a standard tensile test bar. The test is carried out by stressing the tensile bar specimen to a known stress level and immediately exposing it to a chemical environment. This is accomplished by either spraying the chemical onto the specimen or continuously wetting the specimen by a wick. The specimen is exposed to the chemical for 1 min and is examined by any sign of crazing with the naked eye. If no such crazing is evident, the experiment is repeated at a higher level of stress using a fresh specimen each time until crazing is observed. The material is considered safe to use in that particular chemical environment if no crazing is observed at the yield point of the material.

One of the disadvantages of the calibrated solvent test is that it requires a large number of specimens to determine the critical stress level. One other factor is how long the specimen is exposed to chemicals. It is quite possible, that the chemical may attack the polymer if exposed for a long-time period. Since it is not practical to expend a long time for such visual testing, an accelerated method of testing must be developed. This is generally accomplished by carrying out the test at elevated temperature and high stress. As always, there is no substitute for testing an actual part by simulating the service condition; however, this test does provide some useful information regarding the behavior of the polymer exposed to a chemical environment at different stress levels. The critical stress value established for a particular polymer–solvent combination is very useful in determining the level of molded-in stresses in a part. This is further discussed in Chapter 15.

An alternate method for measuring solvent stress-cracking, developed several years ago, has a few advantages over the previous method (8). This method employs a specimen of size 4 × 1 × 0.03 in. strapped to an elliptical jig. The entire assembly, as shown in Figure 9.1, is immersed in a reagent. Because of the elliptical design of the jig, the stress at the high end of the jig is extremely low. Conversely the stress at the low end of the jig is extremely high. The level of stress in the specimen at different points on the jig can be calculated. After 1 min,

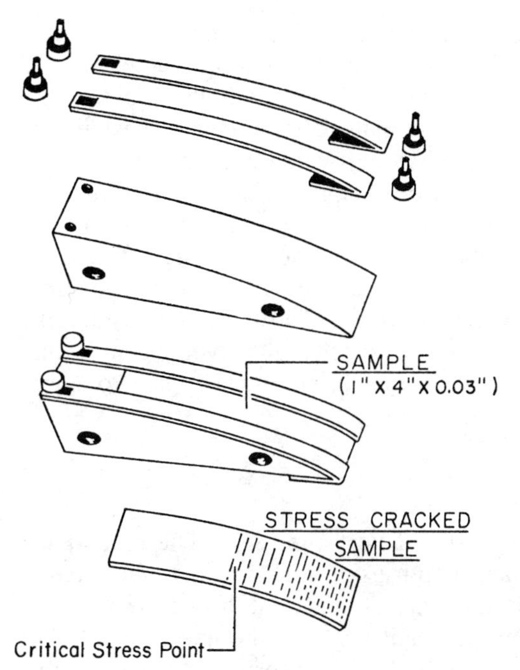

Figure 9-1. Jig for solvent stress-cracking test. (Reprinted by permission of Wiley–Interscience.)

the specimen is observed for crazing. The point at which the crazing stops is considered critical stress point. The critical stress value at this point is determined from a previously calculated value. If no crazing is observed after 1 min, the test is continued for several hours. The test may also be carried out at elevated temperatures to accelerate the stress-cracking process. The biggest advantage of this method is that one can look at the stress-cracking process over the entire range of stress values using only one specimen.

9.5. ENVIRONMENTAL STRESS-CRACKING RESISTANCE (ASTM D 1693)

Environmental stress-cracking is the failure in surface-initiated brittle fracture of a polyethylene specimen or a part under polyaxial stress in contact with a medium in the absence of which fracture does not occur under the same conditions of stress. Combinations of external and/or internal stresses may be involved, and the sensitizing medium may be gaseous, liquid, semisolid, or solid.

There are several conditions necessary for environmental stress-cracking to occur. First of all, the presence of a "stress riser" or a "notch" is a very important factor. The need for some type of stress, "molded-in" or external is inevitable. Finally, without the presence of an external sensitizing agent environmental stress-cracking is impossible (9). Environmental stress-cracking should not be confused with other types of stress cracking, such as solvent stress-cracking and thermal stress-cracking. Environmental stress-cracking describes the tendency of polyethylene products to prematurely fail in the presence of detergents, water, sunlight, oil, or other active environments, usually under conditions of relatively high strain. It is a purely physical phenomenon that involves no swelling or similar mechanical weakening of the material. Polyethylene products are most suceptible to such cracking or crazing under load when exposed to certain chemicals and environments. This phenomenon was first recognized in polyethylene-coated wire which often was lubricated with surface active materials to facilitate installation in conduits. Under these conditions, polyethylene, which appeared to perform satisfactorily in laboratory, very rapidly developed severe cracks that propagated completely through to the conductor (10). Stress-cracking resistance of polyethylene can be improved by increasing molecular weight, reducing stresses by proper fabrication practices, and by incorporating elastomers in the formulation. It is further observed that narrow molecular weight distributions considerably improve the resistance of a polymer of given density and average molecular weight. Large crystalline structures and molecular orientations appear to aggravate the problem (11).

9.5.1. Test Procedure

The test specimens of size 1.5 × $\frac{1}{2}$ in. are cut very precisely. The rectangular specimen is nicked to a fixed length and depth using a sharp blade mounted in the nicking jig (Figure 9.2). The nicked specimen is then bent through 180° so that the nick is on the outside of the bend and right angle to the line of bend. The samples are mounted onto the holder. The holder is inserted in the test tube. Immediately after that, the test tube is filled with fresh reagent to submerge the samples. The reagent can be a surface active agent, soap, or any other liquid organic substance. One of the most commonly used reagent is Igepal CO-630, manufactured by GAF Corporation. The tube is placed in a constant temperature bath maintained at 50 ± 0.5°C or 100.0 ± 0.5°C depending upon the conditions selected for the test. Test specimens are removed after a specified time and observed for crazing. Figure 9.3 illustrates test specimen, specimen holder, and test assembly.

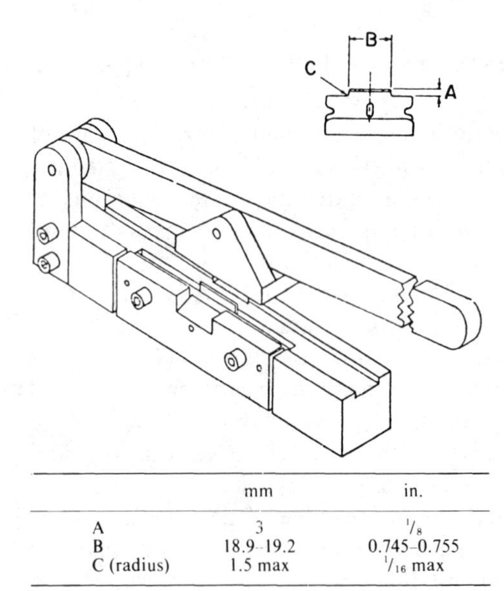

	mm	in.
A	3	$^1/_8$
B	18.9–19.2	0.745–0.755
C (radius)	1.5 max	$^1/_{16}$ max

Figure 9-2. Nicking jig. (Reprinted by permission of ASTM.)

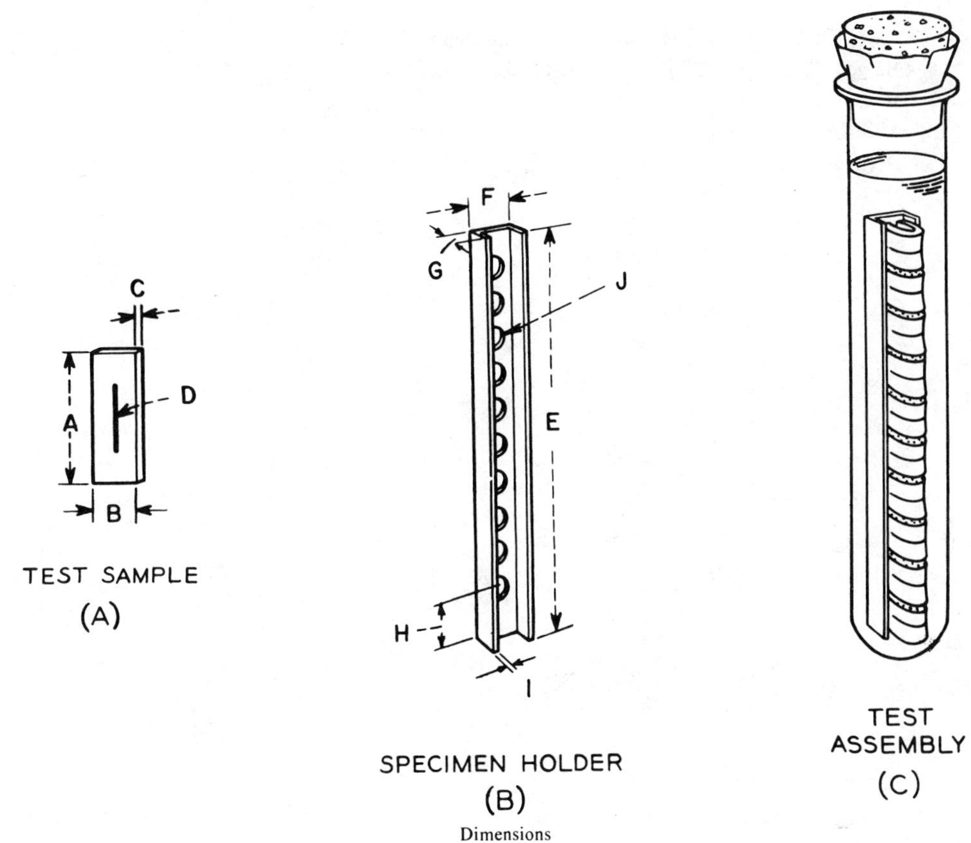

Dimensions

	mm	in.
A	38 ± 2.5	1.5 ± 0.1
B	13 ± 0.8	0.5 ± 0.03
C	See Table 1	
D	See Table 1	
E	165	$6\,^1/_2$
F (outside)	16	$^5/_8$
(inside)	11.75 ± .05	0.463 ± 0.002
G	10	$^3/_8$
H	15	$^{37}/_{64}$
I	2	0.081 (12 B & S)
J	Ten 5-mm holes, 15-mm centers	Ten $^3/_{16}$-in. holes, $^{19}/_{32}$-in. centers

Figure 9-3. Test equipment. (Reprinted by permission of ASTM.)

REFERENCES

1. Richardson, T. A., *Modern Industrial Plastics*, Howard W. Sams and Co., Indianapolis, 1974, p. 112.
2. Deanin, R. D., *Polymer Structure, Properties and Applications*. Cahners, Boston, 1972, p. 455.
3. *Ibid*, p. 449.
4. *Ibid*, p. 427.
5. Borg–Warner Corporation, "Chemical Resistance." *Tech. Bull. Design Tip* No. 6, Parkersburg, W.Va.
6. Smith, W. M., *Manufacture of Plastics*, Vol. 1, Reinhold, New York, 1964, p. 443.

7. Baer, E., *Engineering Design for Plastics*, Reinhold, New York, 1964, p. 778.
8. Bergen, R. L., Jr., "Stress Cracking of Rigid Thermoplastics," *SPE ANTEC*, **8** (1962).
9. Baer, E., Reference 7, p. 772.
10. *Modern Plastics Encyclopedia*, McGraw-Hill, New York, 1967, p. 238.
11. Brydson, J. A., *Plastics Materials*, Reinhold, New York, 1970, p. 117.

CHAPTER 10

Analytical Tests

10.1. INTRODUCTION

Analytical tests are important to material suppliers and processors. These tests provide very basic information that is necessary for characterizing and qualifying the material. Analytical tests, such as density and specific gravity tests, are used as a means of assuring product uniformity. Very few plastics are sold today without additives and modifiers. These additives and modifiers tend to alter the physical properties of the base material depending upon the amount and type used. Compounders of such additives make specific gravity value part of their product specification.

Plastics, unlike metals and ceramics, absorb water. The amount of water absorption depends upon the specific type of plastic. The key properties, such as mechanical, electrical and optical properties, are seriously affected. Water also tends to act as a plasticizer and lowers the softening temperature of the part (1). Plastic materials that absorb a large amount of water normally affect the dimensional stability of the product. The plastic product designer must take into account the water absorption characteristics of the plastic materials to avoid premature failures.

Another important and frequently used test throughout the industry is the moisture analysis test. This simple but effective test provides useful information regarding the processibility of plastic materials. Excessive moisture can cause many processing and visual problems such as splay marks. Tests such as bulk density tests and sieve analysis tests also help to predict the material's behavior during mixing, compounding, and processing. Chapter 7 discusses analytical tests often used in material characterization.

10.2. SPECIFIC GRAVITY (ASTM D 792)

Specific gravity is defined as the ratio of the weight of the given volume of a material to that of an equal volume of water at a stated temperature. The temperature selected for determining the specific gravity of plastic parts is 23°C.

Specific gravity values represent the main advantage of plastics over other materials, namely, light weight. All plastics are sold today on a cost per pound basis and not on a cost per unit volume basis. Such a practice increases the significance of the specific gravity considerably in both purchasing and production control. Two basic methods have been developed to determine specific gravity of plastics depending upon the form of plastic material. Method A is used for a specimen in forms such as sheet, rods, tubes, or molded articles. Method B is developed mainly for material in the form of molding powder, flakes or pellets.

10.2.1. Method A

This method requires the use of a precision analytical balance equipped with a stationary support for an immersion vessel above or below the balance pan. A corrosion-resistant wire for suspending the specimen and a sinker for lighter specimens with a specific gravity of less than 1.00 are employed. A beaker is used as an immersion vessel. A typical set-up for the specific gravity test is shown in Figure 10-1. The test specimen of any convenient size is weighed in air. Next, the specimen is suspended from a fine wire attached to the balance and immersed completely in distilled water. The weight of a specimen in water (and sinker, if used) is determined. The specific gravity of the specimen is calculated as follows:

$$\text{Specific gravity} = \frac{a}{(a + w) - b}$$

where a = weight of specimen in air; b = weight of specimen (sinker, if used)

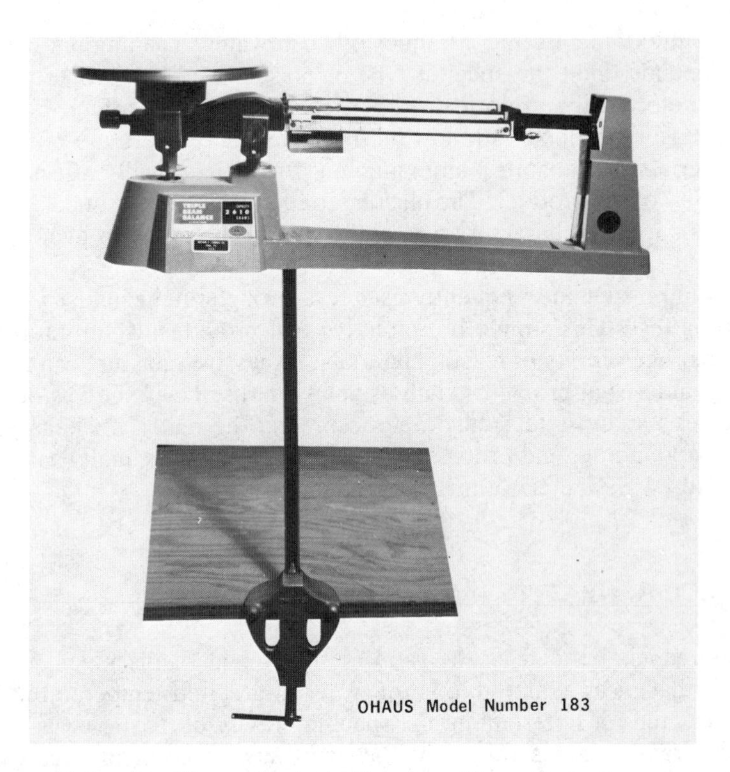

Figure 10-1. Specific gravity test. (Courtesy Ohaus Scale Corporation.)

and wire in water; w = weight of totally immersed sinker (if used) and partially immersed wire.

10.2.2. Method B

This method, suitable for pellets, flakes, or powder, requires the use of an analytical balance, a pycnometer, a vacuum pump, and a vacuum desiccator. The test is started by first weighing the empty pycnometer. The pycnometer is filled with water and placed in a water bath until temperature equilibrium with the bath is attained. The weight of the pycnometer filled with water is determined. After cleaning and drying the pycnometer, 1–5 g of material is added and the weight of the specimen plus the pycnometer is determined. The pycnometer is filled with water and placed in a vacuum desiccator. The vacuum is applied until all the air has been removed from between the particles of the specimen. Lastly, the weight of the pycnometer filled with water and the specimen is recorded. The specific gravity is calculated as follows:

$$\text{Specific gravity} = \frac{a}{(b + a - m)}$$

where a = weight of the specimen; b = weight of the pycnometer filled with water; m = weight of the pycnometer containing the specimen and filled with water.

If the water is substituted by another suitable immersion liquid, the specific gravity of the immersion liquid must be determined and taken into account in calculating the specific gravity.

10.3. DENSITY BY DENSITY GRADIENT TECHNIQUE (ASTM D 1505)

The density of plastic materials is defined as the weight per unit volume and is expressed in g/cm^3 or lb/ft^3. The test method, developed to determine the density of plastics very accurately, is based on observing the level to which a test specimen sinks in a liquid column exhibiting a density gradient in comparison with standard specimens of known density. A number of calibrated glass floats of precisely known density are introduced into the density gradient and allowed to sink in the column to a point where the glass floats' density matches that of the solution. A series of such floats of differing densities within the range of the column serves as a means of calibrating the column (2). The float position versus float density is plotted on a chart large enough to be read accurately to ± 1 mm to obtain a calibration line. When a specimen of unknown density is introduced into the column, the measurement of its position upon reaching equilibrium, when referred to the calibration line gives an accurate measurement of its density.

An alternate method of density determination requires numerical calculation. Table 10-1 lists a number of liquid systems recommended for use in density gradient columns. Figure 10-2 illustrates a typical commercially available density gradient column. A number of papers have been presented on this subject (3–6).

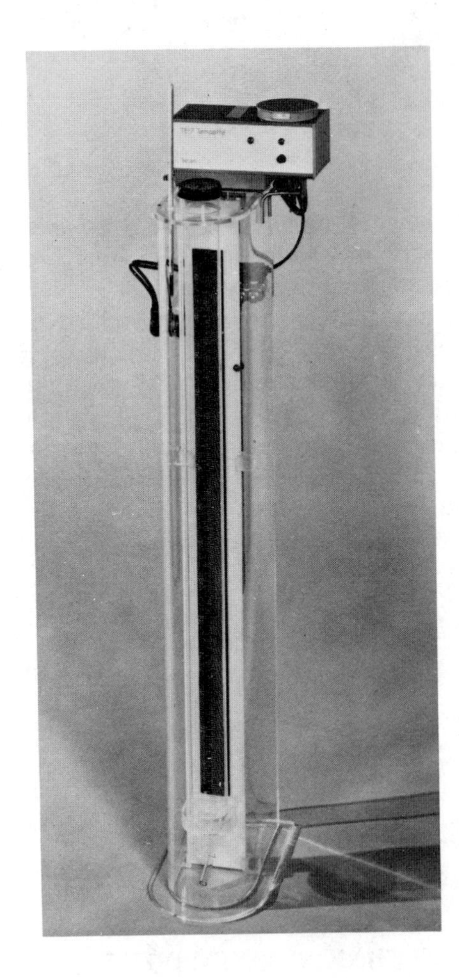

Figure 10-2. Density gradient column. (Courtesy Techne, Inc.)

Table 10-1. Liquid Systems Recommended for Use in Density Gradient Columns[a]

System	Density range (g/ml)
Methanol–benzyl alcohol	0.80–0.92
Isopropanol–water	0.79–1.00
Isopropanol–diethylene glycol	0.79–1.11
Ethanol–carbon tetrachloride	0.79–1.59
Ethanol–water	0.79–1.00
Toluene–carbon tetrachloride	0.87–1.59
Water–sodium bromide	1.00–1.41
Water–calcium nitrate	1.00–1.60
Zinc chloride–ethanol–water	0.80–1.70
Carbon tetrachloride–1,3-dibromopropane	1.60–1.99
1,3-Dibromopropane–ethylene bromide	1.99–2.18
Ethylene bromide–bromoform	2.18–2.89
Carbon tetrachloride–bromoform	1.60–2.89
Tetrachloroethylene–bromoform	1.55–2.70

[a] Reprinted courtesy of Techne, Inc.

10.4. BULK (APPARENT) DENSITY TEST (ASTM D 1895)

Apparent density is a measure of the fluffiness of a material. Bulk density is defined as the weight per unit volume of a material, including voids inherent in the material as tested. Bulk density is commonly used for materials such as molding powders. The test method to determine bulk density has been discussed in detail in Chapter 7.

10.5. WATER ABSORPTION (ASTM D 570)

The tendency of plastics to absorb moisture simply cannot be overlooked since even the slightest amount of water can significantly alter some key mechanical, electrical, or optical property. Water absorption characteristics of plastics depend largely upon the basic type and final composition of a material. For example, materials containing only hydrogen and carbon such as polyethylene and polystyrene, are extremely water-resistant, whereas plastics having oxygen or oxy–

Table 10-2. Water Absorption of Common Plastics[a]

Plastic material	Percent absorption
ABS	0.20–0.45
Acetal	0.22–0.25
Alkyd	0.50–0.25
Acrylic	0.30–0.40
Cellulose acetate	2.00–7.00
Cellulose acetate butyrate	0.90–2.20
Cellulose propionate	1.20–2.80
CTFE	0.00
Epoxy (unfilled)	0.08–0.15
FEP	0.01
Nylon	
Type 6	1.30–1.90
Type 66	1.50–2.0
Type 610	0.40
Type 612	1.5
Type 11	1.10
Polycarbonate	0.15–0.35
Polyester (thermoplastic)	0.8–0.38
Polyethylene	0.010
PPO (Noryl)	0.06–0.07
Polypropylene	0.010
Polysulfone	0.22
Polystyrene	0.03–0.6
SAN	0.2–0.3
TFE	0.01
Urea formaldehyde (cast)	0.02–1.50
PVC	0.07–0.75

[a] From R. Milby, *Plastics Technology*. Reprinted by permission of McGraw–Hill Book Company.

Table 10-3. Immersion Temperatures and Periods

Immersion period	Immersion temperature	Comments
24 hr	23°C	Average materials
2 hr	23°C	Materials with relatively high rate of absorption
Long term	23°C	Test continued until specimen saturation
2 hr	100°C (boiling water)	Water absorption at elevated temperature
½ hr	100°C (boiling water)	Materials with relatively high rate of absorption
Cyclic immersion	23°C 100°C	For special applications such as dinnerware and washing machine agitators

hydrogen groups are very susceptible to water absorption. Cellulose acetate and nylons are good examples of the preceding type. Materials containing chlorine, bromine, or fluorine are water-repellant. Fluorocarbon, such as PTFE, is one such type of water-repellant material (7). Water absorption characteristics of plastic materials are altered by the addition of additives such as fillers, glass fibers, and plasticizers. These additives show a greater affinity to water, especially when they are exposed to the outer surface of the molded article. Some plastics absorb very little water at room temperature, but at higher temperatures, they tend to take in a considerable amount of water and lose properties rapidly. Washing machine agitators, plastic dinnerware, irrigation valves, and sprinklers are examples of applications requiring low-water absorption. Table 10-2 lists typical water absorption values of some common plastics.

The test to determine the water absorption of plastics is relatively simple. Only two pieces of equipment are required—an analytical balance and an oven capable of maintaining a uniform temperature. The test specimen may be a molded disk or a piece cut from a sheet, rod, or tube. Dimensions vary according to the type of specimen. A special conditioning procedure must be followed before actual testing. The specimens are dried in an oven at a specified temperature for a predetermined time and followed by a cooling period in a desiccator and immediately weighed. Table 10-3 shows commonly used immersion temperatures and periods.

Percent increase in weight during immersion is calculated as follows:

$$\text{Increase in weight percent} = \frac{\text{Wet weight} - \text{Conditioned weight}}{\text{Conditioned weight}} \times 100$$

10.6. MOISTURE ANALYSIS

The hygroscopic nature of plastic materials causes processing as well as dimensional stability problems. Materials like ABS and polycarbonate must be dried thoroughly before processing in order to avoid splay marks on molded parts, loss of impact, and loss of other properties. The presence of moisture also tends to produce a weak weld or knit lines, further weakening the molded part. Many

Six Simple Steps to the Resin Moisture Test

Equipment needed consists of: 1) hot plate capable of maintaining surface temperature of 525 F ±25° F, 2) 75 × 25 mm glass microscope slides, 3) tweezers capable of

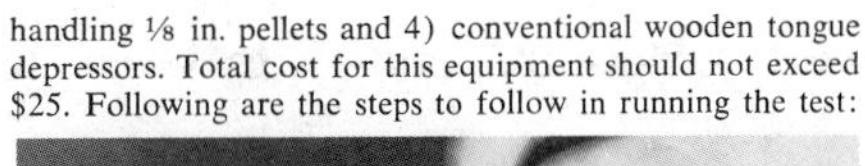
handling ⅛ in. pellets and 4) conventional wooden tongue depressors. Total cost for this equipment should not exceed $25. Following are the steps to follow in running the test:

1. Plug in hot plate (be sure surface is clean) and calibrate it to a surface temperature of 525±25° F. Place two glass slides on surface for 1-2 min.

2. By this time the glass surface temperature should have reached 450-500 F. Use your tweezers to place four or five pellets on one of the glass slides.

3. Now place a second hot slide over the first one to sandwich the pellets between them.

4. Press a tongue depressor on the top of the sandwich until the pellets flatten out to about ½ in. dia.

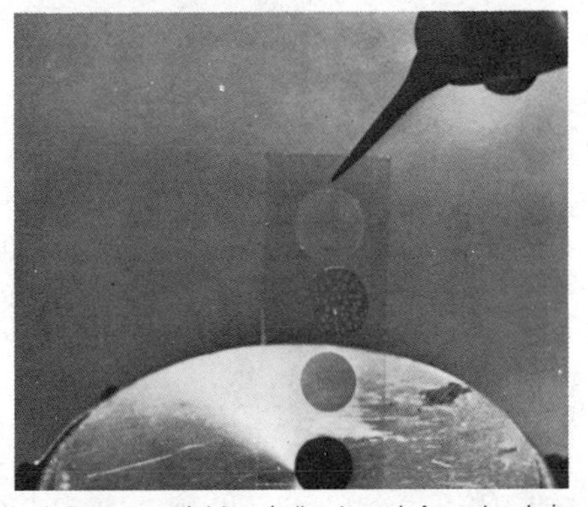

5. Remove sandwich and allow to cool. Amount and size of bubbles indicate percentage of moisture as indicated in photo on facing page, correlating bubbles with moisture.

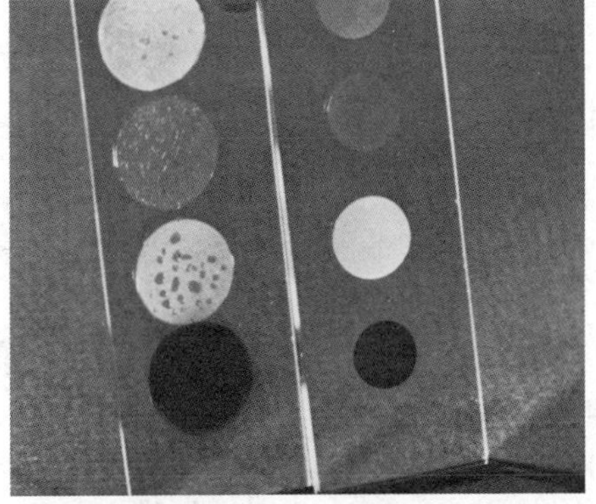

6. Here are typical results. Slide at right indicates dry material; slide at left indicates moisture-laden material. One or two bubbles may be only trapped air.

Figure 10-3. G.E. moisture test. (Reprinted by permission of *Plastic Technology*, March 1969.)

processors conduct routine moisture analysis tests on materials prior to processing. Two basic methods have been developed and are most frequently used.

10.6.1. Moisture Analysis by Oven Drying

This method is used to determine the moisture content of plastic materials supplied in powder or pellet form. The test requires the use of an oven, a weighing pan, and an extremely sensitive scale capable of weighing up to 0.001 g accurately. The test is carried out by weighing a small quantity of material in the weighing pan and placing it in an oven at a specified temperature for 1–2 hr. The pan with the material is removed from the oven at the end of the test period and placed in a desiccator for 30 min and allowed to cool. The pan is reweighed to the nearest 0.001 g. Percent moisture is calculated as follows:

$$\text{Percent moisture} = \frac{A - C}{A - B} \times 100$$

where A = weight of the pan and material; B = weight of the empty pan; C = weight of the pan and the material after drying.

10.6.2. TVI Drying Test

This test was developed by a General Electric engineer and is called the Thomasetti Volatile Indicator (TVI) test. This is a low-cost, quick, and simple method for determining the readiness of moisture-sensitive thermoplastic materials for processing. The test, however, does not determine the moisture content of the material but indicates the absence or presence of moisture in the material. A hot plate capable of maintaining up to 600 ± 25°F, glass microscope slides, tweezers, and a wooden tongue depressor are required for the test. Figure 10-3 shows the step-by-step procedure for conducting this test. This test cannot be used to determine the presence or absence of moisture in glass-reinforced thermoplastics.

10.7. SIEVE ANALYSIS (PARTICLE SIZE) TEST (ASTM D 1921)

The particle size and particle size distribution are important since these two characteristics of materials have a great effect on compounding, processing, and bulk handling. Large and fairly uniform particles are easier to handle and process. Fine particles are difficult to handle and difficult to process. Fine particles, when mixed with large particles, tend to cause uneven melting and hence, nonuniform mold filling, orange peel, and other surface problems. In the case of PVC dry blending operations, fine particles do not allow the plasticizer to be absorbed evenly throughout the batch. Conversely, oversized particles are unable to absorb the plasticizer in sufficient amounts resulting in poor fusing and creating the possibility of gels in the end-products (8). Particle size and particle distribution of dispersion resins affect the viscosity and stability of plastisols and organosols. Larger particles in plastisol compounds fuse more slowly and therefore yield poor physical properties (9). The large particle size of certain fillers, such as calcium carbonate,

Figure 10-4. Sieve analysis. (Courtesy Fisher Scientific Company.)

tends to increase the wear on the extruder screws and barrel and to reduce the physical properties of the end-products (10).

The test method used to determine the particle size and particle size distribution employs a series of sieves with various opening sizes. The material is simply poured from the top and allowed to pass through a series of sieves and is collected at the bottom. The quantity of material retained on each sieve is determined by weighing the sieves before and after the test. A shaker is employed to facilitate separation of various sized particles. Figure 10-4 illustrates a typical set-up for sieve analysis.

REFERENCES

1. Levy, S., and DuBois, J. H., *Plastics Product Engineering Handbook,* Reinhold, New York, 1977, p. 211.
2. Techne Inc., Princeton, N.J., *Tech. Bull.: Density Gradient Column,* Techne Catalog No. 202.
3. Boyer, R. F., Spencer, R. S., and Wiley, R. M., "Use of Density Gradient Tube in the Study of High Polymers," *J. Polym. Sci.,* **1** (1946), p. 249.
4. Tung, L. H., and Taylor, W. C., "An Improved Method of Preparing Density Gradient Tubes," *J. Polym. Sci.,* **21** (1956), p. 144.
5. Mills, J. M., "A Rapid Method of Construction of Linear Density Gradient Columns," *J. Polym. Sci.,* **21** (1956), p. 585.
6. Wiley, R. E., "Setting up Density Gradient Laboratory," *Plast. Tech.* **8**(3) (1962), p. 31.
7. Milby, R., *Plastics Technology*. McGraw-Hill, New York, 1973, pp. 534–536.
8. Schoengood, A. A., "PVC Primer," *Plast. Eng.* (Dec. 1973), p. 29.
9. *Ibid,* p. 30.
10. Prust, R. S., "Quality Control in PVC Compounding," *Plast. Compounding* **1**(3) (May–June 1978), p. 25.

CHAPTER 11

Conditioning Procedures

11.1. CONDITIONING (ASTM D 618)

A true material comparison is possible only when property values are determined by identical test methods under identical conditions (1). Generally speaking, physical and electrical properties of plastics and electrical insulating materials are affected by temperature and humidity. Plastic materials tested above room temperature will yield relatively higher impact strength and lower tensile strength and modulus. High humidity tends to alter the electrical property test results. Obviously, in order to make reliable comparisons of different materials and test results obtained by different laboratories, it is necessary to establish standard conditions of temperature and humidity.

Conditioning is defined as the process of subjecting a material to a stipulated influence or combination of influences for a stipulated period of time (2). Three basic reasons for conditioning of specimens are:

1. To bring the material into equilibrium with normal or average room conditions.
2. To obtain reproducible results regardless of previous history or exposure.
3. To subject the material to abnormal conditions of temperature and humidity in order to predict its service behavior.

STANDARD LABORATORY TEMPERATURE: Standard laboratory temperature is defined as 23°C (73.4°F) with a standard tolerance of ±2°C (±3.6°F).

STANDARD LABORATORY ATMOSPHERE: Standard laboratory atmosphere is defined as an atmosphere having a temperature of 23°C (73.4°F) and a relative humidity of 50 percent with a standard tolerance of ±2°C (±3.6°F) and ±5 percent, respectively.

11.2. DESIGNATION FOR CONDITIONING

Conditioning of a test specimen is designated as:

$$A/B/C$$

where A = a number indicating duration of conditioning (hrs); B = a number

Table 11-1. Conditioning Procedures

Conditioning procedure	Specimen thickness[a] (in.)	Duration (hr)	Temperature (°C)	Humidity (percent)	Special requirement	Application
A	*X*	40	23 ± 2	50 ± 5	None	Majority of tests
	Y	88	23 ± 2	50 ± 5	None	
B	*X*	48	50 ± 2	—	Cool to room temperature for 5-hrs in desiccator over anhydrous calcium chloride	Thermosetting materials
	Y	48	50 ± 2	—	Cool to room temperature for 15-hrs in desiccator over anhydrous calcium chloride	
C	—	96 ± 2	35 ± 1	50 ± 2	—	Studying effect of severe atmospheric moisture
D	—	24 ± $\frac{1}{2}$	23 ± 1	—	Immersion of specimen in distilled water	Electrical and mechanical tests
E	—	48 ± $\frac{1}{2}$	50 ± 1	—	Cool specimen to 23 °C by immersing in distilled water for 1-hr	Electrical and mechanical tests
F		—	23 ± 1	96 ± 1	Time as specified in applicable material specification	—

[a] X = 0.250 and less than 0.250-in.; Y = 0.250-in. and over.

indicating conditioning temperature (°C); C = a number indicating relative humidity (percent or a word) to indicate immersion in liquid.

A sequence of condition is denoted by use of plus(+) sign between successive conditions.

Example. 40/23/50 indicates conditioning for 40 hrs at 23°C at 50 percent R.H.
48/50 + Des indicates conditioning for 48 hrs at 50°C followed by desiccation.

Table 11-1 summarizes conditioning procedures.

REFERENCES

1. Borg-Warner Tech. Rept.: Measurement, Reporting and Interpretation of Thermoplastic Properties, Parkersburg, W. Va., Report No: P-127.
2. Lever, A. E., and Rhys, J. A., *The Properties and Testing of Plastic Materials,* Temple Press, Feltham, England, 1968, p. 7.

SUGGESTED READING

Schmitz, J. V. (Ed.), *Testing of Polymers,* Vol. 1, Interscience, New York, 1965, pp. 41–85.

CHAPTER 12

Miscellaneous Tests

12.1. TORQUE RHEOMETER TEST

The torque rheometer is one of the most versatile pieces of equipment for research and development, production control, and quality control work on polymeric materials. It is a laboratory tool often used to predict processing behavior and to simulate realistic-use conditions. Some of the typical applications include determining melt flow values, stability of polymers and degradation time at varying shear rates, studying rheological properties, studying pigment dispersion, characterizing different formulations, and observing the effects of changing ingredients and temperatures.

The torque rheometer is a torque-measuring rheometer based on the dynamometer principle. A sample of material to be tested is placed in the mixing head where it is subjected to shear by means of two rotating blades. The sample material is also subjected to high temperatures. The dynamometer is suspended freely between two bearing blocks, while it drives the rotors of the measuring head. The shear rate, measured by the angular velocity of the rotors, is set according to a tachometer. The measuring head rotors encounter a resistance torque from the test material that causes the dynamometer to rotate in the opposite direction. The reaction torque is balanced out through the lever system against the torque indicator scale, simultaneously recording on a strip chart recorder. An oil dashpot dampens the movement of the lever system. By sliding the weight on the arm, the zero can be suppressed several times, thus increasing the range without influencing its sensitivity setting. The measuring head is either electrically or oil heated. Figure 12-1 illustrates a torque rheometer schematically.

The torque rheometer is used most extensively in the PVC compounding operation. A great deal of useful information can be derived from a simple fusion test. The fusion test is carried out by weighing a predetermined amount of PVC compound and introducing the charge into the preheated mixing head with rotor blades rotating at a specified rpm. The material is allowed to flux and reach fusion point and continue to the point of degradation. The torque is recorded as a function of time on a strip chart recorder. Figure 12-2 illustrates a typical fusion chart showing fusion point, time to flux, maximum torque at fusion, and total stability of the compound. By altering the quantity of the different additives such as lubricants and stabilizers, one can observe the effect of these variations in terms

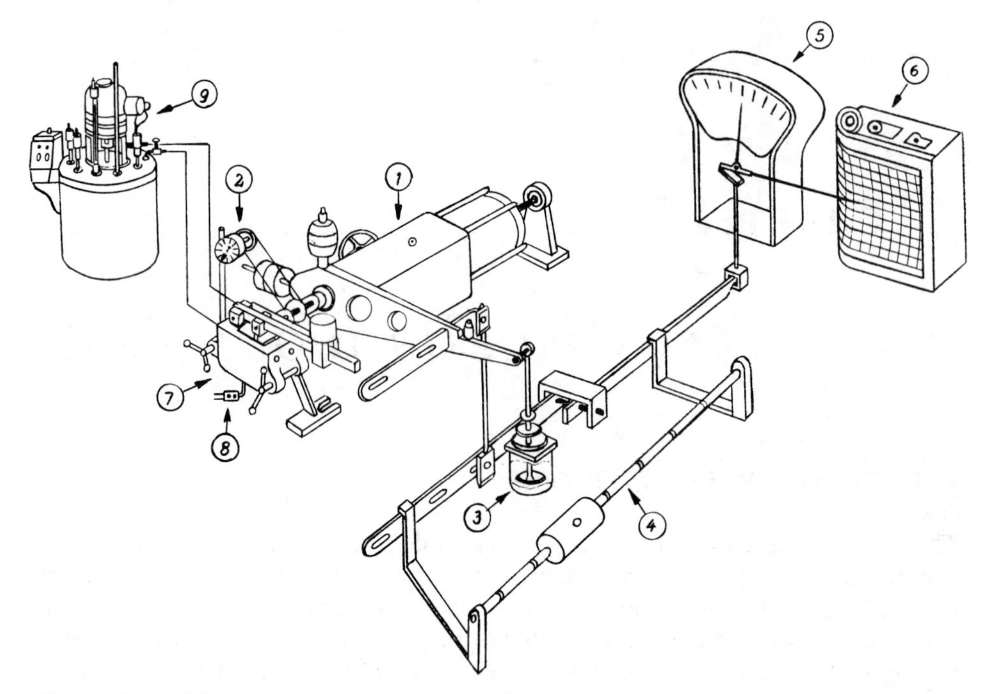

Figure 12-1. Schematic torque rheometer. (Courtesy C. W. Brabender Instruments, Inc.)

of fusion torque, fusion time, and total stability. The correlation between the values obtained from the torque rheometer experiment and the actual manufacturing processes such as extrusion and blow molding can be established. The torque rheometer can also be used to perform capillary flow analysis by simply attaching a small extruder to the dynamometer in place of the mixing head (1). Figure 12-3 illustrates a commercially available torque rheometer.

12.2. PLASTICIZER ABSORPTION TESTS

The ability of polyvinyl chloride (PVC) resin to absorb a plasticizer is of considerable interest to a PVC compounder. The amount of plasticizer that can be added to PVC resin depends upon the type of compound formulated. The amount may vary anywhere from 20 PHR in the case of a flexible extrusion compound to 80 PHR in the case of plastisols. In dry blending a flexible profile formulation for extrusion, it is imperative that the plasticizer added to the PVC resin gets fully absorbed, yielding a dry compound. A semidry powder blend can cause processing as well as conveying problems. The plasticizer absorption efficiency is related to the rate of heating of the resin–plasticizer mix, the type of resin, the particle size and distribution, the surface–volume ratio, and the type of plasticizer. Additives such as filler and impact modifiers also have an effect on plasticizer absorption.

Three basic methods have been developed to study plasticizer absorption characteristics:

1. Plasticizer absorption, Burette method.
2. Plasticizer absorption using a torque rheometer.
3. Plasticizer absorption under applied centrifugal force.

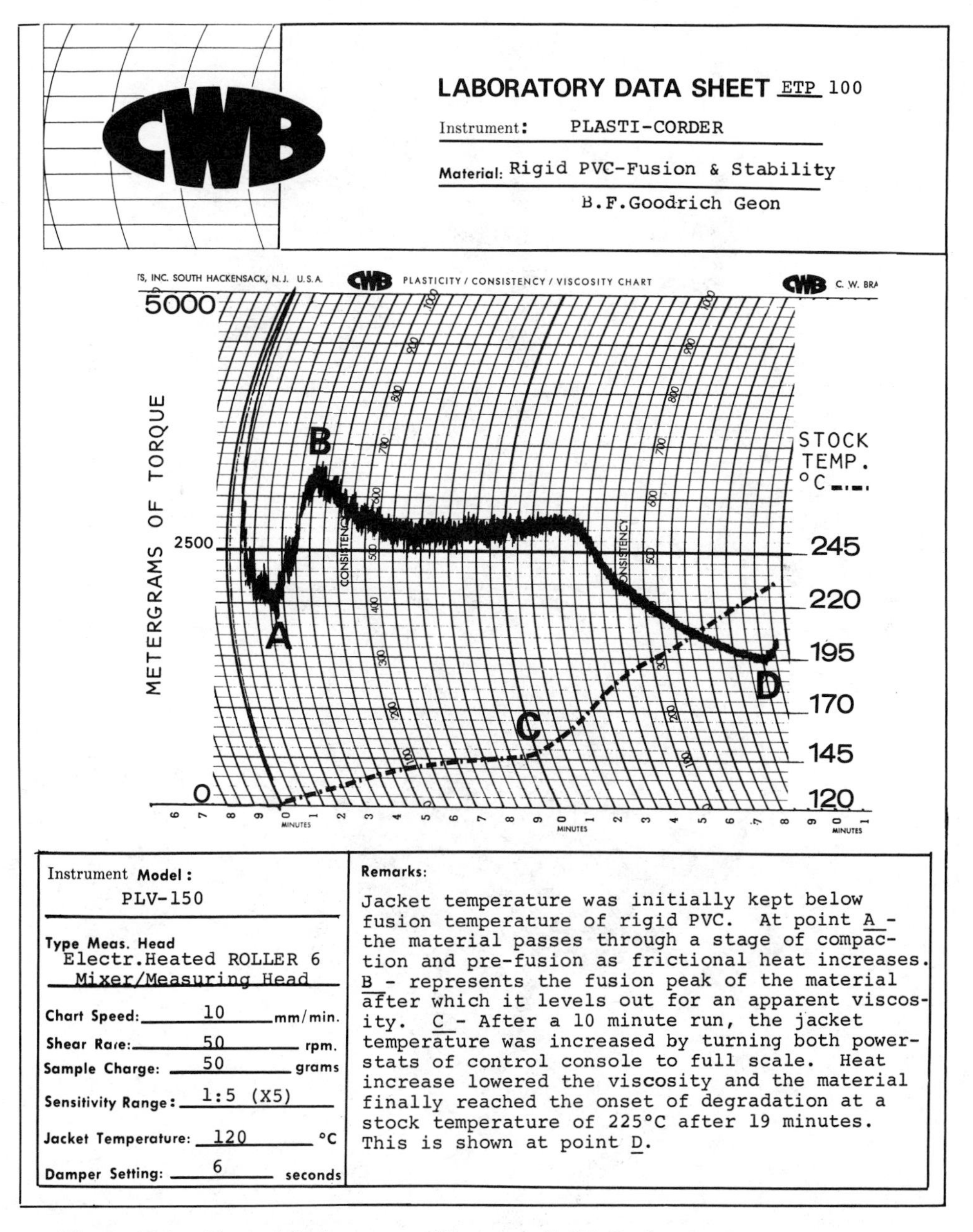

LABORATORY DATA SHEET ETP 100

Instrument: PLASTI-CORDER

Material: Rigid PVC-Fusion & Stability
B.F.Goodrich Geon

Instrument Model: PLV-150

Type Meas. Head: Electr.Heated ROLLER 6 Mixer/Measuring Head

Chart Speed: 10 mm/min.

Shear Rate: 50 rpm.

Sample Charge: 50 grams

Sensitivity Range: 1:5 (X5)

Jacket Temperature: 120 °C

Damper Setting: 6 seconds

Remarks:

Jacket temperature was initially kept below fusion temperature of rigid PVC. At point A - the material passes through a stage of compaction and pre-fusion as frictional heat increases. B - represents the fusion peak of the material after which it levels out for an apparent viscosity. C - After a 10 minute run, the jacket temperature was increased by turning both powerstats of control console to full scale. Heat increase lowered the viscosity and the material finally reached the onset of degradation at a stock temperature of 225°C after 19 minutes. This is shown at point D.

Figure 12-2. Typical fusion curve. (Courtesy C. W. Brabender Instruments, Inc.)

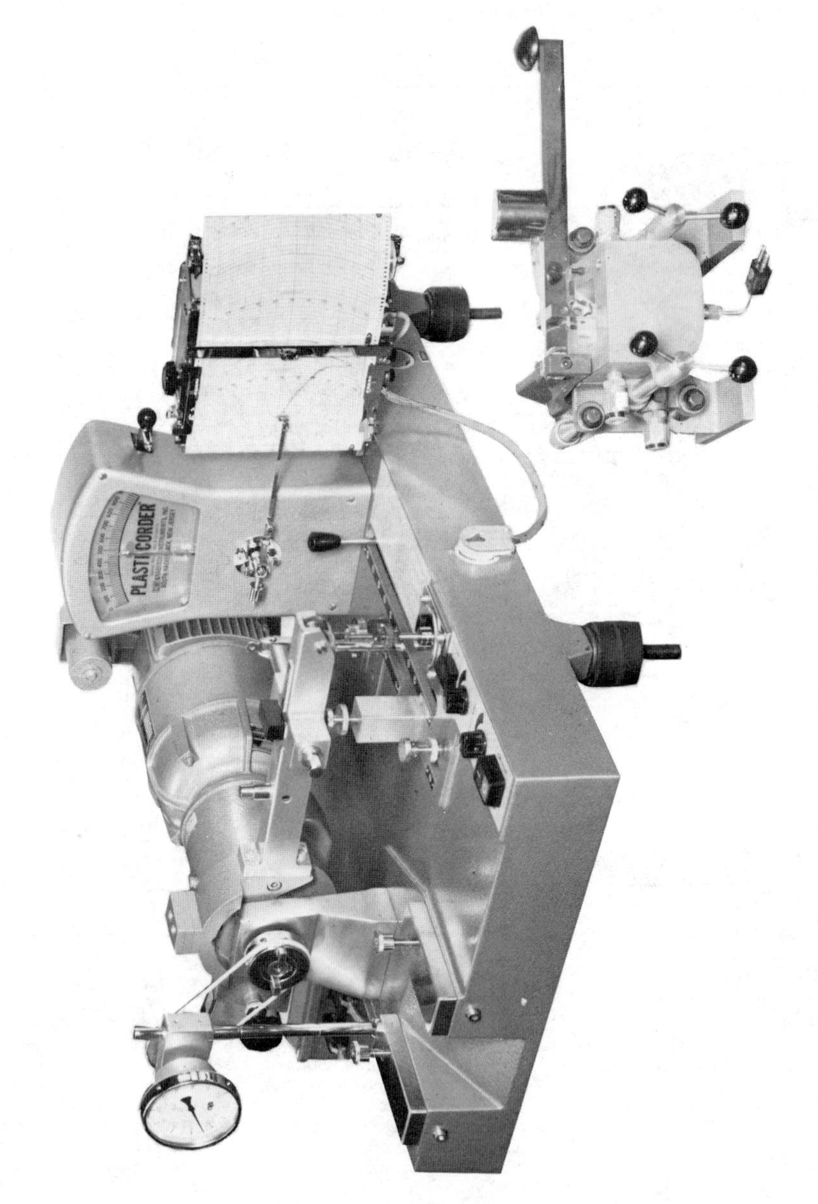

Figure 12-3. Brabender torque rheometer. (Courtesy C. W. Brabender Instruments, Inc.)

12.2.1. Plasticizer Absorption, Burette Method (ASTM D 1755)

This is a quick and simple method for determining the ability of resin to absorb a plasticizer in the standard laboratory atmosphere. A burette with a shortened tip to increase the rate of flow, a titration stand with a glazed tile base, a spatula, and a balance are employed for the test. The burette is filled with commercial grade DOP (di-2 ethylhexyl-phthalate). The next step is to weigh out 5 ± 0.01 g of resin accurately and place it under the burette on the glazed tile. A small amount of plasticizer is added to the resin slowly and distributed throughout the resin using the spatula. This dropwise addition of plasticizer is carried out until the flow point. The flow point is the point at which the resin plasticizer mixture will flow off the spatula. The entire test is repeated to verify the reproducibility. The calculation is carried out as follows:

$$\text{Plasticizer (PHR)} = \frac{\text{cc Plasticizer} \times 100 \times \text{density of plasticizer}}{5}$$

12.2.2. Plasticizer Absorption Using a Torque Rheometer (ASTM D 2396)

This method determines the powder mixing characteristics of the polyvinyl chloride resin. The function of the torque rheometer is described in Section 12.1.

The test requires the use of a torque rheometer equipped with a sigma-style mixer measuring head as shown in Figure 12-4. The mixer is heated either electrically or by circulating heat-transfer oil through the jacket. In order to obtain

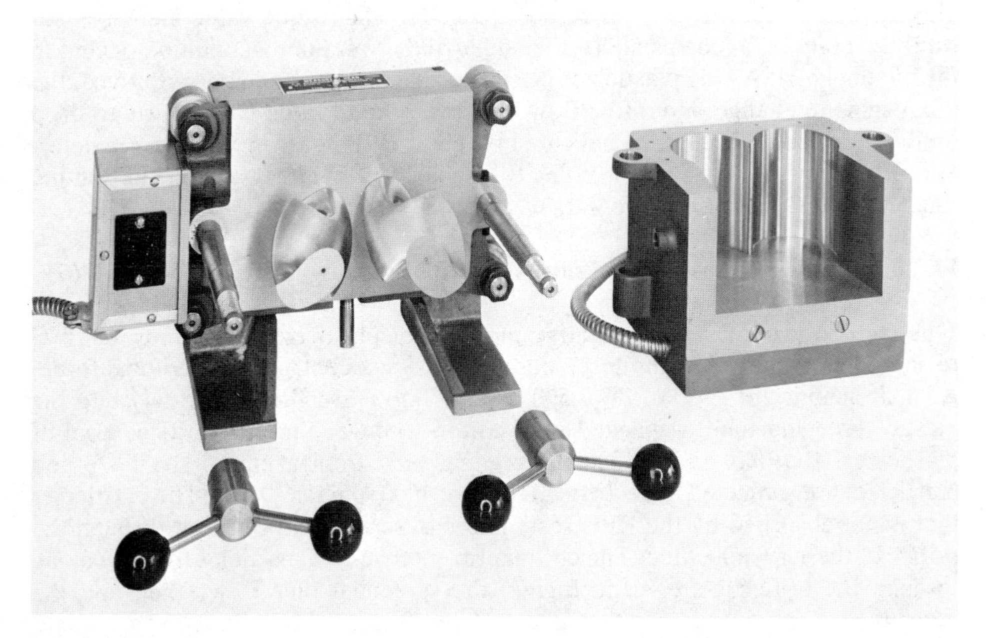

Figure 12-4. Sigma-style mixer for plasticizer absorption test. (Courtesy C. W. Brabender Instruments, Inc.)

consistency, a standard formulation is established as follows:

PVC resin		225.00 ± 0.1 g
Clay		15.75 ± 0.1 g
Basic lead carbonate		22.50 ± 0.1 g
DOP plasticizer		117.00 ± 0.1 g
	Total	380.25 g

Because of lot-to-lot variations in the quality of plasticizer, clay, and basic lead carbonate, it is recommended that the laboratory maintain a large enough inventory of these additives to establish control standards. A standard powder–mix curve should also be generated using standard additives and kept on file for comparison purposes.

The following standard test conditions are used:

Temperature	88 ± 1°C
Mixer speed	63 ± 1 rpm
Material weight	380 ± 10 g

All ingredients except the plasticizer are weighed into the container and mixed thoroughly. The mixer is preheated and allowed to run at a specified speed for 30 min to obtain equilibrium conditions. All dry additives are added to the mixer and allowed to mix for 5 min. Next, the plasticizer is poured quickly into the mixer and mixing is continued for 10 min. beyond the dry point. Figure 12-5 illustrates the entire mixing process graphically. The torque value increases abruptly at point A, as the plasticizer is added and a wet lumpy condition occurs in the mixing bowl. As the plasticizer gets absorbed into the resin and additives, the mix begins to change into a free-flowing powder and torque value starts to drop until the dry point occurs, as indicated by point B. Powder mixing time is determined by drawing two lines at point B as shown in Figure 12-5 and subtracting time at point B from time at point A.

12.2.3. Plasticizer Absorption Under Applied Centrifuge Force (ASTM D 3367)

This method provides a quantitative measure of plasticizer absorption of PVC resin under standard temperature conditions using a controlled centrifugal force. A small quantity of PVC resin (0.500 ± 0.0050 g) is weighed accurately into the plastic screening tube prepacked with cotton to cover the tube orifice. 1 ml of plasticizer (DOP) is added to the screening tube from a pipet. The PVC and plasticizer are subjected to centrifugal force of 3000 rpm for 40 min. The plasticizer that is not absorbed by the PVC resin is removed by centrifugation through the orifice of the screening tube. The cotton prevents the PVC particles from escaping through the orifice. After centrifuging, the screening tube is weighed and the percentage plasticizer absorption is calculated from the difference in the weight of the resin–plasticizer mix.

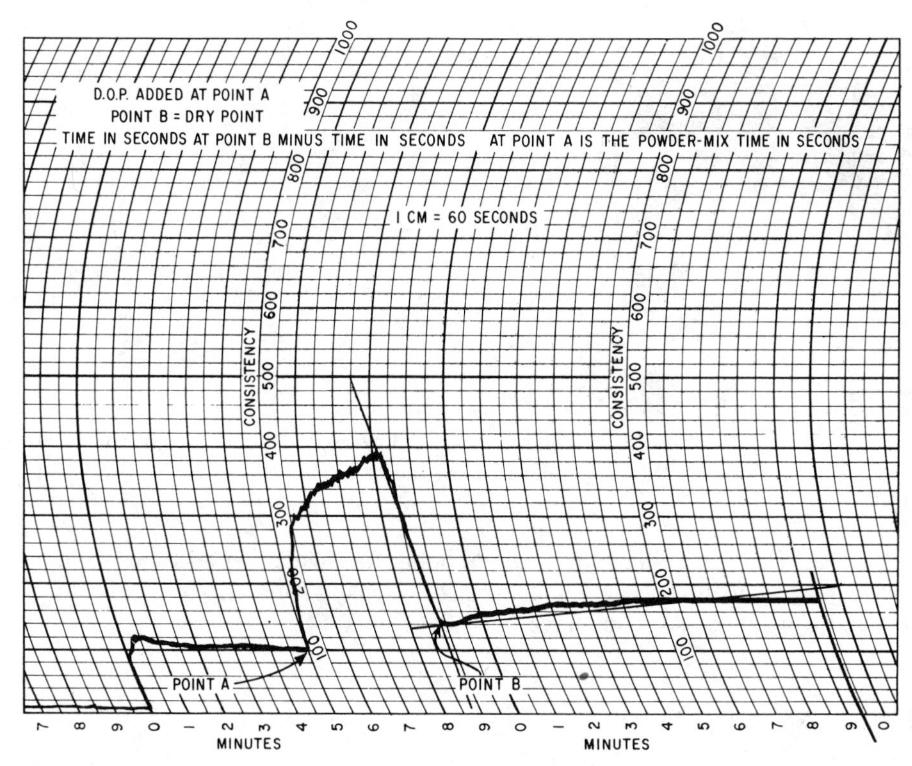

Figure 12-5. Powder mixing process. (Reprinted by permission of ASTM.)

12.3. CUP VISCOSITY TEST

As the name implies, cup viscosity tests employ a cup-shaped gravity device that permits the timed flow of a known volume of liquid through an orifice located at the bottom of the cup. Under ideal conditions, this rate of flow would be proportional to the kinematic viscosity that is dependent upon the specific gravity of the draining liquid. However, the conditions in a simple flow cup cannot be considered ideal for true measurements of viscosity. Cup viscosity tests, however imprecise, are practical, easy-to-use instruments for making flow comparisons under strictly comparable conditions (2,3).

In the plastisol hot dipping operation, a preheated mold is dipped into the plastisol for a predetermined amount of time. At the end of dwell time, the mold is withdrawn and the part is removed. As the process continues, the viscosity of the plastisol reduces considerably. Such a change in the flow properties of plastisol can have a significant effect on the appearance and thickness of the fused coating. The viscosity of the plastisol is conveniently measured by using a flow cup and any necessary adjustments in the viscosity are made.

A Zahn-type viscosity cup such as the one shown in Figure 12-6 is most commonly used. The test is carried out by simply dipping the cup in plastisol or other liquid to be measured and measuring the time interval in sec from the moment of withdrawal until the stream of material flowing from the cup orifice breaks. Zahn

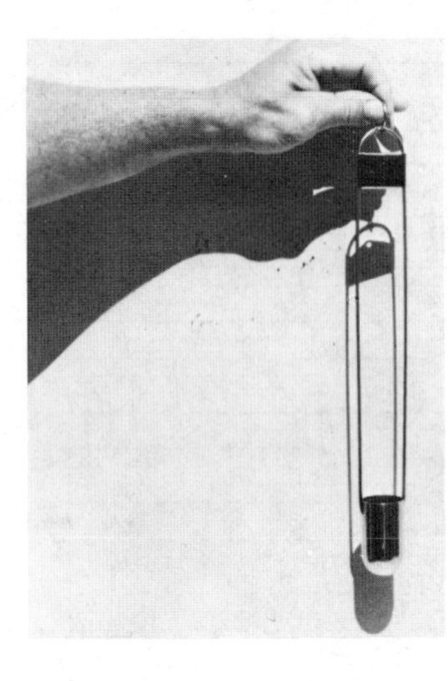

Figure 12-6. Zahn viscosity cup. (Courtesy Gardner Laboratories, Inc.)

cups of varying orifice diameters are available for measuring all types of liquids with varying viscosities. Many other types of viscosity cups such as Shall viscosity cups and Ford viscosity cups have been developed (4).

12.4. BURST STRENGTH TEST

Plastics are used in a variety of applications requiring internal stress applied by the transporting fluid. Plastic pipes, fittings, valves, tanks, and containers are some of the typical examples of pressure vessels.

Two basic tests of primary interest are:

1. Quick-burst strength test.
2. Long-term burst strength test.

12.4.1. Quick-Burst Strength Test (ASTM D 1599)

This method was developed to determine the ability of a plastic pressure vessel to resist rupturing when it is pressurized for a short period of time. Surging is a common phenomena in a fluid transfer system. Surging is a pressure rise in a pipeline caused by a sudden change in the rate of flow or stoppage of flow in the line.

In such cases, pressure vessels are subjected to very high internal pressures for a relatively short period. The short-time rupture strength of pressure vessels is determined by continuously increasing internal hydraulic pressure while immersed in a controlled temperature environment until rupture occurs.

A hydraulic burst-strength tester such as the one shown in Figure 12-7 is employed. A pressure intensifier such as the one shown in Figure 12-8 can also be used. This latter device is relatively simple and only requires the use of shop air and water or other suitable fluid. The test is carried out by simply pressurizing the specimen and uniformly increasing the pressure until the failure occurs. ASTM

D 1599 requires the time to failure for all specimens to be between 60 and 70 sec. The system must be bled thoroughly to avoid entrapment or air bubbles prior to commencing each test. The specimen is considered to have failed when it develops a leak, crack, or ruptures. The hoop stress can be calculated as follows:

$$S = \frac{P(D - t)}{2t} \quad \text{or} \quad S = \frac{P(d + t)}{2t}$$

where S = Hoop stress (psi); P = Internal pressure (psi); D = Average outside diameter (in.); d = Average inside diameter (in.); t = Minimum wall thickness (in.).

Hoop stress is defined as the circumferential stress in a material of cylindrical form subjected to internal or external pressure.

12.4.2. Long-Term Burst Strength Test (ASTM D 1598)

The long-term burst strength of plastic pressure vessels is determined by subjecting the pressure vessels to constant internal pressure and observing time-to-

Figure 12-7. Quick-burst test apparatus. (Courtesy Applied Test Systems, Inc.)

Figure 12-8. Pressure intensifier.

failure. This test is a static pressure test as opposed to the dynamic quick-burst test described earlier.

A hydrostatic pressure tester such as the one shown in Figure 12-9 is employed. It consists of a pressurizing system capable of continuously applying constant internal pressure on the specimen. The apparatus is equipped with a pressure gauge and an individual timing device that is capable of measuring time-to-failure accurately. The specimens are filled with test fluid or gas, and pressure is applied to produce the desired loading. The timers are started immediately after reaching the desired pressure. A constant temperature system may be employed if so desired. The test must be carried out in a standard laboratory atmosphere since any variation in temperature and humidity can cause results to change drastically.

The specimen failure is marked by continuous loss of pressure, bursting, abnormal ballooning, and leakage. The hoop stress in the specimens is calculated by using the formula described in Section 12.4.1.

12.4.3. Developing Long-Term Hydrostatic Design Stress Data and Pressure Rating

The Plastic Pipe Institute of the Society of the Plastics Industry has developed a method of obtaining a long-term hydrostatic design stress and pressure rating a thermoplastic pressure pipe. Hydrostatic design stress is defined as the estimated maximum tensile stress in the wall of the pipe in the circumferential orientation due to internal hydrostatic pressure that can be applied continuously with high degree of certainty that failure of the pipe will not occur. Pressure rating is the estimated maximum pressure that the medium in the pipe can exert continuously with a high degree of certainty that failure of the pipe will not occur.

Long-term hydrostatic design stress is obtained by essentially extrapolating the stress–time regression line on data obtained in Section 12.4.2. The following is a summary of procedures used to obtain long-term hydrostatic design stress (5).

1. Specimens of plastic pipe are subjected to constant internal water pressure at different levels of pressure and the time to rupture is measured. Stress on each specimen is calculated by means of a formula applicable to plastic pipe in the $\frac{1}{2}$- to 48-in. range.

$$S = \frac{Pd}{2t}$$

where S = Hoop stress (psi); P = Pressure (psi); d = Mean diameter (in.); t = Average wall thickness (in.).

The specimens are tested for 10,000 hrs under specified conditions and a linear plot of hoop stress versus time to rupture on log–log coordinates is generated. One such plot is shown in Figure 12-10.

2. The stress rupture data is analyzed by statistical regression to generate a hoop stress versus time equation. This equation is extrapolated mathematically one decade of time to 100,000 hrs (approximately 11.4 yr) to obtain a 100,000 hr design stress.

Figure 12-9. Hydrostatic pressure tester. (Courtesy Applied Test Systems, Inc.)

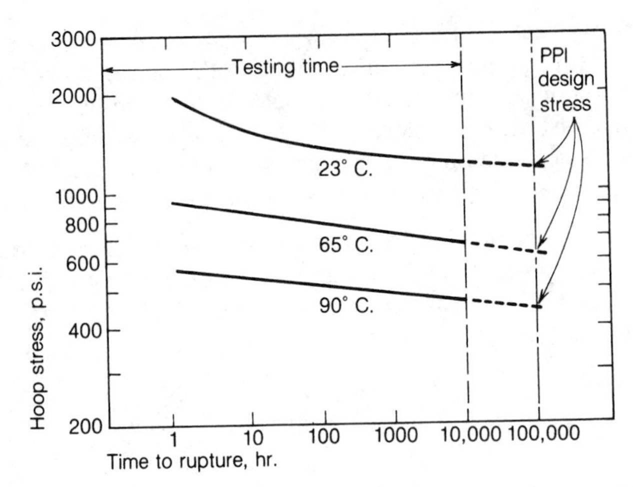

Figure 12-10. Hoop stress versus time to rupture. (Reprinted by permission of *Modern Plastics Encyclopedia*.)

3. Next, the entire design stress scale is divided into continuous increments, each of which is approximately 25 percent larger than the one below it. These increments in psi are 800, 1200, 1600, 2000, 2500, 3200, 4000 and so on. Each material is arbitrarily assigned the threshold value of the increment in which its 100,000 hr design stress falls. The working stresses are calculated from this fundamental stress called hydrostatic design basis.

4. A safety factor, depending upon temperature and type of service, is applied to the hydrostatic design basis to obtain a working stress. This value is then substituted in the pipe design equation to obtain pressure rating or required wall thickness. Typical safety factors are:

Water Service at 23°C–0.5
Water Service at 38°C–0.4
Natural gas, class location 1–0.32

Although this method was primarily developed for plastic pressure pipe, it is not limited to plastic pipe and can be applied to obtain pressure ratings of other materials and pressure vessels. ASTM D 2837, ASTM D 2992, and PPI TR3 describe this method in full detail.

12.5. CRUSH TEST

Parts made from plastic materials are often subjected to compressive loads. The compressive strength values for plastic materials obtained by testing standard test specimens (see Section 2.4) are simply not good enough for determining relative crush resistance of molded articles. The ability of a molded article to resist compressive loading depends upon several factors such as molded-in stress, part design, and processing conditions.

Figure 12-11. Crush tester. (Courtesy Testing Machines, Inc.)

A simple test was devised to study load–deflection characteristics of molded and extruded articles under parallel plate loading. A crush tester such as the one illustrated in Figure 12-11 is employed. It consists of a variable speed drive, two parallel plates, one of which is stationary, and a load cell to measure applied force. A dial indicator is used to measure deflection. More sophisticated compression testers such as the one shown in Figure 2-21 can also be used.

The test is carried out by simply placing a specimen between the parallel plates and applying load until failure occurs. The specimen is considered to have failed if it cracks or fractures. Quite often requirements such as "the part shall not crack or fracture when deflected 10 percent of its original dimension" are placed on the part drawing. The crush test is used as a routine quality control procedure. ASTM D 2412 discusses the external loading properties of plastic pipe by parallel plate loading.

12.6. ACETONE IMMERSION TEST (ASTM D 2152)

This method was developed to determine the quality of rigid PVC pipe and fittings as indicated by their reaction to immersion in anhydrous acetone. An unfused PVC compound attacked by anhydrous acetone causes the material to swell, flake, or completely disintegrate. A properly fused PVC compound is impervious to anhydrous acetone and only a minor swelling, if any, is observed.

The test is carried out by placing a small specimen cut from the molded or extruded article in reagent grade acetone and observing the effect of acetone immersion after 20 min. The presence of water in acetone reduces its effectiveness and therefore acetone must be dried by shaking it with anhydrous calcium sulfate which is removed from the acetone by filtering. Acetone immersion test is used as an on-going quality control test by many PVC pipe and fittings manufacturers.

12.7. ACETIC ACID IMMERSION TEST (ASTM D 1939)

This method evaluates the residual stresses in extruded or molded ABS parts by immersing them in glacial acetic acid and observing the effect of immersion. The presence of excessive residual stresses in ABS parts is indicated by the cracking of the specimen upon immersion in glacial acetic acid. This test is very useful in determining residual stresses in ABS parts that are going to be plated. The plating process requires the parts to be stress-free and therefore many processors of ABS plastics have adopted this method as a production control test.

The specimen, regardless of the size, is immersed into reagent grade glacial acetic acid for 30 sec. Immediately after the immersion period, the specimen is removed, rinsed in running water, and dried. The specimen is examined for cracking. The same specimen is reimmersed for an additional 90 sec or a new specimen for 2 min. After rinsing and drying, the specimen is again examined for cracks. The time taken to develop cracks and the degree of cracking indicates the magnitude of residual stress in the specimen.

12.8. END-PRODUCT TESTING

In spite of numerous field failures and ever-increasing product liability problems, processors of plastic products continue to neglect end-product testing. Too much emphasis is placed on testing raw materials and blaming raw material suppliers for providing substandard material, and not enough importance is given to end-product testing. All major and minor raw material suppliers are well equipped with sophisticated quality control equipment and generally adhere to strict quality standards of their own. Plastics are polymerized under specified conditions in controlled environments. Plastics are not processed under controlled conditions. In the injection molding process, for example, there are at least six major variables and numerous minor variables that can affect the quality of the molded part. Material temperature, injection pressure, mold temperature, regrind/virgin mix, injection speed, and packing are the six major variables. Too high a melt temperature can cause material to degrade and consequently lose physical properties very rapidly. Even the best quality material cannot save the product from failing if the product is improperly processed. The test data supplied with the raw material are derived by testing certain sized specimens molded under a controlled environment. Unfortunately, plastic parts are not molded under the exact same ideal conditions and they vary in size. Therefore, such data are basically of little or no value to a processor from the standpoint of end-product testing. The quality of the molded product cannot be assessed from raw material data provided by the material supplier without proper consideration of molding variations. One other factor to consider is the molded-in stresses that are usually present in all parts depending upon the part geometry, mold design, and molding practices. These stresses, unless relieved by annealing, can cause warpage and premature failure.

It is very clear from the foregoing discussion that end-product testing is imperative if one is to control the quality of the product going out in the field. End-product testing offers numerous advantages. First, it protects consumers from premature product failure and resulting possible injuries. The manufacturer is equally protected by not having to worry about product liability suits. Second, it verifies the manufacturing process for any possible mishaps. More importantly, it reduces unnecessary static between the custom processor and the buyer of the product. An end-product testing specification generated by a buying party helps the custom processor in understanding the product requirement and assures the buyer of receiving a quality product. As we all know, just because the part looks right visually, this does not necessarily mean it has adequate physical properties. Such properties can only be verified by actual testing of the product.

There are many ways to perform end-product testing. The following is a partial list of common end-product tests.

Simulated actual use (functional) testing.
Impact testing.
Torque (shear) test.
Crush test.
Pull test.
Chemical test.

Many companies devise their own end-product tests that simulate actual use conditions. This requires designing the proper fixtures and test equipment and training personnel. Energy-to-break test is the best indication of long-term mechanical performance of a product. This can be accomplished by testing the end-product for impact as well as crush resistance. The impact test, however, is preferred over the crush test. The torque test is often used to verify shear strength of threaded components. The test is fairly simple and requires only a torque wrench and suitable fixtures. A simple pull test is employed to determine the force required to pull apart two components. Figures 12-12 and 12-13 illustrate a torque test apparatus and an inexpensive pull tester. Chemical tests such as acetic acid and acetone immersion tests (see Sections 12.6 and 12.7) are useful in verifying processing conditions.

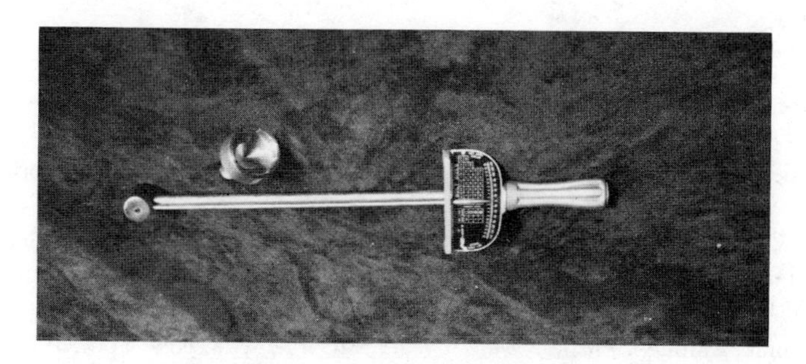

Figure 12-12. Torque tester.

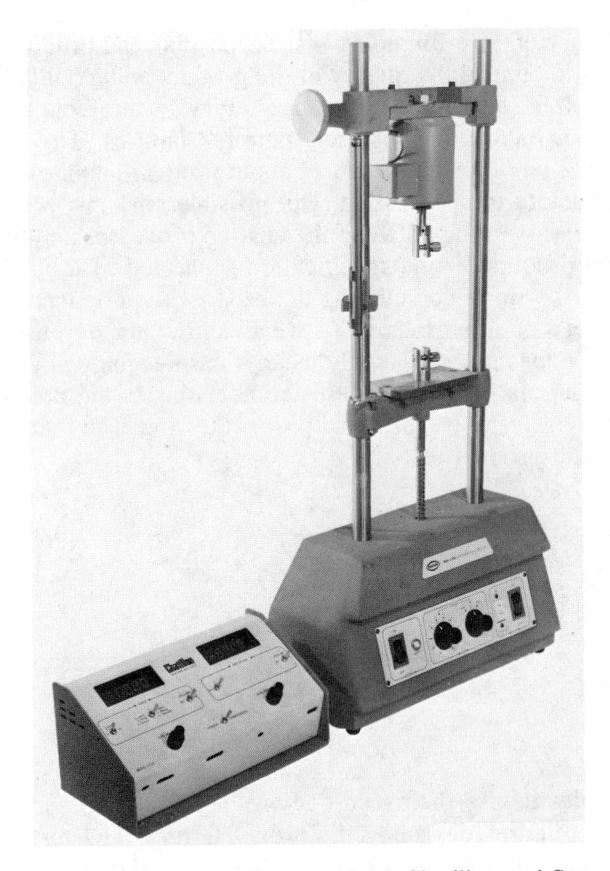

Figure 12-13. Pull tester. (Courtesy John Chatillon and Sons, Inc.)

Many end-product tests are self-devised and so is the equipment. Care must be taken in devising such tests to ensure that they are reliable and the test results are reproducible. Whenever possible, commercially available testing equipment should be employed and guidelines and procedures must be established and followed.

REFERENCES

1. Mentovay, L. W. and Yasenchak, L. P., "Capillary Flow Analysis," *Plast. Design and Processing* (March 1973), p. 18.
2. Technical literature, *Catalog Section C, Rheology,* Gardner Laboratory, Silver Spring, Md., Sept. 1976.
3. Sward, G. G., "Paint Testing Manual," *Am. Soc. for Testing and Materials* Philadelphia, Pa., 1972, pp. 181–185.
4. *Ibid,* pp. 184–185.
5. "Design Guide," *Modern Plastics Encyclopedia,* McGraw–Hill, New York, 1979–1980.

GENERAL REFERENCES

Technical Literature on *Torque Rheometer,* C. W. Brabender Instruments, Inc. South Hackensack, N.J.

Park, R. A., "Characterizing Fluid Plastics by Torque Rheometer," *Plast. Eng.* (Nov. 1976), p. 59.

Allen, E. O. and Willium, R. F., "Prediction of Polymer Processing Characteristics Using C. W. Brabender Plasticorder Torque Rheometer," *SPE ANTEC* (1971), p. 587.

CHAPTER 13

Identification Analysis of Plastic Materials

13.1. INTRODUCTION

Plastic products are manufactured using a variety of processing techniques and materials. It is practically impossible to identify a plastic material or product by a visual inspection or a simple mechanical test. There are many reasons that necessitate the identification of plastics. One of the most common reasons is the need to identify plastic materials used in competitive products. Defective products returned from the field are quite often put through rigorous identification analysis. Sometimes it is necessary to identify a finished product at a later date in order to verify the material used during its manufacture. The custom compounders of reprocessed materials may also need to identify already processed material purchased from different sources. Quite often, processors find substantial quantities of plastic material, hot stamp foils, and decals in the warehouse without any labels to identify the particular type. A little knowledge of the identification process can save time and money.

On a rare occasion, the buyers of molded parts may choose to verify the material specified in the product by performing a simple identification analysis. The development of new material is another reason for such analysis.

There are two ways plastic materials can be identified. The first technique is simple, quick, and inexpensive. It requires very few tools and little knowledge of plastic materials. The second approach is to perform a systematic chemical or thermal analysis. The latter technique is very complex, time-consuming, and expensive. The results can only be interpreted by a person well-versed in polymer chemistry. Plastic materials are often copolymerized, blended, and modified with filler or compounded with different additives such as flame retardants, blowing agents, lubricants, and stabilizers. In such cases, simple identification techniques will not yield satisfactory results. The only true means of positive identification is a complex chemical or thermal analysis.

The first technique is laid out in a flowchart for easy step-by-step identification by process of elimination. This is shown in the Plastics Identification Chart. There are some basic guidelines one must follow in order to simplify the procedure. The first step is to determine whether the material is thermoplastic or thermoset. This

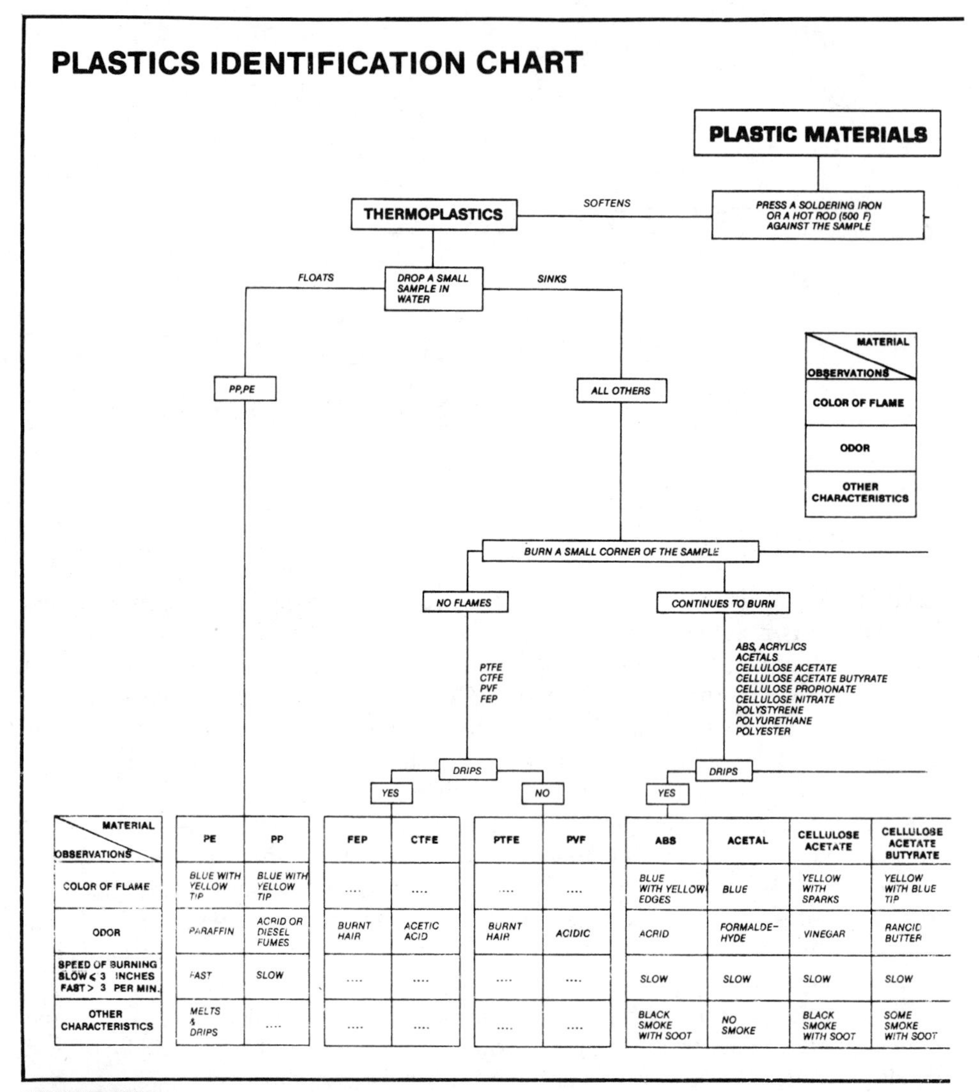

MATERIAL / OBSERVATIONS	PE	PP	FEP	CTFE	PTFE	PVF	ABS	ACETAL	CELLULOSE ACETATE	CELLULOSE ACETATE BUTYRATE
COLOR OF FLAME	BLUE WITH YELLOW TIP	BLUE WITH YELLOW TIP	....	....	....	....	BLUE WITH YELLOW EDGES	BLUE	YELLOW WITH SPARKS	YELLOW WITH BLUE TIP
ODOR	PARAFFIN	ACRID OR DIESEL FUMES	BURNT HAIR	ACETIC ACID	BURNT HAIR	ACIDIC	ACRID	FORMALDE-HYDE	VINEGAR	RANCID BUTTER
SPEED OF BURNING SLOW < 3 INCHES FAST > 3 PER MIN.	FAST	SLOW	....	....	....	....	SLOW	SLOW	SLOW	SLOW
OTHER CHARACTERISTICS	MELTS & DRIPS	....	....	....	....	....	BLACK SMOKE WITH SOOT	NO SMOKE	BLACK SMOKE WITH SOOT	SOME SMOKE WITH SOOT

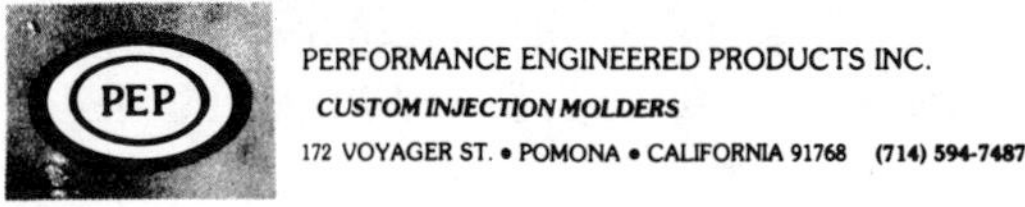

distinction is made by simply probing the sample with a soldering iron or a hot rod heated to approximately 500°F. If the sample softens, the material is thermoplastic. If not, it is thermoset. The next step is to conduct a flame test. It is desirable to use a colorless Bunsen burner. A match stick can also be used in place of a Bunsen burner. However, care must be taken to distinguish between the odor of the materials used in the match and the odor given off by burning plastic materials. Before commencing the burning test, it is advisable to be pre-

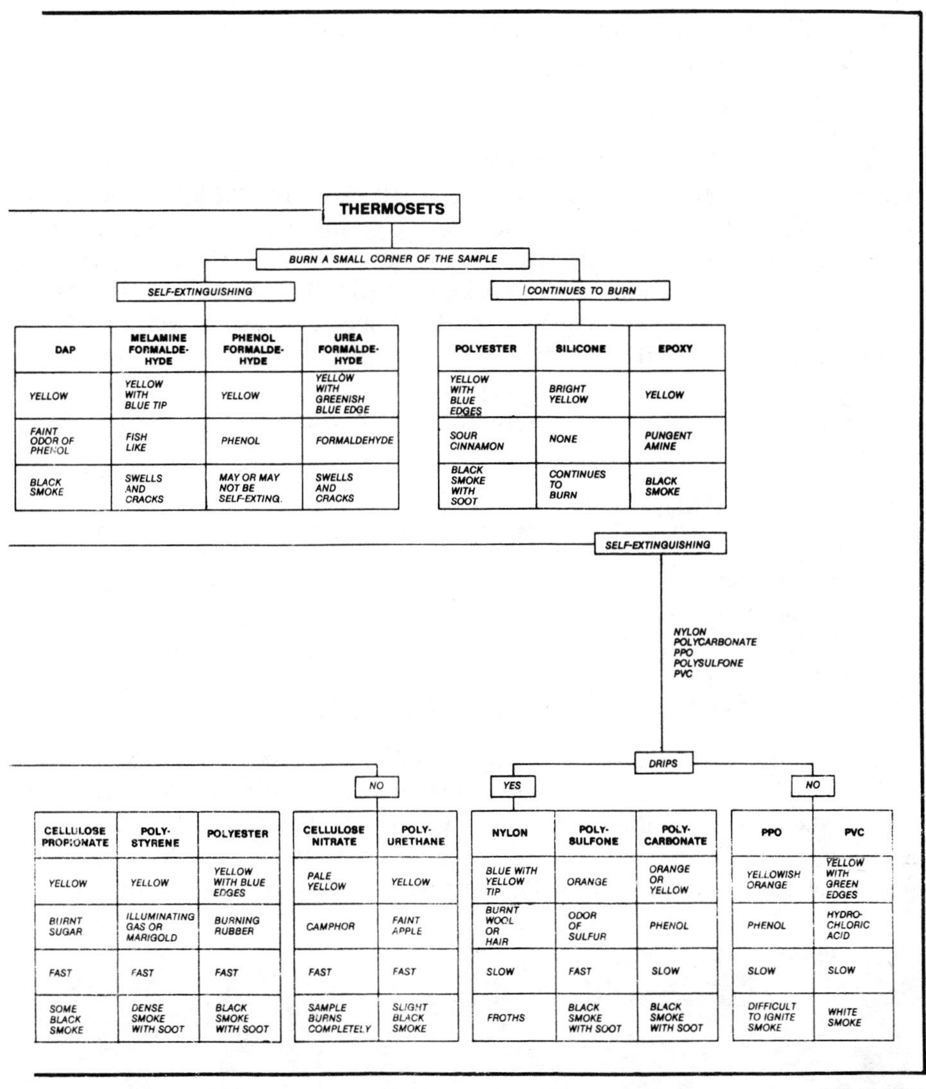
THERMOSETS
BURN A SMALL CORNER OF THE SAMPLE
SELF-EXTINGUISHING
CONTINUES TO BURN
DAP
MELAMINE FORMALDE-HYDE
PHENOL FORMALDE-HYDE
UREA FORMALDE-HYDE
YELLOW
YELLOW WITH BLUE TIP
YELLOW
YELLOW WITH GREENISH BLUE EDGE
FAINT ODOR OF PHENOL
FISH LIKE
PHENOL
FORMALDEHYDE
BLACK SMOKE
SWELLS AND CRACKS
MAY OR MAY NOT BE SELF-EXTING.
SWELLS AND CRACKS
POLYESTER
SILICONE
EPOXY
YELLOW WITH BLUE EDGES
BRIGHT YELLOW
YELLOW
SOUR CINNAMON
NONE
PUNGENT AMINE
BLACK SMOKE WITH SOOT
CONTINUES TO BURN
BLACK SMOKE
SELF-EXTINGUISHING
NYLON
POLYCARBONATE
PPO
POLYSULFONE
PVC
DRIPS
NO
YES
NO
CELLULOSE PROPIONATE
POLY-STYRENE
POLYESTER
YELLOW
YELLOW
YELLOW WITH BLUE EDGES
BURNT SUGAR
ILLUMINATING GAS OR MARIGOLD
BURNING RUBBER
FAST
FAST
FAST
SOME BLACK SMOKE
DENSE SMOKE WITH SOOT
BLACK SMOKE WITH SOOT
CELLULOSE NITRATE
POLY-URETHANE
PALE YELLOW
YELLOW
CAMPHOR
FAINT APPLE
FAST
FAST
SAMPLE BURNS COMPLETELY
SLIGHT BLACK SMOKE
NYLON
POLY-SULFONE
POLY-CARBONATE
BLUE WITH YELLOW TIP
ORANGE
ORANGE OR YELLOW
BURNT WOOL OR HAIR
ODOR OF SULFUR
PHENOL
SLOW
FAST
SLOW
FROTHS
BLACK SMOKE WITH SOOT
BLACK SMOKE WITH SOOT
PPO
PVC
YELLOWISH ORANGE
YELLOW WITH GREEN EDGES
PHENOL
HYDRO-CHLORIC ACID
SLOW
SLOW
DIFFICULT TO IGNITE SMOKE
WHITE SMOKE
VISHU SHAH
© 1981

pared to write down the following observations:

1. Does the material burn?
2. Color of flame.
3. Odor.
4. Does the material drip while burning?

5. Nature of smoke and color of smoke.
6. The presence of soot in the air.
7. Self-extinguishes or continues to burn.
8. Speed of burning—fast or slow.

To identify the material, compare the actual observations with the ones listed in the flowchart. The accuracy of the test can be greatly improved by performing similar tests on a known sample. While performing the identification tests, one must not overlook safety factors. The drippings from the burning plastic may be very hot and sticky. After extinguishing the flame, inhale the smoke very carefully. Certain plastics like acetals give off a toxic formaldehyde gas that may cause a severe burning sensation in the nose and chest.

The results of the simple identification technique can be further confirmed by the following tests:

1. Melting point test.
2. Solubility test.
3. Copper wire test.
4. Specific gravity test.

13.1.1. Melting Point Determination (ASTM D 2117; ASTM D 795)

Two basic methods are used for melting point determination. These methods are fully described in the ASTM Standards Manual. For the first method (ASTM D 795), a Fisher–Johns melting point apparatus as shown in Figure 13-1 is most commonly used. The apparatus consists of a rheostatically controlled heated block, a thermometer, and a viewing magnifier. A small pellet or a sliver of the

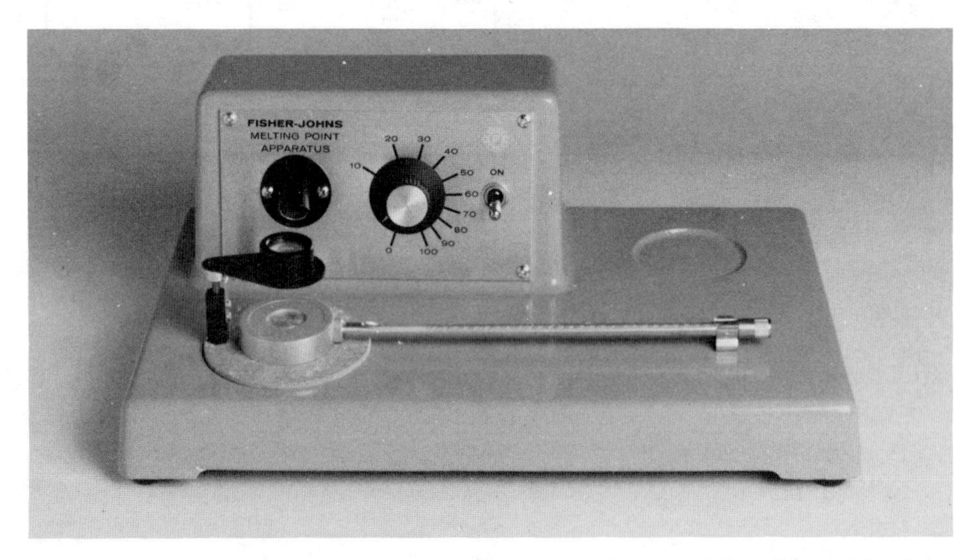

Figure 13-1. Fisher–Johns melting point apparatus. (Courtesy Fisher Scientific Company.)

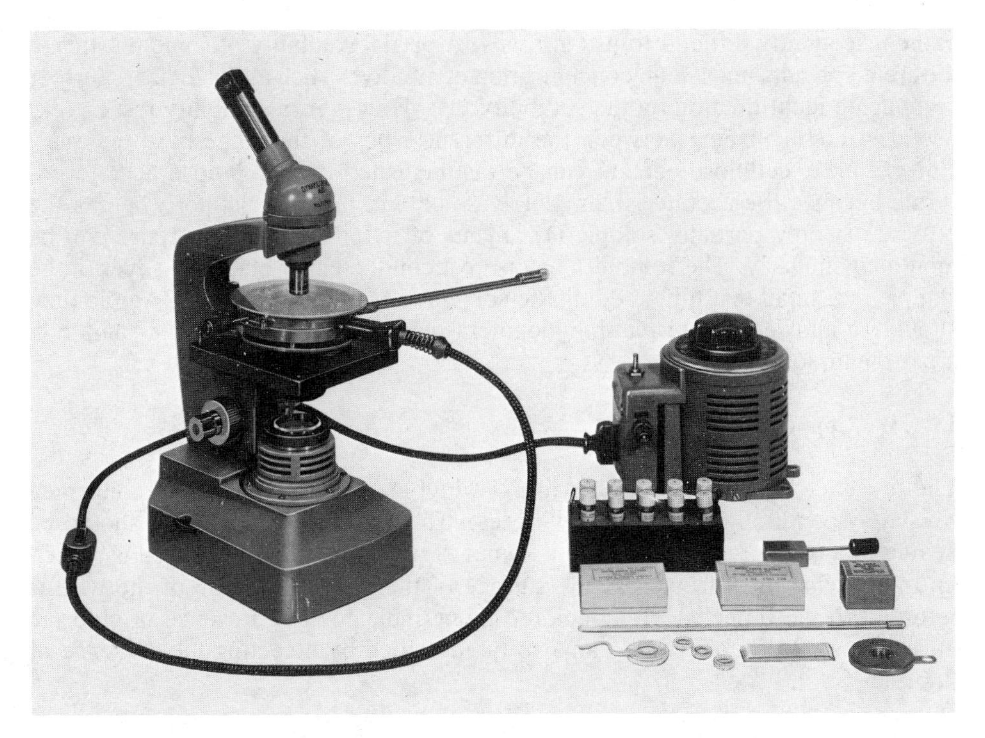

Figure 13-2. Kofler method, melting point determination apparatus. (Courtesy Arthur H. Thomas Company.)

plastic material to be tested is placed on the electrically heated block along with a few drops of silicone oil. A cover glass is placed over the material and the heat is gradually increased until the sample material melts or softens enough to deform. The meniscus formed by the oil is viewed through the magnifier. The temperature at which the meniscus moves is considered the melting point. The expected accuracy of the test is within ±5°F of the published literature value. This method can be used for both crystalline and amorphous plastics. All crystalline plastics have a sharp melting point and the transition is much easier to detect. In contrast, amorphous plastics melt over a wide range and an exact melting point is difficult to determine.

The second method (ASTM D 2117), known as the Kofler method, is used only for semicrystalline polymers. It consists of heating the sample by a hot-stage unit mounted under a microscope and viewing it between crossed polarizers. When crystalline material melts, the characteristic double refraction from the crystalline aggregates disappears. The point at which the double refraction or birefringence (typically a rainbow color) completely disappears is taken as the melting point of the polymer. Figure 13-2 shows a commercially available apparatus. The use of a control sample for comparison is particularly helpful in both methods.

13.1.2. Solubility Test

The behavior of plastic materials in various organic solvents often indicates the type of material. The solubility data found in the literature is of a general nature and consequently difficult to use at times. A partial solubility of some plastics in different solvents and a high concentration of additives such as plasticizers further complicate identification by the solubility test. However, a solubility test is very useful in distinguishing between the different types of the same base polymer. For example, cellulose acetate can be distinguished from cellulose acetate butyrate because the acetate is completely soluble in furfuryl alcohol whereas the butyrate is only partially soluble (1). Types of nylons and polystyrenes can be identified similarly. The solubility test is best conducted by placing a sliver of the sample in a small test tube, adding the solvent, and gently stirring it. Ample time should be allowed before passing judgment regarding solubility of the sample in a particular solvent.

13.1.3. Copper Wire Test

The presence of chlorine such as in polyvinyl chloride can be easily confirmed by simply conducting the copper wire test. The tip of the copper wire should be heated to a red-hot temperature in a flame. A small quantity of material is picked up by drawing the wire across the surface of the sample. The tip of the wire is returned to the flame. A green-colored flame indicates the presence of chlorine in the material. Fluorcarbons can also be identified by detecting the presence of fluorine.

13.1.4. Specific Gravity Test

The increasing use of plasticizers, fillers, reinforcing agents, and other additives makes the identification of plastics by the specific gravity test very difficult. The test is described in detail in Chapter 10.

13.2. CHEMICAL AND THERMAL ANALYSIS FOR IDENTIFICATION OF POLYMERS

The following techniques are those most commonly used today to positively identify plastic materials and additives. Table 13-1 lists the identification technique along with the proper application.

1. Infrared spectroscopy.
2. Gas chromatography (GC).
3. Thermogravimetric analysis (TGA).
4. Differential scanning calorimetry (DSC).
5. Thermo-mechanical analysis (TMA).
6. NMR.
7. X-ray analysis.
8. Pyrolysis.
9. Liquid and gel permeation chromatography.

Table 13-1. Identification Techniques for Polymer and Additives (2)

Technique	Identification
LC/GPC	Polymer molecular weight distribution. Phenols, phosphites, plasticizers, lubricants
GC	Residual monomers
	Nonpolymeric compounds
	Plasticizers
IR	Polymer type
	Additives
Thermal	Fillers
	Lubricants
	Polymer molecular weight
X-ray	Fillers
	Flame retardants
	Stabilizers
NMR	Polyesters
	Silicones
	Phenols
Wet Chemistry	Lubricants
	Flame retardants
	Catalysts

Infrared spectroscopy is by far the most popular method of identification of plastic materials.

13.3. IDENTIFICATION OF PLASTIC MATERIALS

13.3.1. Thermoplastics

A. ABS

ABS is an amorphous terepolymer with a specific gravity of 1.04. It is soluble in toluene and ethylene dichloride. It burns with a yellow flame and an acrid odor, and continues to burn after the removal of the flame source with black smoke with soot and drippings. Infrared spectroscopy is the best method for the positive identification of ABS. The same technique can also be used to further identify a percentage of each polymer in the terepolymer composition.

B. Acetal

Acetal is available as a homopolymer and a copolymer. Both types burn with a blue flame and give off a toxic formaldehyde odor. They are both crystalline with a specific gravity of 1.41. The homopolymer can be distinguished from the copolymer by a Fisher–John melting point test. The copolymer acetal has a lower melting point than the homopolymer. Both types are soluble in hexafluoroacetone sesquihydrate. However, due to the extreme toxicity of this solvent, it is rarely used for identification purposes. A differential thermal analysis (DTA) can also be used to distinguish a homopolymer from a copolymer.

C. *Acrylic*

Acrylic is an amorphous polymer with a specific gravity of 1.18. Acrylic is one of the few transparent plastics used in outdoor applications. It burns with a blue flame with a yellow tip, gives off a fruity odor, and does not produce smoke while it burns. Acrylic is soluble in acetone, benzene, and toluene. Infrared spectroscopy is the best means of identification of this polymer.

D. *Cellulose Acetate*

Cellulose acetate is an amorphous material with a specific gravity of 1.30. It burns very slowly with a yellow flame, drips, gives off an odor of vinegar, and has black smoke with soot. It is soluble in acetone, furfuryl alcohol, and acetic acid. Cellulose acetate is positively identified by infrared spectroscopy.

E. *Cellulose Acetate Butyrate*

This polymer burns with a yellow flame with a blue tip, drips, and gives off an odor of rancid butter. The specific gravity is 1.24 and is soluble in acetone and trichloromethane. Cellulose acetate butyrate is also one of the few weatherable transparent plastics.

F. *Cellulose Propionate*

The cellulose propionate is also an amorphous material like other celluloids with a specific gravity of 1.20. It burns with a dark yellow flame and an odor of burnt sugar. The material drips while burning and the drippings also burn like other celluloids. Acetone, carbon tetrachloride, and trichloromethane are the most common solvents.

G. *Fluorocarbons (FEP, CTFE, PTFE, PVF)*

Fluorocarbon plastics do not actually burn when exposed to a flame. They can be easily identified by a copper wire test that indicates the presence of fluorine by a bright green-colored flame. The fluoroplastics have a very high melting point. PTFE has a waxy surface and a specific gravity of 2.15. Fluoroplastics are practically impossible to dissolve in any chemical.

H. *Nylons*

There are many types of nylons with a specific gravity ranging from 1.04–1.17. All types burn with blue flames and yellow tips and give off a burnt wool or burnt hair odor. Nylon self-extinguishes on removal of the flame. Phenol, *m*-cresol, and formic acid are the most common solvents. Different types of nylons can be identified by the Fisher–Johns melting point test (ASTM D 789). Solubility and specific gravity tests are also used to differentiate between the types of nylons. However, infrared spectroscopy is the best method for positive identification.

I. Polycarbonate

Polycarbonate is one of the toughest transparent thermoplastics. It has a specific gravity of 1.2, burns with a yellow or orange flame with the odor of phenol, and gives off black smoke with soot in the air. Polycarbonate is soluble in methylene dichloride and ethylene dichloride. Infrared spectroscopy is the best method of identification for polycarbonate.

J. Thermoplastic Polyester

Thermoplastic polyester burns with a yellow flame with a blue edge, and as it burns it drips and gives off black smoke with soot. The characteristic odor is one of burning rubber.

K. PVC

This self-extinguishing amorphous plastic has a specific gravity of 1.2–1.7. It burns with a yellow flame and a green tip, and gives off the odor of hydrochloric acid and white smoke. Tetrahydrofuran and methyl ethyl ketone are common solvents for PVC. PVC can be easily identified by the copper wire test which indicates the presence of chlorine by a bright green flame. There are many different types of vinyl polymers such as polyvinyl acetate, polyvinylidene chloride, and ethylene vinyl acetate. The type of vinyl can be positively identified by infrared spectrometric techniques.

L. Polyethylene

Polyethylene is one of the very few crystalline plastics that will float on water. The specific gravity ranges from 0.91–0.96. It burns quickly with a blue flame with a yellow tip. Polyethylene drips while it burns and gives off a paraffin odor similar to a burning candle. Polyethylene is impervious to most common solvents. However, it can be dissolved in hot toluene or hot benzene. Infrared spectroscopy is used to confirm the identity of polyethylene.

M. Polypropylene

This polyolefin family polymer has a specific gravity of 0.88 and floats on water. It burns with a blue flame with a yellow tip and gives off an acrid odor similar to diesel fumes. The material is soluble in hot toluene. Polypropylene is positively identified by infrared spectrometer.

N. Polystyrene

Polystyrene is an amorphous plastic with a specific gravity of 1.09. It burns with a yellow flame and gives off the odor of illuminating gas or marigolds. The material burns and drips, creating a dense black smoke with soot in the air, also dripping continues to burn. Polystyrene is soluble in acetone, benzene, toluene, and ether. Polystyrene is often modified with rubber to improve its impact strength. These different types of polystyrenes can be identified by infrared spectroscopy.

O. Polyphenylene Oxide (PPO)

PPO has a specific gravity of 1.06. It burns with a yellow–orange flame without dripping and gives off an odor of phenol or burning gas. It exhibits self-extinguishing characteristics. PPO is soluble in toluene and dichloroethylene. Like other polymers, PPO can be identified by infrared spectroscopy.

P. Polysulfone

This self-extinguishing amorphous thermoplastic has a specific gravity of 1.24. It burns with an orange flame, gives off the pungent odor of sulfur, and produces black smoke with soot in the air while it burns. Polysulfone is soluble in methylene chloride.

Q. Polyurethane (Thermoplastic)

This generally easy-to-ignite plastic burns with a yellow flame with a faint odor of apple. It produces black smoke. Thermoplastic polyurethane is soluble in tetrahydrofuran and dimethyl formamide. The specific gravity of this material is 1.2.

13.3.2. Thermosetting Plastics

A. Diallyl Pthalate (DAP)

DAP compound's specific gravity ranges from 1.30–1.85 depending upon the type of filler. It burns with a yellow flame, produces black smoke, and gives off the faint odor of phenol. It is self-extinguishing in nature and is difficult to ignite.

B. Epoxy

The specific gravity of epoxy compounds range from 1.10–2.10. It burns with a yellow flame and gives off a black smoke that gives off a pungent amine odor. Different types of epoxies can be characterized by infrared spectroscopy and thermal analysis.

C. Phenol Formaldehyde

Phenol formaldehyde burns with a yellow flame and gives off the odor of phenol. Depending upon the type of filler, it may or may not be self-extinguishing. Phenolic compounds are only available in dark colors. The specific gravity ranges from 1.30–1.90. Phenol formaldehyde is soluble in acetone and acetic acid.

D. *Urea Formaldehyde*

Urea formaldehyde burns with a yellow flame with a greenish blue edge. It gives off a strong odor of formaldehyde and the material usually swells and cracks as it continues to burn.

E. *Melamine Formaldehyde*

This self-extinguishing thermoset has a specific gravity of 1.47–1.8. It burns with a yellow flame with a blue edge and gives off a fishlike odor. The material swells, cracks, and turns white at the edges of a burned section. Melamine formaldehyde is soluble in acetone and cyclohexanone.

F. *Polyesters*

Polyesters burn with a yellow flame with blue edges and produce black smoke with soot in the air. They continue to burn without dripping after the removal of the ignition source and give off a sour cinnamon odor. The specific gravity ranges from 1.30–1.50.

G. *Silicones*

Silicones burn with a bright yellow flame without odors and continue to burn after the removal of the flame source. The specific gravity ranges from 1.05–2.82. The positive identification is made by an infrared spectrometer.

REFERENCES

1. E. I. Dupont Co., *Tech. Bull.: Identification of Thermoplastic Materials,* Wilmington.
2. Coe, G. R., "Instrumental Methods of Polymer Analysis," *SPE ANTEC* **23,** (1977), pp. 496–499.

GENERAL REFERENCES

"Identification of Plastics," *Modern Plastics Encyclopedia,* McGraw–Hill, New York, 1950, pp. 992–1001.

"How to Identify Plastics," *Western Plast. Mag.* (March 1966).

"Identification Chart," *Canad. Plast. Mag.* (1971).

Richardson, T. A., *Modern Industrial Plastics,* Howard W. Sams and Co., Inc., Indianapolis, 1974, pp. 84–90.

Lever, A. E. and Rhys, J. A., *The Properties and Testing of Plastic Materials,* Temple Press, Feltham, England, 1968, pp. 269–275.

Haslam, J. and Willis, H. A., *Identification and Analysis of Plastics,* Iliffe Books, London, England, 1965.

Rodriguez, F., *Principles of Polymer Systems,* McGraw–Hill, New York, 1970, Chapter 15, pp. 464–475.

Kline, G. M., *Analytical Chemistry of Polymers,* Parts 1–3, Interscience, New York, 1962.

Beck, R. D., *Plastics Product Design,* Van Nostrand Reinhold, New York, 1970, pp. 432–440.

CHAPTER 14

Testing of Foam Plastics

14.1. INTRODUCTION

Cellular material (foam) is a generic term for materials containing many cells (either open, closed, or both) dispersed throughout the mass. The increasing use and popularity of foam plastics has created a need for developing test methods particularly suited to foamed plastics. For a long time, the standard test methods developed for solid plastics were employed to determine the properties of cellular plastics. The test methods had to be modified to a degree because of the lower overall strength of the cellular plastic materials. Such modifications and changes created numerous nonstandard test methods which in turn created more confusion among the designers and users of cellular plastics. Through a painstaking effort of ASTM Committees, SPI, and material suppliers, many standard test methods have been developed. The majority of the test methods for cellular plastics are very similar to the ones already developed for noncellular plastics. However, some tests are developed to suit particular needs for cellular plastics and are unique in that respect. The porosity test to measure the open-cell content of rigid foam plastics is one unique test developed especially for foam plastics.

14.2 RIGID FOAM TEST METHODS

14.2.1. Density (ASTM D 1622)

The density of foam plastics is of considerable interest to part designers since many important physical properties are related to foam density. The procedure to determine the density of cellular plastics is very simple. Basically, it requires conditioning a specimen of a shape whose volume can be easily calculated. The specimen is weighed on a balance or a scale. Next, the specimen volume is calculated by measuring length, width, and height using a micrometer, a dial gauge, or a caliper. The density is calculated as follows:

$$\text{Density (lbs/ft}^3\text{)} = \frac{\text{Weight of specimen (lbs)}}{\text{Volume of specimen (ft}^3\text{)}}$$

14.2.2. Cell Size

There is no standard technique to measure cell size. Cell size is the average cell diameter usually measured in microns (μm). One of the earlier techniques involves preparing thin slices of foam. These thin slices are mounted on glass slides and used as negatives, which are projected on a screen using an enlarger or a slide projector. The average cell size is determined from the projection.

14.2.3. Open Cell Content (ASTM D 2856)

The porosity or percentage of open cells in a cellular plastic is determined by this method. The knowledge of open cell content is of extreme importance in specifying cellular plastics for floatation applications, where excessive porosity or high open cell content can adversely affect its floatation characteristics. A high percentage of closed cells, on the other hand, prevents escape of gases and improves insulation characteristics by promoting low thermal conductivity.

Since any conveniently sized specimen can only be obtained by some cutting operation, a fraction of closed cells will be opened during sample preparation and will be included as open cells. Three basic procedures are established to cope with this problem.

Procedure A. Corrects for cells opened during the sample preparation by measuring the cell diameter and allowing for the surface volume.

Procedure B. Corrects for cells opened in sample preparation by cutting and exposing surface areas equal to the surface area of the original sample dimensions.

Procedure C. Does not correct for cells opened during sample preparation and gives good accuracy on highly open-celled materials. The accuracy decreases as the closed cell content increases and as the cell size increases.

The method is based on Boyle's law, which states that at a constant temperature, an increase in volume of a confined gas results in a proportional decrease in pressure. If a chamber size is increased equally with or without material present in the specimen chamber, the pressure drop will be less for the empty chamber. The extent of this difference and the actual volume of the material is a measure of percentage of closed cells.

The apparatus used in the test is called an air pycnometer, schematically illustrated in Figure 14-1. It consists of two cylinders of equal volume with a specimen chamber provided in one of them. Pistons in both of them permit volume changes. When the volumes of both cylinders are altered, the volume change for the specimen-containing cylinder is smaller than for the empty chamber because of the presence of the sample. The extent of this difference is measured and a calculation is carried out to determine the open cell content of the foam sample. A commercially available air pycnometer is shown in Figure 14-2.

14.2.4. Compressive Properties (ASTM D 1621)

The test to determine compressive strength and compressive modulus of rigid cellular plastics is somewhat similar to the one developed for noncellular plastics

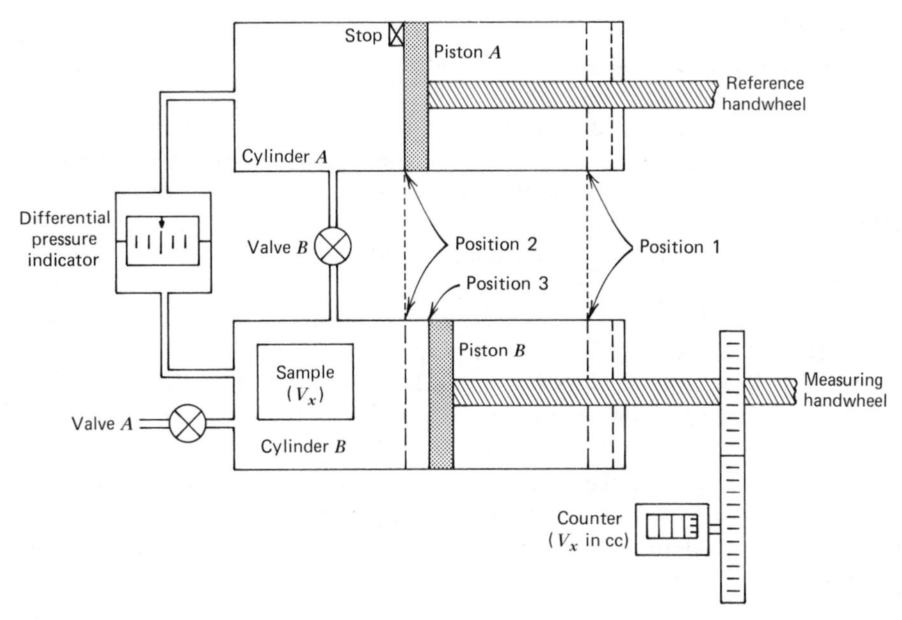

Figure 14-1. Schematic, air pychnometer. (Reprinted by permission of ASTM.)

Figure 14-2. Air pychnometer. (Courtesy Beckman Instruments.)

in Chapter 2. The test is very useful in comparing the compressive strengths of various foam plastic formulations. It provides a standard method of obtaining data for research and development, quality control, and verifying specifications. However, the test is not a direct indication of how the cellular plastic will behave under actual load over a period of time. Further tests such as creep, fatigue, and impact resistance tests must be conducted when designing a cellular part for load-bearing applications.

The two basic procedures that are most commonly used are:

Procedure A. In this procedure, crosshead motion is employed to determine compressive properties. The specimen is compressed 10 percent of its original thickness. The stress at 10 percent deformation is measured. If the yield point occurs before 10 percent deformation, the stress at the yield point is measured. This is shown graphically in Figures 14-3 and 14-4.

Procedure B. This procedure requires the use of a strain measuring device to be mounted on the specimen. The specimen is compressed to a strain of 2 percent. A load–deflection curve is usually plotted using the automatic recorder and the compressive modulus is calculated.

14.2.5. Tensile Properties (ASTM D 1623)

This method for determining tensile strength, modulus, and elongation of rigid cellular plastics is very similar to the one used for noncellular plastics described

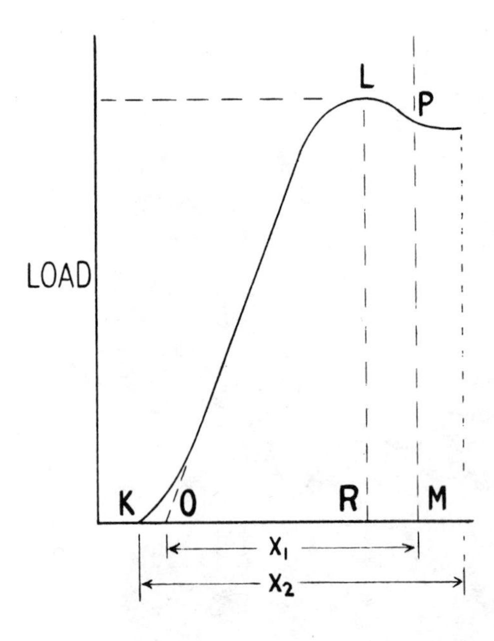

Figure 14-3. Load versus deformation. (Reprinted by permission of ASTM.)

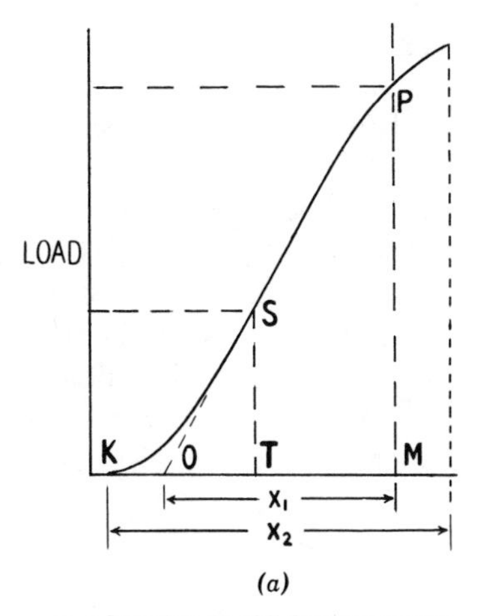

Figure 14-4. Load versus deformation. (Reprinted by permission of ASTM.)

in Chapter 2. The only variation is the shape of the specimen and the method of preparation. Two basic types of test specimens are used: Type A specimen is preferred in cases where enough material exists to form the necessary specimen. Type B is used where only smaller specimens such as sandwich panels are available.

14.2.6. Shear Properties (ASTM C 273)

This test was developed specifically for studying the stress–strain behavior of sandwich constructions or cores when loaded in shear parallel to the plane of the facings. The test advice illustrated in Figure 14-5 can be used for the shear testing of the complete sandwich panel or core materials alone. The load is applied to the ends of steel plates either in tension or compression in such a way that the load is distributed uniformly across the width of the specimen. The maximum load is applied for 3–6 min. A stress–strain curve is plotted and from the initial slope of the stress–strain curve, the shear modulus of the sandwich as a unit or core alone is calculated.

14.2.7. Flexural Properties (ASTM D 790)

Flexural strength at break, flexural yield strength, and modulus of elasticity of rigid cellular plastics can be determined by the same procedures described in Chapter 2.

14.2.8. Dimensional Stability (ASTM D 2126)

In recent years, foam plastics have found numerous applications in the aerospace, electronics, and construction industries. It is important to know how foam plastics will behave under various conditions of temperature and humidity. Many insulation applications demand a high degree of stability for an extended period of time. The test developed for studying the response of rigid cellular plastics to thermal and aging provides a maximum use temperature as well as stability data at the specified temperature and humidity conditions. The test requires the use of a balance, an oven, a cold box, and gauges for accurate measurements of specimens. The test specimens are machined to specified dimensions and accurately measured. After appropriate conditioning, the specimens are subjected to one of the test conditions specified in Table 14-1. Final test measurements are determined after the specimens recover to room temperature. Visual examination along with dimensional check is made. The maximum use temperature is determined by subjecting the specimens to successively higher temperatures and recording dimensional changes until an unacceptable dimensional change is measured.

14.2.9. Water Absorption (ASTM D 2842)

Cellular plastics are used extensively in floatation applications because of their ability to maintain a very low buoyancy factor. The buoyancy factor is directly affected by the amount of water a particular foam plastic will absorb. The test

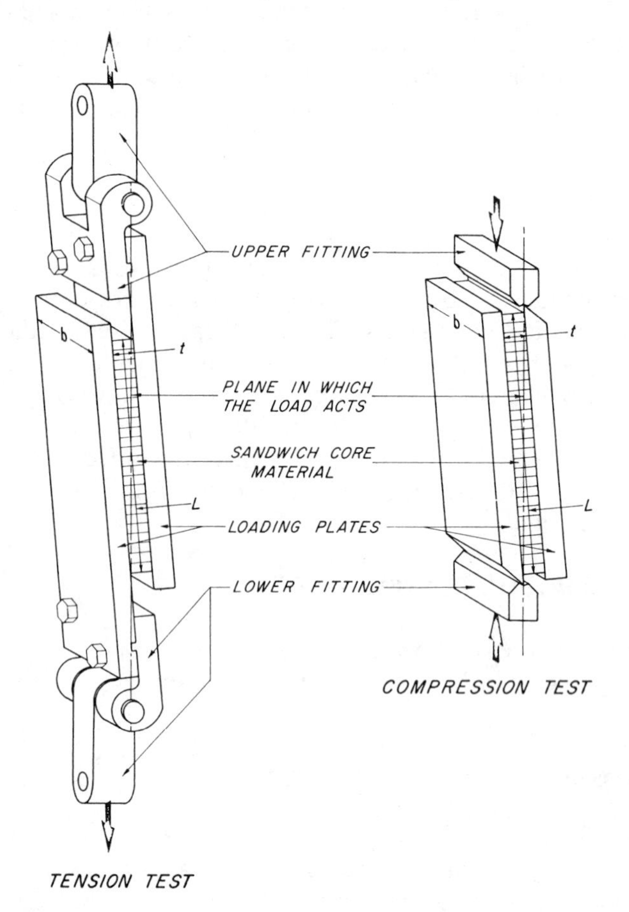

Figure 14-5. Test apparatus—shear property. (Reprinted by permission of ASTM.)

Table 14-1. Temperature and Relative Humidity Conditions[a]

Procedure	Temperature (°C)	Relative Humidity
A	23 ± 2	50 ± 5
B	38 ± 2	97 ± 3
C	70 ± 2	97 ± 3
D	To be selected	95 ± 3
E	−40 ± 3	Ambient
F	−73 ± 3	Ambient
G	70 ± 2	Ambient
H	100 ± 2	Ambient
I	150 ± 2	Ambient
J	200 ± 2	Ambient
X	Temperatures and humidities selected for individual needs	

[a] From ASTM D 2126. Reprinted with permission of ASTM.

method developed to determine water absorption of rigid cellular plastics is fully described in the ASTM Standards Manual. Basically, the test consists of determining the volume of initial dry weight of the object and calculating the initial buoyancy force. The object is then immersed in water and at the end of the immersion period, the final buoyancy force is measured with an underwater weighing assembly. The difference between the initial and final buoyancy force is the weight of water absorbed. This difference is expressed in terms of water absorbed per unit of specimen volume. The test results are seriously affected if proper steps are not followed closely and variables are not controlled carefully.

14.2.10. Water Absorption of Core Materials for Structural Sandwich Constructions (ASTM C 272)

This method of determining the water absorption of core materials for structural sandwich constructions gives basic information about water absorption characteristics and helps to determine the effect of water absorption on mechanical, thermal, and electrical properties. A 3 × 3 in. specimen is cut from a sandwich panel and its edges smoothed. After properly conditioning and weighing the specimen, it is immersed completely in water at a specified temperature and time. After the immersion period, the specimen is removed, dried, and weighed. Water absorption is reported as water gained/cc of specimen tested, or as percentage weight gain. The test is also used as a control test for product uniformity.

14.2.11. Water Vapor Transmission (ASTM C 355)

The water vapor transmission test is used to determine values of water vapor transfer through permeable materials. Two basic methods, the desiccant method and the water method, have been developed. The desiccant method generally yields lower results than the water method.

The rate of water vapor transmission (WVT) is defined as the time rate of water vapor flow of a body between two specified parallel surfaces normal to the surfaces, under steady conditions through unit area, under conditions of test. Water vapor permeance is defined as the ratio of the WVT of a body between two specified parallel surfaces to the vapor pressure difference between the two surfaces. An accepted unit of permeance is perm. Water vapor permeability is defined as the product of permeance and thickness. The unit for permeability is perm-in.

In the desiccant method, the specimen is sealed to the open mouth of a test dish containing desiccant and the assembly is placed in a controlled atmosphere. The rate of water vapor transmission through the specimen into the desiccant is determined by periodic weighings of the dish assembly. A distilled water is used in the water method. The rate of water movement through the specimen is determined by periodically weighing the dish containing the distilled water and specimen.

14.2.12. Weathering Properties

The effect of the outdoor environment on foam plastics is not clearly known yet. This is due to a lack of data available from either the material suppliers or the end-users. In general, the resistance of foamed plastics to the outdoor environment is usually similar to that of the base polymer. The test methods described in Chapter 5 are applicable to foam plastics.

14.2.13. Thermal Conductivity (ASTM C 177-45)

Foam plastics have the lowest thermal conductivity or *K* factor of any insulating material available today. This superior insulating ability depends upon many variable factors. Thermal conductivity of foam plastics is greatly influenced by the blowing agent, cell size, closed cell content, and density of the foam. Temperature and moisture are other factors that affect the thermal conductivity of foam plastics considerably. The method used to determine the thermal conductivity of foam plastics is the same one used for noncellular plastics. These techniques are described in detail in Chapter 3.

14.2.14. Flammability

Because of their superior insulating properties, foam plastics have found numerous applications in the construction and packaging industries. Along with increased use, a considerble amount of concern has been expressed in regard to the flammability of foam plastics. A detailed discussion of various tests used in industry to determine the flammability of cellular and noncellular plastics and their limitations has been included in Chapter 8.

14.2.15. Dielectric Constant and Dissipation Factor (ASTM D 1673)

The foam plastics have unique advantages over the conventional plastics in regard to electrical properties. This is due to the fact that part of the plastic material has been replaced by a gas that has a lower dielectric constant. Thus, foam plastics have an improved dielectric constant, a distinct advantage in wire and cable insulation applications. Another advantage of foam plastics is their low dissipation factor.

This method, although quite similar to the methods used for noncellular plastics, was developed especially for cellular plastics for several reasons. One of the reasons is that foam plastics have surfaces that preclude the use of conventional electrodes such as metal foils attached by petrolatum and similar adhesives. The greater thickness of the foam plastic specimen does not lend itself to the attachment of conventional electrodes.

Dielectric constant and dissipation factor of flat sheets or slabs of both rigid and flexible foam plastics can be determined by this method, at frequencies from 60–100 MHz. The basic apparatus consists of a bridge and a resonant circuit. Since foam plastics do not have surfaces suitable for the attachment of conventional electrodes, prefabricated rigid metal plate electrodes must usually be employed for dielectric constant and dissipation factor tests. Such electrode systems may be of the direct contact type or the noncontacting type.

The specimens are conditioned using standard conditioning procedures. It is preferred that the specimens are free of surface skin unless otherwise specified or agreed upon, since the surface skin may affect the results considerably. A more detailed discussion of the procedures and factors affecting the test results has been presented in Chapter 4.

Because of the basic nature of cellular plastics, inconsistency in thickness and foam densities, the reproducibility of the test results by these methods is considerably poorer than those expected from measurements on noncellular types of insultating materials.

14.3. FLEXIBLE FOAM TEST METHODS

14.3.1. Introduction

The test methods developed for testing flexible foams are quite different than those developed for rigid foams. For rigid foams, separate test methods were developed for specific properties. No such separate test methods relating to specific properties are developed for flexible foams. Instead, a series of test procedures that describe a variety of physical properties of a particular type of material are commonly used to test flexible foams.

In this section, an attempt is made to briefly describe all types of basic and suffix tests developed for flexible foams by the joint efforts of ASTM and the Society of the Plastics Industry. The individual tests for different forms and types of materials are also listed along with a list of basic and suffix tests for each material.

Only a few of these tests apply to each different type of material. The test title, the ASTM designation number, and the applicable test are listed in Table 14-2.

14.3.2. Steam Autoclave Test

The steam autoclave test consists of exposing the foam specimens to a low-pressure steam autoclave at a prescribed temperature and time, and observing the changes in the physical properties of the specimen. The test specimens are exposed to steam following specified preconditioning. After the exposure period, specimens are properly dried in a dry-air oven. The compression load deflection

Table 14-2. Tests for Flexible Cellular Materials

	Slab urethane foams	Molded urethane foams	Slab, bonded, and molded urethane foams	Vinyl polymers and copolymers
ASTM designation	D 1564	D 2406	D 3574	D 1565
Steam autoclave test	X	X	X	
Constant deflection compression set test	X	X	X	X
Load–deflection tests Methods A and B	X	X	X	X
Air flow test	X		X	
Compression load–deflection test	X	X	X	
Dry heat test	X	X	X	
Fatigue test static and dynamic	X	X	X	X
Density test	X	X	X	X
Tear resistance test	X	X	X	
Tension test	X	X	X	
Resilience test	X		X	

is obtained before and after exposure and percent change from original value is reported. The steam autoclave compression set value is also calculated.

14.3.3. Constant Deflection Compression Set Test

This test consists of deflecting the specimen under specified conditions of time and temperature and observing the effect on the thickness of the specimen. The test for a compression set using a constant load has not yet been developed. The test procedure requires the specimen to be compressed between two or more parallel plates to a specified deflection thickness. The entire assembly is placed in a mechanically convected air oven for a specified time and temperature. Following this exposure, the specimen is removed from the apparatus and the recovered thickness is measured. The constant deflection compression set, expressed as a percentage of the original thickness is calculated.

14.3.4. Load Deflection Test

The load deflection test is performed two different ways:

1. Method A: Indentation to specified deflections.
2. Method B: Indentation to specified loads.

Method A: Indentation load deflection (ILD) consists of measuring the load necessary to produce deflection (generally 25 percent –65 percent) in the foam product. The test is carried out by pushing an indentor down into the foam specimen and measuring the force on the foot at various compression amounts.

The test is widely used in the cushioning and bedding industry. Higher IDL values indicates a stiffer foam.

Method B: Indentation residual gauge load (IRGL), consists of deflecting the specimen to a constant load rather than a constant thickness as in the case of ILD method. The specimen is compressed with a specified load for 1 min and the resulting thickness is noted. IRGL value at the specified load is reported. IRGL values at different loads may be obtained. Here, a higher IRGL value indicates a stiffer foam. This test is useful in determining how thick the padding should be under an average person sitting on a seat cushion.

14.3.5. Air Flow Test

The air flow test measures the ease with which air passes through a cellular structure. The resistance to air flow exhibited by the open cells in a flexible foam may be used as an indirect measurement of cell structure characteristics. The test consists of placing a flexible foam specimen in a cavity over a vacuum chamber and creating a specified constant air-pressure differential. The rate of flow of air required to maintain this pressure differential is the air flow value. The test is conducted two different ways—air flow parallel to foam rise and air flow perpendicular to foam rise. Air flow values are proportional to porosity in flexible foam.

14.3.6. Compression Load Deflection Test (CLD)

Compression load deflection test (CLD) consists of measuring the load necessary to produce a 25 percent compression over the entire top area of the foam specimen. This test differs from the ILD test described previously in that the flat compression foot used in the test is larger than the specimen. The compression foot is brought into contact with the specimen. The specimen is compressed 25 percent of its original thickness and after 1 min, the final load is observed. The compression load deflection is recorded as the load required for a 25 percent compression/in.^{2} of specimen area.

14.3.7. Dry Heat Test

The dry heat test consists of exposing foam tensile specimens in an air-circulating oven and observing the effect on tensile properties of the foam. The tensile properties are determined using the same procedure described in the tension test for flexible foams.

14.3.8. Fatigue Test

This test is conducted two ways:

1. Procedure A: Static fatigue at constant deflection.
2. Procedure B: Dynamic fatigue by the roller shear at constant load.

Procedure A helps to determine:

1. A loss of load-bearing properties.
2. A loss of thickness.
3. Structural breakdown by visual examination.

The specimen is placed between the plates with the spacer bars to provide a 75 percent deflection. The plates are clamped and held at 75 percent deflection for 17 hrs at standard test conditions. Thirty minutes after the completion of the test, the specimen thickness and indentation load deflection (ILD) are measured. The percentage loss of thickness and the percentage loss of load deflection are calculated and reported. Results of the visual examination indicating possible breakdown of cellular structure are also noted.

Procedure B determines:

1. A loss of load-bearing properties.
2. A loss of thickness.
3. Possible structure breakdown by visual examination.

This procedure tests the sample dynamically, at a constant load, deflecting the material both vertically and laterally. The apparatus consists of a roller that moves over the specimen along with standad deflection measuring devices. The specimen is tested under a variety of pressures and directions of the roller. The final fatigue value is expressed as a total loss number. The total loss number is equal to the sum of percent losses at each different load. The visual examination results are also included in the report.

14.3.9. Density Test

The density test determines the density of uncored foam by calculating the weight and volume of a regularly shaped specimen. The specimen is weighed on a balance and its dimensions are measured using calipers. The density of the specimen is calculated by dividing the weight by the volume of the specimen.

14.3.10. Tear-Resistance Test

This test was developed to determine the tear-resistance of foam. The tear-resistance test, also known as the block method, uses a block specimen described in Figure 14-6. The block specimen is mounted by spreading the block so that

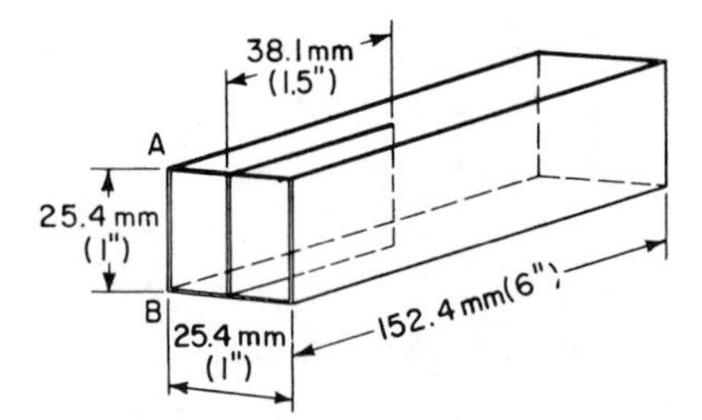

Figure 14-6. Block specimen—tear-resistance test. (Reprinted by permission of ASTM.)

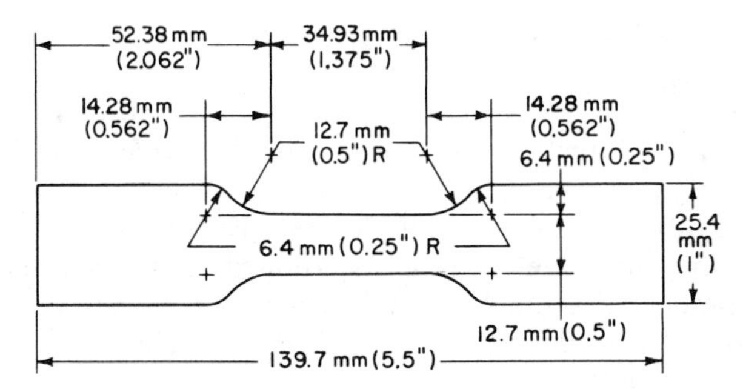

Figure 14-7. Die for cutting dumbbell-shaped specimen for tension test. (Reprinted by permission of ASTM.)

each tab is held in the jaw of the tensile testing machine. The load is applied by separating the jaws at a specified speed. The force required to rupture the specimen is recorded. The tear-resistance is calculated from the maximum load recorded and the average thickness of the specimen. The tear-resistance is reported in pounds-force per linear inch of thickness.

14.3.11. Tension Test

This method is similar to the method described in Chapter 2 for the tensile testing of plastics. The test determines the effect of the application of tensile load to a foam specimen. Tensile stress, tensile strength, and ultimate elongation are also obtained. The test specimens are dumbbell-shaped, cut from a flat sheet with a die such as is shown in Figure 14-7. The tensil strength is determined by placing the specimen in the jaws of the tensile testing machine and pulling it apart at a specified speed. The load at break divided by the original cross-sectional area of the specimen is the tensile strength.

Figure 14-8. Resilience test apparatus. (Courtesy Custom Scientific Instruments, Inc.)

14.3.12. Resilience Test (ASTM D 1564)

This test, also known as the ball rebound test, is very useful in evaluating the resilience of foams. The test consists of dropping a steel ball onto a foam specimen and noting the height of rebound. The apparatus as shown in Figure 14.8, consists of a clear plastic tube with a series of circles inscribed on it to measure the rebound height directly in percentage. The ball is dropped so that it does not strike the tube on the drop or the rebound. The resilience value is the ratio of the rebound height to the original height expressed as a percent. The higher value of the resilence indicates more "lively" foam.

14.3.13. Raw Material Tests

Raw material tests were developed specifically for the purpose of checking the uniformity of the two main ingredients used in the manufacture of urethane foams. Test methods for both toluene diisocyanate and polyol are fully described in Part 36 of the ASTM Standard Manual. Only a very brief description of each test is covered in this chapter.

A. *Test Methods for Urethane Foam Isocyanate Raw Materials (ASTM D 1638)*

This method covers the procedures for testing isocyanate raw materials used in preparing urethane foam. Two types of isocyanates are covered by this method:

1. Toluene diisocyanate (purified).
2. Modified or crude isocyanates.

The tests used to quality each type of isocyanate are long, complex, and different.

Tests for Toluene Diisocyanate (Purified)

1. Assay (the percentage by weight of TDI present in the sample).
2. Isomer content.
3. Total chlorine content.
4. Hydrolyzable chlorine.
5. Acidity.
6. Freezing point.
7. Specific gravity.
8. Color.

Tests for Crude or Modified Isocyanates

1. Acidity.
2. Amine equivalent.
3. Brookfield viscosity.

The toluene diisocyanate is most often checked by foam producers for assay and acidity.

B. *Test Methods for Urethane Foam Polyol Raw Materials (ASTM D 2849)*

These methods cover the testing of polyol raw materials used in the manufacture of urethane foams including both polyester and polyethers. The following tests are employed to quality polyols.

1. Sodium and potassium concentrations.
2. Acid and alkalinity numbers.
3. Hydroxyl number.
4. Unsaturation.
5. Water content.
6. Suspended matter.
7. Specific gravity.
8. Viscosity.
9. Color.

Hydroxyl number and water content are two of the most often performed tests by foam producers to check the uniformity of polyols.

14.3.14. Processibility Test for Foams

These types of tests are most often performed by manufacturers of foam products to insure themselves against variations that may be encountered during processing. Processability tests provide valuable information regarding the behavior of materials during processing.

Rate-of-Rise (Volume Increase) Properties of Urethane Foaming Systems (ASTM D 2237)

This method covers the determination of the rate at which the volume of a foaming system changes under standard conditions. A rise rate curve is plotted to obtain the extrapolated initiation time (EIT) and extrapolated rise time (ERT).

The apparatus consists of a cylindrical container and a float connected to a very sensitive height measuring device. One such device is illustrated in Figure 14-9. The entire system, including the apparatus and components, is conditioned to 23 ± 1°C. The components of the foaming system are mixed and placed in the bottom of the constant cross section cylindrical container. The mixing and stopwatch are started simultaneously. The height gauge is moved over to the foaming system and lowered into the container. The height versus time data is continuously recorded. After the foam has cured and cooled, a nominal 2-in. cube is cut from the core of the foam to inspect cell structure and calculate density. The change in height is calculated as a percentage of the total change in height using the obtained density value. A change in height versus time curve is plotted on semilogarithmic graph paper as shown in Figure 14-10. A linear extrapolation of the straight line section of the rise rate curve to intersect both 0 percent and 100 percent change in height is made. The 0 percent extrapolation value is considered the extrapolated initiation time and the 100 percent extrapolation value is con-

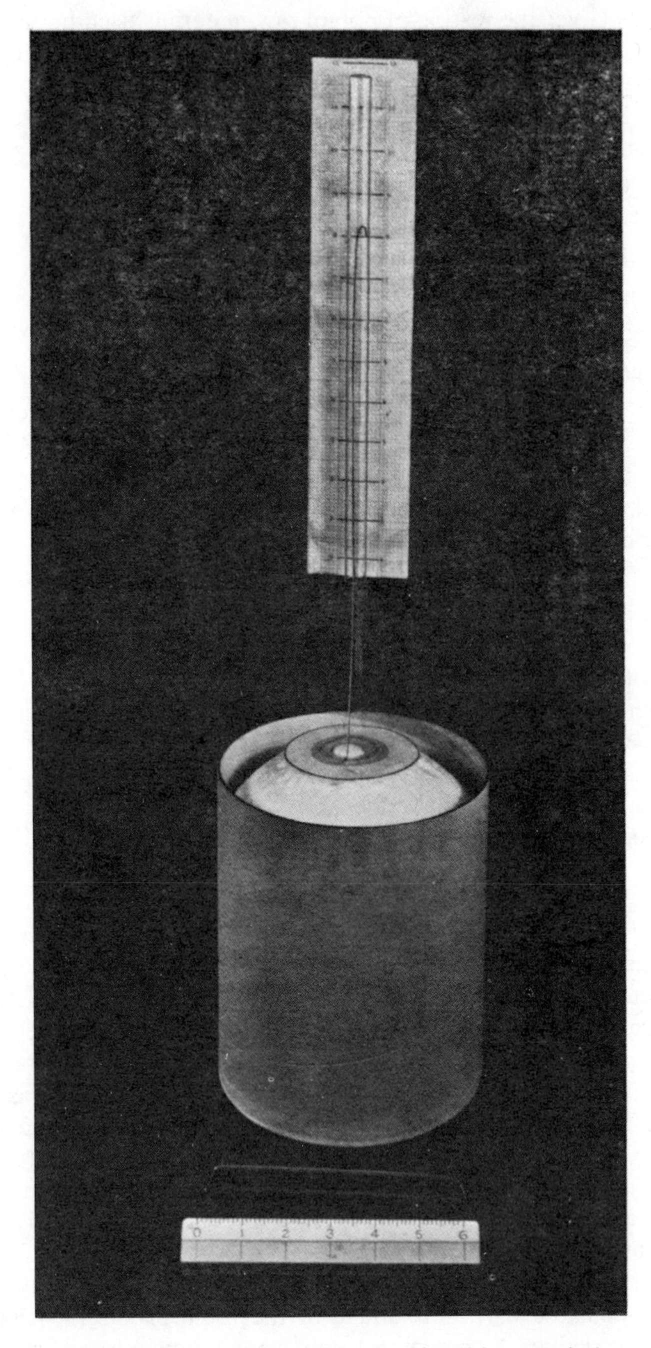

Figure 14-9. Rate-of-rise test apparatus. (Reprinted by permission of ASTM.)

Table 14-3. Foam Properties Chart[a]

Type	Density (lb/ft^3)	Tensile strength (psi) (ASTM D 1623)	Compressive strength at 10 percent deflection (psi) (ASTM D 1621)	Maximum service temperatire (°F) Dry	Maximum service temperatire (°F) Wet	Thermal conductivity (Btu/ft^2) (ASTM D 2326)	Coefficient of linear expansion (10^{-5} in./in./F) (ASTM D 696)	Dielectric constant (ASTM D 1673)	Dissipation factor at 28°C. and 1 meg.	WVT (perm-in.) (ASTM C 355)	Water absorption, percent by volume (96 hr.) (ASTM D 2842)
ABS (acrylonitrile-butadiene-styrene) Injection molding type Pellets	40–56	2000–4000	2300–3700	176–180	—	0.58–2.1	3.7–9.5	—	—	—	0.4–0.6
Flame-retarded Pellets	45–55	1500–3000		180	—		5.1	—	—	—	—
Acrylic Boards	2.6–6.2	—	67–320[b]	220–230	—	0.22	2.9–3.2	1.90	0.0036	—	11.8–13.3
Cellulose acetate Boards and rods (rigid, closed cell foam)	6.0–8.0	170	125	350	—	0.31	2.5	1.12	—	—	13–17 at 100% R.H. 1.9–2.5 at 50% R.H.
Epoxy Rigid closed cell Precast blocks, slabs, sheet	5.0	51	90	350	—	0.26	—	1.19–1.08 at 10^6 and 10^{10}	—	—	—
	10.0	180	260	350	—	0.28	—	1.36–1.24 at 10^6 and 10^{10}	—	—	—
	20.0	650	1080	350	—	0.32	—	1.55–1.41 at 10^6 and 10^{10}	—	—	—
Syntactic One-part free flowing powder	14.0–20.0	—	125	350–400	—	<0.5	—	1.38	0.006	—	—
Two-part moldable like damp sand	20.0	—	1000	—	—	0.38	—	1.45	0.01	—	—
Rigid sheet or pack-in-place	23.0	—	2100	500	—	0.36	—	—	—	—	1.8
Spray-applied Two-package systems (liquid)	1.8–2.0	26–31	13–17	160	—	0.11–0.12	—	—	—	—	—
Foam-in-place Two-package systems (liquid)	2–2.3	—	20–26	200	—	0.11–0.13	—	—	—	—	—
Ethylene copolymer Extruded sheet	35.0	600–800	—	130	—	—	—	—	—	—	<0.5
Ionomer Sheet, rod	2.0–20.0	57–105	24–15.2	150–155	—	0.27–0.34	—	1.5 at 10^6	0.003	0.34–2.00	0.40–1.0

Table 14-3. (*Continued*)

Type	Density (lb/ft³)	Tensile strength (psi) (ASTM D 1623)	Compressive strength at 10 percent deflection (psi) (ASTM D 1621)	Maximum service temperature (°F) Dry	Wet	Thermal conductivity (Btu/ft²) hr/°F/in (ASTM D 2326)	Coefficient of linear expansion (10^{-5} in./in./F) (ASTM D 696)	Dielectric constant (ASTM D 1673)	Dissipation factor at 28°C. and 1 meg.	WVT (perm-in.) (ASTM C 355)	Water absorption, percent by volume (96 hr.) (ASTM D 2842)
Phenolic	$\frac{1}{3}$–$1\frac{1}{2}$	3–17	2–15	—	—	0.21–0.28	—	—	—	—	—
Foam-in-place											
Liquid resin	2–5	20–54	22.85	Continuous service at 300		0.20–0.22	0.5	—	—	2 lb./ft.³ 5 lb./ft.³ 2074 g. 1844 g. per day/m.²	13–51 at 100% R.H. 1–4 at 50% R.H.
	7–10	80–130	158–300			0.24–0.28	—	1.19–1.2	0.028–0.031	—	10–15 at 100% R.H.
	10–22	—	300–1200	—	—	—	—	1.19–1.2	0.028–0.031	—	1–5 at 50% R.H.
Syntactic castable	50–60	>1000	8000–13,000	275	—	1.0	10	2.1	0.03	—	>0.5
Two-component: liquid and paste											
Phenylene oxide-based foamable resin	50	3300	5500	200	—	0.86[c]	3.8×10^{-5}	216	0.0017	—	—
Pellets											
Polybenzimidazole	3–6	44–125	30–125	600	—	0.216	—	—	—	—	—
Blown											
Slabs											
Syntactic	15–38	200–2000	700–4310	600	—	0.42–0.78	—	1.82–2.40	—	—	—
Molding powder, slabs											
Polycarbonate	50	5500	7500	270	—	1.05[c]	2.5	220	0.001	—	—
Pellets											
Polyethylene	1.3–2.6	20–30	5	160–180		0.28–0.40	9.5–2.3	1.05 at 10^6	0.0002 at 10^9	0.40	<0.50
Low-density foam											
Planks, rods, round, ovals, net											
Noncrosslinked sheet	2.1–3.3	35–100	3	160–180	—	0.28–0.34	2.3	1.05 at 10^6	0.0002 at 10^9	0.20–<0.40	<0.50
Noncrosslinked tubing	2.1–3.3	—	—	160–180	—	0.26–0.28 at 40°F. mean	2.3	—	—	0.13	<0.50
Crosslinked sheet, rolls, cord	1.6–2.4	40–70	2.9–3.0	175–200	—	0.25–0.28					0.1 (24 hr)
Intermediate density foam	3.6–4.4	70–110	4.3–14	180–200	—	0.30	—	1.06 at 10^6	0.0002 at 10^9	<0.40	<0.50
Crosslinked plank, sheet, rolls	5.5–7	110–210	2–18	180–200	—	0.32–0.34	—	1.07 at 10^6	0.0002 at 10^9	<0.40	<0.50
	9.0–10.5	210–300	15–30	180–200	—	0.34–0.40	—	1.15 at 10^6	0.0002 at 10^9	<0.40	<0.50

High-density foam Molded parts and shapes with solid integral skin	25.0–50.0	1200	1300	230	230	0.92	4.18	—	—	—	0.22
Crosslinked foam Rolls, sheets	0.9–12.5	46–210	2.0–18.5	180	—	0.27–0.4	<13	1.1–1.55	0.002–0.0007	—	0.1–0.5
Polypropylene Low-density foam Roll, sheets	0.6	20–40	0.7	250		0.27		1.02 at 10^4	0.00006 at 10^4	—	0.02 (3 hr)
High-density foam Molded parts and shapes with solid integral skin	35.0	1600	2100								
—	—	4.2	—	—	—	—	—				
Crosslinked foam Sheets	3.0	118–147	175–1200	275	275	0.27	—	—	—	—	0.5
Polystyrene	1.0	21–28	13–18	165–175	—	0.26	3.0–4.0	1.06–1.02 at 10^3 to 10^6 cycles	0.0001–0.0007 at 10^2 to 10^6 cycles	1.2–6.0	2–6
Shapes, boards, and billets Molded from expandable sheets	2.0	42–68	35–45	165–175	—	0.24 at 70°F mean temp.	3.0–4.0			0.6–1.2	2–4
	5.0	148–172	85–130	165–175	—	0.246	3.0–4.0			0.4–0.6	2–4
Extruded boards and billets	1.5–20	55–70	25–55 at 5%			0.21–0.29[d] 0.23–0.26[c]					
	2.0–2.6	60–105 at 5%	25–60	165–175 (aged)	—	0.17–0.19[d] 0.18–0.21[e]	3.0–4.0	<1.05 at 10^2 to 10^8 cycles	<0.0004 at 10^3 to 10^8 cycles	0.3–1.1	1.0
	2.0–5.0	180–200	100–180 at 5%								
Extruded film and sheet	6.0	300–500	42.5 52.5	170–175 (aged)	—	0.24 at 70°D mean temperature	—	1.27	0.00011	1.50	
	8.0	400–700		175	—		—	—	—	1.25	
	10.0	600–1000	68.0	175	—		—	1.28	0.00015	1.00	
Polyurethane Rigid (closed cell)	1.3–3.0	15–95	15–60	180–250	—	0.11–0.17[f,h] 0.21	4–8	1.05	—	0.6–4.0	1.0–5.0
Molded parts; boards, blocks, slabs; pipe covering; one-shot, two- and three-package	4–8	90–290	70–275	200–250	—	0.15–0.21[f,h] 0.21–0.29[g,h]	4	1.10	0.0018	0.9–2.0	0.6–2.0
systems for foam-in-place; for spray, pour, or froth-pour techniques	9–12	230–450	290–550	250–275	—	0.19–0.25[f,h] 0.31–0.35[g,h]	4	1.2	0.0032	—	
	13–18	475–700	650–1100	250–300	—	0.26–0.34[f,h] 0.36–0.40[g,h]	4	1.3	0.0055	—	
	19–25	775–1300	1200–2000	250–300	—	0.34–0.42[e,h] 0.42–0.52[f,h]	4	1.4	—	—	0.2
	26–40	1350–2500	2100–4000	275–300	—	—	4	1.5	—	—	—
	41–70	3000–8000	5000–15000	300	—	0.57[g]	4	—	—	0.08	—
Flexible—free rise (slabstock) Slabs, sheets, blocks, custom shapes	0.9–8.0	8–45	0.2–2.0 at 25%	150–175	—	0.2–0.25 at 2 lb./cu. ft. density	—	1.0–1.5	—	—	—

Table 14-3. (*Continued*)

Type	Density (lb/ft³)	Tensile strength (psi) (ASTM D 1623)	Compressive strength at 10 percent deflection (psi) (ASTM D 1621)	Maximum service temperatire (°F) Dry	Wet	Thermal conductivity (Btu/ft²) hr/°F/in (ASTM D 2326)	Coefficient of linear expansion (10^{-5} in./in./F) (ASTM D 696)	Dielectric constant (ASTM D 1673)	Dissipation factor at 28°C. and 1 meg.	WVT (perm-in.) (ASTM C 355)	Water absorption, percent by volume (96 hr.) (ASTM D 2842)
Flexible—molded (foam-in-place)	1.2–20.0	10–1350	0.25–100 at 25%	150	—	0.3 at 2 lb./cu. ft. density	—	1.1	—	—	—
Two- and three-package or one-shot systems for mixing on job											
Isocyanurate foams	1.5–3.0	15–50	20–60	300	—	0.105–0.17	4	—	—	3	1.5
Bun											
Laminate	2.0–3.5	20–60	30–70	300	—	0.11–0.16	4	—	—	3	1.5
Pour	1.5–3.0	25–75	20–80	300	—	0.11–0.17	4	—	—	3	1.5
Spray	2.0–3.0	20–60	20–65	300	—	0.115–0.17	4	—	—	3	1.5
Semirigid (foamed-in-place)	1.2–26.0	20–1350	20–2100	175–350	—	0.11–0.30	—	2.2	—	—	—
Integral skin molded (flexible)	25–65 (skin)	20–1350	1–5 at 25%	150–175	—	—	—	—	—	—	—
Two- and three-package systems	5–20 (core)										
Skinned molded (rigid)	25–65 (skin)	100–2700	40–3000	150–250	—	0.12–0.80	4	2.5	—	—	—
Two- and three-package systems	3–30 (core)										
Microcellular elastomer	20–70	50–2500 at 50%	50–2500 at 25%	200	150	—	—	—	—	—	—
Two- and three-package systems		5–100 at 10%									
Polyvinyl chloride	13–60	—	—	150–175	—	Depends on density	—	—	—	—	—
Plastisol mechanically frothed											
Liquid or paste											
Plastisol with blowing agent	3–60	50–3000	—	150–175	—	Depends on density	—	—	—	—	—
Liquid or paste											
Flexible open cell	10–up	10–200	—	125–225	—	Depends on density	—	—	—	—	—
Sheets and rolls; cored cusions; other molded shapes											
Flexible closed cell	4.0–11	50–150 (D412)	0.5–40.0	130–150	—	0.24–0.28 at 70°F.	—	—	—	0.20 max.	Nil
Sheets, tubes, and molded shapes											
Rigid closed cell	2–4	1000 and up	—	—	—	2.0 at 70°F.	4.0–6.0	—	—	—	—
Boards and billets											
Silicone	9.6 (unrestricted)	—	—	400	—	—	—	1.42 at 10^5	0.001 at 10^5	—	—
Liquid (10% closed cell)											
Liquid (closed cell)	21.0–31.0	100–150	10 at 75%	500–650	—	0.36	—	1.3–1.4	<0.01	—	0.1
Sheet (open cell)	10.0	45	5 at 75%	500	—	0.6	—	1.2	0.007	—	—
Sheet (closed cell)	25–34	80–100	1.5–2.9	450	—	—	—	—	—	—	Maximum 5%[i]

Styrene-acrylonitrile	0.5	—	1.5 at 5%	—	—	0.32	—	—	—	—	—
Products or shapes molded from expandable	0.8	20	6.0 at 5%	170–190	—	0.29	—	—	—	2.0–4.0	Nil
beads; finished boards	1.0	30	6.0 at 5%	—	—	0.29	—	—	—	—	—
Urea formaldehyde	0.8–1.2	Poor	5	120	—	0.18–0.21	—	—	—	28–35	1.9
Block, shred, foam-in-place											

[a] Reprined with permission of McGraw–Hill Book Company, New York.

[b] At 5 percent.

[c] ASTM C 177.

[d] At 40°F.

[e] At 70°F.

[f] Blow with fluorocarbon.

[g] Blown with CO_2.

[h] First number in each sequence is the value for unaged material; the number is the value after aging.

[i] ASTM D 1056.

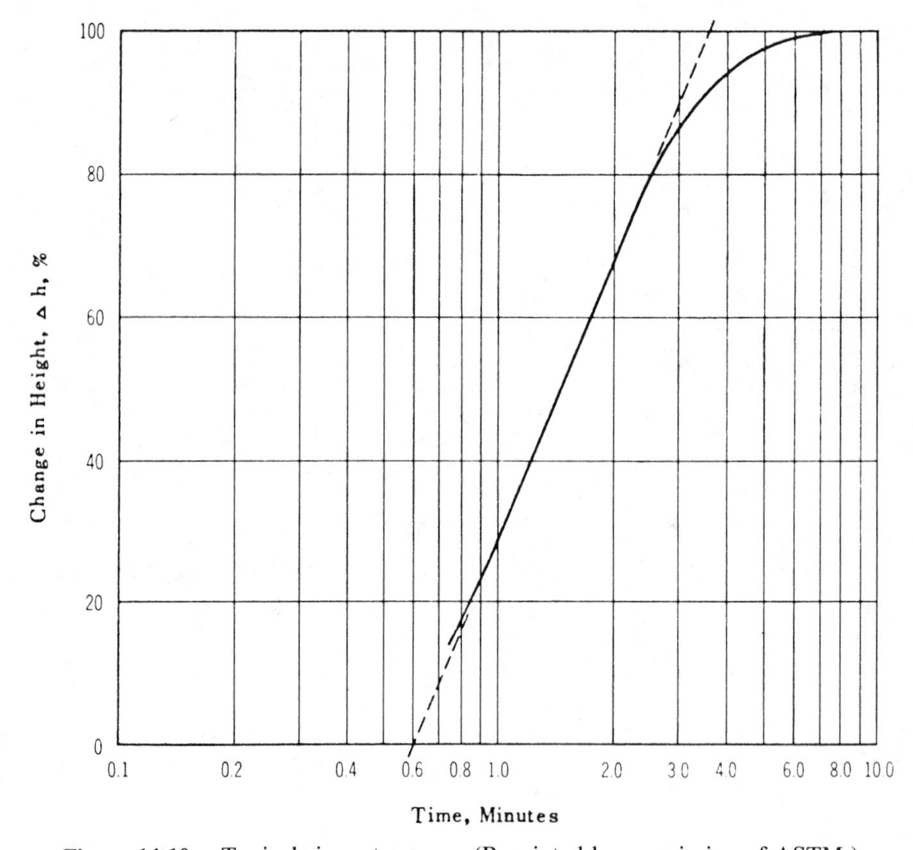

Figure 14-10. Typical rise rate curve. (Reprinted by permission of ASTM.)

sidered the extrapolated rise time. To improve the accuracy of the test results, the experimental values are corrected to 23°C through the use of temperature coefficients for ERT and EIT.

The data obtained by this method cannot be directly applied to manufacturing since the method of mixing, time–temperature relation of the mold surface during foaming, and shape and cross section of a mold cavity can seriously affect the rate of rise value. However, if properly interpreted, the rate-of-rise data obtained by this method can be helpful in predicting foaming behavior under nonstandard conditions.

14.4. FOAM PROPERTIES

Table 14-3 lists and compares important properties for both rigid and flexible foams.

GENERAL REFERENCES

Frisch, K., and Saunders, J., *Plastics Foams*, Marcel Dekker, New York, 1972.

Benning, C. J., *Plastics Foams*, Vols. I and II, Wiley-Interscience, New York, 1969.

Remington, W. J. and Priser, R., "A New Apparatus for Determining the Cell Structure of Cellular Materials," *Rubber World*, **183**, (1958), pp. 261–264.

Harding, R. H., "Determination of Average Cell Volume in Foamed Plastics," *Mod. Plast.*, **37**(10), (1960), pp. 156–160.

Benning, C. J. and Nutter, J. I., "Filled Polyethylene Foams," *22nd SPE ANTEC*, Montreal, Mar 7–11, 1966 Reprints.

Benning, C. J., "Polyethylene Foam: I. Modified PE Foam Systems," *J. Cellular Plast.*, **3**(2), (Feb. 1967), pp. 62–72.

Benning, C. J., "Polyethylene Foam, II. Mechanical Properties of Polyethylene Foams Prepared at High and Low Rates of Extrusion," *J. Cellular Plast.*, **3**(3), (March 1967), pp. 125–137

CHAPTER 15

Failure Analysis

15.1. INTRODUCTION

Whether you are an engineer, designer, purchasing agent, or production manager, the subject of failure analysis is bound to catch up with you. Just about everyone gets directly or indirectly involved with the complex task of failure analysis. The situation is usually one of the utmost urgency and nervousness. Parts are returned from the field and the reputation of the manufacturer is at stake. Fear of losing an entire product line or even going out of business is not totally unreal. All these factors and the lack of sufficient information regarding failure analysis techniques make this task very complex. In this chapter we discuss a step-by-step analysis of a defective product returned from the field.

At this point, it is important to examine the reasons behind part failure. Part failure is generally related to one of four key factors—material, process, design, and user abuse. Failures arising from hasty material selection are not uncommon in plastics or any other industry. In an application that demands high-impact resistance, a high-impact material must be specified. If the material is to be used outdoors for a long period, a UV-resistant material must be specified. For proper material selection, careful planning, a thorough understanding of plastic materials, and reasonable prototype testing are required. But, proper material selection alone will not prevent a product from failing. While designing a plastic product, the designer must use the basic rules and guidelines provided by the material supplier for designing a particular part in that material. One must remember that with the exception of a few basic rules in designing plastic parts, the design criteria changes from material to material as well as from application to application. Today, design-related failures are by far the most common type of failure.

After proper material selection and design, the responsibility shifts from the designer to the plastic processor. The most innovative design and a very careful material selection cannot make up for poor processing practices. Molded-in stresses, voids, weak weld lines, and moisture in the material are some of the most common causes for premature product failures. Lastly, one other factor over which no one has any control is consumer abuse of the product. In spite of the built-in safety factor, warning labels, and user's instructions, failure arising from consumer abuse is quite common in the plastics industry.

15.2 TYPES OF FAILURES

15.2.1. Mechanical Failure

Mechanical failure arises from the applied external forces which when they exceed the yield strength of the material cause the product to deform, crack, or break into pieces. The force may have been applied in tension, compression, and impact for a short or a long period of time at varying temperatures and humidity conditions.

15.2.2. Thermal Failure

Thermal failures occur from exposing products to an extremely hot or extremely cold environment. At abnormally high temperatures the product may warp, twist, melt, or even burn. Plastics tend to get brittle at low temperatures. Even the slightest amount of load may cause the product to crack or even shatter.

15.2.3. Chemical Failure

Very few plastics are totally impervious to all chemicals. Failure occurring from exposing the products to certain chemicals is quite common. Residual or molded-in stress, high temperatures, and external loading tend to aggravate the problem.

15.2.4. Environmental Failure

Plastics exposed to outdoor environments are susceptible to many types of detrimental factors. Ultraviolet rays, humidity, microorganisms, ozone, heat, and pollution are major environmental factors that seriously affect plastics. The effect can be anywhere from a mere loss of color, slight crazing and cracking, to a complete breakdown of the polymer structure.

15.3. ANALYZING FAILURES

The first step in analyzing any type of failure is to determine the cause of the failure. Before proceeding with any elaborate tests, some basic information regarding the product must be gathered. If the product is returned from the field, have the district manager or consumer give you basic information, such as the date of purchase, date of installation, date when the first failure encountered, geographic location, types of chemicals used with or around the product, whether the product was used indoors or outdoors. All this information is very vital if one is to analyze the defective product proficiently. For example, if the report from the field along with the defective product indicates a certain type of chemical was used with the product, one can easily check the chemical compatibility of the product or go one step further and simulate the actual-use condition using that same chemical. Recordkeeping also simplifies the task of failure analysis. A simple date code or cavity identification number will certainly enhance the traceability. Four basic methods are employed to analyze product failure.

1. Visual examination.
2. Identification analysis.
3. Stress analysis.
4. Microtoming.

By zeroing in on the type of failure, one can easily select the appropriate method of failure analysis.

15.3.1. Visual Examination

A careful visual examination of the returned part can reveal many things. Excessive splay marks indicate that the materials were not adequately dried before processing. The failure to remove moisture from hygroscopic materials can lower the overall physical properties of the molded article and in some cases even cause them to become brittle. The presence of foreign material and other contaminants is also detrimental and could have caused the part to fail. Burn marks on molded articles are easy to detect. They are usually brown streaks and black spots. These marks indicate the possibility of material degradation during processing causing the breakdown of molecular structure leading to overall reduction in the physical properties. Sink marks and weak weld lines, readily visible on molded parts, represent poor processing practices and may contribute to part failure.

A careful visual examination will also reveal the extent of consumer abuse. The presence of unusual chemicals, grease, pipe dope, and other substances may give some clues. Heavy marks and gouges could be the sign of excessively applied external force.

The defective part should also be cut in half using a sharp saw blade. The object here is to look for voids caused by trapped gas and excessive shrinkage, especially in thick sections during molding. A reduction in wall thickness caused by such voids could be less than adequate for supporting compressive or tensile force or withstanding impact load and may cause part to fail. Lastly, if the product has failed because of exposure to UV rays and other environmental factors, a slight chalking, microscopic cracks, large readily visible cracks, or loss of color will be evident.

15.3.2. Identification Analysis

One of the main reasons for product failure is simply use of the wrong material. When a defective product is returned from the field, material identification tests must be carried out to verify that the material used in the defective product is, in fact, the material specified on the product drawing. Chapter 13 discusses in detail simple material identification techniques. However, identifying the type of material is simply not enough. Since all plastic materials are supplied in a variety of grades with a broad range of properties, the grade of material must also be determined. A simple technique such as the melt index test (see Chapter 7) can be carried out to confirm the grade of a particular type of material. The percentage of regrind material mixed with virgin material has a significant effect on the physical properties. Generally, the higher the level of regrind material mixed with virgin, the lower the physical properties. If during processing, higher than recommended temperature and long residence time is used, chances are that the material is degraded. This degraded material, when reground and mixed with virgin material, can cause a significant reduction in overall properties. Unfortunately, the percentage of regrind used with virgin is almost impossible to determine by performing tests on the molded parts. However, a correlation between the melt index value and the part failure rate can be established by conducting a series of tests to determine the minimum or maximum acceptable melt index value (1).

Part failures due to impurities and contamination of virgin material are quite common. Material contamination usually occurs during processing. A variety of purging materials are used to purge the previous material from the extruder barrel before using the new material. Not all of these purging materials are compatible. Such incompatibility can cause the loss of properties, brittleness, and delamination. In the vinyl compounding operation, failure to add key ingredients, such as an impact modifier, can result in premature part failure. Simple laboratory techniques cannot identify such impurities, contamination, or the absence of a key ingredient. More sophisticated techniques, such as infrared (IR) analysis and gel permeation chromatography (GPC) must be employed. These methods can not only positively identify the basic material, but also point out the type and level of impurities in most cases. Chapter 7 discusses the application of GPC and IR in detail.

15.3.3. Stress Analysis

Once the part failure resulting from poor molding practices or improper material usage through visual examination and material identification is ruled out, the next logical step is to carry out an experimental stress analysis. Experimental stress analysis is one of the most versatile methods for analyzing parts for possible failure. The part can be externally stressed or can have residual or molded-in stresses. External stresses or molded-in stresses or a combination of both can cause a part to fail prematurely. Experimental stress analysis can be conducted to determine the actual levels of stress in the part. Four basic methods are used to conduct stress analysis.

1. Photoelastic method.
2. Brittle coatings method.
3. Strain gauge method.
4. Chemical method.

1. Photoelastic Method

The photoelastic method for experimental stress analysis is quite popular among design engineers and has proved to be an extremely versatile, yet simple technique. Photoelastic properties are discussed in detail in Chapter 6.

If the parts to be analyzed are made out of one of the transparent materials, stress analysis is simple. All transparent plastics, being birefringent, lend themselves to photoelastic stress analysis. The transparent part is placed between two polarizing mediums and viewed from the opposite side of the light source. The fringe patterns are observed without applying external stress. This allows the observer to study the molded-in or residual stresses in the part. High fringe order indicates the area of high stress level whereas low fringe order represents an unstressed area. Also, close spacing of fringes represents a high stress gradient. A uniform color indicates uniform stress in the part. (2) Next, the part should be stressed by applying external force and simulating actual-use conditions. The areas of high stress concentration can be easily pinpointed by observing changes in fringe patterns brought forth by external stress. Figure 15-1 illustrates a typical stress pattern in a part.

Another technique known as the photoelastic coating technique can be used to photoelastically stress-analyze opaque plastic parts. The part to be analyzed is coated with a photoelastic coating, service loads are applied to the part, and coating is illuminated by polarized light from the reflection polariscope (3). Molded-in or residual stresses cannot be observed with this technique. However, the same part can be fabricated using one of the transparent plastic materials. In summary, photoelastic techniques can be used successfully for failure analysis of a defective product.

2. *Brittle-Coating Method*

The brittle-coating method is yet another technique of conveniently measuring the localized stresses in a part. Brittle coatings are specially prepared lacquers that are usually applied by spraying on the actual part. The part is subjected to stress

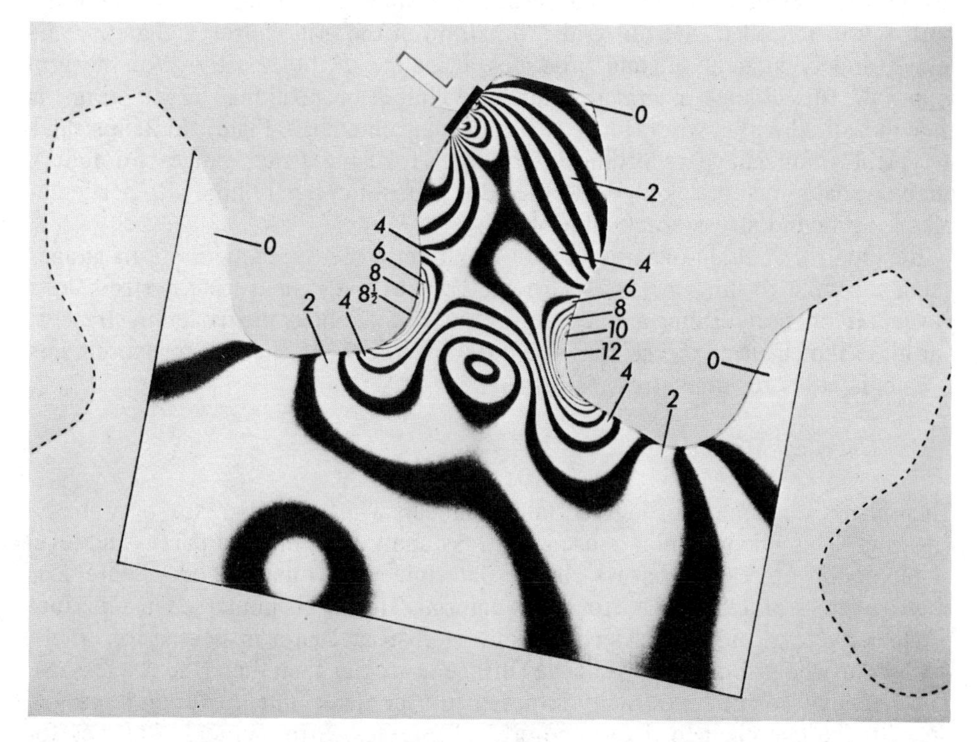

Figure 15-1. Typical photoelastic stress pattern. (Courtesy Measurements Group, Vishay Inc.)

after air drying the coating. The location of maximum strain and the direction of the principle strain is indicated by the small cracks that appear on the surface of the part as a result of external loading. Thus, the technique offers valuable information regarding the overall picture of the stress distribution over the surface of the part. The data obtained from the brittle coating method can be used to determine the exact areas for strain gauge location and orientation, allowing precise measurement of the strain magnitude at points of maximum interest. They are also useful for the determination of stresses at stress concentration points that are too small or inconveniently located for installation of strain gauges (4). The brittle-coating technique, however, is not suitable for detailed quantitative analysis like photoelasticity. Sometimes it is necessary to apply an undercoating prior to the brittle coating to promote adhesion and to minimize compatibility problems. Further discussion on this subject is found in the literature. (5)

3. *Strain Gauge Method*

The electrical resistance strain gauge method is the most popular and widely accepted method for strain measurements. The strain gauge consists of a grid of strain-sensitive metal foil bonded to a plastic backing material. When a conductor is subjected to a mechanical deformation, its electrical resistance changes proportionally. This principle is applied in the operation of a strain gauge. For strain measurements, the strain gauge is bonded to the surface of a part with a special adhesive and then connected electrically to a measuring instrument. When the test part is subjected to a load, the resulting strain produced on the surface of the part is transmitted to the foil grid. The strain in the grid causes a change in its length and cross section, and produces a change in the resistivity of the grid material. This change in grid resistance, which is proportional to the strain, is then measured with a strain gauge recording instrument (6). Figure 15-2 illustrates a typical strain gauge recording instrument. In using strain gauges for failure analysis, care must be taken to test the adhesives for compatibility with particular plastics to avoid stress-cracking problems.

Residual or molded-in stresses can be directly measured with strain gauges using the hole drilling method. This method involves measuring a stress at a particular location, drilling a hole through the part to relieve the frozen-in stresses, and then remeasuring the stress. The difference between the two measurements is calculated as residual stress (7).

4. *Chemical Method*

Most plastics, when exposed to certain chemicals while under stress, show stress-cracking. This phenomenon is used in stress analysis of molded parts. The level of molded-in or residual stress can be determined by this method. Figure 15-3 shows a stress-cracking curve for ABS material. The part is immersed in a mixture of glacial acetic acid and water for 2 min at 73°F and later inspected for cracks that occur where tensile stress at the surface is greater than the critical stress (8). The part may also be externally stressed to a predetermined level and sprayed on with the chemical to determine critical stresses. Stress-cracking curves for many types of plastics have been developed by material suppliers. If a defective

Figure 15-2. Typical strain gauge recording instrument. (Courtesy Measurements Group, Vishay Inc.)

product returned from the field appears to have stress-cracked, similar tests should be carried out to determine molded-in stresses as well as the effect of external loading by simulating end-use conditions. Failures of such types are quite common in parts where metal inserts are molded-in or inserted after molding.

15.3.4. Microtoming

Microtoming is a technique of slicing an ultrathin section from a molded plastic part for microscopic examination. This technique has been used by biologists and metallurgists for years but only in the last decade has this technique been used successfully as a valuable failure analysis tool.

Microtoming begins with the skillful slicing of an 8–10 μm-thick section from a part and mounting the slice on a transparent glass slide. The section is then examined under a light transmission microscope equipped with a polarizer for photoelastic analysis. A high power (1000×) microscope which will permit photographic recording of the structure in color is preferred. By examining the microstructure of a material, much useful information can be derived. For example, microstructural examination of a finished part that is too brittle may show that

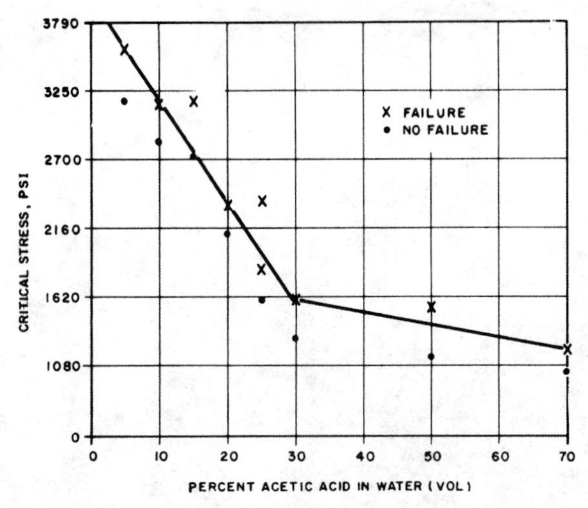

Stress cracking curve for Cycolac-T ABS — data received from the Borg Warner Corp.

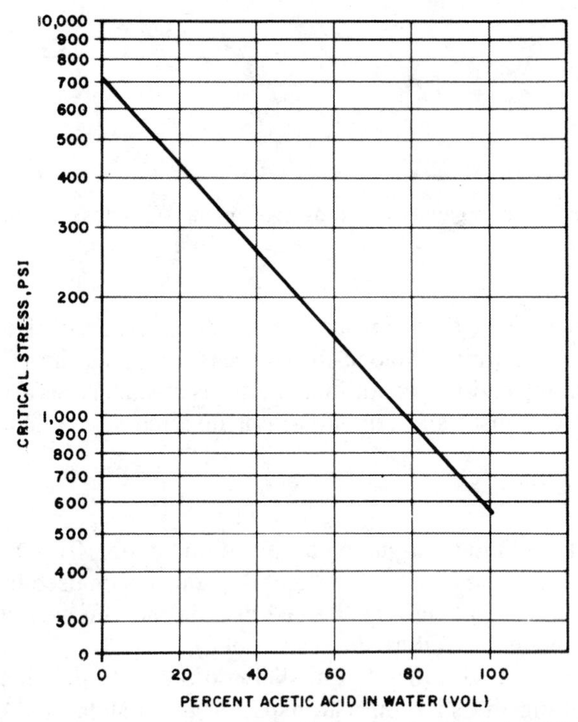

Stress cracking curve for Kralastic-MH ABS — data receiv.ed from the Uniroyal Chemical Div.

Figure 15-3. Stress-cracking curve for ABS. (Reprinted by permission from *Plastics Design and Processing*.)

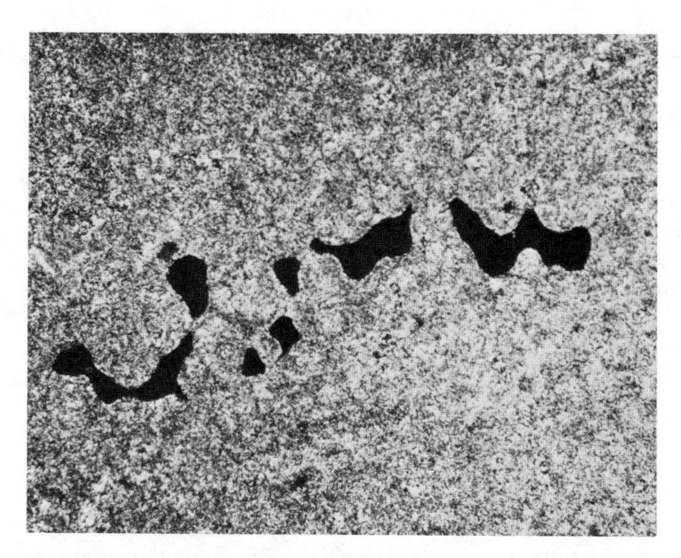

Figure 15-4. Shrinkage voids created by insufficient time and pressure to freeze the gates during injection molding process. (Courtesy *Engineering Design,* published by E. I. DuPont de Nemours and Co., Inc.)

the melt temperature was either nonuniform or too low. The presence of unmelted particles is usually evident in such cases. Another reason for frequent failures of the injection molded part is failure to apply sufficient time and pressure to freeze the gates. This causes the parts to be underpacked which creates center-wall shrinkage voids. Figure 15-4 illustrates such shrinkage voids. Voids tend to reduce the load-bearing capabilities and toughness of a part through the concentration of stress in a weak area. Contamination, indicated by abnormality in the microstructure, almost always creates some problems. Contamination caused by the mixing of different polymers can be detected through such analysis by carefully studying the differences in polymer structures. Quite often, a poor pigment dispersion also causes parts to be brittle. This is readily observable through the microtoming technique. In order to achieve optimum properties, additives such as glass fibers and fillers must disperse properly. Microtoming a glass fiber-reinforced plastic part reveals the degree of bonding of the glass fiber to the resin matrix as well as the dispersion and orientation of glass fibers. Molded-in stresses as well as stresses resulting from external loading are readily observed under cross polarized light because of changes in birefringence when the molecular structure is strained. Microtoming technique can also be applied to check the integrity of spin and ultrasonic or vibration welds. (9)

REFERENCES

1. Vogt, J. P., "Testing for Mechanical Integrity Assures Service Life of Plastic Parts," *Plast. Design and Processing* (March 1976), p. 13.
2. "Introduction to Stress Analysis by the Photoelastic Coating Technique," *Tech. Bull. IDCA-1*, Photoelastic Division, Vishay Intertechnology, Inc.

3. "Reflection Polariscope," *Tech. Bull. S-103-A*, Photoelastic Division, Vishay Intertechnology, Inc.
4. Holman, J. P., *Experimental Methods for Engineers*, McGraw–Hill, New York, 1971, pp. 333–334.
5. Durelli, A. J., Phillips, E. A., and Tsao, C. H., *Introduction to the Theoretical and Experimental Analysis of Stress and Strain*, McGraw–Hill, New York, 1958.
6. Measurements Group, Tech. Bull., Vishay Intertechnology, Inc.
7. Morita, D. R. "QC Tests That Can Help Pin Point Material or Design Problems," *Plastics Design Forum*, (May–June 1980), pp. 51–55.
8. Vogt, J. P., "Testing for Mechanical Integrity Assures Service Life of Plastic Parts," *Plast. Design and Processing*, (March 1976), p. 12.
9. Sessions, M. L., *Microtoming, Engineering Design with Dupont Plastics*, E. I. Dupont Co., Wilmington, Spring 1977, p. 12.

GENERAL REFERENCES

Dally, J. W. and Riley, W. F., *Experimental Stress Analysis*, McGraw–Hill, New York, 1965.

Hetenyi, M., *Handbook of Experimental Stress Analysis*, John Wiley and Sons, New York, 1950.

Kuske, A., and Robertson, G., *Photoelastic Stress Analysis*, John Wiley and Sons, New York, 1974, fpp. 263–274.

Bell, R. G., Cook, D. C., "Microtoming, an Emerging Tool for Analyzing Polymer Structures," *Plast. Eng.* (Aug. 1979), p. 18.

Frocht, M. M., *Photoelasticity*, John Wiley and Sons, New York, 1941.

Leven, M. M., *Photoelasticity*, Pergamon Press, New York, 1963.

CHAPTER 16

Quality Control

16.1. INTRODUCTION

Quality, as defined by the *American Heritage Dictionary,* is a characteristic or attribute of something; property or a feature. According to Juran (1), "Quality is a universal concept, applicable to all goods and services, more appropriately described as fitness for use." Fitness for use is the extent to which the product successfully serves the purpose of the user during usage. Quality control is the regulatory process through which we measure actual quality performance, compare it with standards, and act on the difference (2).

Quality control is the means by which every step in the production process receives the attention required to assure that all parts and end-products meet the desired specifications. Some parts are so complicated that if a production machine introduces a minor variance, the end-unit cannot be assembled. This means that the manufacturer must make a determination of every critical step of the manufacturing process. Then, he must decide at which point inspections will be made and what controls will be established in order to produce parts and end-products with all of the desired specifications.

In these days of rules and regulations, the only way any manufacturer can survive and be profitable is to have a firm grasp and a clear understanding of the science of quality control. For a plastics manufacturer or processor, the challenge is unique. The majority of materials are newly developed and are not precise in their composition. Manufacturing processes and procedures are different from conventional techniques; products made from such new materials have no previous history. Finally, the rapid growth of the plastics industry has created severe problems in training new people.

A well-established quality control system serves many useful purposes. First, it keeps the present customer happy, which in turn attracts new business. Second, it allows one to meet all regulatory and contractual obligations. More important, the system acts as a signaling device for any unforeseen problems and thereby reduces costly rejects. In the case of plastics, controlling the quality of the product is not a simple matter of inspecting and testing the product as it comes off the machine or assembly line. Many variables and unknowns, such as post-mold shrinkage, play an important role in controlling the ultimate quality of the product. Figure 16-1 graphically illustrates the interdependence of the major variables and

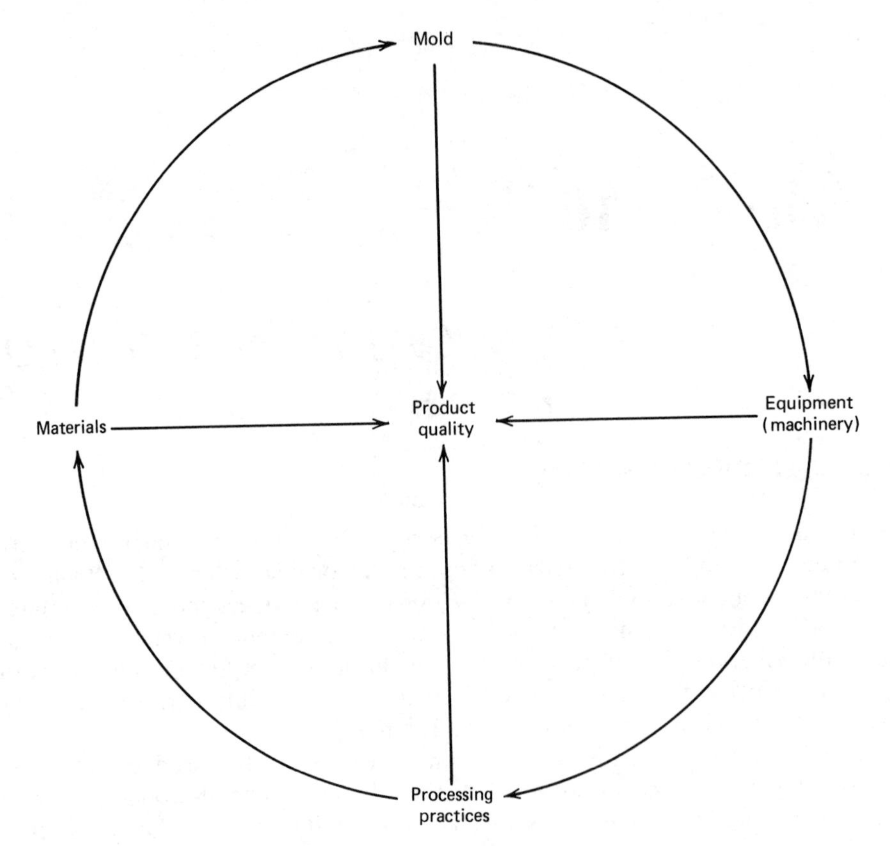

Figure 16-1. Interdependence of major variables in controlling product quality.

their effect on product quality. In this chapter, we discuss statistical quality control, an important aspect of quality control, and the quality control system in general.

16.2 STATISTICAL QUALITY CONTROL

Statistical quality control (SQC) is relatively new. The original concepts were developed by Walter A. Shewart of Bell Telephone Laboratories in the early 1930s. Before the development of SQC methods, the only way to assure the quality of the product was to inspect 100 percent of the goods produced. Thus, the quality was "inspected into" the product. No consideration was given to the concept of controlling the process to produce a good part. The statistical quality control method emphasizes controlling the process by examining the product rather than controlling the product alone. Simply stated, the more adequately the manufacturing process is controlled, the less the probability of scrap and rejects (3).

There are two main facets of statistical quality control. One of them is the use of the process control charts for in-process manufacturing operations. These charts, also referred to as "variables control charts" or "attributes control charts,'' are aimed at evaluating present as well as future performance. The other facet of statistical quality control is acceptance inspection or acceptance sampling. This technique forms the basis for scientifically evaluating past performance and accepting or rejecting the product.

16.2.1. Process Control Charts

Ideally, every manufacturer would like to see the manufacturing process controlled to a degree where he could produce identical parts day-in and day-out. In the real world, this is not possible. The difficulties in controlling the process arise from inherent variations caused by equipment and tool wear, operator skill, and manufacturing environment. Statistical quality control techniques, through the use of control charts, allow us to achieve the following:

1. To statistically determine whether the process is in or out of control.
2. If the process begins to drift out of control, it signals a need for investigation and correction.
3. To pinpoint the cause of the process variation.
4. To reduce inspection cost.
5. To improve the quality of the product and reduce the reject and scrap rate.
6. To establish a stable, smooth-running, and predictable process.
7. To study the capability of a particular process.
8. To check suitability of specifications.

In recent years, the growing trend in the plastics industry has been towards automation and high-speed production. Many new high-speed, fast-cycling injection molding machines and high-throughput extruders have been developed. A 64 cavity mold in a machine operating at a 3-sec cycle and producing over 70,000 parts/hr is not uncommon. In contrast, a single cavity mold in a 3000-ton machine producing a part weighing over 100 lbs is also not uncommon. Some large extruders which produce 8- and 10-in.-diameter pipe can consume as much as 2000 lbs/hr. It is obvious that in such a high production environment, one cannot allow the process to drift out of control or high scrap. Statistical quality control plays a very important role in controlling such processes in a highly competitive market where a high scrap and reject rate is totally unacceptable.

A. *Variables Control Charts ($\bar{x}$ and R charts)*

Before we discuss how control charts are used in the plastics manufacturing operation, it is necessary to understand a few basic terms.

$\bar{x}$ (AVERAGE, ARITHMETIC MEAN). The mean value of a sample of observations of the variable x.

$$\bar{x} = \frac{\Sigma x}{n}$$

For example, the individual weights of bottle caps are 7, 3, 9, 5, and 6 g. To compute the average ($\bar{x}$), sum the weights (x) and divide the sum by the total number (n) of bottle caps.

$$\bar{x} = \frac{7 + 3 + 9 + 5 + 6}{5} = 6$$

R (RANGE). The range is the difference between the highest and the lowest individual values in the group. The range calculated from the previous example would be six. $\bar{R}$ (pronounced R-bar) is the average of a series of R values.

Figure 16-2 illustrates a typical variables control chart ($\bar{x}$ and R chart). The center line of the $\bar{x}$ chart represents the average of a series of $\bar{x}$ values. This value is symbolized by $\bar{\bar{x}}$ (pronounced x double-bar). The center line of the R chart represents the average of the ranges of the samples. The top and the bottom lines of both charts represent the upper control limit (UCL) and the lower control limit (LCL), respectively. Generally, the control limits are three standard deviations (3σ) above and below the center line. This means that the probability of measurements falling between ± 3 standard deviation is 99.73 percent of the observed values. Or stated another way, between ± 3 standard deviations from the mean, one expects to find 99.73 percent of all the observed values. The values represented in Figure 16-2 are derived from the following example.

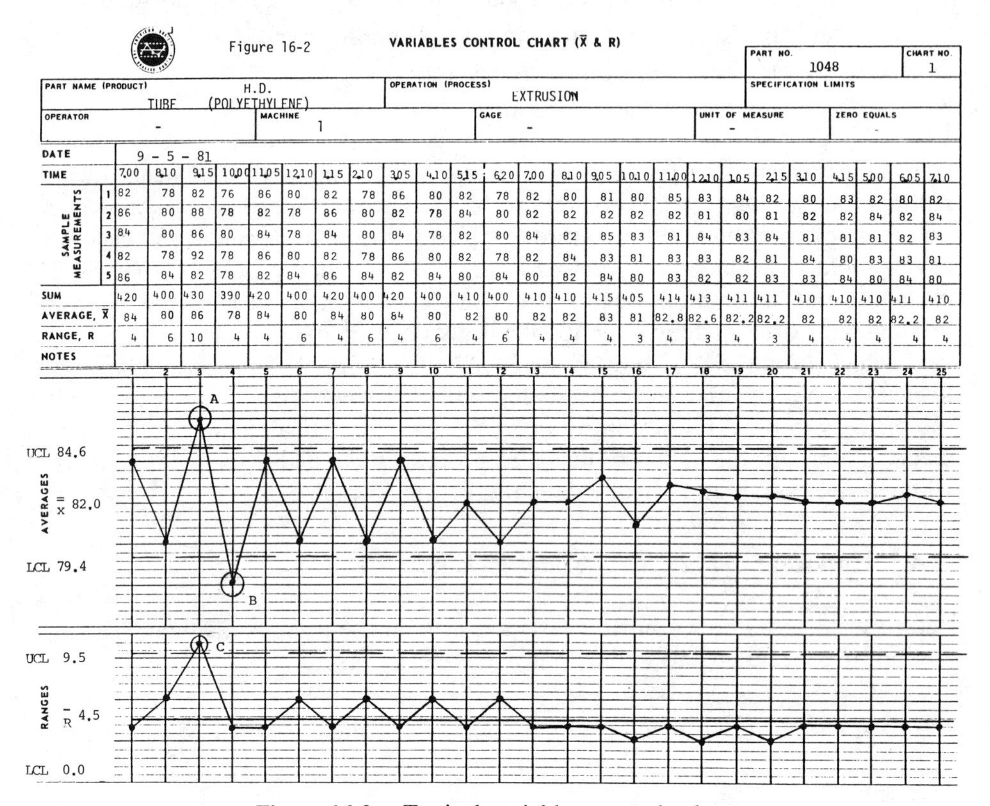

Figure 16-2

VARIABLES CONTROL CHART ($\bar{X}$ & R)

PART NAME (PRODUCT)	OPERATION (PROCESS)	PART NO.	CHART NO.
H.D. TUBE (POLYETHYLENE)	EXTRUSION	1048	1

OPERATOR	MACHINE	GAGE	UNIT OF MEASURE	ZERO EQUALS	SPECIFICATION LIMITS
-	1	-	-	-	

DATE: 9 - 5 - 81

	1	2	3	4	5	6	7	8	9	10	11	12
TIME	7.00	8.10	9.15	10.00	11.05	12.10	1.15	2.10	3.05	4.10	5.15	6.20
SAMPLE MEASUREMENTS 1	82	78	82	76	86	80	82	78	86	80	82	78
2	86	80	88	78	82	78	86	80	82	78	84	80
3	84	80	86	80	84	78	84	80	84	78	82	80
4	82	78	92	78	86	80	82	78	86	80	82	78
5	86	84	82	78	82	84	86	84	82	84	80	84
SUM	420	400	430	390	420	400	420	400	420	400	410	400
AVERAGE, $\bar{X}$	84	80	86	78	84	80	84	80	84	80	82	80
RANGE, R	4	6	10	4	4	6	4	6	4	6	4	6
NOTES												

	13	14	15	16	17	18	19	20	21	22	23	24	25
TIME	7.00	8.10	9.05	10.10	11.00	12.10	1.05	2.15	3.10	4.15	5.00	6.05	7.10
SAMPLE MEASUREMENTS 1	82	80	81	80	85	83	84	82	80	83	82	80	82
2	82	82	82	82	82	81	80	81	82	82	84	82	84
3	84	82	85	83	81	84	83	84	81	81	81	82	83
4	82	84	83	81	83	83	82	81	84	80	83	83	81
5	80	82	84	80	83	82	82	83	83	84	80	84	80
SUM	410	410	415	405	414	413	411	411	410	410	410	411	410
AVERAGE, $\bar{X}$	82	82	83	81	82.8	82.6	82.2	82.2	82	82	82	82.2	82
RANGE, R	4	4	4	3	4	3	4	3	4	4	4	4	4
NOTES													

Figure 16-2. Typical variables controls chart.

Example. Let us assume that we have just installed an extruder to make a 2-in.-inside-diameter tubing. The nominal wall thickness of 82 mils is to be maintained. As in the case of any manufacturing process, variations are expected. These variations are expected due to a number of factors such as variations in material, melt temperature, normal wear in the die, extruder speed, and tube puller speed. The quality control inspector is required to cut five pieces of tubing approximately every hour, measuring wall thickness and computing the average weight. Figure 16-2 also lists the wall thickness of the samples taken on Monday morning soon after starting the extruder. The wall thicknesses are recorded in mils for the next 24 hrs. The following calculations are carried out so that the variables control chart for $\bar{x}$ and R can be plotted.

$$\bar{\bar{x}} = \frac{\text{Sum of the means of the samples}}{\text{Sum of the number of sample means}}$$

$$= \frac{\Sigma\bar{x}}{n}$$

$$= \frac{2050}{25}$$

$$= 82$$

$$R = \frac{\text{Sum of sample ranges}}{\text{Sum of the number of sample ranges}}$$

$$= \frac{\Sigma R}{n}$$

$$= \frac{113}{25}$$

$$= 4.52$$

The upper control limit (UCL) and lower control limit (LCL) of the $\bar{x}$ chart is computed by:

$$\text{UCL} = \bar{x} + A_2\bar{R}$$

$$= 82 + 0.577\ (4.52)$$

$$= 84.6$$

$$\text{LCL} = \bar{\bar{x}} - A_2\bar{R}$$

$$= 82 - 0.577\ (4.52)$$

$$= 79.4$$

A_2 is a factor used in calculating the control limits for the $\bar{x}$ chart. Table 16-1 lists the factors for calculating the control limits for the average and range charts. The n in the table refers to the number of items in the sample. In this example, the number of items in the sample is five.

The upper control limit and the lower control limit for the range chart is calculated in a very similar manner.

$$\begin{aligned} \text{UCL} &= D_4\overline{R} \\ &= 2.115(4.52) \\ &= 9.55 \\ \text{LCL} &= D_3\overline{R} \\ &= 0(4.52) \\ &= 0 \end{aligned}$$

Here again, the D_4 and D_3 values derived from Table 16-1 are the factors used to compute the control limits for the range chart. From this information we now can chart $\overline{x}$ and R values as shown in Figure 16-2.

INTERPRETATION. By surveying the average chart we can safely conclude that if a sample of five tubes is randomly picked and measured, the arithmetic mean wall thickness will fall between 79.4 and 84.6 mils about 99.73 percent of the time. A similar conclusion regarding the range chart can also be made. The average chart also reveals that at two points marked "A" and "B" the process went outside the limits of the expected variability. This variation was attributed to the instability in the process due to the start-up. As is evident from the chart, the

Table 16-1. Factors for Control Charts[a]

Number of items in sample, n	Chart for averages: Factors for control limits A_2	Chart for ranges: Factors for central line d_2	Chart for ranges: Factors for control limits D_3	Chart for ranges: Factors for control limits D_4
2	1.880	1.128	0	3.267
3	1.023	1.693	0	2.575
4	0.729	2.059	0	2.282
5	0.577	2.326	0	2.115
6	0.483	2.534	0	2.004
7	0.419	2.704	0.076	1.924
8	0.373	2.847	0.136	1.864
9	0.337	2.970	0.184	1.816
10	0.308	3.078	0.223	1.777
11	0.285	3.173	0.256	1.744
12	0.266	3.258	0.284	1.716
13	0.249	3.336	0.308	1.692
14	0.235	3.407	0.329	1.671
15	0.223	3.472	0.348	1.652

[a] From B. Mason, *Statistical Techniques in Business and Economics*. Reprinted with permission of Richard D. Irvin, Inc.

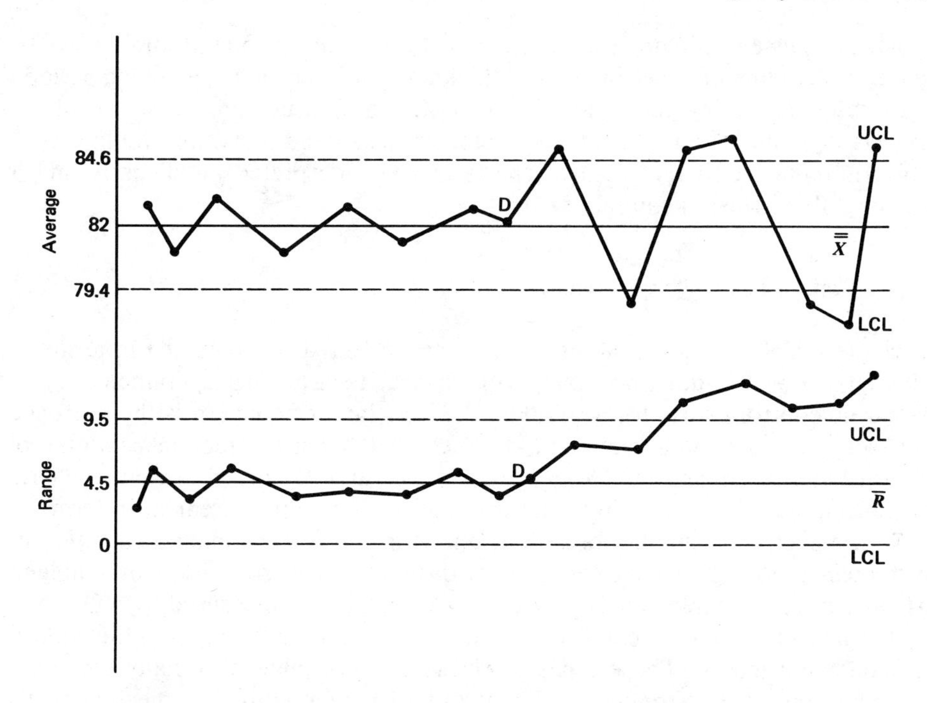

Figure 16-3. Chart showing process out of control due to material variation.

process stabilized itself with time and variations minimized. By looking at the range chart, we can see that the process was out of control at point "C" (during start-up), but stabilized later on.

Figure 16-3 shows the control chart of the same process a few days later. Up to point "D" in the chart, the process was well within the control limits. Beyond point "D," the process seems to have gone completely out of control. A little investigation revealed that a high-density polyethylene material (with a slightly different melt index value) from a new supplier was introduced at point D. The difference in the extrusion characteristics of this new material accounted for such gross variation. Figure 16-4 illustrates a typical $\bar{x}$ chart showing a trend. The

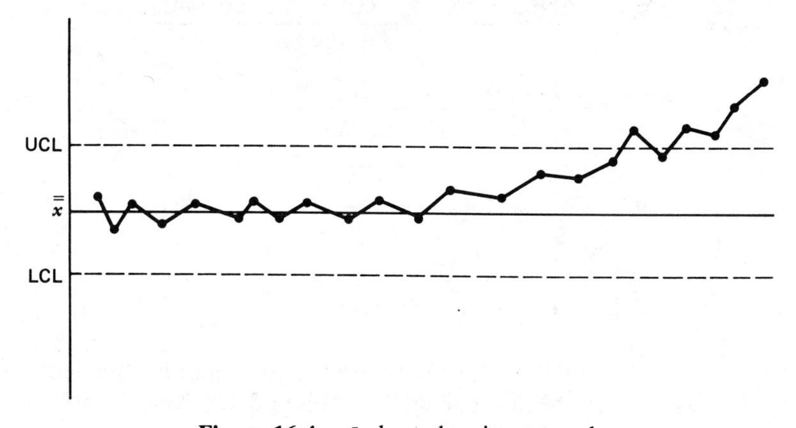

Figure 16-4. $\bar{x}$ chart showing a trend.

assignable cause producing such a trend was traced down to the tooling wear. A steady but gradual increase in the wall thickness was recorded over a long period.

From the foregoing discussion, it is clear that a thorough understanding of the process equipment, equipment operators, process conditions, and tooling condition is required so that control charts can be interpreted intelligently and a meaningful conclusion can be drawn.

B. Control Charts for Attributes

Attribute is defined as a characteristic or property that is appraised in terms of whether it does or does not exist with respect to a given requirement. (4) In preparing control charts for attributes, the only item of concern is the presence or absence of a given characteristic or defect. The most common inspection tool for attributes is Go and No-Go gauges. Unlike variable charts, attributes charts do not require actual measurements. Instead, a simple measurement in terms of percent defective or number of defects/unit is made. Some common examples of attributes are the presence or absence of flash, sink marks or shorts in a molded article, molded-in threads acceptable or unacceptable as measured by a Go–No-Go thread gauge, and the color of the molded article visually matches the control or it does not match. These types of charts are inevitable, especially when the item is produced in extremely large quantities, making individual measurements impractical, time-consuming, and very expensive. Control charts for attributes are also exployed when the measurement technique is destructive. One of the biggest advantages of such charts is their simplicity and ease-of-understanding by all levels of personnel.

P CHARTS. One of the most commonly used control charts for attributes is the "percent defective" or "*P*" chart. *P* charts are prepared by obtaining a series of appropriately sized samples. A sample size of 50 or 100 is the most convenient. Next, the number of defective samples is counted and recorded. From this information, the percent defective in a sample (P), average (mean) percent defective ($\overline{P}$), upper control limit, and lower control limit can be calculated as follows:

$$P = \frac{\text{Number of defective units in sample}}{\text{Sample size}}$$

$$\overline{P} = \frac{\text{Sum of percent defective}}{\text{Total number of samples}}$$

$$\text{UCL} = \overline{P} + 3\sqrt{\frac{\overline{P}(1 - \overline{P})}{n}}$$

$$\text{LCL} = \overline{P} - 3\sqrt{\frac{\overline{P}(1 - \overline{P})}{n}}$$

If a negative value (less than zero) of the lower control limit is obtained, it should be considered as zero on the chart since the percent defective cannot be less than zero.

C CHARTS. The C chart is also known as the "C-bar" chart and is a special type of attributes control chart. Unlike the P chart which portrays *percent defective,* the C chart uses the *number of defects per unit.* For example, a black speck and a deep sink mark on a molded part are two defects per part for C chart calculation but the part is one defective for P chart purposes.

The upper and lower control limits of a C chart is calculated as follows:

$$\text{Control limit} = \overline{C} \pm 3\sqrt{\overline{C}}$$

$\overline{C}$ is the center line of the control chart.

16.2.2. Acceptance Sampling

Acceptance sampling is a widely used and accepted statistical quality control technique. The acceptance sampling technique calls for selecting a sample randomly from a lot and deciding whether to accept or reject the lot, based on the number of defective items found in the sample.

Contrary to the belief that 100 percent inspection means 100 percent quality, 100 percent inspection is not only impractical but also unreliable, ineffective, and impossible in the case of destructive testing. Acceptance sampling drastically reduces the cost of inspection and in most cases provides better quality accurance than 100 percent inspection due to lower inspector fatigue and boredom. At this point, it is important to understand that the acceptance sampling technique does not "control quality" but it helps in determining the course of action. Based on the acceptance sampling plan, one may decide to accept or reject a particular lot.

A. *Sampling Theory*

The need for an acceptance sampling plan arises from another need to compromise between the consumer's demand for a perfect quality product and the manufacturer's inability to provide a perfect quality product due to process limitations and variations. Since 0 percent defective product quality is not possible, a compromise in terms of some absolute value of quality greater than 0 percent defective must be made. Theoretically, an ideal sampling plan is one that rejects all lots worse than the standard and accepts all lots equal to or better than the standard. Figure 16-5 illustrates one such ideal sampling plan diagram. According to this plan, all groups of products greater than 5 percent defective would be rejected and less than 5 percent defective would be accepted. Such an ideal sampling plan which can distinguish between acceptable and rejectable lots 100 percent of the time is not possible. Even 100 percent inspection is not capable of achieving perfect discrimination between good and bad lots. In other words, there is always a chance that good lots may be rejected and bad lots accepted with any type of sampling plan. The smaller the sample size, the greater the risk of making an erroneous judgement. The probability of acceptance is high with a good quality product and it becomes less and less so as the product gets worse. This fact forms the basis for a curve known as the operating characteristics curve. The curve is also commonly referred to as the OC curve. An OC curve is a plot of the quality of an incoming lot in percent defective against the probability that a lot will be accepted when sampled according to the plan. The lots that contain 0 percent

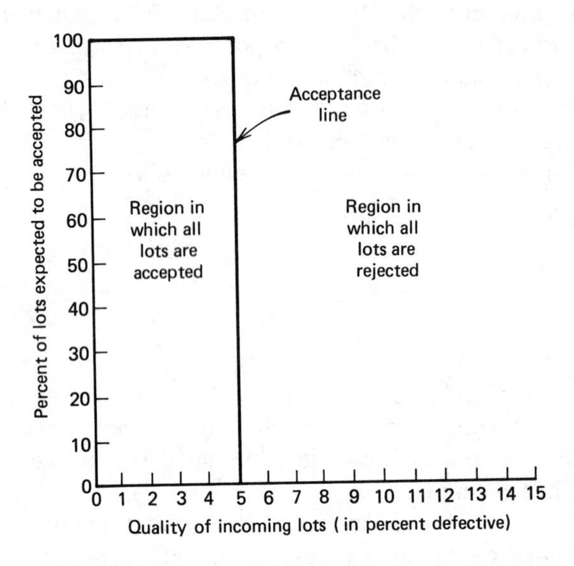

Figure 16-5. Ideal sampling plan.

defective will be accepted 100 percent of the time; conversely, the lots that are 100 percent defective will never be accepted. Depending upon the sampling plan and the quality of the incoming lots, the probability of acceptance (or rejection) increases or decreases. A typical OC curve of a sample size (n) of 100 and acceptance number (c) of 2 is illustrated in Figure 16-6. In this case, $n = 100$ and $c = 2$ means that for a sampling plan that calls for a sample size of 100, the maximum allowable number of defects in the sample is two. The OC curve in Figure 16-6 also indicates that if a lot is 1 percent defective, the lot will be accepted under this sampling plan ($n = 100$, $c = 2$) 92 percent of the time. Also, if the lot is 5 percent defective, it will be accepted under this sampling plan 12 percent of the time. Obviously, a distinctly different OC curve for each sampling plan with its own characteristic risk pattern can be plotted. There are two points on the OC

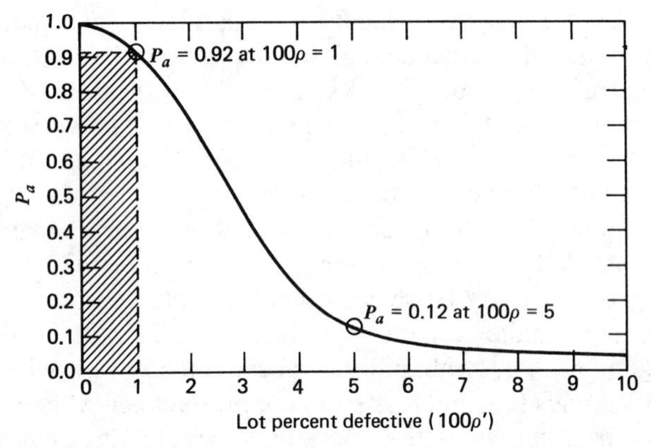

Figure 16-6. A typical operating characteristic curve. (Reprinted with permission from Richard D. Irwin, Inc.)

curve that describe relatively, the acceptable region and the nonacceptable region. These two are known as producer's risk (α) and consumer's risk (β). Producer's risk is defined as the risk or probability of rejecting a good lot. Every producer would like to keep this risk as small as possible. The consumer's risk is defined as the risk or probability of accepting a poor lot. This is the probability of accepting a poor lot which is unsatisfactory and sending it out to the consumer. The consumer may reject the lot quite often if the consumer's risk of a sampling plan is kept high. A balance between the producer's risk and the consumer's risk must be attained for economics reasons.

B. *Types of Sampling Plans*

The three basic types of sampling plans that are most frequently used are:

1. Single sampling plan.
2. Double sampling plan.
3. Multiple sampling plan.

SINGLE SAMPLING PLAN. This type of sampling plan is used when the results of a single sample from an inspection lot are conclusive in determining its acceptability. The lot is accepted if the number of defectives found in the sample is equal to or less than the acceptance number AC or C. Similarly, the lot is rejected if the number of defectives found in the sample is equal to or greater than the rejection number RE or r.

DOUBLE SAMPLING PLAN. A double sampling plan is used if the inspection of the first sample leads to the decision to accept or reject or to take a second sample. After inspection of the second sample, if necessary, a decision to accept or reject is made. Figure 16-7 shows a typical double sampling plan. Figure 16-8 illustrates a switching plan for sampling inspection.

MULTIPLE SAMPLING PLAN. A multiple sampling plan is an extension of a double sampling plan. As long as the number of defectives falls between acceptance and rejections numbers, the inspection is continued.

C. *Classification of Sampling Plans*

Sampling plans are classified into three major classes:

AQL sampling plan.
LTPD sampling plan.
AOQL sampling plan.

A detailed discussion of all three classes of sampling plans is beyond the scope of this book. Our discussion will be confined to the AQL sampling plan.

Lot Size	Sample	Sample size	Cumulative sample size	1.0%	
				AC	RE
2–8	First	2	2	0	1
9–15	First	3	3	0	1
16–25	First	5	5	0	1
26–50	First	8	8	0	1
51–90	First	13	13	0	1
91–150	First	20	20	0	1
151–280	First Second	20 20	20 40	0 1	2 2
281–500	First Second	32 32	32 64	0 1	2 2
501–1200	First Second	50 50	50 100	0 3	3 4
1201–3200	First Second	80 80	80 160	1 4	4 5
3201–10,000	First Second	125 125	125 250	2 6	5 7
10,001–35,000	First Second	200 200	200 400	3 8	7 9
35,001–150,000	First Second	315 315	315 630	5 12	9 13

Figure 16-7. Double sampling plan for normal inspection MIL-STD-105 D Level II.

D. AQL Sampling Plan

AQL (acceptable quality level) is defined as the maximum percent defective that for acceptance sampling, can be considered acceptable as a process average. A 4 percent AQL sampling plan will regularly accept 4 percent defective product. AQL sampling plans are designed to protect the supplier from having good lots rejected. The consumer's risk of accepting bad lots can only be determined by studying the OC curve for the AQL sampling plans. The specified producer's risk in an AQL plan is generally 5 percent, meaning a plan that will accept lots of AQL quality with a stated probability of 95 percent. For example, 2 percent AQL plan will reject 2 percent defective material 5 percent of the time or accept it 95 percent of the time.

AQL sampling plans are the most widely accepted and used plans throughout the industry and government. The best known published source of such AQL sampling plans is the government publication titled, *Sampling Procedures and Tables for Inspection by Attributes MIL-STD-105 D.* This standard covers the range of AQL's from 0.15 percent defective to 10 percent defective.

E. Use of Sampling Tables

The following procedure is recommended for using the sampling tables:

1. Determine the lot size.
2. Select the inspection level as specified. If the inspection level is unspecified, normal inspection level II should be used. Some companies use past quality history as a guideline for selecting the inspection level. For example, if the supplier has mostly submitted "good" lots in the past, level I (reduced inspection) is used. Level III (tightened inspection) is used if lots have been mostly "bad." Special inspection levels *S*-1 through *S*-4 may also be used for expensive as well as destructive-type inspections or for parts made with repetitive-type processes.

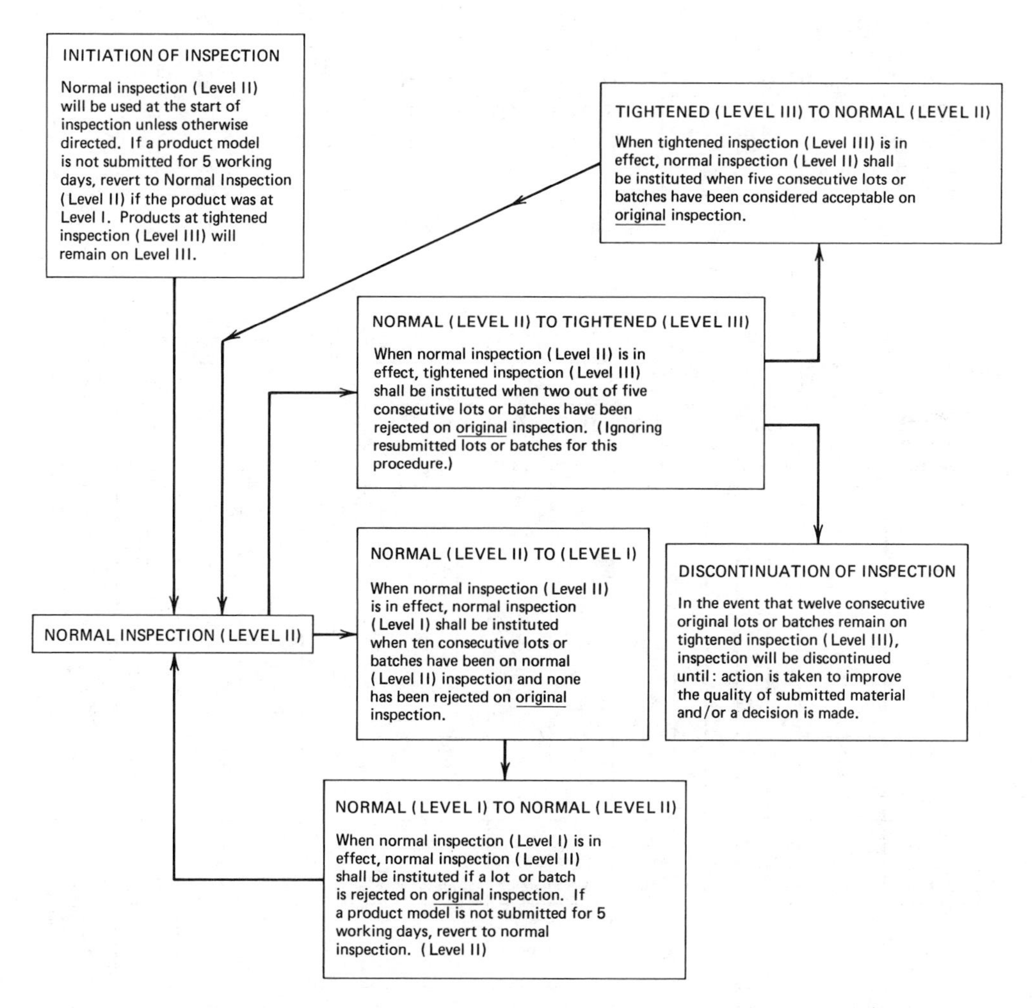

Figure 16-8. Switching plan for quality control/quality audit statistical sampling. (Courtesy Rainbird Corporation.)

Table 16-2. Sample Size Code Letters

Lot or batch size	Special inspection levels				General inspection levels		
	S-1	S-2	S-3	S-4	I	II	III
2–8	A	A	A	A	A	A	B
9–15	A	A	A	A	A	B	C
16–25	A	A	B	B	B	C	D
26–50	A	B	B	C	C	D	E
51–90	B	B	C	C	C	E	F
91–150	B	B	C	D	D	F	G
151–280	B	C	D	E	E	G	H
281–500	B	C	D	E	F	H	J
501–1,200	C	C	E	F	G	J	K
1,201–3,200	C	D	E	G	H	K	L
3,201–10,000	C	D	F	G	J	L	M
10,001–35,000	C	D	F	H	K	M	N
35,001–150,000	D	E	G	J	L	N	P
150,001–500,000	D	E	G	J	M	P	Q
500,001 and over	D	E	H	K	N	Q	R

Table 16-3. Single Sampling Plans for

Sample size code letter	Sample Size	Acceptable Quality											
		0.010	0.015	0.025	0.040	0.065	0.10	0.15	0.25	0.40	0.65	1.0	1.5
		AC RE	AC RE	AC RE	AC RE	AC RE	AC RE	AC RE	AC RE	AC RE	AC RE	AC RE	AC RE
A	2	↓	↓	↓	↓	↓	↓	↓	↓	↓	↓	↓	↓
B	3	↓	↓	↓	↓	↓	↓	↓	↓	↓	↓	↓	↓
C	5	↓	↓	↓	↓	↓	↓	↓	↓	↓	↓	↓	↓
D	8	↓	↓	↓	↓	↓	↓	↓	↓	↓	↓	↓	0 1
E	13	↓	↓	↓	↓	↓	↓	↓	↓	↓	↓	0 1	↑
F	20	↓	↓	↓	↓	↓	↓	↓	↓	↓	0 1	↑	↓
G	32	↓	↓	↓	↓	↓	↓	↓	↓	0 1	↑	↓	1 2
H	50	↓	↓	↓	↓	↓	↓	↓	0 1	↑	↓	1 2	2 3
J	80	↓	↓	↓	↓	↓	↓	0 1	↑	↓	1 2	2 3	3 4
K	125	↓	↓	↓	↓	↓	0 1	↑	↓	1 2	2 3	3 4	5 6
L	200	↓	↓	↓	↓	0 1	↑	↓	1 2	2 3	3 4	5 6	7 8
M	315	↓	↓	↓	0 1	↑	↓	1 2	2 3	3 4	5 6	7 8	10 11
N	500	↓	↓	0 1	↑	↓	1 2	2 3	3 4	5 6	7 8	10 11	14 15
P	800	↓	0 1	↑	↓	1 2	2 3	3 4	5 6	7 8	10 11	14 15	21 22
Q	1250	0 1	↑	↓	1 2	2 3	3 4	5 6	7 8	10 11	14 15	21 22	↑
R	2000	↑	↑	1 2	2 3	3 4	5 6	7 8	10 11	14 15	21 22	↑	↑

↓ = Use first sampling plan below arrow. If sample size equals, or exceeds, lot or batch size, do 100 percent inspection.

↑ = Use first sampling plan above arrow.

AC = Acceptable number.

RE = Rejection number.

3. Determine the sample size code letter from the lot size information using Table 16-2.

4. Select the sampling plan as specified. If unspecified, the single sampling plan should be used.

5. From the information regarding the inspection level, sample size code letter, sampling plan, and specified AQL value, determine the acceptance and rejection number. If the AQL value is unspecified, start with a 2.5 percent defective AQL value.

Example. Given the following information, determine the sample size, acceptance and rejection number using (1) single and (2) double sampling plans.

Lot size: 1000
Inspection level: Normal
AQL: 1.0

Solution. From Table 16-2, note that the 1000 piece lot size corresponds to code letter "J" for normal inspection level II.

1. Using Table 16-3 (single sampling plan for normal inspection), we find that for code letter "J" and AQL of 1.0, the sample size is 80 with acceptance number AC = 2 and rejection number RE = 3. Thus, the course of action

Normal Inspection (Master Table)

Levels (normal inspection)													
2.5	4.0	6.5	10	15	25	40	65	100	150	250	400	650	1000
AC RE	AC RE	AC RE	AC RE	AC RE	AC RE	AC RE	AC RE	AC RE	AC RE	AC RE	AC RE	AC RE	AC RE
↓	↓	0 1	↓	↓	1 2	2 3	3 4	5 6	7 8	10 11	14 15	21 22	30 31
	0 1	↑		1 2	2 3	3 4	5 6	7 8	10 11	14 15	21 22	30 31	44 45
0 1	↑	↓	1 2	2 3	3 4	5 6	7 8	10 11	14 15	21 22	30 31	44 45	↑
↑	↓	1 2	2 3	3 4	5 6	7 8	10 11	14 15	21 22	30 31	44 45	↑	
↓	1 2	2 3	3 4	5 6	7 8	10 11	14 15	21 22	30 31	44 45	↑		
1 2	2 3	3 4	5 6	7 8	10 11	14 15	21 22	↑	↑	↑			
2 3	3 4	5 6	7 8	10 11	14 15	21 22	↑						
3 4	5 6	7 8	10 11	14 15	21 22	↑							
5 6	7 8	10 11	14 15	21 22	↑								
7 8	10 11	14 15	21 22	↑									
10 11	14 15	21 22	↑										
14 15	21 22	↑											
21 22	↑												
↑													

Table 16-4. Double Sampling Plans for

Sample size code letter	Sample	Sample size	Cumulative sample size	Acceptable										
				0.010	0.015	0.025	0.040	0.065	0.10	0.15	0.25	0.40	0.65	1.0
				AC RE	AC RE	AC RE	AC RE	AC RE	AC RE	AC RE	AC RE	AC RE	AC RE	AC RE
A				│	│	│	│	│	│	│	│	│		│
B	First	2	2	│	│	│	│	│	│	│	│	│		│
	Second	2	4	│	│	│	│	│	│	│	│	│		│
C	First	3	3	│	│	│	│	│	│	│	│	│		│
	Second	3	6	│	│	│	│	│	│	│	│	│		│
D	First	5	5	│	│	│	│	│	│	│	│	│		│
	Second	5	10	│	│	│	│	│	│	│	│	│		↓
E	First	8	8	│	│	│	│	│	│	│	│	│	·	
	Second	8	16	│	│	│	│	│	│	│	│	│	↓	
F	First	13	13	│	│	│	│	│	│	│	│	↑		
	Second	13	26	│	│	│	│	│	│	│	│	↓		
G	First	20	20	│	│	│	│	│	│	│	│	·	↑	│
	Second	20	40	│	│	│	│	│	│	│	↓		│	↓
H	First	32	32	│	│	│	│	│	│	│	·	↑	│	0 2
	Second	32	64	│	│	│	│	│	│	↓		│	↓	1 2
J	First	50	50	│	│	│	│	│	│	·	↑	│	0 2	0 3
	Second	50	100	│	│	│	│	│	↓		│	↓	1 2	3 4
K	First	80	80	│	│	│	│	│	·	↑	│	0 2	0 3	1 4
	Second	80	160	│	│	│	│	↓		│	↓	1 2	3 4	4 5
L	First	125	125	│	│	│	│	·	↑	│	0 2	0 3	1 4	2 5
	Second	125	250	│	│	│	↓		│	↓	1 2	3 4	4 5	6 7
M	First	200	200	│	│	│	·	↑	│	0 2	0 3	1 4	2 5	3 7
	Second	200	400	│	│	↓		│	↓	1 2	3 4	4 5	6 7	8 9
N	First	315	315	│	│	·	↑	│	0 2	0 3	1 4	2 5	3 7	5 9
	Second	315	630	│	↓		│	↓	1 2	3 4	4 5	6 7	8 9	12 13
P	First	500	500	│	·	↑	│	0 2	0 3	1 4	2 5	3 7	5 9	7 11
	Second	500	1000	↓		│	↓	1 2	3 4	4 5	6 7	8 9	12 13	18 19
Q	First	800	800	·	↑	│	0 2	0 3	1 4	2 5	3 7	5 9	7 11	11 16
	Second	800	1600		│	↓	1 2	3 4	4 5	6 7	8 9	12 13	18 19	26 27
R	First	1250	1250	↑	│	0 2	0 3	1 4	2 5	3 7	5 9	? 11	11 16	↑
	Second	1250	2500	│	│	1 2	3 4	4 5	6 7	8 9	12 13	18 19	26 27	│

↓ = Use first sampling plan below arrow. If sample size equals, or exceeds lot or batch size, do 100 percent inspection.
↑ = Use first sampling plan above arrow.
AC = Acceptance number.
RE = Rejection number.
· = Use corresponding single sampling plan (or alternatively, use double sampling plan below, where available).

should be accept—if the number of defective is 2 or less. Reject—if the number of defective is 3 or more.

2. Using Table 16-4 (for double sampling plan for normal inspection), we find that for the code letter "J" and AQL of 1.0, the first sample size is 50 pieces and acceptance number AC = 0 and rejection number RE = 3.

Normal Inspection (Master Table)

Quality Levels (normal inspection)														
1.5	2.5	4.0	6.5	10	15	25	40	65	100	150	250	400	650	1000
AC RE	AC RE	AC RE	AC RE	AC RE	AC RE	AC RE	AC RE	AC RE	AC RE	AC RE	AC RE	AC RE	AC RE	AC RE
		↓	·		↓	·	·	·	·	·	·	·	·	·
	↓	·	↑	↓	0 2 1 2	0 3 3 4	1 4 4 5	2 5 6 7	3 7 8 9	5 9 12 13	7 11 18 19	11 16 26 27	17 22 37 38	25 31 56 57
↓	·	↑	↓	0 2 1 2	0 3 3 4	1 4 4 5	2 5 6 7	3 7 8 9	5 9 12 13	7 11 18 19	11 16 26 27	17 22 37 38	25 31 56 57	↑
·	↑	↓	0 2 1 2	0 3 3 4	1 4 4 5	2 5 6 7	3 7 8 9	5 9 12 13	7 11 18 19	11 16 26 27	17 22 37 38	25 31 56 57	↑	
↑	↓	0 2 1 2	0 3 3 4	1 4 4 5	2 5 6 7	3 7 8 9	5 9 12 13	7 11 18 19	11 16 26 27	17 22 37 38	25 31 56 57	↑		
↓	0 2 1 2	0 3 3 4	1 4 4 5	2 5 6 7	3 7 8 9	5 9 12 13	7 11 18 19	11 16 26 27	↑	↑	↑			
0 2 1 2	0 3 3 4	1 4 4 5	2 5 6 7	3 7 8 9	5 9 12 13	7 11 18 19	11 16 26 27	↑						
0 3 3 4	1 4 4 5	2 5 6 7	3 7 6 9	5 9 12 13	7 11 18 19	11 16 26 27	↑							
1 4 4 5	2 5 6 7	3 7 8 9	5 9 12 13	7 11 18 19	11 16 26 27	↑								
2 5 6 7	3 7 8 9	5 9 12 13	7 11 18 19	11 16 26 27	↑									
3 7 8 9	5 9 12 13	7 11 18 19	11 16 26 27	↑										
5 9 12 13	7 11 18 19	11 16 26 27	↑											
7 11 18 19	11 16 26 27	↑												
11 16 26 27	↑													
↑														

The second sample size is 50 pieces and the acceptance number (for both samples) is AC = 3. The rejection number (for both samples) is RE = 4. The following procedure should be followed.

Select 50 pieces from the lot, and Accept—if the number of defective is 0; Reject—if the number of defective is 3 or more. If the number of defective falls between 0 and 3, let's say 2, then select another 50 pieces and Accept—if the number of defective in both samples is 3 or less. Reject—if the number of defective in both samples is 4 or more.

F. LTPD Sampling Plan

The lot tolerance percent defective (LTPD) sampling plan is defined as an allowable percentage defective, a figure which may be considered a borderline distinction between a satisfactory lot and an unsatisfactory one. A 2 percent LTPD sampling plan will regularly reject 2 percent defective product. The specified consumer's risk in an LTPD sampling plan is generally 10 percent, meaning a plan that will reject lots of LTPD quality with a stated probability of 90 percent. For example, a 6.0 LTPD plan will accept 6 percent defective material 10 percent of the time and reject it 90 percent of the time.

G. AOQL Sampling Plan

Average outgoing quality (AOQ) is defined as the average quality of outgoing products including all accepted lots plus all rejected lots after the rejected lots have been effectively 100 percent inspected and all defectives replaced by non-defectives. Average outgoing quality limit (AOQL) is defined as the maximum of AOQ for all possible incoming qualities for a given sampling inspection plan. In other words, AOQL is the worst average quality that can exist in the long run in the outgoing product. For example, a 2.5 AOQL plan assures us that in the long run the accepted material will not be more than 2.5 percent defective. AOQL sampling plans are designed to protect the consumer with specified risk. They offer a low probability of acceptance if the product quality exceeds the required AOQL.

16.3 QUALITY CONTROL SYSTEM

16.3.1. Raw Material Quality Control

Any well-established quality control system begins with control of purchased material. Such a system assures ones that the purchased material, in fact, meets the specified requirements. In most cases, processors rely on material suppliers to provide the same quality material time after time. If the end-product or the particular process employed to make this end-product is sensitive to changes in material quality and uniformity, such reliance on material suppliers may prove costly in the long run. The steps involved in setting up a good raw material quality control system are:

1. *Supplier Selection.* The first step in setting up such a system is to select a reputable supplier of material. Items to check for are past history, industry reputation, and future commitment. The supplier's ability to verify the quality of the material he is supplying should be investigated. One must also look at the quality and the type of supplier's manufacturing and test facility, the frequency

of testing, quality control procedures and quantity, and more important, the quality of personnel. These considerations are of the utmost importance when a purchase of material from a custom compounding house or from a totally unknown supplier is considered. If the product liability risk is high, it may behoove you to consider requiring material suppliers to certify the material, that is, it meets the minimum requirements specified in the material specifications.

2. *Receiving Inspection.* Many types of tests have been devised for testing raw materials. Depending upon the severity of the need for inspection, the types of tests selected may vary from being very basic and simple to very sophisticated and complex. Some of the most common basic tests are the melt index test, specific gravity, bulk density, spiral flow test, and viscosity tests. Gel permeation chromatography, infrared analysis, thermal analysis, and rheometry are some of the more elaborate raw material quality control tests. These tests are discussed in detail in Chapter 7. Some processors also choose to mold test bars from a small sample of raw material and conduct physical tests such as tensile, impact, and flexural and then evaluate the results to see if the results meet the pre-established specifications.

16.3.2. Process Quality Control

In-process quality control serves the basic purpose of providing assurance that the product continues to meet the specified requirements. By employing the process control chart techniques of statistical quality control, we are able to continuously monitor the process and determine whether the process is in or out of control. Patrol or floor inspection gives the inspector an opportunity to verify the visual and dimensional conformity of the processed parts.

16.3.3. Product Quality Control

There are two major areas of interest in product quality control. One of them is receiving inspection, where a product manufactured by an outside vendor is inspected when it is received. The other one is the outgoing lot inspection in which the product manufactured in-house is inspected prior to shipping. Here again, the principles of statistical quality control are applied. A sampling plan is selected based on the requirements and the AQL, LPTD, or AOQL value is specified. The product quality control involves visual inspection, measurement inspection, and in some cases, actual product testing. More and more emphasis is being put on end-product testing. Preferably, the test will simulate actual use, since a part which is aesthetically appealing and well within the specified tolerance only gives a partial indication of overall part quality.

16.3.4. Visual Standards

One of the reasons for the tremendous success of plastic products in the consumer market is that the products made from plastic materials are aesthetically more appealing in terms of color and feel than products made from other materials. The majority of quality control systems fail to recognize the importance of visual standards or guidelines. Quite often, too much emphasis is placed on measurement and testing of the product and not enough on the visual standards.

Visual defects, such as orange peel, sinks, and cold flow, are quite common among fabricated plastic parts. These defects are not usually encountered in parts made from other materials. Furthermore, the terminology that prevails in the plastics industry to describe visual defects is totally different than the terminology used for conventional materials. Identifying visual defects is not only necessary for controlling aesthetic quality of fabricated parts but is also necessary in assessing the overall quality and strength of the part. For example, a visual defect such as splay marks on the part indicate the presence of moisture in the material, brown streaks indicate the beginning of material degradation, both of which can lower the overall properties of the material.

A visual standards manual must include the basic definitions and explanations of recurring visual defects along with proper illustrations. Since it is difficult to qualify the visual defects in terms of actual measurements, such as gauging a diameter or wall thickness, some guidelines and accept–reject criteria must be established. Figures 16-9 and 16-10 illustrate typical pages from a visual standards manual. A typical visual defects summary chart is shown in Figure 16-11.

16.3.5. Mold (Tool) Control

The quality of the molded part is only as good as the mold that produces that part. New equipment, skilled operators, or good molding practices cannot make up for a defective or worn out mold. In spite of this proven fact, the majority of manufacturers often fail to recognize the importance of effective mold control systems.

A good mold control system starts with proper inventory control and adequate mold storage facilities. Documentation is the key word. From the inception of the mold, every little detail regarding the particulars of the mold must be logged. A mold information form such as the one shown in Figure 16-12 can be used. Once the mold is released for production, the first article should be performed, all dimensions carefully logged, and samples retained for future reference. A mold history record card such as the one shown in Figure 16-13 should be generated and kept up-to-date. Any mold modification, however minor, must be recorded along with routine maintenance and repair data on the history card. The mold service schedule should be based upon the number of parts produced from the mold.

16.3.6. Workmanship Standards

Workmanship standards are essential to the smooth and successful operation of any quality control system. They are considered the best means of achieving quality workmanship.

Workmanship standards are nothing more than a simplified guide, explaining through the use of drawings, sketches, and photographs, the proper method of carrying out the specified task. The task may consist of simply deburring or hot stamping the parts or assembling them using solvent cementing techniques. The

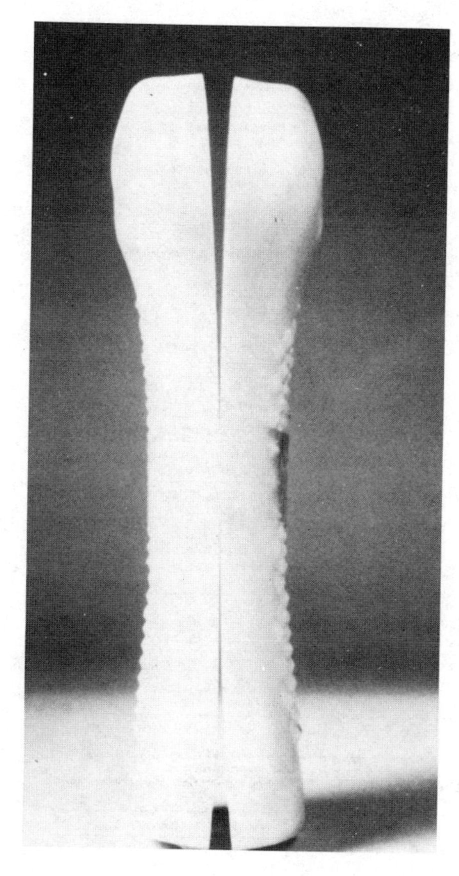

Figure 16-9. Visual defect. Wax page. (Courtesy Rainbird Corporation.)

ORANGE PEEL

Preferred

1. No uneven surface texture on molded part resembling orange peel.

Acceptable

1. A slight orange peel on inside of an external part.
2. A slight orange peel on outside of an internal part.
3. Orange peel not to exceed 1/4" in diameter on the part.

Reject

1. Orange peel on sealing surfaces.
2. Excessive orange peel, greater than 1/4" in diameter on molded part.

Figure 16-10. Criteria for accepting or rejecting the part. (Courtesy Rainbird Corporation.)

VISUAL DEFECT	PREFERRED	ACCEPTABLE	REJECT
Splay	None	Slight—internal part $\leq$ 5% of part area	$>$5% of part area, external part, sealing surface
FLASH Small	None at parting line, holes threads	Thin flash $\leq$ 0.007 Thick flash $\leq$ 0.005	Flash in holes, water passages, sealing surfaces & threads $>$ 0.007
FLASH Medium	None at parting line, holes threads	Thin flash $\leq$ 0.015; thick flash $\leq$ 0.010 Thin flash in holes and threads	Thick flash in holes & threads $>$ 0.015, any flash on sealing surface
FLASH Large (Part Size)	None at parting line, holes threads	Thin flash $\leq$ 0.020; thick flash $\leq$ 0.015 Thin flash in holes and threads	Thick flash in holes and threads $>$ 0.020 any flash on sealing surface
Weld Line	Almost invisible fine line, not possible to feel with fingernail	Slightly visible, medium visible yet unable to feel with fingernail	Heavy, broken surface. Easy to feel with fingernail
Orange Peel	Parts with smooth surface	Slight orange peel on inside of external part. $\leq$ 1/2" in diameter	On sealing surfaces, $>$ 1/4" in diameter
Cold Slug	None	Slight, $\leq$ 1/4" in diameter	Cold slug in threads, cold slug $>$ 1/4" in diameter
Contamination	None	Slight, $\leq$ 10% of total area	Heavy, $>$ 10% of total area. Contamination on external part.
Sink Marks	None	$\leq$ 0.010 deep	Sink mark on sealing surface, $>$ 0.010 deep
Short Shot	Completely filled part	Slight on any size part	Shorts in threads, med. or heavy short, short on sealing surfaces
Burn Mark	None	Slight, $\leq$ 0.015 in diameter, slight brown streak	Burn mark on sealing surfaces on threads, $>$ 0.015 in diameter. Medium or heavy burn marks.
Voids	None	Slight void	Any voids close to water passages or voids breaking into the water passage. Medium or heavy voids
Ovality Out of Roundness (O. O. R.)	Perfect round holes	$\leq$0.003 O. O. R. up to 1" diameter hole $\leq$0.005 O. O. R. up to 4" diameter hole $\leq$0.010 O. O. R. 6" diameter hole	$>$0.003 O. O. R., 2" diameter hole $>$0.005 O. O. R., 4" diameter hole $>$0.010 O. O. R., 6" diameter hole
Gate Marks	Almost invisible, none on sealing surface	$\leq$0.005 on any size part	$>$0.005 on any part, gage mark on sealing surface
Jetting	None	Slight jetting	Medium or heavy jetting
Mold Deposit	Part free of mold deposit	Slight	Medium or heavy, mold deposit on sealing surface
Ejector pin marks	No indentation or protrusion created by ejector pin mark None on sealing surface	$\leq$0.005 long protrusion $\leq$0.010 deep indentation	$>$0.005 long protrusion $>$0.010 deep indentation, marks on sealing surface
Color	Matches with color chip	Slight variation	Heavy (wide) variation on any part.

Slight—visible at close scrutiny.
Medium—visible at arms length.
Heavy—obvious at first glance or to untrained eye.

Figure 16-11. Visual defect summary chart. (Courtesy Rainbird Corporation.)

majority of workmanship standards provide preferred, acceptable, and reject criteria. Figure 16-14 illustrates a typical workmanship standard.

Some of the most obvious advantages of such workmanship standards are a lower reject rate, elimination of unnecessary rework, early detection of defects, and reduced risk of rejecting good parts. Proper implementation of workmanship standards can improve quality and reliability by eliminating the time lost in rework because of varied interpretation and personal opinions.

16.3.7. Documentation

Documentation is the heart of the quality control system. The data compiled through the documentation of test results, dimensional measurements, process capability studies, and sampling inspection can be used for statistical analysis. Many private and governmental agencies require that proper documents and records be retained for certain minimum time periods. For example, the nuclear industry requires that records be maintained for 40 years (5). Retention of such records can also provide necessary proof in the courtroom in case of product liability suits.

A "good" documentation system is one that is easy to implement, easy to understand, and easy to maintain. The records documenting inspection must indicate the characteristics observed, number of observations made, number and types of discrepancies, final dispositions, inspector identification, and most important, date of documentation.

Lastly, without proper records it would be practically impossible to trace the reason for product failure. By carefully studying the dimensional measurement records of a part, one can also identify the equipment as well as the tool wear.

16.3.8. Quality Assurance Manual

A quality control system without a quality assurance manual describing in detail a quality assurance program cannot function adequately. The sole purpose of a quality assurance manual is to provide clear and precise written instructions and procedures so that there can be no misunderstanding and confusion between different organizations within and outside the company. Whenever possible, oral instruction should be avoided, since it can only result in misinterpretation and gross distortion of the message. The following is a broad outline of a typical quality assurance manual.

1. An organization chart describing the responsibilities of each individual in the organization.
2. Function and responsibility of the quality control organization.
3. Material review board (MRB) function and corrective action procedures.
4. Receiving inspection procedures.
5. In-process inspection procedures (first article inspection procedures)
6. Shipping inspection procedures.
7. Disposition guidelines.
8. Quality audit program. (Under this program, the Quality Manager or Supervisor periodically makes unannounced checks on inspected units, accuracy of gauges and test equipment, etc. to verify the accuracy as well as adequacy of quality assurance systems.)
9. Procedures for handling customer returns.
10. Gauge and test equipment calibration and maintenance procedure.
11. Mold control (tool and die) program.
12. Miscellaneous test procedures.
13. Method of recording inspection data and exhibit of sample forms.
14. Retention of records and documents.
15. Defective material rework and reinspection procedures.
16. Visual standards.
17. Workmanship standards.

Mold Information

Mold Number: ____________ Company: ____________

Mold Maker: ____________

Date Mfg.: ____________

No.	Part Number	Model No. Desc.	Material	Supplier/Grade

☐ 2 Plate ☐ 3 Plate ☐ Stacked ☐ Hot Runner

☐ Hot Manifold ☐ Insulated Runner ☐ Std. Unit Die ☐ Other

Mold Base Type: ____________ H

Mold Base Size: ______ L X ______ W X ______ H

Insert Size: ______ L X ______ W X ______ H

Mold Material: ____________

Insert Material: ____________

Number of Cavities: ____________

Mold Hardness: ____________

Machine Size		Nozzle Radius	Locating Ring Diameter	Sprue Bushing Size
Tons	OZ's			

Figure 16-12. Mold information form. (Courtesy Performance Engineered Products, Inc.)

16.4. GENERAL

16.4.1. Quality Control and Machine Operators

Although not a part of the quality control team, machine operators usually play a very important and direct role in controlling the quality of the parts. A smart quality control manager places part of the burden of inspecting the quality of the parts on machine operators. This also helps to lower the inspection cost by reducing the work load of inspectors and improving the overall quality. Such a

Type of Gate: ______

Type of KO: ______

Unwinding: ______

Core Pull: ______

Limit Switches: ______

Mold Transducer: ______

Mold Finish: ______

Plating Information: ______

Hot Sprue Bushing: ______

Hot Runner: ______

Insulated Runner: ______

Top Half	GPM	Bottom Half	GPM
Cavity:		Cavity:	
Core:		Core:	
Other:		Other:	

Mold Accesory Information: ______

Scheduled Maintenance Record

Maintenance	Date

Gaging

Part Number	Gage Type	Serial Number

Figure 16-12. (*Continued*)

program of machine operator participation in quality control of the product cannot be successful without the proper training of the participants. Machine operators must be trained to look for visual defects such as flow marks, shorts, flash, to trim the gate without leaving protrusions from the part or gouging into the part, to use a go–no-go gauge, or perform a simple test between the cycles. An illustration or sample of preferred maximum acceptable and unacceptable parts should

Mold History Record Card

Mold No: ____________________ Print No: ____________________

Part No: ____________________ Mold Maker: ____________________

Company: ____________________ Date Mfg: ____________________

Mold Movement Record

Date	From	To	Authorization	Remarks

Mold Modification Record

Date	Modification No.	Reason

Critical Variables Data Record

Dimension	Tolerance		Measured	Remarks	Date
	+	−			

Maintenance/Repair History

Date	P. O. No.	No. of Parts Run	Description of Repair/Maintenance	Cost	Mold Shop

Figure 16-13. Mold history record card. (Courtesy Performance Engineered Products, Inc.)

workmanship
standards

SOLDERLESS CONNECTIONS
Crimped Pins - Multi-Wire

MAGNIFICATION 5X

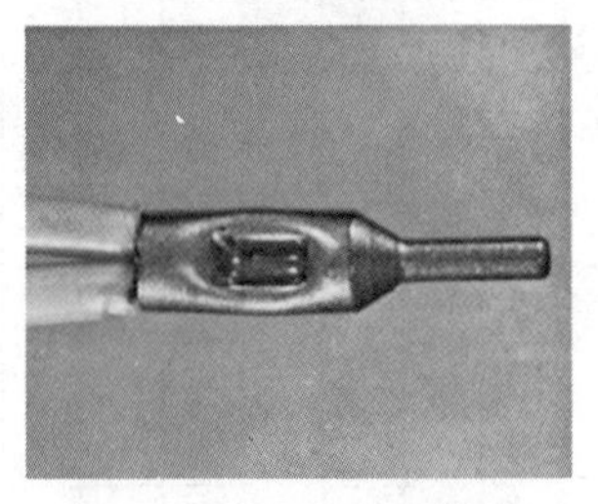

PREFERRED

1. Insulation has been stripped evenly and terminates flush with the rear of the contact barrel.
2. Conductors are bottomed in the support well.
3. Crimp indent is well formed and centered.

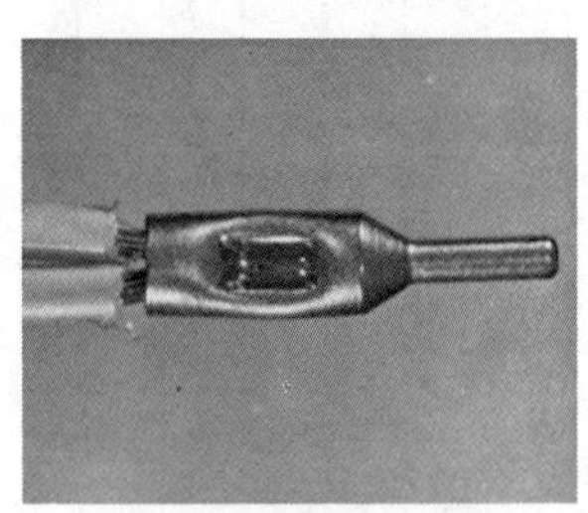

ACCEPTABLE

1. Insulation terminates within 1/16 inch of the rear of the contact barrel.
2. Insulation trim is slightly uneven.

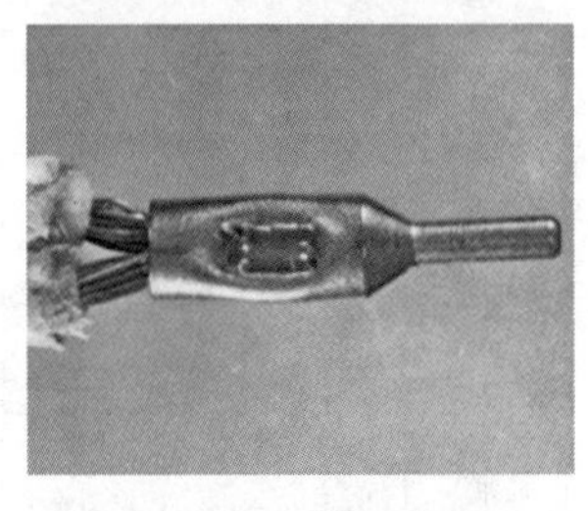

ACCEPTABLE MINIMUM

1. Exposed bare wire is maximum but does not exceed 1X insulated wire diameter plus 1/16 inch or a maximum gap of 1/8 inch, whichever is less, from the rear of the contact barrel. (Above tolerance includes allowance for latent shrinkage of the wire insulation).

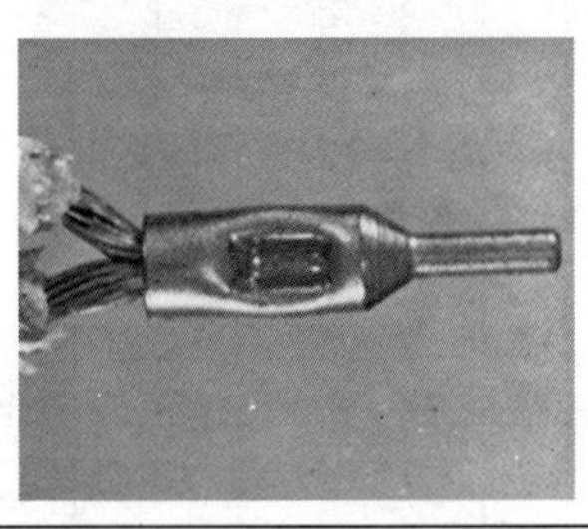

REJECT

1. Exposed bare wire exceeds maximum tolerance specified above.
2. Conductors are not bottomed in the barrel.
3. Crimp indent is too low.

Figure 16-14. A typical workmanship standard. (Courtesy Martin–Marietta Company.)

be provided to the machine operator for the sake of better understanding and comparison purposes.

REFERENCES

1. Juran, J. M., *Quality Control Handbook,* McGraw–Hill, New York, 1974, p 2-2.
2. *Ibid,* p. 2-11.

3. Debing, L. M., *Quality Control for Plastics Engineers,* Reinhold, New York, 1957, p. 1.
4. "Quality Assurance Terms and Definitions," *Military Standard Mil-Std-109B,* U.S. Govt. Printing Office, Washington, D.C., 1969.
5. "Nondestructive Inspection and Quality Control," *Metals Handbook,* **11,** American Society for Metals, Philadelphia p. 408.

GENERAL REFERENCES

Grant, E. L., *Statistical Quality Control,* McGraw–Hill, New York, 1974.

Juran, J. M., *Quality Control Handbook,* McGraw–Hill, New York, 1974.

Mason, R. D., *Statistical Techniques in Business and Economics,* Irvin, Homewood, IL, 1974.

Dodge, H. F., and Romig, H. G., *Sampling Inspection Tables—Single and Double Sampling,* John Wiley and Sons, New York, 1944.

Kenney, C. W., Andrew, D. E., *Inspection and Gaging,* Industrial Press, New York, 1977.

Statistical Quality Control Handbook, Western Electric Company, New York, 1956.

"Sampling Plans," *Military Standard Mil-Std-105D,* U.S. Govt. Printing Office, Washington, D.C.

CHAPTER 17

Product Liabilities and Testing

17.1. INTRODUCTION

Between 1970 and 1980, one phrase caught the attention of the manufacturers and suppliers more than any other phrase: product liability. The Consumer Product Safety Commission reports that the number of product liability suits increased from about 50,000 in 1960 to more than 500,000 in 1976 (1). The number is expected to reach over 1 million lawsuits by 1983. The cost of liability insurance has increased substantially in the last five years. Virtually every industry has been plagued with liability suits and the plastics industry is certainly no exception.

The entire concept of product liability suits has emerged from the total lack of concern regarding the product safety by product designers and manufacturers. The Occupational Safety and Health Administration (OSHA) reports the loss of over 10,000 lives each year and over 2 million disabling injuries annually (2). Recently, a material supplier and a plastic fittings manufacturer and distributor were sued by an angry consumer because plastic fittings failed prematurely, flooding the entire building and ruining expensive furnishings and carpeting. In another case, a molding machinery manufacturer was sued heavily because a machine operator lost his right hand while trying to free the part from the mold and accidently tripped the switch that closed the mold. Yet another classic illustration of a plastic product liability case involved a small manufacturer of PVC handles that were incapable of handling high heat. The handle was softened by high temperature, exposing live electrical contacts and electrocuting a person.

From the foregoing discussion it is clear that product liability and product safety are interrelated. The majority of manufacturers, especially the smaller ones, allow themselves to become the target of such product liability suits by thinking that they are impervious to product liability suits. The fact of the matter is that the product liability involves everyone: material manufacturers, product designers, fabricators, sellers, and installers. The manufacturer may be held liable if:

1. The product is defective in design and is not suitable for its intended use.
2. The product is manufactured defective and proper testing and inspection was not carried out.

3. The product lacks adequate labeling and warnings.
4. The product is unsafely packaged.
5. The proper records of product sale, distribution, and manufacturer are not kept up-to-date.
6. The proper records of failure and customer complaints are not maintained (3).

The questions we all face are what can a machine manufacturer do to avoid expensive lawsuits? What are the steps a manufacturer of a product must take before placing the product into the hands of somewhat novice customers? How can a molder who is merely providing a service to the industry protect himself from unknowingly getting involved in such product liability problems? How many ways can a design engineer design a product with all the safeguards built-in without affecting the product's originality, cost, and his or her creativity? How can a material supplier prevent getting sued because a product made from his material failed because of product design problems and not material quality?

Obviously, there is no single answer to all of these questions. The following is a general guideline everyone should follow to steer clear of unwarranted product liability suits.

17.2. PRODUCT/EQUIPMENT DESIGN CONSIDERATIONS

The product or equipment design engineer is often considered a prime mover of product safety. The key factors to be considered in designing a safe product or equipment are consumer ignorance, manufacturing mishaps, and deliberate misuse. A team of designers must review the design individually and collectively from different viewpoints and all possible angles. If a company is not large enough to staff a team of designers, an outside consulting firm should be allowed to review the design. Product insurance representatives should also be consulted during preliminary design since many insurance companies have engineers on staff as safety consultants. The design engineer should also be familiar with all standards and regulations concerning his or her product. Some of the other minor design considerations are selecting components with a high degree of reliability, designing systems to permit ready access for operating, repairing, and replacing components, designing equipment that takes into consideration the capabilities and limitations of operators, designing components that are incapable of being revised or improperly installed, and anticipating all possible environmental and chemical hazards (4).

17.3. PACKAGING CONSIDERATIONS

Packaging in a broad sense is defined as the outer shell of the product. Since plastics are used extensively as an encloser material for products such as appliances, electrical equipment, liquid chemicals, food, and beverages, the product safety and liability considerations are of extreme interest. The product enclosures should be designed to be tamperproof to prevent the insides from being exposed to persons unfamiliar with the potential hazards. The packaging material should be tested for toxicity, chemical compatibility, and environmental resistance. Identification labels indicating product name, model number, serial and lot number, date code, and manufacturing code have been found useful in making products safer.

17.4. INSTRUCTIONS, WARNING LABELS, AND TRAINING

One of the major lines of defense against the product liability suits is providing adequate instructions, warning labels, and training to the consumer as well as to the installer and service persons. All products cannot be made 100 percent safe. There will always be some degree of risk involved in handling certain products or machinery. Therefore, designers of the product or equipment must take into account the safety aspect and come up with a systematic procedure to deal with well-designed but inherently hazardous products. First of all, a clear, concise, but easy to read instruction manual is in order. The writer of such an instruction manual must take into account the possible misinterpretation by the reader and must be aware of the consequences in case this happens.

The machinery or product manufacturer should not only comply with government or industry regulations regarding warning labels, but also place warning labels on his own wherever it is deemed necessary to prevent accidents. The warnings should not only "warn" but also indicate the consequences of disregarding the warning (5). Machinery manufacturers as well as fabricators can prevent the majority of accidents by implementing a proper training program for the machine operators, installers, maintenance personnel, and foremen. Developing a safety training program on specific machines that the operators will be working with, having safety refresher courses, and distributing safety bulletins are a few of the most useful suggestions that have been proven very successful (6).

17.5. TESTING AND RECORDKEEPING

One of the most powerful weapons any manufacturer trying to steer clear of product liability suits can have is the comprehensive testing and recordkeeping program. In many cases, quality control records may be the only defense the manufacturer may have in a courtroom.

The quality control testing should start with an inspection of the raw material and components as they are received. Whenever possible, the supplier should be required to meet military or other industry specifications. If the raw material or components are to be used in a potentially hazardous product, the supplier should be requested to provide certification along with each shipment. In-process testing is equally important. Here again, the quality control and destructive or nondestructive testing requirements should be set in accordance with the industry standards. In the case of machine manufacturers, a thorough preshipment inspection should help eliminate any unexpected surprises. In some cases, it is advisable to retain an independent testing laboratory. The data generated by the independent test laboratories is often found to be more useful and convincing than self-generated data in courtroom defense. Four good reasons outlined below make a strong case for independent testing for product liability (7).

1. *Objectivity.* A manufacturer may be too close to his own products to maintain an impartial, unbiased viewpoint regarding their safety features. An outside safety engineer can look at the products impassionately, pointing out unsafe features that may escape an internal review.

2. *Exposure.* Independent engineering and testing laboratories make it their business to keep abreast of current specifications and safety laws as well as proposed safety legislation.

3. *Independence.* Since accountability is becoming a more important aspect of product liability, a documented report containing solid evidence of testing and fail–safe analysis by an independent firm carries much weight in establishing the intent of the manufacturer to design and make safe products before they reach the market.

4. *Anticipation.* Product testing before injury and/or product liability suits is good preventive medicine.

The other important task for a manufacturer is recordkeeping. A well-organized recordkeeping policy accomplishes many objectives. First, it establishes that the company is taking reasonable precautions to produce safe products. Second, in the event of an accident or injury, it allows the manufacturer to backtrack and pinpoint the cause of failure. Third, with the help of records, the manufacturer can prove that his product did in fact meet the minimum requirements. It is advisable to keep records of design, manufacturing, inspection, quality control, and testing procedures and results on file for at least five years. The retention period should be based upon individual product need. The records should also include material specification, suppliers, serial and lot numbers, and customer names (8). Other useful documents include company's safety policy manual, design changes, failure reports, marketing and shipping records, and advertisement records.

17.6. SAFETY STANDARDS ORGANIZATIONS

Appendix E lists the organizations responsible for setting standards for safety.

REFERENCES

1. Zavita, W. E. and Hauser, C. M., "The Case for Independent Testing for Product Safety," *Plast. Engin.* (Dec. 1977), p. 33.
2. Kolb, J., and Ross, S. S., *Product Safety and Liability,* McGraw–Hill, New York, 1980, p. 4.
3. *Ibid,* p. 12.
4. *Ibid,* p. 330.
5. *Ibid,* p. 208.
6. Allchin, T., "Product Liability in Plastics Industry," *SPE Pacific Tech. Conf. (PACTEC),* **57,** 1975, p. 57.
7. Zavita, Reference 1, p. 34.
8. Fountas, N., "Product Liability—Prepare Now for Judgement Day," *Plast. World* (Feb. 1978), 68.

CHAPTER 18

Nondestructive Testing

18.1. INTRODUCTION

The term nondestructive testing is applied to tests or measurements carried out without harming or altering the properties of the part. Quite often, in order to make measurements or to study certain characteristics of a part, it becomes necessary to destroy the integrity of the part. For example, it is practically impossible to measure wall thickness of certain areas of 6 in. PVC pipe fitting without actually cutting the part. Similarly, shrinkage voids in a 3-in.-diameter extruded Teflon* rod can only be found by cutting the rod in several places. Such destructive techniques are not only very expensive but also time-consuming. Nondestructive testing methods allow one to determine flaws, imperfections, and nonuniformities without destroying the part. Nondestructive tests range from simple visual examination, weighing, and hardness measurements to more complex electrical and ultrasonic tests. The discussion on nondestructive testing in this chapter is confined to ultrasonic measurement and testing techniques. Visual examination, hardness measurements, and electrical measurement techniques have been discussed in other chapters of this book.

18.2. ULTRASONIC TESTING

Ultrasonic testing is one of the most widely used methods for nondestructive inspection. In plastics, the primary application is the detection of discontinuities and measurement of thickness. Ultrasonic techniques can also be used for determining moisture content of plastics, studying the joint integrity of a solvent welded plastic pipe and fittings, and testing welded seams in plastic plates (1).

The term ultrasonic, in a broad sense, is applied to describe high-frequency sound with a frequency above 20,000 cycles/sec. Commercial ultrasonic testing equipment generally employs the testing frequency in the range from 0.75–20 MHz. To provide a basis for understanding the ultrasonic system and how it

* Teflon is a registered trademark of the Dupont Company.

operates, it is necessary to introduce the following terms:

FREQUENCY GENERATOR: A device that imposes a short burst of high frequency alternating voltage on a transducer.

TRANSDUCER: A transducer or a probe is a device that emits a beam of ultrasonic waves when bursts of alternating voltage are applied to it. An ultrasonic transducer is comprised of piezoelectric material. Piezoelectric material is material that vibrates mechanically under a varying electric potential and develops electrical potentials under mechanical strain, thus transforming electrical energy into mechanical energy and vice versa (2). As the name implies, an electrical charge is developed by a piezoelectric crystal when pressure is applied to it and reverse is also true. The most commonly encountered piezoelectric materials are quartz, lithium sulfate, and artifical ceramic materials such as barium titanate.

Many different types of ultrasonic transducers are available, differing in diameter of the probe, frequency, and frequency bandwidth. Each transducer has a characteristic resonant frequency at which ultrasonic waves are most effectively generated and received. Narrow band width transducers are capable of penetrating deep, as well as detecting small flaws. However, these transducers do a poor job of separating echos. Broad bandwidth transducers exhibit excellent echo separation but poor flaw detection and penetration (3). Transducers of a frequency range of 2–5 MHz are most common. For plastic materials, transducers in the range of 1–2 MHz seem to yield the best results.

COUPLANTS: Air, being one of the worst transmitters of sound waves at high frequencies and due to a lack of impedance matching between air and most solids, must be replaced by a suitable coupling agent between the transducer and the material being tested. Many different types of liquids have been used as coupling agents. Glycerine seems to have the highest acoustic impedance. However, oil is the most commonly used couplant. Grease, petroleum jelly, and pastes can also be used as couplants, although a wetting agent must be added to increase wettability as well as viscosity. Some couplants have a tendency to react with the test specimen material and therefore chemical compatibility of the couplant should be studied prior to application. Couplants that are difficult to remove from test specimens should also be avoided. Any type of contamination between the test specimen and the transducer can seriously affect the thickness measurements, especially in the case of thin films. Therefore, it is absolutely necessary to remove these contaminants before applying the couplant. The basic sequence of operation in any ultrasonic measurement system is:

1. Generation of ultrasonic frequency by means of a transducer.
2. Use of a coupling agent (couplant), such as oil or water, to help transmit the ultrasonic waves into the material.
3. Detection of the ultrasonic energy after it has been modified by the material.
4. Displaying of the energy by means of a recorder, cathode-ray tube or other devices.

The three basic ultrasonic measurement techniques most widely used today are:

1. Pulse–echo.
2. Transmission.
4. Resonance.

18.2.1. Pulse–Echo Technique

The pulse–echo technique is the most popular of the three basic ultrasonic, nondestructive testing techniques. The pulse–echo technique is very useful in detecting flaws as well as thickness measurement.

Figure 18-1 shows the principle of the pulse–echo technique. The initial pulse of ultrasonic energy from a transducer is introduced into the test specimen through the couplant. This sound wave travels through the thickness of the specimen until a reflecting surface is encountered, at which time the sound wave reflects back to the transducer. This is called the back wall echo. If the wave encounters a flaw in its path, the flaw acts as a reflecting surface and the wave is reflected back to the transducer. The echo in this case is referred to as a flaw echo. In both cases, the reflected wave travels back to the transducer, causing the transducer element to vibrate and induce an electrical energy that is normally amplified and displayed onto a CRT or other such devices. The echo wave coming from the back wall of the specimen is marked by its transit time from the transducer to the backwall and return. Similarly, the transit time for the flaw echo can also be determined by this technique. Since transit time corresponds to the thickness of the specimen, it is quite possible to calculate the thickness of the specimen using simple computer logic. Figure 18-2 illustrates a typical commercially available pulse–echo instrument. One other technique known as the immersion test technique has generated tremendous interest among the manufacturers who are in favor of automated inspection techniques. In the immersion technique, as shown in Figure 18-3, the

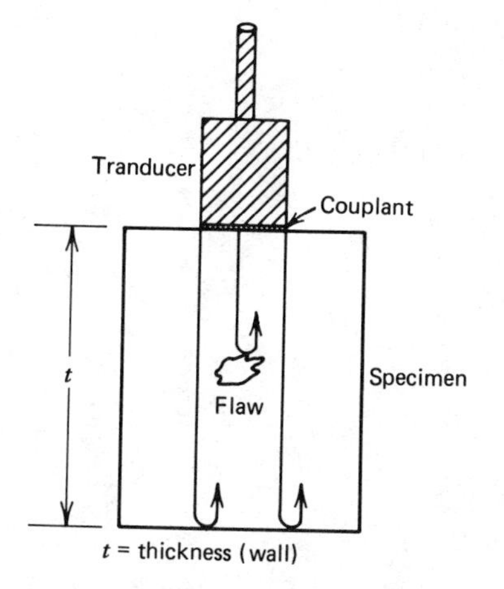

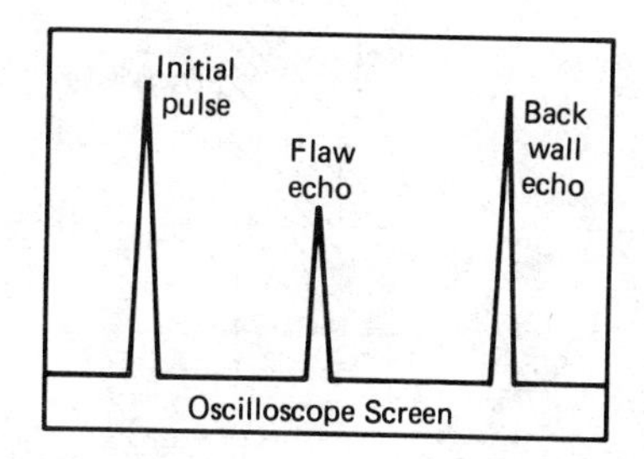

Figure 18-1. Pulse–echo technique.

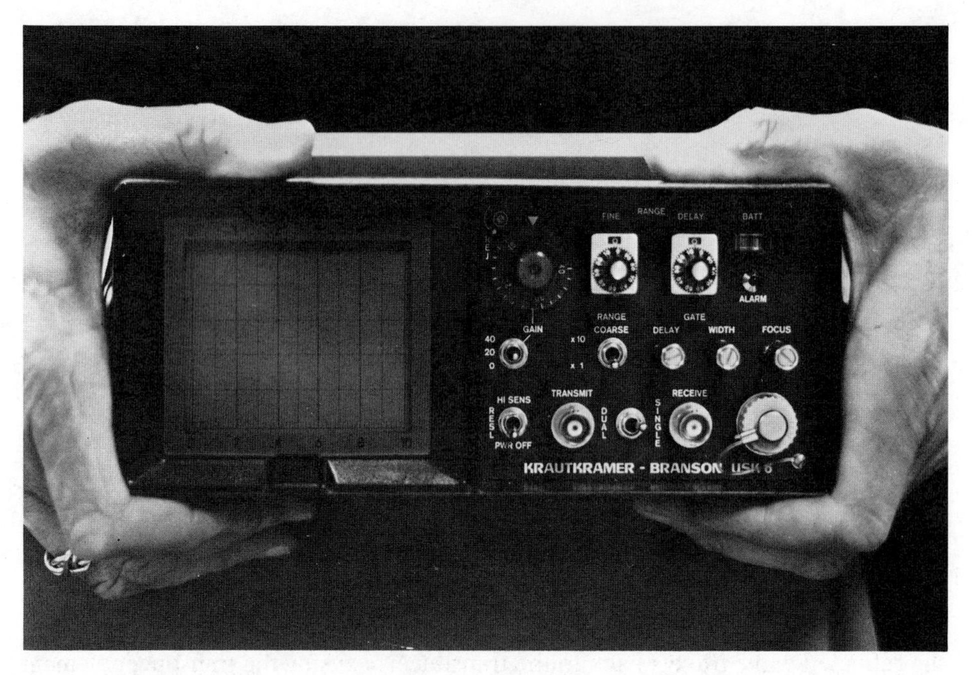

Figure 18-2. Instrument for detecting flaws using pulse–echo technique. (Courtesy Krautkramer–Branson Inc.)

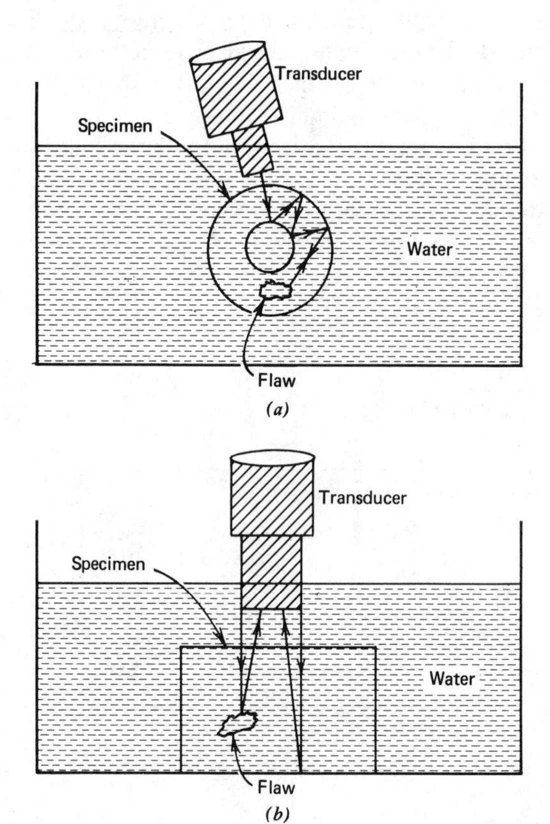

Figure 18-3. Immersion technique. (*a*) Angle beam pulse–echo inspection. (*b*) Straight beam pulse–echo inspection.

specimen is completely immersed in the liquid (usually water) which in turn acts as a couplant. The transducer is mounted in a fixture that is moved across the specimen or the specimens such as pipe or sheet pass continuously under the fixture. The sound beam can be directed either perpendicular or at an angle to the surface of the test piece. The water between the transducer and the specimen acts as a couplant. In this manner, a uniform and nonabrasive coupling is achieved (5). The immersion technique is very useful in the automatic inspection of pipe, sheet, rods, and plates by detecting flaws as the pieces are being extruded under water.

18.2.2. Transmission Technique

In this technique, the intensity of ultrasound is measured after it has passed through the specimen (6). The transmission technique requires two transducers, one to transmit the sound waves and one to receive them. Figure 18-4 illustrates the basic concept of the transmission technique and flaw detection using the technique. The transmission testing can be done either by direct beams or reflected beams. In either case, the flaws are detected by comparing the intensity of ultrasound transmitted through the test specimen with the intensity transmitted through a reference standard made of the same material. The best results are achieved by using the immersion technique since this technique provides uniform and efficient coupling between transducers and test specimen. The main application of the transmission technique is in detecting flaws in laminated plastic sheets.

18.2.3. Resonance Technique

This method is primarily useful for measuring the thickness of the specimen. This is accomplished by determining the resonant frequencies of a test specimen. The detailed discussion of this technique is found in the literature (7).

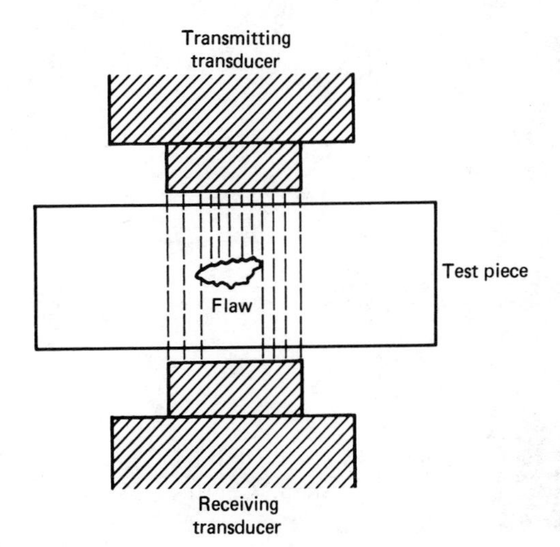

Figure 18-4. Transmission technique.

18.3. APPLICATION OF ULTRASONIC NDT IN PLASTICS

Ultrasonic nondestructive testing (NDT) has gained popularity in the past decade along with the growth of the plastic industry and along with an increasing emphasis placed on automation and material saving. Two major areas in which ultrasonic testing concepts are applied extensively are flaw detection and thickness measurement. The pulse–echo technique is used to detect a flaw such as voids and bubbles in an extruded rod of rather expensive materials such as Teflon and nylon. The flaw detection unit and other auxilliary equipment can be programmed so that the specific portion of the rod with a flaw is automatically cut off and discarded without disturbing the continuous extrusion process. The transmission technique is commonly used to detect flaws in laminates. Thickness measurement by ultrasonic equipment is simple, reliable, and fast. This NDT technique simplifies wall thickness measurement of parts with hard-to-reach areas and complex part geometry. Automated wall thickness measurement and control of large diameter extruded pipe is accomplished by using the immersion technique. An ultrasonic sensing unit is placed in a cooling tank to continuously monitor wall thickness. In the event of an out of control condition, a closed-loop feed back control system is activated and corrections are made to bring the wall thickness closer to the set point. Many such systems are commercially available. Figure 18-5 illustrates one such system based on gamma backscatter principal. Also illustrated in Figure 18-6 is a wall thickness measurement instrument. The ultrasonic NDT technique is used extensively by gas companies to examine the integrity of plastic pipe socket joints after they have been solvent cemented together (8). Ultrasonic measurements can also be used for determining the moisture content of plastics. In ma-

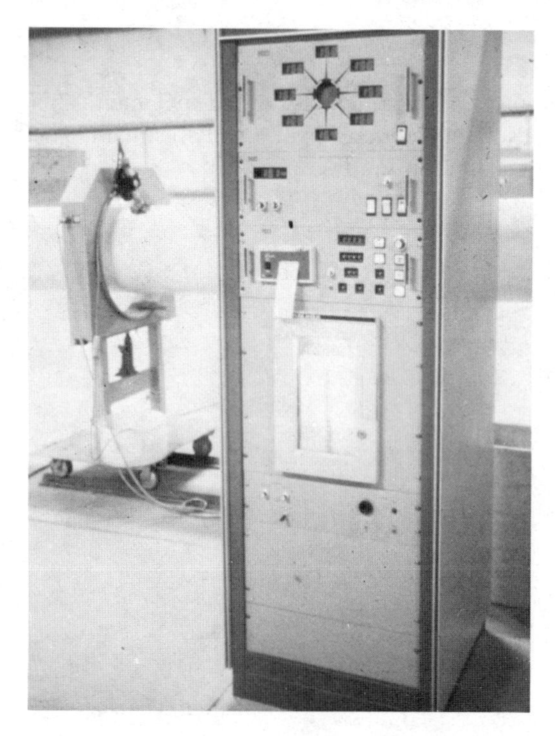

Figure 18-5. Wall thickness measurement system. (Courtesy NDC Systems.)

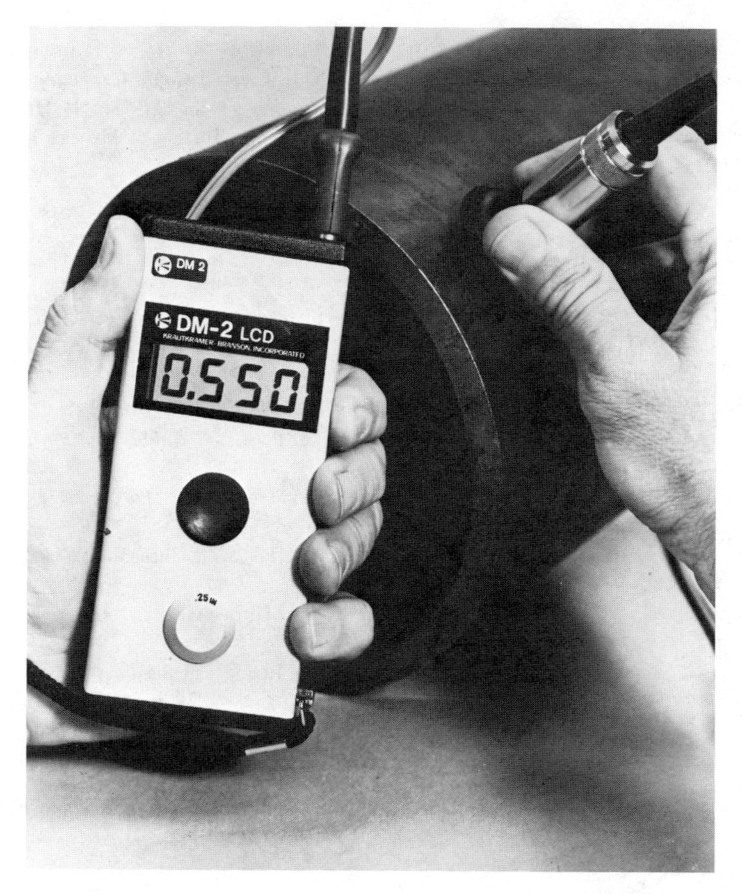

Figure 18-6. Thickness tester. (Courtesy Krautkramer–Branson Inc.)

terials like nylons, the attenuation and the acoustic velocity change with the change in moisture content (9). The use of ultrasonics in testing reinforced plastics (10) and the testing of missiles and rockets (11) have been discussed.

REFERENCES

1. Krautkramer, J. and Krautkramer, H., *Ultrasonic Testing of Materials*, Springer–Verlag, New York, 1969.
2. Ostrofsky, B., "Ultrasonic Inspection of Welds," *Welding J.* (March 1965), p. 97–5.
3. "Nondestructive Inspection and Quality Control," *Metals Handbook*. Vol. II, American Society for Metals, Metals Park, OH, 1976, p. 179.
4. Krautkramer, Reference 1, p. 152.
5. *Ibid*, p. 238.
6. *Ibid*, p. 141.
7. *Ibid*, p. 131.
8. *Nondestructive Examination of Plastic Pipe Socket Joints*, NDT Application Report. No J-714, Branson Instrument Company, Stamford, Connecticut.

9. Krautkramer, Reference 1, p. 436.
10. Hastings, C. H., Lopilato, S. A., and Lynnworth, L. C., *Ultrasonic Inspection of Reinforced Plastics and Resin–Ceramic Composites. Nondestructive Test*, **19** (1961), pp. 340–346.
11. "Symposium on Recent Developments in Nondestructive Testing of Missiles and Rockets," *ASTM Spec. Tech. Pub. No. 350* (1963).

SUGGESTED READING

Lamble, J. H., *Principle and Practice of Nondestructive Testing*, Wiley–Interscience, New York, 1962.

LeGrand, R., "Nondestructive Testing Methods," *Machinist* (Sept. 1946), p. 893.

"Symposium on Nondestructive Testing," *ASTM Spec. Tech. Publ. No. 149* (1953).

Hitt, W. C. and Ramsey, J. B., "Ultrasonic Inspection of Plastics," *Rubber and Plastics Age*, **44**(4) (Apr 1963), p. 411.

Seaman, R. E., "Ultrasonic Inspection by Pulsed Transmissions," *Br. Plast.*, **29**(7), (July 1956) p. 262.

Baumeister, G. B., "Production Testing of Bonding Materials with Ultrasonics," *ASTM Bull. No. 204*, (Feb. 1955), p. 50.

Miller, N. B., and Boruff, V. H., "Adhesive Bonds Tested Ultrasonically," *Adhesives Age* **6**, (June 1963) p. 32.

Zurbrick, J. R., "Nondestructive Testing of Glass–Fiber Reinforced Plastics: Key to Composition Characterization and Design Properties Prediction," *S.P.E. J.*, **24**(9), (Sept. 1968), p. 56.

Hatfield, P., "Ultrasonic Measurements in High Polymers," *Research*, **9**, (Oct. 1956) p. 388.

Coggeshall, A. D., "Nondestructive Quality Control Tests on Finished Reinforced Plastic Parts," *Plast. Tech.* (Dec. 1969), p. 43.

CHAPTER 19

Professional and Testing Organizations

19.1. AMERICAN NATIONAL STANDARDS INSTITUTE (ANSI)

In 1918, when ANSI was founded, standardization activities were just beginning in the United States. Many groups were developing standards and their interests and activities overlapped. The standards they produced often duplicated or conflicted with each other. The result was the waste of manpower, money, and considerable confusion. Five professional/technical societies and three government departments decided a coordinator was needed and created ANSI to handle the job.

ANSI is a federation of standards competents from commerce and industry, professional, trade, consumer, and labor organizations and government. ANSI, in cooperation with these federation participants,

1. Identifies the needs for standards and sets priorities for their completion.
2. Assigns development work to competent and willing organizations.
3. Sees to it that public interests, including those of the consumer, are protected and represented.
4. Supplies standards writing organizations with effective procedures and management services to ensure efficient use of their manpower and financial resources and timely development of standards.
5. Follows up to assure that needed standards are developed on time.

Another role is to approve standards as American National Standards when they meet consensus requirements. It approves a standard only when it has verified evidence presented by a standards developer that those affected by the standard have reached substantial agreement on its provisions. ANSI's other major roles are to represent U.S. interests in nongovernmental international standards work, to make national and international standards available, and to inform the public of the existence of these standards.

19.2. AMERICAN SOCIETY FOR TESTING AND MATERIALS (ASTM)

The American Society for Testing and Materials was founded in 1898. It is a scientific and technical organization formed for "the development of standards on characteristics and performance of materials, products, systems and services and the promotion of related knowledge." ASTM is the world's largest source of voluntary consensus standards. The society operates through more than 135 main technical committees with 1550 subcommittees. These committees function in prescribed fields under regulations that ensure balanced representation among producers, users, and general interest participants. The society currently has 28,000 active members, of whom approximately 17,000 serve as technical experts on committees, representing 76,200 units of participation.

Membership in the society is open to all concerned with the fields in which ASTM is active. An ASTM standard represents a common viewpoint of those parties concerned with its provisions, namely, producers, users, and general interest groups. It is intended to aid industry, government agencies, and the general public. The use of an ASTM standard is purely voluntary. It is recognized that, for certain work or in certain regions, ASTM specifications may be either more or less restrictive than needed. The existence of an ASTM standard does not preclude anyone from manufacturing, marketing, or purchasing products, or using products, processes, or procedures not conforming to the standard. Because ASTM standards are subject to periodic reviews and revision, it is recommended that all serious users obtain the latest revision.

A new edition of the Book of Standards is issued annually. On the average, about 30 percent of each part is new or revised. In 1980, the annual book of ASTM standards, which consisted of 48 parts and over 48,000 pages, included over 6000 ASTM standards and tentatives.

19.3. FOOD AND DRUG ADMINISTRATION (FDA)

The Food and Drug Administration, first established in 1931, is an U.S. government agency of the Department of Health and Human Services. The FDA's activities are directed toward protecting the health of the nation against impure and unsafe foods, drugs, and cosmetics and other potential hazards.

The plastics industry is mainly concerned with the Bureau of Foods which conducts research and developes standards on the composition, quality, nutrition, and safety of foods, food additives, colors and cosmetics, and conducts research designed to improve the detection, prevention, and control of contamination. The FDA is concerned about indirect additives. Indirect additives are those substances capable of migrating into food from a contacting plastic materials. Extensive tests are carried out by the FDA before issuing safety clearance to any plastic material that is to be used in food contact applications. Plastics used in medical devices are tested with extreme caution by the FDA's Bureau of Medical Devices which develops FDA policy regarding safety and effectiveness of medical devices.

Field operations for the enforcement of the laws under the jurisdiction of the FDA are carried out by 11 regional field offices, 22 district offices, and 124 resident inspection posts.

19.4. NATIONAL BUREAU OF STANDARDS (NBS)

The National Bureau of Standards was established by act of Congress in March 1901. The bureau's overall goal is to strengthen and advance the nation's science and technology and to facilitate their effective application for public benefit.

The bureau conducts research and provides a basis for the nation's physical measurement system, scientific and technological services for industry and government, a technical basis for increasing productivity and innovation, promoting international competitiveness in American industry, maintaining equity in trade and technical services, promoting public safety. The bureau's technical work is performed by the National Measurement Laboratory, the National Engineering Laboratory, and the Institute for Computer Sciences and Technology.

19.5. NATIONAL ELECTRICAL MANUFACTURERS ASSOCIATION (NEMA)

The National Electrical Manufacturers Association was founded in 1926. This 560 member association consists of manufacturers of equipment and apparatus for the generation, transmission, distribution, and utilization of electric power. The membership is limited to corporations, firms, and individuals actively engaged in the manufacture of products included within the product scope of NEMA product subdivisions.

NEMA develops product standards covering such matters as nomenclature, ratings, performance, testing, and dimensions. NEMA is also actively involved in developing National Electrical Safety Codes and advocating their acceptance by state and local authorities. Along with a monthly news bulletin, NEMA also publishes manuals, guidebooks, and other material on wiring, installation of equipment, lighting, and standards. The majority of NEMA standardization activity is in cooperation with other national organizations. The manufacturers of wires and cables, insulating materials, conduits, ducts, and fittings are required to adhere to NEMA standards by state and local authorities.

19.6. NATIONAL FIRE PROTECTION ASSOCIATION (NFPA)

The National Fire Protection Association was founded in 1896 with the objective of developing, publishing, and disseminating standards intended to minimize the possibility and effect of fire and explosion. NFPA's membership consists of individuals from business and industry, fire service, health care, insurance, educational, and government institutions. NFPA conducts fire safety education programs for the general public and provides information on fire protection and prevention. Also provided by the association is the field service by specialists on flammable liquids, electricity, gases, and marine problems.

Each year, statistics on causes and occupancies of fires and deaths resulting from fire are compiled and published. NFPA sponsors seminars on the Life Safety Codes, National Electrical Code, industrial fire protection, hazardous materials, transportation emergencies, and other related topics. NFPA also conducts research programs on delivery systems for public fire protection, arson, residential fire sprinkler systems, and other subjects. NFPA publications include *National Fire Codes Annual, Fire Protection Handbook, Fire Journal*, and *Fire Technology*.

19.7. NATIONAL SANITATION FOUNDATION (NSF)

The National Sanitation Foundation, more commonly known as NSF, is an independent, nonprofit environmental organization of scientists, engineers, technicians, educators, and analysts. NSF frequently serves as a trusted neutral agency for government, industry, and consumers, helping them to resolve differences and unite in achieving solutions to problems of the environment.

At NSF, a great deal of work is done on the development and implementation of NSF standards and criteria for health-related equipment. Standard No. 1, concerning soda fountain and luncheonette equipment, was adopted in 1952. Since then, many new standards have been developed and successfully implemented. The majority of NSF standards relate to water treatment and purification equipment, products for swimming pool applications, plastic pipe for potable water as well as drain, waste, and vent (DWV) uses, plumbing components for mobil homes and recreational vehicles, laboratory furniture, hospital cabinets, polyethylene refuse bags and containers, aerobic waste treatment plants, and other products related to environmental quality.

Manufacturers of equipment, materials, and products that conform to NSF standards are included in official listings and these producers are authorized to place the NSF seal on their products. Representatives from NSF regularly visit the plants of manufacturers to make certain that products bearing the NSF seal do indeed fulfill applicable NSF standards.

19.8. PLASTICS TECHNICAL EVALUATION CENTER (PLASTEC)

The Plastics Technical Evaluation Center, more commonly known as PLASTEC, is one of 20 information analysis centers sponsored by the Department of Defense to provide the defense community with a variety of technical information services applicable to plastics, adhesives, and organic matrix composites. For the last 21 years, PLASTEC has served the defense community with authoritative information and advice in such forms as engineering assistance, responses to technical inquiries, special investigations, field trouble shooting, failure analysis, literature searches, state-of-the-art reports, data compilations, and handbooks. PLASTEC has also been heavily involved in standardization activities. In recent years, PLASTEC has been permitted to serve private industry.

The significant difference between a library and technical evaluation center is the quality of the information provided to the user. PLASTEC uses its database "library" as a means to an end to provide succinct and timely information which has been carefully evaluated and analyzed. Examples of the activity include recommendation of materials, counseling on designs, and performing trade-off studies between various materials, performance requirements, and costs. Applications are examined consistent with current manufacturing capabilities, and the market availability of new and old materials alike is considered. PLASTEC specialists can reduce raw data to the user's specifications and supplement them with unpublished information that updates and refines published data. PLASTEC works to spin-off the results of government-sponsored R & D to industry and similarly to utilize commercial advancements to the government's goal of highly sought technology transfer. PLASTEC has a highly specialized library to serve the varied needs of their own staff and customers. The most useful part of the library is the file of more than 30,000 carefully selected documents. This file, which is made up primarily of significant reports and conference papers, is without question the best of its kind in the United States. After careful review, the documents are indexed, abstracted, and processed into a database. About 3000 new documents are added to the collection each year.

PLASTEC offers a great deal of information and assistance to the design engineer in the area of specifications and standards on plastics. PLASTEC has complete visual search microfilm file and can display and print the latest issues of specifications, test methods, and standards from Great Britain, Germany, Japan, U.S.A., and International Standards Organization. Military and Federal specifications and standards and industry standards, such as ASTM, NEMA, and UL are on file and can be quickly retrieved.

19.9. SOCIETY OF PLASTICS ENGINEERS (SPE)

The Society of Plastics Engineers was founded in 1942 with the objective of promoting scientific and engineering knowledge relating to plastics. SPE is a professional society of plastics scientists, engineers, educators, students, and others interested in the design, development, production, and utilization of plastics materials, products, and equipment. SPE currently has over 22,000 members scattered among its 80 sections. The individual sections as well as the SPE main body arranges and conducts monthly meetings, conferences, educational seminars, and plant tours throughout the year. SPE also publishes *Plastics Engineering, Polymer Engineering and Science, Plastics Composites*, and the *Journal of Vinyl Technology*. The society presents a number of awards each year encompassing all levels of the organization, section, division, committee, and international. SPE divisions of interest are color and appearance, injection molding, extrusion, electrical and electronics, thermoforming, engineering properties and structure, vinyl plastics, blow molding, medical plastics, plastics in building, decorating, mold making and mold design.

19.10. SOCIETY OF PLASTICS INDUSTRY (SPI)

The Society of Plastics Industry is a major society, whose membership consists of manufacturers and processors of plastics materials and equipment. The society has four major operating units consisting of the Eastern Section, the Midwest Section, the New England Section, and the Western Section. SPI's Public Affairs Committee concentrates on coordinating and managing the response of the plastics industry to issues like toxicology, combustibility, solid waste, and energy. The Plastic Pipe Institute is one of the most active divisions, promoting the proper use of plastic pipes by establishing standards, test procedures, and specifications. Epoxy Resin Formulators Division has published over 30 test procedures and technical specifications. Risk management, safety standards, productivity, and quality are a few of the major programs undertaken by the machinery division. SPI's other divisions include Expanded Polystyrene Division, Fluoropolymers Division, Furniture Division, International Division, Plastic Bottle Institute, Machinery Division, Molders Division, Mold Makers Division, Plastic Beverage Container Division, Plastic Packaging Strategy Group, Polymeric Materials Producers Division, Polyurethane Division, Reinforced Plastic/Composites Institute, Structural Foam Division, Vinyl Siding Institute, Vinyl Formulators Division.

The National Plastics Exposition and Conference, held every three years by the Society of Plastic Industry, is one of the largest plastic shows in the world. SPI works very closely with other organizations such as ASTM and ANSI to develop new test methods, standards, and specifications.

19.11. UNDERWRITERS LABORATORIES (UL)

Underwriters Laboratories, founded in 1894, is chartered as a not-for-profit organization to establish, maintain, and operate laboratories for the investigation of materials, devices, products, equipment, constructions, methods, and systems with respect to hazards affecting life and property.

There are five testing facilities in the U.S. and over 200 inspection centers. More than 700 engineers and 500 inspectors conduct tests and follow-up investigations to insure that potential hazards are evaluated and proper safeguards provided. UL has six basic services it offers to manufacturers, inspection authorities, or government officials. These are product listing service, classification service, component recognition service, certificate service, inspection service and fact finding and research.

UL's Electrical Department, which is the largest of six departments, is in charge of evaluating individual plastics and other products using plastics as components. Electrical Department evaluates consumer products such as TV sets, power tools, appliances, and industrial and commercial electrical equipment and components. In order for a plastic material to be recognized by UL, it must pass a variety of UL tests including the UL 94 flammability test and the UL 746 series, short and long-term property evaluation tests. When a plastic material is granted Recognized Component Status, a yellow card is issued. The card contains precise identification of the material including supplier, product designation, color, and its UL 94 flammability classification at one or more thicknesses. Also included are many of the property values such as temperature index, hot wire ignition, high-current arc ignition, and arc resistance. These data also appear in the recognized component directory.

UL publishes the names of the companies who have demonstrated the ability to provide a product conforming to the established requirements, upon successful completion of the investigation and after agreement of the terms and conditions of the listing and follow-up service. Listing signifies that production samples of the product have been found to comply with the requirements, and that the manufacturer is authorized to use the UL's listing mark on the listed products which comply with the requirements.

UL's consumer advisory council was formed to advise UL in establishing levels of safety for consumer products, to provide UL with additional user field experience and failure information in the field of product safety, and to aid in educating the general public in the limitations and safe use of specific consumer products.

APPENDIX A

Index of Test Equipment Manufacturers

Accelerated Weathering
Atlas Electric Devices Co.
Q-Panel Company
American Ultraviolet Company

Arc Resistance
Amprobe Instruments
Beckman Instruments, Inc.
Custom Scientific Instrument, Inc.
Testing Machines, Inc.

Bubble Viscometer
Gardner Laboratories

Burst Strength Tester
Testing Machines, Inc.
Rutherford Research Products Co.
Applied Test Systems, Inc.

Charpy Impact Test
Testing Machines, Inc.
Tinius Olsen Testing Machine Co.
Custom Scientific Instruments, Inc.

Colorimeter
Macbeth Corp.
Gardner Laboratories
Applied Color System, Inc.
Testing Machines, Inc.

Compressive Strength and Modulus
Tinius Olsen Testing Machine Co.
Instron Corp.
Dillon W.C. and Co.
Custom Scientific Instruments, Inc.
Testing Machines, Inc.
Acco Industries, Inc.
MTS Systems Corp.

Creep Properties
Custom Scientific Instruments, Inc.
Applied Test Systems, Inc.
Acco Industries, Inc.
Instron Corporation
Tinius Olsen Testing Machine Company

Cup Viscosity Test
Gardner Laboratories
Fisher Scientific Co.
VWR Scientific Co.

Dielectric Strength, Dielectric Constant
Biddle Co.
Hipotronics
Custom Scientific Instruments, Inc.

Drop Impact Test
Tinius Olsen Testing Machine Co.
Gardner Laboratories
Kayeness, Inc.
T.M. Long Co.
Custom Scientific Instruments, Inc.

DSC
Perkin–Elmer Corp.
Dupont Co.

Durometer Hardness
Testing Machines, Inc.
Rex Gauge Co.
Shore Instrument and Mfg. Co., Inc.

Elongation
Tinius Olsen Testing Machine Co.
Instron Corp.
Testing Machines, Inc.
Dillon W.C. and Co.

Environmental Stress Cracking
Custom Scientific Instruments, Inc.

Fatigue Failure
Fatigue Dynamics, Inc.
Instron Corp.
Acco Industries, Inc.
Tinius Olsen Testing Machine Co.

Flammability
Custom Scientific Instrument, Inc.
Arapahoe Chemicals, Inc.
Testing Machines, Inc.
United States Testing Co., Inc.

Flexural Modulus/Strength
Tinius Olsen Testing Machine Co.
Instron Corp.
Testing Machines, Inc.
Dillon W.C. and Co.
MTS Systems Corp.

Gel Point/Gel Time
Shyodu Precision Instrument Co.
Sunshine Scientific Instrument, Inc.
Techne, Inc.
Testing Machines, Inc.

Glossmeter
Gardner Laboratories

GPC
Waters Associates, Inc.

Hardness
Shore Instruments & Mfg. Co.
Acco Industries, Inc.
Ames Precision Machine Works
K. J. Law Engineering, Inc.
Barber–Colman Co.

Haze
Gardner Laboratories

HDT
Tinius Olsen Testing Machines, Inc.
Custom Scientific Instruments, Inc.
Testing Machines, Inc.

Impact Strength
Tinius Olsen Testing Machines, Inc.
Gardner Laboratories
Testing Machines, Inc.
Rheometrics, Inc.
Custom Scientific Instruments, Inc.
T.M. Long Co.
Kayeness, Inc.

Inherent/Intrinsic Viscosity
Fisher Scientific Co.

IR Analysis
Foxboro Analytical Div. of the Foxboro Co.
Infrared Industries, Inc.
Digilab

Melt Index
Kayeness, Inc.
Instron Corp.
Tinius Olsen Testing Machine, Inc.
Testing Machines, Inc.
Custom Scientific Instruments, Inc.
Monsanto Co.

Oxygen Index
Custom Scientific Instruments, Inc.
Testing Machines, Inc.
MKM Machine Tool Co.

Peak Exothermic Temperature
Testing Machines, Inc.
Gardner Laboratories
Techne, Inc.
Shyodu Precision Instrument Co.
Sunshine Scientific Instrument Co.

Polarizer/Photoelasticity
Kayeness, Inc.
Photolastic Div. Measurement Group

Rheometer
Instron Corp.
Rheometrics, Inc.
Tinius Olsen Testing Machine Co.
Monsanto Co.

Specific Gravity
Techne, Inc.
Testing Machines, Inc.
Ohaus Scale Corp.

Stress Relaxation
Instron Corp.

Stress/Strain
Tinius Olsen Testing Machine Co.
Instron Corp.
Photolastic Div.

Tensile Impact Strength
Testing Machines, Inc.
Custom Scientific Instruments, Inc.
Tinius Olsen Testing Machine Co.

Tensile Strength
Tinius Olsen Testing Machine Co.
Instron Corp.
W.C. Dillon and Co.
Custom Scientific Instruments, Inc.
Testing Machines, Inc.
MTS Systems Corp.

Torsion Tester
Acco Industries, Inc.
Testing Machines, Inc.
Custom Scientific Instruments, Inc.
Tinius Olsen Testing Machine Co.
Instron Corp.

TGA
Dupont Co.
Perkin–Elmer Corp.
Harrop Laboratories

TMA
Dupont Co.
Harrop Laboratories
Perkin–Elmer Corp.

Vicat Softening Point
Tinius Olsen Testing Machine Co.
Testing Machines, Inc.
Custom Scientific Instruments, Inc.

Viscosity
Brookfield Engineering Laboratories, Inc.
Gardner Laboratories
Fisher Scientific
VWR Scientific
Testing Machines, Inc.
Brabender C.W. Instruments, Inc.
Haake, Inc.

Weathering
Atlas Electric Devices Co.
Q-Panel Co.
American Ultraviolet Co.
Testing Machines, Inc.
Blue M Electric Co.

ALPHABETICAL INDEX OF COMPANIES AND ADDRESSES

ACCO Industries, Inc.
929 Connecticut Avenue
Bridgeport, Connecticut 06602
(203) 335-2511

American Ultraviolet Co.
195 Commerce St.
Chatham, New Jersey 07928
(201) 635-8355

Atlas Electric Devices Co.
4114 N. Ravenswood
Chicago, Illinois 60613
(312) 327-4520

Applied Test Systems, Inc.
Saxonburg Blvd.
Saxonburg, Pennsylvania 16056
(412) 265-3300

Arapahoe Chemical, Inc.
2075 N. 55th St.
Boulder, Colorado 80031
(303) 442-1926

Ames Precision Machine Works
5270 Gaddes Road
Ann Arbor, Michigan 48105

Blue M Electric Co.
138th and Chatham St.
Blue Island, Illinois 60406

Brabender, C.W. Instruments, Inc.
50 E. Wesley St.
S. Hackensack, New Jersey 07606
(201) 343-8425

Brookfield Engineering Laboratories, Inc.
240 Cushing St.
Stoughton, Massachusetts 02072
(617) 344-4310

Biddle Co.
Plymouth Meeting, Pennsylvania 19462
(215) 646-9200

Barber–Colman Co.
Motor Division
Dept. S
14353 Rock St.
Rockford, Illinois 61101
(815) 877-0241

Custom Scientific Instruments, Inc.
Box A 13 Wing Drive
Wippany, New Jersey 07981
(201) 538-8500

Dillon W.C. and Co.
14620 Keswick St.
Van Nuys, California 91405
(213) 786-8812

Digilab
237 Putnam Avenue
Cambridge, Massachusetts 02139
(617) 868-4330

Dupont De Nemours and Co.
Dupont Instrument Div.
Quillen Bldg. Concord Plaza
Wilmington, Delaware 19898
(302) 772-5500

Fisher Scientific
711 Forbes Avenue
Pittsburg, Pennsylania 15219
(412) 562-8300

Fatigue Dynamics, Inc.
P.O. Box 2533
Dearborn, Michigan 48123
(313) 273-8270

Foxboro Analytical
Division of the Foxboro Co.
140 Water Street
Box 449
S. Norwalk, Connecticut 06856
(203) 853-1616

Gardner/Neotec Div., Pacific Scientific
2431 Linden Lane
Silver Spring, Maryland 20910
(301) 495-7000

Haake, Inc.
244 Saddle River Road
Saddle Brook, New Jersey 07662
(201) 843-2320

Hipotronics, Inc.
P.O. Drawer A
Brewster, New York 10509
(914) 279-8031

Harrop Laboratories
3470 E. 5th Avenue
Columbus, Ohio 43219
(614) 236-0291

Instron Corp.
100 Royal St.
Canton, Massachusetts 02021
(617) 828-2500

Infrared Industries, Inc.
P.O. Box 989
Santa Barbara, California 93102
(805) 684-4181

Kayeness, Inc.
RD 3, Box 30
Honeybrook, Pennsylvania 19344
(215) 273-3711

K.J. Law Engineering, Inc.
23660 Research Drive
Farmington Hills, Michigan 48024

Long T.M. Co., Inc.
40 S. Bridge St.
Somerville, New Jersey 08876
(201) 369-3362

MKM Machine Tool Co., Inc.
P.O. Box 309
Jaffersonville, Indiana 47130
(812) 282-6627

Monsanto Co.
947 W. Waterloo Road
Akron, Ohio 44314
(216) 745-0302

MTS Systems Corp.
Box 24012
Minnespolis, Minnesota 55424
(612) 944-4000

Ohaus Scale Corp.
29 Hanover Rd.
Florham Park, New Jersey 07932
(201) 377-9000

Perkin–Elmer Corp.
Main Ave.
Norwalk, Connecticut 06856
(203) 762-1000

Photolastic Division
Measurements Group
Vishay Corp.
P.O. Box 27777
Raleigh, North Carolina 27611
(919) 365-3800

Q-Panel Co.
15610 Industrial Pkwy.
Cleveland, Ohio 44135
(216) 267-9200

Rex Gauge Co.
P.O. Box 46 Q
Glenview, Illinois 60025
(312) 724-6668

Rheometrics Inc.
2438 U.S. Hwy. #22
Union, New Jersey 07083
(201) 687-4838

Rutherford Research Products Co.
Box 249
Rutherford, New Jersey 07070
(201) 933-2091

Shore Instruments and Mfg. Co.
90-35 Van Wyck Expwy
Jamaica, New York 11435
(212) 526-4089

Shyodu Precision Instrument Co.
197 King St.
Brooklyn, New York 11231
(212) 858-7700

Sunshine Scientific Instruments, Inc.
1810 Grant Ave.
Philadelphia, Pennsylvania 19115
(215) 673-5600

Techne, Inc.
3700 Brunswick Pike
Princeton, New Jersey 08540
(609) 452-9275

Testing Machines, Inc.
400 Bayview Avenue
Amityville, New York 11701
(516) 842-5400

Tinius Olsen Testing Machine Co., Inc.
Easton Road
Willowgrove, Pennsylvania 19090
(215) 675-7100

United States Testing Co., Inc.
1415 Park Avenue
Hoboken, New Jersey 07030
(201) 792-2400

APPENDIX B

Abbreviations: Polymer Materials

ABS	Acrylonitrile–butadiene–styrene
AN	Acrylonitrile
CA	Cellulose acetate
CAB	Cellulose acetate butyrate
CAP	Cellulose acetate propionate
CN	Cellulose nitrate
CP	Cellulose Propionate
CPE	Chlorinated polyethylene
CPVC	Chlorinated polyvinyl chloride
CTFE	Chlorotrifluoroethylene
DAP	Diallyl phthalate
EC	Ethyl cellulose
ECTFE	Poly(ethylene–chlorotrifluoroethylene)
EP	Epoxy
EPDM	Ethylene–propylene–diene monomer
EPR	Ethylene propylene rubber
EPS	Expanded polystyrene
ETFE	Ethylene/tetrafluoroethylene copolymer
EVA	Ethylene–vinyl acetate
FEP	Perfluoro(ethylene–propylene) copolymer
FRP	Fiberglass-reinforced polyester
HDPE	High-density polyethylene
HIPS	High-impact polystyrene
HMWPE	High-molecular-weight polyethylene
LDPE	Low-density polyethylene
MF	Melamine–formaldehyde
PA	Polyamide
PAPI	Polymethylene polyphenyl isocyanate

PB	Polybutylene
PBT	Polybutylene terephthalate (thermoplastic polyester)
PC	Polycarbonate
PE	Polyethylene
PES	Polyether sulfone
PET	Polyethylene terephthalate
PF	Phenol–formaldehyde
PFA	Polyfluoro alkoxy
PI	Polyimide
PMMA	Polymethyl methacrylate
PP	Polypropylene
PPO	Polyphenylene oxide
PS	Polystyrene
PSO	Polysulfone
PTFE	Polytetrafluoroethylene
PTMT	Polytetramethylene terephthalate (thermoplastic polyester)
PU	Polyurethane
PVA	Polyvinyl alcohol
PVAC	Polyvinyl acetate
PVC	Polyvinyl chloride
PVDC	Polyvinylidene chloride
PVDF	Polyvinylidene floride
PVF	Polyvinyl fluoride
TFE	Polytetrafluoroethylene
SAN	Styrene–acrylonitrile
SI	Silicone
TPE	Thermoplastic Elastomers
TPX	Polymethylpentene
UF	Urea formaldehyde
UHMWPE	Ultrahigh-molecular-weight polyethylene
UPVC	Unplasticized polyvinyl chloride

APPENDIX C

Glossary

A

ACCELERATED AGING: A test procedure in which conditions are intensified in order to reduce the time required to obtain a deteriorating effect similar to one resulting from normal service conditions.

ACCELERATED WEATHERING: A test procedure in which the normal weathering conditions are accelerated by means of a device (machine).

AGING: The process of exposing plastics to natural or artificial environmental conditions for a prolonged period of time.

ARC RESISTANCE: The ability of a plastic material to resist the action of a high voltage electrical arc, usually stated in terms of time required to render the material electrically conductive.

APPARENT DENSITY: (Bulk Density). The weight of the unit volume of material including voids (air) inherent in the material as tested.

ABRASION RESISTANCE: The ability of a material to withstand mechanical action such as rubbing, scraping, or erosion that tends to progressively remove material from its surface.

AMORPHOUS POLYMERS: Polymeric materials that have no definite order or crystallinity. The polymer molecules are arranged in completely random fashion. Examples of amorphous plastics are polystyrene, PVC, PMMA, etc.

B

BULK FACTOR: The ratio of the volume of any given quantity of the loose plastic material to the volume of the same quantity of the material after molding or forming. Bulk factor is a measure of volume change that may be expected in fabrication.

BULK DENSITY: See apparent density.

BRITTLE FAILURE: The failure resulting from the inability of a material to absorb energy, resulting in instant fracture upon mechanical loading.

BIREFRINGENCE (DOUBLE REFRACTION): Birefringence is the difference between index of refraction of light in two directions of vibration.

BRITTLENESS TEMPERATURE: The temperature at which plastics and elastomers exhibit brittle failure under impact conditions.

BROOKFIELD VISCOMETER: The Brookfield viscometer is the most widely used instrument for measuring the viscosity of plastisols and other liquids of a thixotropic nature. The instrument measures shearing stress on a spindle rotating at a definite, constant speed, while immersed in the sample. The degree of spindle lag is indicated on a rotating dial. This reading, multiplied by a conversion factor based on spindle size and rotational speed, gives a value for viscosity in centipoises. By taking measurements at different rotational speeds, an indication of the degree of thixotropy of the sample is obtained.

BUBBLE VISCOMETER: In a bubble viscometer, a transparent liquid streams downward in the ring-shaped zone between the glass wall of a sealed tube and a rising air bubble. The rate at which the air bubbles rises, under controlled conditions and within certain limits, is a direct measure of kinematic viscosity of streaming liquids.

BURST STRENGTH: The internal pressure required to break a pressure vessel such as a pipe or fitting. The pressure (and therefore the burst strength) varies with the rate of pressure build-up and the time during which the pressure is held.

C

CAPILLARY RHEOMETER: An instrument for measuring the flow properties of polymer melts. It is comprised of a capillary tube of specified diameter and length, means for applying desired pressures to force the molten polymer through the capillary, means for maintaining the desired temperature of the apparatus, and means for measuring differential pressures and flow rates. The data obtained from capillary rheometers is usually presented as graphs of shear stress against shear rate at constant temperature.

CELLULAR PLASTICS: See foamed plastics.

CHARPY IMPACT TEST: A destructive test of impact resistance, consisting of placing the specimen in a horizontal position between two supports, then striking the specimen with a pendulum striker swung from a fixed height. The magnitude of the blow is increased until the specimen breaks. The result is expressed in in-lb or ft-lb of energy.

CHALKING: A whitish, powdery residue on the surface of a material caused by material degradation (usually from weather).

CHROMA (SATURATION): The attribute of color perception that expresses the degree of departure from gray of the same lightness.

CIE (COMMISSION INTERNATIONALE DE L'ECLAIRAGE): The international commission on illuminants responsible for establishing standard illuminants.

COEFFICIENT OF THERMAL EXPANSION: The fractional change in length or volume of a material for unit change in temperature.

COMPRESSIVE STRENGTH: The maximum load sustained by a test specimen in a compressive test divided by the original cross section area of the specimen.

CONDITIONING: Subjecting a material (or test specimens) to standard environmental and/or stress history prior to testing.

COLORIMETER: An instrument for matching colors with results approximately the same as those of visual inspection, but more consistently.

CONTINUOUS USE TEMPERATURE: The maximum temperature at which material may be subjected to continuous use without fear of premature thermal degradation.

CRAZING: An undesirable defect in plastic articles, characterized by distinct surface cracks or minute frostlike internal cracks, resulting from stresses within the article. Such stresses may result from molding shrinkage, machining, flexing, impact shocks, temperature changes, or action of solvents.

CRYSTALLINITY: A state of molecular structure in some resins attributed to the existence of solid crystals with a definite geometric form. Such structures are characterized by uniformity and compactness.

CROSSLINKING: Applied to polymer molecules, the setting up of chemical links between the molecular chains. When extensive, as in most thermosetting resins, crosslinking makes one infusible super molecule of all the chains. Crosslinking can be achieved by irradiation with high energy electron beams or by chemical crosslinking agents such as organic peroxides.

CREEP: Due to its viscoelastic nature, a plastic subjected to a load for a period of time tends to deform more than it would from the same load released immediately after application, and the degree of this deformation is dependent on the load duration. Creep is the permanent deformation resulting from prolonged application of stress below the elastic limit. Creep at room temperature is sometimes called cold flow.

CREEP MODULUS (APPARENT MODULUS): The ratio of initial applied stress to creep strain.

CREEP RUPTURE STRENGTH: Stress required to cause fracture in a creep test within specified time.

CUP FLOW TEST: Test for measuring the flow properties of thermosetting materials. A standard mold is charged with preweighed material, and the mold is closed using sufficient pressure to form a required cup. Minimum pressures required to mold a standard cup and the time required to close the mold fully are determined.

CUP VISCOSITY TEST: A test for making flow comparisons under strictly comparable conditions. The cup viscosity test employs a cup-shaped gravity device that permits the timed flow of a known volume of liquid passing through an orifice located at the bottom of the cup.

D

DENSITY: Weight per unit volume of a material expressed in grams per cubic centimeter, pounds per cubic foot, etc.

DIELECTRIC STRENGTH: The electric voltage gradient at which an insulating material is broken down or "arced through" in volts per mil of thickness.

DIELECTRIC CONSTANT (PERMITTIVITY): The ratio of the capacitance of a given configuration of electrodes with a material as dielectric to the capacitance of the same electrode configuration with a vacuum (or air for most practical purposes) as the dielectric.

DISSIPATION FACTOR: The ratio of the conductance of a capacitor in which the material is dielectric to its susceptance, or the ratio of its parallel reactance to its parallel resistance. Most plastics have a low dissipation factor, a desirable property because it minimizes the waste of electrical energy as heat.

DIMENSIONAL STABILITY: Ability of a plastic part to retain the precise shape in which it was molded, fabricated, or cast.

DROP IMPACT TEST: Impact resistance test in which the predetermined weight is allowed to fall freely onto the specimen from varying heights. The energy absorbed by the specimen is measured and expressed in in.-lb or ft-lb.

DIFFERENTIAL SCANNING CALORIMETRY (DSC): DSC is a thermal analysis technique that measures the quantity of energy absorbed or evolved (given off) by a specimen in calories as its temperature is changed.

DUCTILITY: Extent to which a material can sustain plastic deformation without fracturing.

DUROMETER HARDNESS: Measure of the indentation hardness of plastics. It is the extent to which a spring-loaded steel indentor protrudes beyond the pressure foot into the material.

E

ELONGATION: The increase in length of a test specimen produced by a tensile load. Higher elongation indicates higher ductility.

EMBRITTLEMENT: Reduction in ductility due to physical or chemical changes.

ENVIRONMENTAL STRESS CRACKING: The susceptibility of a thermoplastic article to crack or craze formation under the influence of certain chemicals and stress.

EXTENSOMETER: Instrument for measuring changes in linear dimensions; also called a strain gauge.

EXTRUSION PLASTOMETER (RHEOMETER): A type of viscometer used for determining the melt index of a polymer. It consists of a vertical cylinder with two longitudinal bored holes (one for measuring temperature and one for containing the specimen, the latter having an orifice of stipulated diameter at the bottom and a plunger entering from the top). The cylinder is heated by external bands and weight is placed on the plunger to force the polymer specimen through the orifice. The result is reported in grams/10 min.

F

FAILURE ANALYSIS: The science of analyzing failures (product) employing a step-by-step method of elimination.

FATIGUE FAILURE: The failure or rupture of a plastic article under repeated cyclic stress, at a point below the normal static breaking strength.

FATIGUE LIMIT: The stress below which a material can be stressed cyclically for an infinite number of times without failure.

FATIGUE STRENGTH: The maximum cyclic stress a material can withstand for a given number of cycles before failure.

FADOMETER: An apparatus for determining the resistance of materials to fading by exposing them to ultraviolet rays of approximately the same wavelength as those found in sunlight.

FALLING WEIGHT IMPACT TESTER: See drop impact tester.

FLAMMABILITY: Measure of the extent to which a material will support combustion.

FLEXURAL MODULUS: Within the elastic limit, the ratio of the applied stress on a test specimen in flexure to the corresponding strain in the outermost fiber of the specimen. Flexural modulus is the measure of relative stiffness.

FLEXURAL STRENGTH: The maximum stress in the outer fiber at the moment of crack or break.

FOAMED PLASTICS (CELLULAR PLASTICS): Plastics with numerous cells disposed throughout its mass. Cells are formed by a blowing agent or by the reaction of the constituents.

G

GEL POINT: The stage at which liquid begins to gel, that is, exhibits pseudo-elastic properties.

GEL TIME: Gel time is the interval of time between introduction of the catalyst and the formation of a gel.

GEL PERMEATION CHROMATOGRAPHY (GPC): A newly developed column chromatography technique employing a series of columns containing closely packed rigid gel particles. The polymer to be analyzed is introduced at the top of the column and then is eluted with a solvent. The polymer molecules diffuse through the gel at rates depending on their molecular size. As they emerge from the columns, they are detected by differential refractometer coupled to a chart recorder, on which a molecular weight distribution curve is plotted.

GLOSSMETER: An instrument for measuring specular gloss at various angles.

H

HARDNESS: The resistance of plastic materials to compression and indentation. Brinnel hardness and shore hardness are major methods of testing this property.

HAZE: The cloudy or turbid aspect of appearance of an otherwise transparent specimen caused by light scattered from within the specimen or from its surfaces.

HEAT DEFLECTION TEMPERATURE (HDT): Temperature at which a standard test bar deflects 0.010 in. under a stated load of either 66 or 264 psi.

HOOKE'S LAW: Stress is directly proportional to strain.

HOOP STRESS: The circumferential stress in a material of cylindrical form subjected to internal or external pressure.

HUE: The attribute of color perception by means of which an object is judged to be red, yellow, green, blue, purple, or intermediate between some adjacent pair of these.

HYSTERESIS: The cyclic noncoincidence of the elastic loading and the unloading curves under cyclic stressing. The area of the resulting elliptical hysteresis loop is equal to the heat generated in the system.

HYGROSCOPIC: Materials having the tendency to absorb moisture from air. Plastics, such as nylons and ABS, are hygroscopic and must be dried prior to molding.

I

IMPACT STRENGTH: Energy required to fracture a specimen subjected to shock loading.

IMPACT TEST: A method of determining the behavior of material subjected to shock loading in bending or tension. The quantity usually measured is the energy absorbed in fracturing the specimen in a single blow.

INDENTATION HARDNESS: Resistance of a material to surface penetration by an indentor. The hardness of a material as determined by the size of an indentation made by an indenting tool under a fixed load, or the load necessary to produce penetration of the indentor to a predetermined depth.

INDEX OF REFRACTION: The ratio of velocity of light in vacuum (or air) to its velocity in a transparent medium.

INFRARED ANALYSIS: A technique frequently used for polymer identification. An infrared spectrometer directs infrared radiation through a film or layer of specimen and measures and records the relative amount of energy absorbed by the specimen as a function of wavelength or frequency of infrared radiation. The chart produced is compared with correlation charts for known substances to identify the specimen.

INHERENT VISCOSITY: In dilute solution viscosity measurements, inherent viscosity is the ratio of the natural logarithm of the relative viscosity to the concentration of the polymer in grams per 100 ml of solvent.

INTRINSIC VISCOSITY: In dilute solution viscosity measurements, intrinsic viscosity is the limit of the reduced and inherent viscosities as the concentration of the polymeric solute approaches zero and represent the capacity of the polymer to increase viscosity.

ISO: Abbreviation for the International Standards Organization.

ISOCHRONOUS (EQUAL TIME) STRESS–STRAIN CURVE: A stress–strain curve obtained by plotting the stress vs corresponding strain at a specific time of loading pertinent to a particular application.

IZOD IMPACT TEST: A method for determining the behavior of materials subjected to shock loading. Specimen supported as a cantilever beam is struck by a weight at the end of a pendulum. Impact strength is determined from the amount of energy required to fracture the specimen. The specimen may be notched or unnotched.

K

K FACTOR: A term sometimes used for thermal insulation value or coefficient of thermal conductivity. Also, see thermal conductivity.

L

LUMINOUS TRANSMITTANCE (LIGHT TRANSMITTANCE): The ratio of transmitted light to incident light. The value is generally reported in percentage of light transmitted.

M

METAMERISM: Metamerism is a phenomenon of change in the quality of color match of any pair of colors as illumination or observer or both are changed.

MELT INDEX TEST: Melt index test measures the rate of extrusion of a thermoplastic material through an orifice of specific length and diameter under prescribed conditions of temperature and pressure. Melt index value is reported in grams per 10 minutes for specific condition.

MODULUS OF ELASTICITY (ELASTIC MODULUS, YOUNG'S MODULUS): The ratio of stress to corresponding strain below the elastic limit of a material.

MONOMER: (*Mono-mer single-unit*) A monomer is a relatively simple compound that can react to form a polymer (multiunit) by combination with itself or with other similar molecules or compounds.

MONOCHROMATIC LIGHT SOURCE: A light source capable of producing light of only one wavelength.

MOLECULAR WEIGHT: The sum of the atomic weights of all atoms in a molecule. In high polymers, the molecular weight of individual molecules varies widely, therefore, they are expressed as weight average or number average molecular weight.

MOLECULAR WEIGHT DISTRIBUTION: The relative amount of polymers of different molecular weights that comprise a given specimen of a polymer.

N

NECKING: The localized reduction in cross section that may occur in a material under stress. Necking usually occurs in a test bar during a tensile test.

NOTCH SENSITIVITY: Measure of reduction in load-carrying ability caused by stress concentration in a specimen. Brittle plastics are more notch sensitive than ductile plastics.

NEWTONIAN BEHAVIOR: A flow characteristic evidenced by viscosity that is independent of shear rate, that is, the shear stress is directly proportional to shear rate. Water, mineral oil, etc. are typical Newtonian liquids.

NON-NEWTONIAN BEHAVIOR: The behavior of liquid that does not satisfy the requirement for a Newtonian liquid as defined. The flow of molten polymers is generally non-Newtonian producing lower viscosities at higher rate of shear.

O

ORIENTATION: The alignment of the crystalline structure in polymeric materials so as to produce a highly uniform structure.

OXYGEN INDEX: The minimum concentration of oxygen expressed as a volume percent, in a mixture of oxygen and nitrogen that will just support flaming combustion of a material initially at room temperature under the specified conditions.

P

PEAK EXOTHERMIC TEMPERATURE: The maximum temperature reached by reacting thermosetting plastic composition is called peak exothermic temperature.

PHOTOELASTICITY: An experimental technique for the measurement of stresses and strains in material objects by means of the phenomenon of mechanical birefringence.

POISSON'S RATIO: Ratio of lateral strain to axial strain in an axial loaded specimen. It is a constant that relates the modulus of rigidity to Young's modulus.

POLARIZER: A medium or a device used to polarize the incoherent light.

POLARIZED LIGHT: Polarized electromagnetic radiation whose frequency is in the optical region.

POLYMER: (*Poly-many, mer-unit*) A polymer is a high molecular weight organic compound whose structure can be represented by a repeated monomeric unit.

POLYMERIZATION: A chemical reaction in which the molecules of monomers are linked together to form polymers.

PROPORTIONAL LIMIT: The greatest stress that a material is capable of sustaining without deviation from proportionality of stress and strain (Hooke's Law).

R

REFRACTIVE INDEX: See index of refraction.

RELATIVE HUMIDITY: The ratio of the quantity of water vapor present in the atmosphere to the quantity that would saturate it at the existing temperature.

It is also the ratio of the pressure of water vapor present to the pressure of saturated water vapor at the same temperature.

RELATIVE VISCOSITY: Ratio of kinematic viscosity of a specified solution of the polymer to the kinematic viscosity of the pure solvent.

RHEOLOGY: The science dealing with the study of material flow.

RHEOMETER: See extrusion plastometer (rheometer).

ROCKWELL HARDNESS: Index of indentation hardness measured by a steel ball indentor. See also indentation hardness.

S

SECANT MODULUS: The ratio of total stress to corresponding strain at any specific point on the stress–strain curve.

SHEAR STRENGTH: The maximum load required to shear a specimen in such a manner that the resulting pieces are completely clear of each other.

SHEAR STRESS: The stress developing in a polymer melt when the layers in a cross section are gliding along each other or along the wall of the channel (in laminar flow).

SHEAR RATE: The overall velocity over the cross section of a channel with which molten or fluid layers are gliding along each other or along the wall in laminar flow.

SHORE HARDNESS: See indentation hardness.

SPE: Abbreviation for Society of Plastics Engineers.

SPECIFIC GRAVITY: The ratio of the weight of the given volume of a material to that of an equal volume of water at a stated temperature.

SPECTROPHOTOMETER: An instrument that measures transmission or apparent reflectance of visible light as a function of wavelength, permitting accurate analysis of color or accurate comparison of luminous intensities of two sources of specific wavelengths.

SPECIMEN: A piece or a portion of a sample used to conduct a test.

SPECULAR GLOSS: The relative luminous reflectance factor of a specimen at the specular direction.

SPI: Abbreviation for Society of Plastics Industry.

SPIRAL FLOW TEST: A method for determining the flow properties of a plastic material based on the distance it will flow under controlled conditions of pressure and temperature along the path of a spiral cavity using a controlled charge mass.

STRAIN: The change in length per unit of original length, usually expressed in percent.

STRESS: The ratio of applied load to the original cross sectional area expressed in pounds per square inch.

STRESS–STRAIN DIAGRAM: Graph of stress as a function of strain. It is constructed from the data obtained in any mechanical test where a load is applied to a material and continuous measurements of stress and strain are made simultaneously.

STRESS OPTICAL SENSITIVITY: The ability of some materials to exhibit double refraction of light when placed under stress is referred to as stress–optical sensitivity.

STRESS CONCENTRATION: The magnification of the level of applied stress in the region of a notch, crack, void, inclusion, or other stress risers.

STRESS RELAXATION: The gradual decrease in stress with time under a constant deformation (strain).

S–N DIAGRAM: Plot of stress (S) against the number of cycles (N) required to cause failure of similar specimens in a fatigue test.

SURGING: Pressure rise in a pipeline caused by a sudden change in the rate of flow or stoppage of flow in the line. These changes of pressure cause elastic deformation of the pipe walls and changes in the density of fluid column.

T

TENSILE STRENGTH: Ultimate strength of a material subjected to tensile loading.

TENSILE IMPACT ENERGY: The energy required to break a plastic specimen in tension by a single swing of a calibrated pendulum.

THERMOGRAVIMETRIC ANALYSIS (TGA): A testing procedure in which changes in the weight of a specimen are recorded as the specimen is progressively heated.

THERMOPLASTIC: A class of plastic material that is capable of being repeatedly softened by heating and hardened by cooling. ABS, PVC, polystyrene, polyethylene, etc. are thermoplastic materials.

THERMOSETTING PLASTICS: A class of plastic materials that will undergo a chemical reaction by the action of heat, pressure, catalysts, etc., leading to a relatively infusible, nonreversible state. Phenolics, epoxies, and alkyds are examples of typical thermosetting plastics.

THERMAL CONDUCTIVITY: The ability of a material to conduct heat. The coefficient of thermal conductivity is expressed as the quantity of heat that passes through a unit cube of the substance in a given unit of time when the difference in temperature of the two faces is 1 degree.

THERMOMECHANICAL ANALYSIS (TMA): A thermal analysis technique consisting of measuring physical expansion or contraction of a material or changes in its modulus or viscosity as a function of temperature.

TORSION: Stress caused by twisting a material.

TORSION PENDULUM: An equipment used for determining dynamic mechanical properties of plastics.

TOUGHNESS: The extent to which a material absorbs energy without fracture. The area under a stress–strain diagram is also a measure of toughness of a material.

TRISTIMULUS COLORIMETER: The instrument for color measurement based on spectral tristimulus values. Such an instrument measures color in terms of three primary colors: red, green, and blue.

U

Ultraviolet: The region of the electromagnetic spectrum between the violet end of visible light and the x-ray region, including wavelengths from 100 to 3900 Å. Photon of radiations in the UV area have sufficient energy to initiate some chemical reactions and to degrade some plastics.

Ultrasonic Testing: A nondestructive testing technique for detecting flaws in material and measuring thickness based on the use of ultrasonic frequencies.

V

Vicat Softening Point: The temperature at which a flat-ended needle of 1 sq. mm circular or square cross section will penetrate a thermoplastic specimen to a depth of 1 mm under a specified load using a uniform rate of temperature rise.

Viscosity: A measure of resistance of flow due to internal friction when one layer of fluid is caused to move in relationship to another layer.

Viscometer: An instrument used for measuring the viscosity and flow properties of fluids.

W

Water Absorption: The amount of water absorbed by a plastic article when immersed in water for a stipulated period of time.

Weatherometer: An instrument used for studying the effect of weather on plastics in accelerated manner using artificial light sources and simulated weather conditions.

Weathering: A broad term encompassing exposure of plastics to solar or ultraviolet light, temperature, oxygen, humidity, snow, wind, pollution, etc.

Y

Yellowness Index: A measure of the tendency of plastics to turn yellow upon long term exposure to light.

Yield Point: Stress at which strain increases without accompanying increase in stress.

Yield Strength: The stress at which a material exhibits a specified limiting deviation from the proportionality of stress to strain. Unless otherwise specified, this stress will be the stress at the yield point.

Young's Modulus: The ratio of tensile stress to tensile strain below the proportional limit.

Z

ZAHN VISCOSITY CUP: A small U-shaped cup suspended from a looped wire, with an orifice of any one of five sizes at the base. The entire cup is submerged in test sample and then withdrawn. The time in seconds from the moment the top of the cup emerges from the sample until the stream from the orifice first breaks is the measure of viscosity.

APPENDIX D

Trade Names

Trade Name	Material	Manufacturer
Abson	ABS	Mobay Corp.
Acrylafil	G.F. SAN	Fiberfil, Inc.
Acryloid	Acrylic modifiers	Rohm and Hass Co.
Araldite	Epoxy resins	Ciba–Geigy Corp.
Bakelite	Polyethylene	Union Carbide Corp.
Beetle	Urea–formaldehyde	American Cyanamid Co.
Celcon	Acetal copolymer	Celanese Co.
Celanex	Thermoplastic polyester	Celanese Co.
Corvel	Powder coating compounds	Polymer Corp.
Cycolac	ABS compounds	Borg–Warner Corp.
Cycoloy	ABS–polycarbonate copolymer	Borg–Warner Corp.
Dacovin	PVC compounds	Diamond Shamrock Co.
Dacron	Polyester fibers	Dupont Co.
Delrin	Acetal homopolymer	Dupont Co.
Dow Corning	Silicones	Dow Corning Corp.
Dylan	Polyethylene	ARCO
Dylen	Polystyrene	ARCO
Dypro	Polypropylene	ARCO
Durez	Phenolic, polyester, diallyl phthalate, alkyd	Hooker Corp.
Epoxylite	Epoxy	Epoxylite Corp.
Estane	Polyurethane	B.F. Goodrich Co.
Ethofil	G.F. polyethylene	Fiberfill, Inc.
Formaldafil	G.F. Acetal	Fiberfill, Inc.
Geon	PVC resins and compounds	B.F. Goodrich Co.
Halar	Fluoropolymer	Allied Corp.
Halon	PTFE Fluorocarbon	Allied Corp.
Hetron	polyester	Ashland Chem.

K-resin	Butadiene–styrene	Phillips Co.
Kydex	Acrylic/PVC	Rohm and Hass Co.
Kynar	Polyvinylidene fluoride	Pennwalt Corp.
Lexan	Polycarbonate	G.E. Co.
Lucite	Acrylic resin	Dupont Co.
Lustran	ABS resin	Monsanto Co.
Lustrex	Polystyrene	Monsanto Co.
Marlex	Polyethylene, polypropylene	Phillips co.
Merlon	Polycarbonate	Mobay Co.
Myler	Polyester film	Dupont Co.
Noryl	Polyphenylene oxide	G.E. Co.
Nylafil	G.F. nylon	Fiberfill, Inc.
Nylatron	Nylon	Polymer Corp.
Petrothene	Polyethylene	U.S.I. Co.
Plaskon	Phenolic, epoxy, DAP	Plaskon Products
Plexiglass	Acrylic	Rohm and Hass Co.
Polypenco	Nylon rods, tubes, sheets	Polymer Corp.
Profax	Polypropylene	Hercules, Inc.
Rovel	SAN	Uniroyal
Saran	Vinylidene chloride copolymers	Dow Co.
Styrafil	G.F. polystyrene	Fiberfill Co.
Styron	Polystyrene	Dow Co.
Surlyn	Ionomer resin	Dupont Co.
Teflon	Tetrafluoroethylene (TFE) and fluorinated ethylenepropylene (FEP)	Dupont Co.
Tenite	Cellulosics	Eastman, Inc.
Tyril	SAN copolymer	Dow Co.
TPR	Thermopolastic elastomer	Uniroyal, Inc.
Ultrathene	Polyethylene	U.S.I. Co.
Versamid	Polyamide resins	General Mills, Inc.
Vespel	Polyimide resins	Dupont Co.
Vydyne	Nylon	Monsanto
Zytel	Nylons	Dupont Co.

APPENDIX E

Safety Standards Organizations

American Gas Association (AGA), Arlington, Virginia

AGA is the association of the gas distribution industry. The association is mainly responsible for research, standardization, and information related to the production, distribution, and utilization of gas.

American National Standards Institute (ANSI), New York, New York

ANSI is a nonprofit organization consisting of members from commerce, industry, professionals, trade consumers, labor organizations, and government. ANSI in cooperation with these groups identifies the needs for standards and sets priorities for their completion, assigns development work, supplies a standards-writing organization with effective procedures and management services, and approves standards as American National Standards.

American Society for Quality Control (ASQC), Milwaukee, Wisconin

ASQC is a technical and professional organization responsible for developing and publishing standards related to Quality Control.

American Society for Testing and Materials (ASTM), Philadelphia, Pennsylvania

ASTM is a scientific and technical organization formed for "The development of standards on characteristics and performance of materials, products, systems and services, and the promotion of related knowledge." ASTM is the world's largest source of Voluntary Consensus Standards.

American Society of Electroplated Plastics, Washington, D.C.

This nonprofit organization is devoted to improving the efficiency of its industry and to promoting the use of electropolated plastics. Standards and guidelines developed by the society mainly deal with subjecs such as parts design, mold design, substrate fabrication requisites, plastic product quality, test procedures and standards, and performance capabilities.

American Society of Safety Engineers, Park Ridge, Illinois

This is a technical society interested in the advancement of the profession and professional development of its members. The organization develops standards for the profession and professional safety engineer and participates in standards policy bodies.

Association of Home Appliance Manufacturers, Chicago, Illinois

The association develops voluntary appliance performance standards and makes safety recommendations to Underwriters Laboratories and the American Gas Association, represents the industry in consumer and government relations, compiles statistics, sponsors certification programs, and provides consumer appliance information, educational materials, and technical aids.

Factory Mutual Engineering and Research, Norwood, Massachusetts

This organization provides property and production loss–prevention engineering service to industrial organizations insured in the four companies of the Factory Mutual System. Many standards dealing with flammability, wind damage, fire sprinklers, etc., have been developed.

Industrial Safety Equipment Association, Arlington, Virginia

This association represents the manufacturers of personal protection equipment and machinery safeguard devices.

International Association of Plumbing and Mechanical Officials (IAPMO), Los Angeles, California

IAPMO sponsors the uniform plumbing code (UPC) that is used in over 2500 jurisdications in the United States and is a mandatory code for 10 states.

Juvenile Products Manufacturer's Association, Moorestown, New Jersey

The association, which represents juvenile furniture manufacturers, is responsible for developing juvenile product safety performance standards.

National Association of Plastic Fabricators, Washington, D.C.

The association consists of suppliers of materials and machinery, and manufacturers of decorative plastic laminated products for residential and commercial use.

National Electrical Manufacturers Association (NEMA), Washington, D.C.

The association has over 200 separate standards publications for electrical apparatus and equipment. The NEMA membership includes over 500 major electrical manufacturing companies in the U.S.A.

National Fire Protection Association (NFPA), Boston, Massachusetts

NFPA is involved in the standards-making field under which codes, standards, and recommended practices are developed as guides to engineering for reducing loss of life and property by fire. Standards are published yearly as national fire codes.

National Machine Tool Builder's Association, McLean, Virginia

The membership of this association consists of companies producing machine tools. The association publishes standards and documents.

National Safety Council, Chicago, Illinois

National Safety Council's objective is to determine and evaluate methods and procedures that prevent accidents and minimize injury and economic loss resulting from accidents. The National Safety Council also provides leadership to expedite the adoption and the use of those methods and procedures that best serve the public interest.

National Sanitation Foundation (NSF), Ann Arbor, Michigan

NSF has published standards and criteria under which testing and certification services are currently extended to over 1600 manufacturers who use the NSF seal on over 25,000 items of equipment or products.

Polyurethane Manufacturer's Association, Chicago, Illinois

The organization is a private nonprofit trade association of companies involved in the manufacture of solid polyurethane thermosetting elastomers and related chemicals and equipment suppliers.

Society of Plastics Industry (SPI), New York, New York

SPI is a trade and technical society of over 1200 companies in all branches of the plastics industry interested in quality standards, research, uniform accounting, wage rate surveys, codes, public relations, informative labeling, safety, fire prevention, food packaging, etc.

Underwriters Laboratories (UL)

UL, an independent organization devoted to testing for public safety, was established to maintain and operate laboratories for examination and testing the safety of devices, systems, and related materials.

APPENDIX F

Trade Publications

PLASTICS

Modern Plastics, McGraw–Hill Publishing Company, 770 Lexington Ave., New York 10021.

Plastics Engineering, Society of Plastics Engineers, 14 Fairfield Drive, Brookfield Center, Connecticut 06805.

**Plastics World*, Cahners Publishing Company, 221 Columbus Ave., Boston, Massachusetts 02116.

**Plastics Technology*, Bill Brothers Publishing Corporation and Communications, Inc., 633 Third Ave., New York, New York 10017.

**Plastics Design Forum*, Industry Media, Inc., 1129 E. 17th Ave., Denver, Colorado 80218.

**Plastics Design and Processing*, Lake Publishing Corporation, Box 159 700 Peterson Road, Libertyville, Illinois 60048.

**Plastics Machinery and Equipment*, Industry Media, Inc., 1129 E. 17th Ave., Denver, Colorado 80218.

**Plastics Compounding*, Industry Media, Inc., 1129 E. 17th Ave., Denver, Colorado 80215.

**Reinforced Plastics and Composite World*, Cahners Publishing Company, 3375 S. Bannock St., Englewood, Colorado 80110.

Plastics, Western Plastics News, Inc., 1704 Colorado Ave., Santa Monica, California 90404.

OTHERS

**Design News*, Cahners Publishing Co., 221 Columbus Ave., Boston, Massachusetts 02116.

**Materials Engineering*, Penton/IPC Publication, 1111 Chester Ave., Cleveland, Ohio 44114.

**Quality*, Hitchcock Publishing Company, 3579 Hitchcock Publishing Co., Hitchcock Building, Wheaton, Illinois 60187.

**Industry Week*, Penton/IPC Publication, 1111 Chester AVe., Cleveland, Ohio 44114.

* These periodicals are circulated without charge to qualified individuals.

Plant Engineering, Technical Publishing, 1301 S. Grove Ave., Barrington, Illinois 60010.

**Instrument and Apparatus News (IAN)*, Chilton Company, P.O. Box 2006, Radnor, Pennsylvania 19089.

**Design Engineering*, Morgan–Grampian Publishing Co., 2 Park Ave., New York, New York 10016.

**Product Design and Development*, Chilton Company, P.O. Box 2000, Radnor, Pennsylania 19089.

*INDUSTRY PUBLICATIONS

Engineering Design, E.I. Dupont de Nemours and Co., 2533 Nemours Building, Wilmington, Delaware 19898.

Dupont Tech–Topics, E.I. Dupont de Nemours and Co. Polymer Products Dept., Wilmington, Delaware 19898.

Elastomers Notebook, E.I. Dupont de Nemours and Co., Elastomer Division, Wilmington, Delaware 19898.

Dupont Magazine, E.I. Dupont de Nemours and Co., Wilmington, Delaware 19898.

Bakelite Review, Union Carbide Corp., Plastics Division, 270 Park Ave., New York, New York 10017.

Materials News, Dow Corning Corp., Midland, Michigan 48640.

APPENDIX G

Independent Testing Laboratories

Battelle Memorial Institute
505 King Avenue
Columbus, Ohio 43201
(614) 424-6424

Delsen Testing Laboratories, Inc.
1031 Flower St.
Glendale, California 91201
(213) 245-8517

Detroit Testing Laboratory
8720 Northend Ave.
Oak Park, Michigan 48238
(313) 398-9880.

Gaynes Testing Laboratories, Inc.
1642 W. Fulton St.
Chicago, Illinois 60612
(312) 421-5257

Ghesquire Plastic Testing, Inc.
20304 Harper Avenue
Harperwoods, Michigan 48225
(313) 885-3535

Hauser Laboratories
P.O. Box G
Boulder, Colorado 80306
(303) 443-4662

Hunter Associates Laboratory, Inc.
11495 Sunset Hills Road
Reston, Virginia 22090
(703) 471-6870

Jamieson Laboratories
Washington Hwy.
RFD #8
Smithfield, Rhode Island 02917
(401) 231-1590

LC Laboratories
1254 Chestnut St.
Newton, Massachusetts 02164
(617) 244-9222

Scientific Process and Research, Inc.
400 Cleveland Ave.
P.O. Box 1268
Highland Park, New Jersey 08904
(201) 846-3477

Skeist Laboratories, Inc.
112 Naylon Avenue
Livingston, New Jersey 07039
(201) 994-1050

South Florida Test Service, Inc.
9200 N.W. 58th St.
Miami, Florida 33178
(305) 592-3170

Springborn Laboratories, Inc.
Water St.
Enfield, Connecticut 06082
(203) 749-8371

Tropical Marine Testers, Inc.
P.O. Box 14036
N. Palm Beach, Florida 33408
(305) 622-7373

United States Testing Co.
1415 Park Avenue
Hoboken, New York 07030
(201) 792-2400

Underwriters Laboratories, Inc.
1285 Walt Whitman Road
Melville, New York 11747
(516) 271-6200

APPENDIX H

Specifications

The following is the index of material specifications. A complete copy of these specifications may be obtained by writing to the agency responsible for publishing the particular specification.

ABRASION RESISTANT MATERIAL
ASTM D700-75
ISO2797-1974
L-P-385C
L-P-385C
L-P-392A
L-P-395C
MIL-HDBK-149A
MIL-M-20693B
MIL-P-46122B(MR)
MIL-P-46124A(MR)
MIL-P-47082(MI)
MIL-P-47136(MI)

ACETAL
ANSI/ASTM D2133-78
ASTM D2948-72
L-P-392A
MIL-HDBK-700(MR)
MIL-P-46137A(MR)

ACRYLIC
ANSI/ASTM D788-78A
BS 3412-1976
L-P-380C
L-P-380C
MIL-HDBK-700(MR)
MIL-I-46058C(4)
MIL-P-19735B

ACRYLIC–SILANE
ANSI/ASTM D2660-70(1)

ACRYLONITRILE
ANSI/ASTM D1788-78A
ANSI/ASTM D2474-78

ACRYLONITRILE CONTENT
ANSI/ASTM D1788-78A
ASTM D1431-78
L-P-399B

ACRYLONITRILE–BUTADIENE–STYRENE
ANSI/ASTM D1788-78A
ANSI/ASTM D3011-74(1)
L-P-1183B
MIL-HDBK-700(MR)

ADHESIVE
AMI12
AMS3368
AMS3369
AMS3370
AMS3371
AMS3372
AMS3737
ASTM D1763-76
L-C-530B(1)

AIRCRAFT APPLICATION
AMS3828
AMS3832
L-P-383(1)
MIL-G-83410(USAF)

Appendix H reprinted with permission from "Specifications for Plastics—Desk Top Data Bank," Cordura Publications, Inc., La Jolla, California, 1979.

MIL-P-25395A(ASG)
MIL-P-25421B
MIL-P-25515C
MIL-P-25515C
MIL-P-25518A(1)(ASG)
MIL-P-25770A(ASG)
MIL-R-25506B(1)(AS)
MIL-R-7575C(2)
MIL-R-81090(Wep)
MIL-R-83309(USAF)
MIL-R-83330
MIL-R-9300B
MIL-S-83474(USAF)

ALKYD
MIL-HDBK-700(MR)
MIL-R-21417A(SH)

AMINO
ISO2112-1977
MIL-HDBK-700(MR)

AMINO–SILANE
ASTM D2408-67(1972)

AMMUNITION
L-P-398B
MIL-C-15567(1)(OS)
MIL-C-20301
MIL-E-10853B
MIL-L-51149(MI)
MIL-M-3165A
MIL-P-13298A(MU)
MIL-P-14536(Ord)
MIL-P-48296(PA)
MIL-R-21931A(OS)
MIL-R-51209(MU)
MIL-R-60671(MU)
MIL-R-82483A(OS)
MIL-R-82657(OS)

ARC-RESISTANT MATERIAL
ASTM D700-75
ASTM D709-78
MIL-M-14G
MIL-P-997D(2)

ASBESTOS
AMS3842A
AMS3843A
AMS3858A
ANSI/ASTM D2853-70(1)
ANSI/ASTM D3011-74(1)
ASTM D2897-72
ASTM D700-75
ASTM D709-78
ASTM D709-78
ISO800-1977
MIL-C-47221(MI)
MIL-M-14G
MIL-M-46891(MI)
MIL-P-25770A(ASG)
MIL-P-25770A(ASG)
MIL-P-47134(MI)

AUTOMATIVE APPLICATIONS
ASTM D2000-77A

B STAGE
AMS3821
AMS3822
AMS3826
AMS3827
AMS3827
AMS3830
AMS3832
AMS3845
AMS3845/1
AMS3845/2
AMS3845/2
AMS3847
AMS3847
AMS3858A
AMS3867
AMS3894B
AMS3903
AMS3906
MIL-G-55636B(1)
MIL-S-83474(USAF)

BISPHENOL-A
ASTM D1763-76
MIL-R-46068(MU)

BLOWING AGENT
MIL-HDBK-700(MR)

BORON FILAMENT
AMS3867
AMS3867/1
AMS3867/2
AMS3867/3

BUTADIENE
ANSI/ASTM D1788-78A
ASTM D1418-79
ASTM D1892-78
MIL-L-47106(MI)

CABLE
BS 3412-1976
L-P-385C

L-P-385C
L-P-390C(1)
MIL-HDBK-139(MU
MIL-HDBK-699A(MR)
MIL-M-0024041B(SHIPS)
MIL-N-18352A(NOrd)
MIL-P-22096B
MIL-P-46122B(MR)
MIL-P-47082(MI)

CALENDERING
ANSI/ASTM D1788-78A
L-P-1183B
MIL-HDBK-139(MU)
MIL-HDBK-149A
MIL-HDBK-700(MR)

CARBOXY–NITROSO RUBBER
MIL-R-83322

CARBOXYL TERMINATED
MIL-L-47106(MI)
MIL-P-23942(1)(AS)
MIL-P-82658(1)(OS)

CASTING RESIN
AMS3571
AMS3740A
ANSI/ASTM D2473-78
ASTM D1763-76
MIL-C-82644(OS)
MIL-HDBK-139(MU)
MIL-HDBK-700(MR)
MIL-P-46124A(MR)

CELL NUMBER
ANSI/ASTM D1755-78
ANSI/ASTM D1784-78
ANSI/ASTM D1788-78A
ANSI/ASTM D2287-78
ANSI/ASTM D2473-78
ANSI/ASTM D2474-78
ANSI/ASTM D2853-70(1)
ANSI/ASTM D3011-74(1)
ASTM D1892-78
ASTM D2146-77
ASTM D2647-78
ASTM D2848-71
ASTM D2897-72
ASTM D2948-72
ASTM D2952-71(1977)
ASTM D3013-77
ASTM D3220-73
ASTM D3221-78

CELLULAR MATERIAL
MIL-HDBK-139(MU)
MIL-HDBK-149A
MIL-HDBK-699A(MR)
MIL-HDBK-700(MR)

CELLULOSE
ANSI/ASTM D1201-62
ANSI/ASTM D705-62(19)
ASTM D700-75
ASTM D704-62(1975)
ASTM D709-78
ASTM D709-78
ISO2112-1977
ISO800-1977
MIL-M-14G

CELLULOSE ACETATE
ASTM D706-78
L-P-397C
MIL-HDBK-700(MR)
MIL-M-3165A

CELLULOSE ACETATE BUTYRATE
ANSI/ASTM D707-78
L-P-349C

CELLULOSE NITRATE
MIL-C-15567(1)(OS)

CELLULOSE PROPIONATE
ASTM D1562-72
MIL-P-46074B(MR)

CELLULOSIC

ANSI/ASTM D1201-62
L-P-349C
MIL-HDBK-700(MR)

CARAMIC FIBER
MIL-P-47253(MI)

CHEMICAL RESISTANT MATERIAL
AMI12
AMS3650A
AMS3651C
ANSI/ASTM D1784-78
ASTM D1431-78
L-P-1035A
L-P-385C
L-P-385C
L-P-389A(1)
L-P-392A
L-P-395C
MIL-M-20693B
MIL-P-46109C(MR)

MIL-P-46120A(MR)
MIL-P-46122B(MR)
MIL-P-46124A(MR)
MIL-P-46133A(MR)
MIL-P-46174(MR)
MIL-R-7575C(2)

CHLORINATED POLY(VINYL CHLORIDE)
ANSI/ASTM D1784-78

CHLOROPRENE
ASTM D1418-79
ASTM D2000-77A

CHLOROTRIFLUOROETHYLENE
ANSI/ASTM D1430-78
ASTM D3275-78
L-P-385C

CHOPPED GLASS FIBERS
MIL-F-47018A(MI)
MIL-M-46861(2)(MI)
MIL-M-46862(MI)
MIL-P-43043B(MR)
MIL-P-46069(2)(MU)
MIL-P-46109C(MR)
MIL-P-46892(MI)
MIL-P-82650(OS)

CHROME COMPLEX
ASTM D2410-67(1972)

CHROME–SILANE
AMS3823B

CLOTH
AMS3821
AMS3822
AMS3823B
AMS3824
AMS3825
AMS3826
AMS3827
AMS3830
AMS3835
AMS3835
AMS3837
AMS3845
AMS3845/1
AMS3845/2
AMS3846
AMS3847
AMS3902A
AMS3903
AMS3903/1
AMS3903/2
AMS3903/3
AMS3903/4
AMS3903/5
AMS3903/6
AMS3903/7
AMS3903/8
ANSI/ASTM D2409-67(1)
ANSI/ASTM D2660-70(1)
ASTM D1668-73
ASTM D2150-70(1977)
ASTM D2408-67(1972)
ASTM D2410-67(1972)
ASTM D709-78
L-P-383(1)
MIL-F-46885(MI)
MIL-F-47018A(MI)
MIL-F-47078(MI)
MIL-F-47079(MI)
MIL-G-55636B(1)
MIL-L-52696(ME)
MIL-P-17549C(2)(SHIP)
MIL-P-25395A(ASG)
MIL-P-25421B
MIL-P-25515C
MIL-P-25515C
MIL-P-25518A(1)(ASG)
MIL-P-43043B(MR)
MIL-P-47134(MI)
MIL-P-82540(1)(OS)
MIL-P-997D(2)
MIL-Y-83370A(USAF)

COATING RESIN
AMI12
AMS3684
AMS3737
AMS3750
ASTM D1763-76
ASTM D3222-78
ASTM D3275-78
ASTM D3275-78
BS 3412-1976
L-C-530B(1)
MIL-C-47071(MI)
MIL-C-47272(MI)
MIL-HDBK-139(MU)
MIL-I-46058C(4)
MIL-R-46896(MI)

COMPRESSION MOLDING
AMI12
AMS3650A
ANSI/ASTM D1430-78

ANSI/ASTM D1788-78A
ANSI/ASTM D2287-78
ANSI/ASTM D707-78
ASTM D1562-72
ASTM D1892-78
ASTM D706-78
ASTM D729-78
ASTM D789-78
L-P-1183B
L-P-380C
L-P-380C
L-P-385C
L-P-385C
MIL-C-47027(MI)
MIL-C-47027(MI)
MIL-HDBK-139(MU)
MIL-HDBK-700(MR)
MIL-M-20693B
MIL-M-46861(2)(MI)
MIL-M-46862(MI)
MIL-P-22096B
MIL-P-22985B
MIL-P-3409
MIL-P-46169A(MR)
MIL-P-46892(MI)
MIL-P-47134(MI)

CONTINUOUS FILAMENT
AMS3846
ANSI/ASTM D2409-67(1)
ANSI/ASTM D2660-70(1)
ASTM D2408-67(1972)
ASTM D2410-67(1972)
ASTM D3317-74A
ASTM D709-78
L-P-383(1)
MIL-G-46887(MI)
MIL-G-47024(MI)
MIL-G-47296(MI)
MIL-P-25395A(ASG)
MIL-P-25421B
MIL-P-25515C
MIL-P-25518A(1)(ASG)
MIL-P-46892(MI)
MIL--Y-83370A(USAF)
MIL-Y-83371(USAF)

CORROSION RESISTANT MATERIAL
L-C-530B(1)
L-P-385C
L-P-385C
MIL-M-0024041B(SHIPS)
MIL-P-21347D
MIL-P-46120A(MR)
MIL-P-46174(MR)
MIL-P-47253(MI)
MIL-R-46068(MU)

CREEP
ANSI/ASTM D2953-71
ASTM D1457-78
ASTM D709-78
L-P-395C
MIL-HDBK-149A
MIL-M-0024041B(SHIPS)
MIL-P-21347D
MIL-P-46120A(MR)
MIL-P-46137A(MR)
MIL-P-47298(2)(MI)

CRYOGENIC EQUIPMENT
L-P-385C
L-P-385C
MIL-P-46115B(MR)
MIL-P-46133A(MR

CURING AGENT
AMS3358
AMS3359
AMS3360
AMS3362A
AMS3363B
AMS3364A
AMS3365B
AMS3366B
AMS3368
AMS3369
AMS3370
AMS3371
AMS3372
AMS3571
AMS3734
AMS3735A
AMS3736
AMS3737
AMS3738A
AMS3739A
AMS3740A
AMS3750
ASTM D1763-76
ASTM D3013-77
MIL-C-47003(MI)
MIL-C-47016(MI)
MIL-C-47072(MI)
MIL-C-47171(MI)
MIL-C-47221(MI)
MIL-C-47257(MI)
MIL-C-47272(MI)

MIL-C-81247(WP)
MIL-C-82644(OS)
MIL-I-46865(MI)
MIL-I-46877(1)(MI)
MIL-I-46879(MI)
MIL-I-47151(MI)
MIL-L-47274(1)(MI)
MIL-L-52696(ME)
MIL-M-0024041B(SHIPS)
MIL-P-46892(MI)
MIL-P-47199(MI)
MIL-P-47298(2)(MI)
MIL-P-83455(USAF)
MIL-R-46092(1)(MI)
MIL-R-47252(MI)
MIL-S-83384(USAF)
MIL-S-83474(USAF)

DECORATIVE MATERIAL
AMS3835
AMS3835
L-P-1183B
MIL-HDBK-139(MU)
MIL-T-17171D(SHIPS)

DENSITY
ANSI/ASTM D1248-74
ANSI/ASTM D2581-73
BS 3412-1976
MIL-P-51431(EA)

DESIGNATION
ANSI/ASTM D1248-74
ANSI/ASTM D1755-78
ANSI/ASTM D1765-78A
ANSI/ASTM D1784-78
ANSI/ASTM D1788-78A
ANSI/ASTM D2287-78
ANSI/ASTM D2473-78
ANSI/ASTM D2581-73
ANSI/ASTM D2853-70(1)
ANSI/ASTM D2953-71
ANSI/ASTM D2953-71
ANSI/ASTM D3011-74(1)
ASTM D1418-79
ASTM D1892-78
ASTM D2000-77A
ASTM D2146-77
ASTM D2647-78
ASTM D2848-71
ASTM D2897-72
ASTM D2948-72
ASTM D2952-71(1977)
ASTM D3013-77
ASTM D3220-73
ASTM D3221-78
ASTM D789-78
BS 3412-1976
ISO/R1874-1971
ISO/R1874-1971
ISO1060-1975
ISO1622-1975
ISO1872-1972(E)
ISO2112-1977
ISO2798-1974
J130
MIL-G-55636B(1)
MIL-M-14G

DIALLYL PHTHALATE
ASTM D1636-75A
MIL-M-14G

DIAMETER
AMS3867
AMS3867/1
AMS3867/2
AMS3867/3

DIMENSIONAL STABILITY
ASTM D709-78
L-P-392A
L-P-393A(2)
L-P-395C
MIL-M-14G
MIL-P-21347D
MIL-P-22985B
MIL-P-3409
MIL-P-46137A(MR)
MIL-P-81390
MIL-T-17171D(SHIPS)

EASE OF FABRICATION
ASTM D1457-78
L-P-349C
L-P-397C

ELECTRICAL APPLICATION
AMI12
AMS3358
AMS3359
AMS3360
AMS3361
AMS3362A
AMS3363B
AMS3364A
AMS3365B
AMS3366B
AMS3368

AMS3369
AMS3370
AMS3371
AMS3372
AMS3734
AMS3736
AMS3738A
AMS3739A
AMS3750
AMS3750
AMS3906/5
ANSI/ASTM D1202-62
ANSI/ASTM D1248-74
ANSI/ASTM D1430-78
ANSI/ASTM D2287-78
ANSI/ASTM D2581-73
ANSI/ASTM D2953-71
ARP765
ASTM D2146-77
ASTM D2647-78
ASTM D2952-71(1977)
ASTM D3013-77
ASTM D3159-78A
ASTM D3275-78
ASTM D3307-78
ASTM D700-75
ASTM D700-75
ASTM D704-62(1975)
ASTM D709-78
BS 3412-1976
ISO2112-1977
ISO2113-1972
ISO800-1977
L-P-1035A
L-P-1041A
L-P-349C
L-P-380C
L-P-380C
L-P-383(1)
L-P-385C
L-P-385C
L-P-389A(1)
L-P-392A
L-P-394B
L-P-395C
L-P-396B
L-P-397C
L-P-399B
MIL-C-47027(MI)
MIL-C-47163(MI)
MIL-C-47224(MI)
MIL-C-47233(MI)
MIL-C-47272(MI)
MIL-E-47231(MI)
MIL-F-47078(MI)
MIL-F-47079(MI)
MIL-G-55636B(1)
MIL-HDBK-149A
MIL-I-46058C(4)
MIL-I-46865(MI)
MIL-I-46877(1)(MI)
MIL-I-46879(MI)
MIL-I-47151(MI)
MIL-M-0024041B(SHIPS)
MIL-M-14G
MIL-M-24325(2)(SHIPS)
MIL-M-24519(1)(NAVY)
MIL-M-47137(MI)
MIL-N-18352A(NOrd)
MIL-P-22985B
MIL-P-25421B
MIL-P-3409
MIL-P-43081B(MR)
MIL-P-46109C(MR)
MIL-P-46115B(MR)
MIL-P-46120A(MR)
MIL-P-46122B(MR)
MIL-P-46124A(MR)
MIL-P-46129A(MR)
MIL-P-46131B(MR)
MIL-P-46133A(MR)
MIL-P-46161A(MR)
MIL-P-46169A(MR)
MIL-P-46169A(MR)
MIL-P-46174(MR)
MIL-P-46892(MI)
MIL-P-47082(MI)
MIL-P-47199(MI)
MIL-P-47253(MI)
MIL-P-47298(2)(MI)
MIL-P-81390
MIL-P-83455(USAF)
MIL-P-997D(2)
MIL-R-25042B
MIL-R-25042B
MIL-R-25042B
MIL-R-25506B(1)(AS)
MIL-R-46092(1)(MI)
MIL-R-46896(MI)
MIL-R-7575C(2)

ELECTRONIC ASSEMBLIES

AMS3358
AMS3359
AMS3360
AMS3361

AMS3362A
AMS3363B
AMS3364A
AMS3365B
AMS3366B
AMS3571
AMS3734
AMS3735A
AMS3736
AMS3738A
AMS3739A
AMS3740A
AMS3750
L-P-385C
L-P-385C
MIL-C-46866(MI)
MIL-C-47153(MI)
MIL-C-47163(MI)
MIL-C-47224(MI)
MIL-C-47233(MI)
MIL-C-47272(MI)
MIL-C-82644(OS)
MIL-I-46058C(4)
MIL-I-46877(1)(MI)
MIL-I-47151(MI)
MIL-P-43081B(MR)
MIL-P-47199(MI)
MIL-R-47004(MI)
MIL-S-83384(USAF)

ENCAPSULATION

AMS3358
AMS3359
AMS3360
AMS3361
AMS3362A
AMS3363B
AMS3364A
AMS3365B
AMS3366B
AMS3368
AMS3369
AMS3370
AMS3371
AMS3372
AMS3571
AMS3734
AMS3735A
AMS3736
AMS3738A
AMS3739A
AMS3740A
MIL-C-47153(MI)
MIL-C-47163(MI)
MIL-C-47224(MI)
MIL-C-47233(MI)
MIL-C-47272(MI)
MIL-C-82644(OS)
MIL-I-46865(MI)
MIL-I-47151(MI)
MIL-M-0024041B(SHIPS)
MIL-M-24325(2)(SHIPS)
MIL-P-47199(MI)
MIL-P-47298(2)(MI)
MIL-R-46092(1)(MI)
MIL-R-47004(MI)
MIL-S-83384(USAF)

EPOXY

AMS3734
AMS3735A
AMS3736
AMS3737
AMS3738A
AMS3739A
AMS3740A
AMS3821
AMS3822
AMS3823B
AMS3828
AMS3832
AMS3837
AMS3867
AMS3867/1
AMS3867/2
AMS3867/3
AMS3894/1
AMS3894/2
AMS3894/3
AMS3894/4
AMS3894/5
AMS3894/6
AMS3894/7
AMS3894/8
AMS3894/9
AMS3894B
AMS3903
AMS3903/1
AMS3903/2
AMS3903/3
AMS3903/4
AMS3903/5
AMS3903/6
AMS3903/7
AMS3903/8
AMS3906

AMS3906/1
AMS3906/2
AMS3906/3
AMS3906/4
AMS3906/5
AMS3906/6
AMS3906/7
ASTM D1763-76
ASTM D3013-77
ASTM D709-78
L-C-530B(1)
MIL-C-46866(MI)
MIL-C-47071(MI)
MIL-C-47072(MI)
MIL-C-47153(MI)
MIL-C-47221(MI)
MIL-C-47224(MI)
MIL-C-47233(MI)
MIL-C-47257(MI)
MIL-C-47272(MI)
MIL-C-81247(WP)
MIL-C-82644(OS)
MIL-F-47078(MI)
MIL-G-46887(MI)
MIL-G-55636B(1)
MIL-HDBK-700(MR)
MIL-I-46058C(4)
MIL-I-46865(MI)
MIL-I-46879(MI)
MIL-I-47151(MI)
MIL-L-52696(ME)
MIL-M-24325(2)(SHIPS)
MIL-M-46861(2)(MI)
MIL-M-46862(MI)
MIL-M-47137(MI)
MIL-P-25421B
MIL-P-46069(2)(MU)
MIL-P-46892(MI)
MIL-R-21931A(OS)
MIL-R-46896(MI)
MIL-47252(MI)
MIL-R-82657(OS)
MIL-R-82664(OS)
MIL-R-9300B

EPOXY COMPATIBLE
AMS3823B
AMS3837
ASTM D2408-67(1972)
ASTM D2410-67(1972)
MIL-G-46887(MI)
MIL-G-47296(MI)
MIL-R-60346B
MIL-Y-83370A(USAF)
MIL-Y-83371(USAF)

ETHYL CELLULOSE
ANSI/ASTM D787-78
MIL-E-10853B
MIL-M-3165A
MIL-P-22985B

ETHYLENE
ANSI/ASTM D2853-70(1)
ASTM D2647-78
ASTM D2952-71(1977)
ASTM D3159-78A
ASTM D3275-78
BS 3412-1976

ETHYLENE–PROPYLENE
ASTM D1418-79
ASTM D2000-77A
L-C-530B(1)
MIL-HDBK-149A
MIL-R-83285(USAF)
MIL-R-83412A(USAF)
MIL-T-47111(MI)

EXTRUSION
AMI12
AMS3650A
AMS3651C
ANSI/ASTM D1248-74
ANSI/ASTM D1430-78
ANSI/ASTM D1788-78A
ANSI/ASTM D2116-78
ANSI/ASTM D2133-78
ANSI/ASTM D2287-78
ANSI/ASTM D2473-78
ANSI/ASTM D2581-73
ANSI/ASTM D2853-70(1)
ANSI/ASTM D2874-70(1)
ANSI/ASTM D3011-74(1)
ANSI/ASTM D707-78
ANSI/ASTM D787-78
ANSI/ASTM D788-78A
ASTM D1431-78
ASTM D1457-78
ASTM D1562-72
ASTM D1892-78
ASTM D2146-77
ASTM D2848-71
ASTM D2897-72
ASTM D2948-72
ASTM D3159-78A
ASTM D3220-73

ASTM D3221-78
ASTM D3222-78
ASTM D3275-78
ASTM D3307-78
ASTM D703-78
ASTM D706-78
ASTM D729-78
ASTM D789-78
BS 3412-1976
ISO1622-1975
L-P-1183B
L-P-349C
L-P-380C
L-P-380C
L-P-385C
L-P-385C
L-P-385C
L-P-389A(1)
L-P-390C(1)
L-P-392A
L-P-393A(2)
L-P-395C
L-P-396B
L-P-398B
L-P-399B
MIL-C-47027(MI)
MIL-HDBK-139(MU)
MIL-HDBK-149A
MIL-HDBK-700(MR)
MIL-M-20693B
MIL-M-3165A
MIL-N-18352A(NOrd)
MIL-P-21347D
MIL-P-22096B
MIL-P-22985B
MIL-P-3409
MIL-P-46074B(MR)
MIL-P-46115B(MR)
MIL-P-46115B(MR)
MIL-P-46120A(MR)
MIL-P-46124A(MR)
MIL-P-46129A(MR)
MIL-P-46131B(MR)
MIL-P-46133A(MR)
MIL-P-47082(MI)
MIL-P-51431(EA)

FILAMENT WINDING

AMS3828
AMS3832
AMS3906
MIL-C-47257(MI)
MIL-HDBK-139(MU)
MIL-HDBK-700(MR)
MIL-P-82540(1)(OS)
MIL-R-60346B

FILM-FORMING MATERIAL

AMS3684
L-P-1041A
MIL-HDBK-139(MU)
MIL-P-47136(MI)

FINISH

AMS3823B
AMS3824
AMS3825
AMS3826
AMS3832
AMS3835
AMS3835
AMS3837
AMS3846
AMS3902A
ANSI/ASTM D2409-67(1)
ANSI/ASTM D2660-70(1)
ASTM D2150-70(1977)
ASTM D2408-67(1972)
ASTM D2410-67(1972)
MIL-F-46885(MI)
MIL-G-46887(MI)
MIL-G-47296(MI)
MIL-HDBK-700(MR)
MIL-M-46861(2)(MI)
MIL-P-25395A(ASG)
MIL-P-25515C
MIL-P-25515C
MIL-P-25518A(1)(ASG)
MIL-P-82540(1)(OS)
MIL-R-60346B
MIL-Y-83371(USAF)

FLAME RETARDANT MATERIAL

ASTM D1636-75A
ASTM D3013-77
ASTM D700-75
ASTM D704-62(1975)
ASTM D709-78
MIL-G-55636B(1)
MIL-M-14G
MIL-M-24519(1)(Navy)
MIL-M-47137(MI)
MIL-P-46160(MR)
MIL-P-46161A(MR)
MIL-P-997D(2)

MIL-R-21417A(SH)
MIL-R-21607D(SH)
MIL-T-1717D(SHIPS)

FLEXIBLE MATERIAL
AMS3358
AMS3359
AMS3360
AMS3361
AMS3362A
AMS3363B
AMS3364A
AMS3365B
AMS3366B
AMS3750
ANSI/ASTM D2953-71
J130
MIL-HDBK-699A(MR)
MIL-I-46877(1)(MI)
MIL-L-47274(1)(MI)
MIL-M-0024041B(SHIPS)
MIL-N-18352A(NOrd)
MIL-P-22096B
MIL-P-47136(MI)
MIL-P-47298(2)(MI)
MIL-P-83455(USAF)
MIL-S-83384(USAF)

FLOW TEMPERATURE
ANSI/ASTM D707-78
ASTM D1562-72
ASTM D706-78
L-P-349C

FLUID RESISTANT MATERIAL
ANSI/ASTM D1784-78
ARP765
ASTM D2000-77A
MIL-HDBK-149A
MIL-P-46120A(MR)
MIL-R-25042B
MIL-R-7575C(2)
MIL-R-81090(Wep)
MIL-R-83248(1)
MIL-R-83309(USAF)
MIL-R-83484/1(USAF)
MIL-R-83485/1(USAF)
MIL-R-9300B

FLUORESCENT MATERIAL
MIL-I-46058C(4)

FLUOROCARBON
ANSI/ASTM D2116-78
ASTM D1457-78
ASTM D2897-72
ASTM D3159-78A
ASTM D3275-78
ASTM D3307-78
L-P-389A(1)
MIL-C-47083(MI)
MIL-HDBK-149A
MIL-HDBK-700(MR)
MIL-R-83248(1)
MIL-R-83485/1(USAF)

FLUOROSILICONE
MIL-HDBK-149A
MIL-P-83455(USAF)

FOAM
AMS3751
MIL-HDBK-139(MU)
MIL-HDBK-700(MR)

FUNGUS-RESISTANT MATERIAL
L-C-530B(1)
L-P-393A(2)
MIL-C-47233(MI)
MIL-P-46115B(MR)
MIL-P-46129A(MR)
MIL-P-46131B(MR)
MIL-P-47082(MI)
MIL-P-81390
MIL-S-83384(USAF)

FUSES
MIL-C-15567(1)(OS)
MIL-P-21347D

GASKETS
AMS3651C
AMS3842A
AMS3843A
L-P-385C
L-P-385C
MIL-HDBK-149A
MIL-HDBK-149A
MIL-HDBK-699A(MR)
MIL-P-46122B(MR)

GEARS
ASTM D709-78
L-P-392A
MIL-M-20693B
MIL-P-46174(MR)

GLASS FABRIC
AMS3821
AMS3822

AMS3823B
AMS3824
AMS3825
AMS3826
AMS3827
AMS3830
AMS3835
AMS3835
AMS3837
AMS3845
AMS3845/1
AMS3845/2
AMS3867
AMS3867/1
AMS3867/1
AMS3867/2
AMS3867/2
AMS3867/3
AMS3867/3
ANSI/ASTM D2409-67(1)
ANSI/ASTM D2660-70(1)
ASTM D1668-73
ASTM D2150-70(1977)
ASTM D2408-67(1972)
ASTM D2410-67(1972)
ASTM D709-78
ISO2113-1972
L-P-383(1)
MIL-F-46885(MI)
MIL-F-47018A(MI)
MIL-F-47078(MI)
MIL-F-47079(MI)
MIL-G-55636B(1)
MIL-L-52696(ME)
MIL-P-25395A(ASG)
MIL-P-25421B
MIL-P-25515C
MIL-P-25518A(1)(ASG)
MIL-P-43043B(MR)
MIL-P-82540(1)(OS)
MIL-P-997D(2)
MIL-R-25042B
MIL-R-25042B
MIL-R-25506B(1)(AS)
MIL-R-7575C(2)
MIL-R-9299C
MIL-R-9300B

GLASS FIBER

AMS3906
AMS3906/1
AMS3906/2
AMS3906/3
AMS3906/4
AMS3906/5
AMS3906/6
AMS3906/7
ANSI/ASTM D1201-62
ANSI/ASTM D2853-70(1)
ANSI/ASTM D3011-74(1)
ASTM D1636-75A
ASTM D2897-72
ASTM D2948-72
ASTM D3220-73
ASTM D700-75
ASTM D704-62(1975)
L-P-383(1)
L-P-395C
MIL-C-47083(MI)
MIL-HDBK-700(MR)
MIL-L-52696(ME)
MIL-M-14G
MIL-M-19887A(SHIPS)
MIL-M-24325(2)(SHIPS)
MIL-M-24519(1)(NAVY)
MIL-M-46861(2)(MI)
MIL-M-46862(MI)
MIL-P-17549C(2)(SHIP)
MIL-P-21347D
MIL-P-23943(1)(AS)
MIL-P-25395A(ASG)
MIL-P-25421B
MIL-P-25515C
MIL-P-25515C
MIL-P-25518A(1)(ASG)
MIL-P-43043B(MR)
MIL-P-46069(2)(MU)
MIL-P-46109C(MR)
MIL-P-46131B(MR)
MIL-P-46137A(MR)
MIL-P-46161A(MR)
MIL-P-46169A(MR)
MIL-P-46174(MR)
MIL-P-46892(MI)
MIL-P-47134(MI)
MIL-P-81390
MIL-P-82540(1)(OS)
MIL-P-82650(OS)
MIL-R-21607D(SH)
MIL-R-46068(MU)
MIL-R-60346B

GLASS ROVING

AMS3828
AMS3832
ASTM D2150-70(1977)

ISO2797-1974
MIL-C-47257(MI)
MIL-G-46887(MI)
MIL-G-47024(MI)
MIL-G-47296(MI)
MIL-M-46861(2)(MI)
MIL-P-239431(1)(AS)
MIL-P-43043B(MR)
MIL-P-82540(1)(OS)
MIL-P-82650(OS)
MIL-R-60346B

GRAPHITE
ASTM D2897-72
MIL-P-47134(MI)
MIL-Y-83371(USAF)

GRAPHITE FIBER
AMS3616
AMS3894/1
AMS3894/2
AMS3894/3
AMS3894/4
AMS3894/5
AMS3894/6
AMS3894/7
AMS3894/8
AMS3894/9
AMS3894B
MIL-G-83410(USAF)

HAND LAYUP
AMS3867
AMS3894B
AMS3906
MIL-G-83410(USAF)
MIL-HDBK-139(MU)
MIL-HDBK-700(MR)

HEAT CURABLE MATERIAL
AMS3368
AMS3369
AMS3372
AMS3616
AMS3618
AMS3619
AMS3684
AMS3735A
AMS3821
AMS3822
AMS3827
AMS3828
AMS3832
MIL-C-46866(MI)
MIL-I-46879(MI)
MIL-P-46892(MI)
MIL-S-83474(USAF)

HEAT RESISTANT MATERIAL
AMI12
AMS3358
AMS3359
AMS3360
AMS3361
AMS3362A
AMS3363B
AMS3364A
AMS3365B
AMS3366B
AMS3616
AMS3618
AMS3619
AMS3650A
AMS3651C
AMS3684
AMS3822
AMS3826
AMS3830
AMS3845
AMS3845/1
AMS3845/2
AMS3846
AMS3847
AMS3858A
AMS3894/1
AMS3894/2
AMS3894/6
AMS3894/9
AMS3903/1
AMS3903/2
AMS3903/3
AMS3903/4
AMS3906/3
AMS3906/4
AMS3906/5
ARP765
ASTM D2000-77A
ASTM D2408-67(1972)
ASTM D700-75
ASTM D704-62(1975)
ASTM D709-78
ISO2112-1977
ISO800-1977
L-P-380C
L-P-385C
MIL-C-46866(MI)
MIL-C-47016(MI)

MIL-C-47027(MI)
MIL-C-47071(MI)
MIL-C-47153(MI)
MIL-C-47221(MI)
MIL-C-82644(OS)
MIL-F-46885(MI)
MIL-F-47079(MI)
MIL-G-47024(MI)
MIL-G-83410(USAF)
MIL-I-46879(MI)
MIL-I-47151(MI)
MIL-M-14G
MIL-M-14G
MIL-M-24325(2)(SHIPS)
MIL-M-46891(MI)
MIL-M-47026(MI)
MIL-M-47137(MI)
MIL-P-19735B
MIL-P-25395A(ASG)
MIL-P-25421B
MIL-P-25515C
MIL-P-25515C
MIL-P-25770A(ASG)
MIL-P-3409
MIL-P-43038B(MR)
MIL-P-46115B(MR)
MIL-P-46120A(MR)
MIL-P-46129A(MR)
MIL-P-46133A(MR)
MIL-P-46174(MR)
MIL-P-997D(2)
MIL-R-25042B
MIL-R-25042B
MIL-R-83248(1)
MIL-R-83330
MIL-R-83485/1(USAF)
MIL-R-9300B
ZZ-R-765B(1)

HIGH FREQUENCY APPLICATION
AMS3650A
AMS3651C
MIL-P-3409
MIL-P-997D(2)

HIGH MODULUS FIBER
AMS3867/1
AMS3867/2
AMS3894B
AMS3902A
MIL-Y-83370A(USAF)
MIL-Y-83371(USAF)

HYDRAZINE-RESISTANT
MIL-R-83412A(USAF)

HYDROLYTIC REVERSION
MIL-P-83455(USAF)
MIL-S-83384(USAF)

IMINE
MIL-L-47274(1)(MI)

IMPACT-RESISTANT MATERIAL
AMS3650A
ANSI/ASTM D1201-62
ANSI/ASTM D787-78
ASTM D2146-77
ASTM D700-75
ASTM D704-62(1975)
ISO2112-1977
ISO800-1977
L-P-1035A
L-P-1183B
L-P-393A(2)
L-P-394B
L-P-395C
L-P-397C
L-P-398B
MIL-C-47153(MI)
MIL-M-14G
MIL-M-24325(2)(SHIPS)
MIL-N-18352A(NOrd)
MIL-P-21347D
MIL-P-22096B
MIL-P-46074B(MR)
MIL-P-46124A(MR)
MIL-P-81390

INFRARED SPECTRUM
AMS3616
AMS3619
MIL-C-47221(MI)
MIL-C-81247(WP)
MIL-L-47106(MI)
MIL-L-47274(1)(MI)
MIL-P-46174(MR)
MIL-P-48296(PA)
MIL-P-51431(EA)
MIL-R-47252(MI)

INJECTION MOLDING MATERIAL
AMI12
AMS3650A
ANSI/ASTM D1430-78
ANSI/ASTM D1788-78A
ANSI/ASTM D2116-78

ANSI/ASTM D2133-78
ANSI/ASTM D2287-78
ANSI/ASTM D2853-70(1)
ANSI/ASTM D3011-74(1)
ANSI/ASTM D707-78
ANSI/ASTM D787-78
ASTM D1431-78
ASTM D1562-72
ASTM D1892-78
ASTM D2848-71
ASTM D2897-72
ASTM D706-78
ASTM D729-78
ASTM D789-78
L-P-1183B
L-P-380C
L-P-380C
L-P-385C
L-P-385C
L-P-392A
L-P-397C
L-P-399B
MIL-C-47027(MI)
MIL-HDBK-139(MU)
MIL-HDBK-700(MR)
MIL-M-19887A(SHIPS)
MIL-M-20693B
MIL-M-3165A
MIL-P-21347D
MIL-P-22096B
MIL-P-22985B
MIL-P-3409
MIL-P-46074B(MR)
MIL-P-46109C(MR)
MIL-P-46115B(MR)
MIL-P-46124A(MR)
MIL-P-46129A(MR)
MIL-P-46131B(MR)
MIL-P-46137A(MR)
MIL-P-46160(MR)
MIL-P-46892(MI)

INSULATING ADHESIVE
MIL-C-47171(MI)
MIL-L-47274(1)(MI)
MIL-R-47252(MI)

INSULATION (i.e., ELECTRICAL)
AMS3650A
AMS3651C
ANSI/ASTM D1248-74
ASTM D709-78
BS 3412-1976
L-P-385C
L-P-385C
MIL-C-47153(MI)
MIL-C-47233(MI)
MIL-C-47272(MI)
MIL-F-47018A(MI)
MIL-HDBK-149A
MIL-P-46122B(MR)
MIL-P-46161A(MR)
MIL-P-47082(MI)
MIL-P-997D(2)
MIL-R-46896(MI)

IONOMER
MIL-P-46124A(MR)

ISOCYANATE TERMINATED
MIL-L-47108(MI)

ISOPRENE
ASTM D1418-79

JACKETING MATERIAL
ANSI/ASTM D1248-74
L-P-390C(1)
MIL-HDBK-139(MU)
MIL-M-20693B
MIL-N-18352A(NOrd)
MIL-P-22096B

LAMINATE
AMS3821
AMS3822
AMS3823B
AMS3824
AMS3825
AMS3826
AMS3827
AMS3830
AMS3835
AMS3835
AMS3837
AMS3845
AMS3845/1
AMS3845/2
AMS3846
AMS3847
AMS3858A
AMS3894B
AMS3902A
ANSI/ASTM D2409-67(1)
ANSI/ASTM D2660-70(1)
ASTM D2150-70(1977)
ASTM D2408-67(1972)

ASTM D2410-67(1972)
ASTM D709-78
L-P-383(1)
MIL-C-47027(MI)
MIL-F-46885(MI)
MIL-F-47018A(MI)
MIL-F-47078(MI)
MIL-F-47079(MI)
MIL-G-46887(MI)
MIL-G-83410(USAF)
MIL-HDBK-139(MU)
MIL-P-17549C(2)(SHIP)
MIL-P-25395A(ASG)
MIL-P-25515C
MIL-P-25515C
MIL-P-25518A(1)(ASG)
MIL-P-25770A(ASG)
MIL-P-43038B(MR)
MIL-P-43043B(MR)
MIL-P-997D(2)
MIL-R-21607D(SH)
MIL-R25042B
MIL-R-25042B
MIL-R-25506B(1)(AS)
MIL-R-83330
MIL-R-9300B
MIL-T-1717D(SHIPS)

LAMINATING MATERIAL
AMS3616
AMS3618
AMS3619
MIL-F-47018A(MI)
MIL-HDBK-139(MU)
MIL-HDBK-700(MR)
MIL-L-52696(ME)
MIL-R-3745
MIL-R-83330
MIL-R-9299C

LOW PRESSURE LAMINATING
AMS3821
AMS3827
AMS3867
L-P-383(1)
MIL-G-83410(USAF)
MIL-P-25395A(ASG)
MIL-P-25421B
MIL-P-25518A(1)(ASG)
MIL-P-25770A(ASG)
MIL-P-43038B(MR)
MIL-P-43043B(MR)
MIL-R-21607D(SH)
MIL-R-25042B
MIL-R-25042B
MIL-R-25506B(1)(AS)
MIL-R-7575C(2)
MIL-R-9300B

LOW TEMPERATURE CURE
AMS3906/6
AMS3906/7
MIL-C-47072(MI)
MIL-L-52696(ME)
MIL-R-21607D(SH)

LOW-TEMPERATURE-RESISTANT MATERIAL
AMI12
AMS3358
AMS3359
AMS3360
AMS3361
AMS3362A
AMS3363B
AMS3364A
AMS3365B
AMS3366B
ANSI/ASTM D787-78
ARP765
ASTM D2146-77
L-P-385C
L-P-385C
L-P-394B
MIL-C-46866(MI)
MIL-P-3409
MIL-P-46124A(MR)
MIL-P-46133A(MR)
MIL-P-47082(MI)
MIL-P-47199(MI)
MIL-R-83485/1(USAF)
MIL-S-21923(NOrd)(2)
ZZ-R-765B(1)

MACHINABILITY
AMS3740A
ASTM D709-78
MIL-C-82644(OS)

MARINE APPLICATION
MIL-L-52696(ME)
MIL-M-0024041B(SHIPS)
MIL-P-17549C(2)(SHIP)
MIL-R-21607D(SH)

MAT
ANSI/ASTM D1201-62
L-P-383(1)

MIL-M-46861(2)(MI)
MIL-P-17549C(2)(SHIP)
MIL-P-25395A(ASG)
MIL-P-25421B
MIL-P-25515C
MIL-P-25515C
MIL-P-25518A(1)(ASG)
MIL-P-43043B(MR)

MECHANICAL STRENGTH
AMI12
AMS3822
AMS3826
AMS3828
AMS3845
AMS3845/1
AMS3845/2
AMS3847
AMS3867
AMS3894B
AMS3903
AMS3906
AMS3906
AMS3906/2
AMS3906/4
ANSI/ASTM D1201-62
ASTM D1431-78
ASTM D1457-78
ASTM D709-78
ASTM D709-78
L-P-1035A
L-P-395C
MIL-C-47027(MI)
MIL-C-47224(MI)
MIL-C-47233(MI)
MIL-M-46861(2)(MI)
MIL-M-47026(MI)
MIL-M-47137(MI)
MIL-P-21347D
MIL-P-25515C

MELAMINE
ASTM D709-78
MIL-M-14G

MELAMINE–FORMALDEHYDE
ASTM D704-62(1975)
ISO2112-1977

MELT FLOW RATE
ANSI/ASTM D1248-74
ANSI/ASTM D2116-78
ANSI/ASTM D2133-78
ANSI/ASTM D2581-73
ASTM D1431-78
ASTM D3275-78
BS 3412-1976
ISO 1622-1975
ISO 1872-1972(E)
L-P-390C(1)
L-P-399B
MIL-P-46120A(MR)
MIL-P-46124(MR)

MELT PROCESS
ANSI/ASTM D2116-78
ASTM D3159-78A
ASTM D3222-78
ASTM D3275-78

METHACRYLATE
ANSI/ASTM D788-78A
L-P-380C
L-P-380C
MIL-P-19735B

MINERAL-FILLED
ANSI/ASTM D1201-62
ASTM D1636-75A
ASTM D700-75
ASTM D704-62(1975)
ISO800-1977
MIL-C-47224(MI)
MIL-M-14G
MIL-M-24325(2)(SHIPS)
MIL-M-46891(MI)
MIL-M-47137(MI)

MOISTURE-RESISTANT MATERIAL
ASTM D700-75
ASTM D709-78
L-P-1041A
L-P-385C
MIL-C-46866(MI)
MIL-C-47163(MI)
MIL-C-47233(MI)
MIL-C-47272(MI)
MIL-C-82644(OS)
MIL-HDBK-149A
MIL-I-46877(1)(MI)
MIL-I-46879(MI)
MIL-I-47151(MI)
MIL-M-0024041B(SHIPS)
MIL-M-14G
MIL-M-20693B
MIL-M-47137(MI)
MIL-P-3409

MIL-P-46115B(MR)
MIL-P-47082(MI)

MOLDING

AMI12
AMS3616
AMS3650A
AMS3651C
AMS3821
AMS3822
AMS3858A
ANSI/ASTM D1201-62
ANSI/ASTM D1248–74
ANSI/ASTM D1788-78A
ANSI/ASTM D2116-78
ANSI/ASTM D2133-78
ANSI/ASTM D2287-78
ANSI/ASTM D2473-78
ANSI/ASTM D2581-73
ANSI/ASTM D2853-70(1)
ANSI/ASTM D2874-70(1)
ANSI/ASTM D705-62(19)
ANSI/ASTM D707-78
ANSI/ASTM D787-78
ANSI/ASTM D788-78A
ASTM D1431-78
ASTM D1457-78
ASTM D1562-72
ASTM D1636-75A
ASTM D1892-78
ASTM D2146-77
ASTM D2848-71
ASTM D2948-72
ASTM D3013-77
ASTM D3159-78A
ASTM D3220-73
ASTM D3221-78
ASTM D3222-78
ASTM D3275-78
ASTM D3307-78
ASTM D700-75
ASTM D703-78
ASTM D704-62(1975)
ASTM D706-78
ASTM D729-78
ASTM D789-78
BS 3412-1976
ISO 1622-1975
ISO2112-1977
ISO800-1977
L-P-1035A
L-P-1041A
L-P-1183B
L-P-349C
L-P-380C
L-P-380C
L-P-385C
L-P-385C
L-P-389A(1)
L-P-390C(1)
L-P-392A
L-P-393A(2)
L-P-394B
L-P-395C
L-P-396B
L-P-397C
L-P-398B
L-P-399B
MIL-C-46866(MI)
MIL-C-47027(MI)
MIL-C-47028(MI)
MIL-C-47083(MI)
MIL-C-47221(MI)
MIL-C-47224(MI)
MIL-HDBK-139(MU)
MIL-HDBK--149A
MIL-HDBK--700(MR)
MIL-L-47108(MI)
MIL-M-0024041B(SHIPS)
MIL-M-14G
MIL-M-20693B
MIL-M-24325(2)(SHIPS)
MIL-M-24519(1)(NAVY)
MIL-M-3165A
MIL-M-46861(2)(MI)
MIL-M-46862(MI)
MIL-M-46891(MI)
MIL-M-47026(MI)
MIL-M-47137(MI)
MIL-N-18352A(NOrd)
MIL-P-16617B(OS)
MIL-P-19735B
MIL-P-21347D
MIL-P-22096B
MIL-P-22985B
MIL-P-23943(1)(AS)
MIL-P-3409
MIL-P-43038B(MR)
MIL-P-43043B(MR)
MIL-P-43081B(MR)
MIL-P-46069(2)(MU)
MIL-P-46074B(MR)
MIL-P-46109C(MR)
MIL-P-46115B(MR)

MIL-P-46120A(MR)
MIL-P-46124A(MR)
MIL-P-46129A(MR)
MIL-P-46131B(MR)
MIL-P-46133A(MR)
MIL-P-46137A(MR)
MIL-P-46160(MR)
MIL-P-46161A(MR)
MIL-P-46169A(MR)
MIL-P-46174(MR)
MIL-P-46892(MI)
MIL-P-47082(MI)
MIL-P-47134(MI)
MIL-P-47199(MI)
MIL-P-47298(2)(MI)
MIL-P-51431(EA)
MIL-P-81390
MIL-P-82650(OS)
MIL-R-83322
MIL-R-83485/1(USAF)
MIL-S-21923(NORD)(2)
MIL-S-83474(USAF)

NATURAL RUBBER
ASTM D1418-79

NITRILE
ASTM D1418-79
MIL-HDBK-149A
MIL-R-83309(USAF)

NITROGEN TETROXIDE-RESISTANT
MIL-R-83322

NOMENCLATURE
ANSI/ASTM D1566-78A
ASTM D1418-79
ASTM D1600-75
ASTM D3317-74A

NONCORROSIVE MATERIAL
AMS3362A
AMS3363B
AMS3364A
AMS3365B
AMS3366B
AMS3734
AMS3735A
AMS3736
AMS3737
AMS3738A
AMS3739A
AMS3740A
AMS3750
MIL-G-83410(USAF)
MIL-I-46058C(4)
MIL-P-25770A(ASG)
MIL-R-21931A(OS)
MIL-R-25042B
MIL-R-25506B(1)(AS)
MIL-R-83330

NONMAGNETIC MATERIAL
MIL-P-21347D

NONTOXIC MATERIAL
L-P-1035A

NUCLEAR REACTOR
ANSI/ASTM D2953-71

NYLON
ASTM D2897-72
ASTM D709-78
ASTM D789-78
ISO/R1874-1971
L-P-395C
MIL-HDBK-700(MR)
MIL-M-19887A(SHIPS)
MIL-M-20693B
MIL-N-18352A(NOrd)
MIL-P-22096B

O-RINGS
MIL-HDBK-149A
MIL-HDBK-699A(MR)
MIL-R-83248(1)
MIL-R-83322
MIL-R-83485/1(USAF)

OPTICAL APPLICATION
L-P-349C
L-P-397C
MIL-P-46124A(MR)

ORGANIC FIBER
AMS3902A
AMS3903
AMS3903/1
AMS3903/2
AMS3903/3AMS3903/4
AMS3903/5
AMS3903/6
AMS3903/7
AMS3903/8
ASTM D3317-74A
MIL-Y-83370A(USAF)

ORGANOSOLS
ANSI/ASTM D1755-78

OZONE-RESISTANT MATERIAL
ARP765
MIL-P-47082(MI)
MIL-R-83285(USAF)
MIL-S-21923(NORD)(2)

PAINT
MIL-R-21417A(SH)

PHENOL–FORMALDEHYDE
ASTM D700-75
MIL-P-47134(MI)
MIL-R-3745

PHENOLIC
AMS3823B
AMS3830
AMS3837
AMS3858A
ASTM D700-75
ASTM D709-78
ISO800-1977
MIL-C-47028(MI)
MIL-C-47071(MI)
MIL-F-46885(MI)
MIL-F-47018A(MI)
MIL-G-47024(MI)
MIL-HDBK-700(MR)
MIL-M-14G
MIL-46891(MI)
MIL-M-47026(MI)
MIL-P-16617B(OS)
MIL-P-23943(1)(AS)
MIL-P-25515C
MIL-P-25515C
MIL-P-25770A(ASG)
MIL-P-82650(OS)
MIL-R-15189A
MIL-R-9299C

PHENOLIC COMPATIBLE
AMS3823B
AMS3837
ASTM D2408-67(1972)
ASTM D2410-67(1972)
MIL-F-46885(MI)

PHENYLSILANE
MIL-F-47079(MI)

PIPE
ANSI/ASTM D1248-74
ANSI/ASTM D2874-70(1)
BS 3412-1976
L-P-1035A
MIL-HDBK-139(MU)
MIL-P-46120A(MR)

PLASTISOL
ANSI/ASTM D1755-78
ISO 1060-1975
MIL-HDBK-139(MU)

POLSULFIDE
ASTM D2000-77A

POLY(ARYL SULFONE ETHER)
MIL-P-46133A(MR)

POLY(METHYL METHACRYLATE)
ANSI/ASTM D788-78A

POLY(VINYL CHLORIDE COVINYL ACETATE)
L-P-1035A

POLY(VINYL CHLORIDE)
ANSI/ASTM D1755-78
ANSI/ASTM D1784-78
ANSI/ASTM D2287-78
ISO1060-1975
L-P-1035A
L-P-1041A
MIL-P-47136(MI)

POLY(VINYLIDENE CHLORIDE)
L-P-1041A

POLY(VINYLIDENE FLUORIDE)
ASTM D3222-78
MIL-P-46122B(MR)

POLYACETAL
MIL-C-47027(MI)

POLYAMIDE
AMS3845
ASTM D2897-72
ASTM D789-78
ISO/R1874-1971
MIL-HDBK-700(MR)
MIL-M-19887A(SHIPS)
MIL-M-20693B
MIL-N-18352A(NOrd)
MIL-P-22096B

POLYBUTADIENE
MIL-L-47105(MI)
MIL-P-23942(1)(AS)
MIL-P-82658(1)(OS)

POLYBUTYLENE
ANSI/ASTM D2581-73

POLYCARBONATE
ANSI/ASTM D2473-78
ASTM D2848-71
L-P-393A(2)
MIL-C-47027(MI)
MIL-HDBK-700(MR)
MIL-P-81390
AMS3650A
ANSI/ASTM D1430-78
ASTM D1418-79
L-P-385C
L-P-385C

POLYDICHLOROSYTRENE
MIL-P-3409

POLYESTER
AMS3823B
AMS3826
AMS3827
ANSI/ASTM D1201-62
ASTM D2150-70(1977)
L-P-383(1)
MIL-HDBK-700(MR)
MIL-M-14G
MIL-M-24519(1)(NAVY)
MIL-P-17549C(2)(SHIP)
MIL-P-25395A(ASG)
MIL-P-43038B(MR)
MIL-P-43043B(MR)
MIL-P-46169A(MR)
MIL-P-82540(1)(OS)
MIL-R-21607D(SH)
MIL-R-25042B
MIL-R-25042B
MIL-R-46068(MU)
MIL-R-7575C(2)
MIL-R-82483A(OS)

POLYESTER COMPATIBLE
AMS3823B
AMS3825
AMS3835
AMS3835
ANSI/ASTM D2409-67(1)
ANSI/ASTM D2660-70(1)
ASTM D2150-70(1977)
ASTM D2410-67(1972)
MIL-P-82540(1)(OS)
MIL-R-60346B

POLYETHYLENE
ANSI/ASTM D1248-74
ANSI/ASTM D2853-70(1)
ASTM D1418-79
ASTM D2647-78
ASTM D2952-71(1977)
BS 3412-1976
ISO1872-1972(E)
L-C-530B(1)
L-P-390C(1)
MIL-HDBK-700(MR)
MIL-P-43081B(MR)
MIL-P-51431(EA)

POLYIMIDE
AMS3616
AMS3618
AMS3619
AMS3684
AMS3845/1
AMS3845/2
AMS3847
MIL-G-55636B(1)
MIL-R-83330

POLYIMIDE COMPATIBLE
MIL-Y-83370A(USAF)
MIL-Y-83371(USAF)

POLYISOBUTYLENE
MIL-P-13298A(MU)
MIL-P-14536(Ord)

POLYPHENYLENE OXIDE
ANSI/ASTM D2874-70(1)
MIL-P-46115B(MR)
MIL-P-46129A(MR)
MIL-P-46131B(MR)

POLYPHENYLENE SULFIDE
MIL-P-46174(MR)

POLYPROPYLENE
ANSI/ASTM D2853-70(1)
ASTM D2146-77
L-P-394B
MIL-HDBK-700(MR)
MIL-P-46109C(MR)

POLYSTYRENE
ANSI/ASTM D3011-74(1)
ASTM D703-78
ISO1622-1975
L-P-396B

MIL-HDBK-700(MR)
MIL-P-21347D

POLYSULFIDE
MIL-E-47231(MI)
MIL-L-51149(MI)

POLYSULFONE
MIL-P-46120A(MR)

POLYTEREPHTHALATE
ASTM D3220-73
ASTM D3221-78
MIL-P-46160(MR)
MIL-P-46161A(MR)

POLYTETRAFLUOROETHYLENE
AMS3651C
AMS3842A
AMS3843A
ASTM D1457-78
MIL-P-47253(MI)
MIL-P-48296(PA)

POLYURETHANE
AMS3571
ASTM D1418-79
MIL-HDBK-149A
MIL-I-46058C(4)
MIL-M-0024041B(SHIPS)
MIL-P-47082(MI)
MIL-P-47298(2)(MI)

POTTING RESIN
AMS3358
AMS3359
AMS3360
AMS3361
AMS3362A
AMS3363B
AMS3364A
AMS3365B
AMS3366B
AMS3734
AMS3735A
AMS3736
AMS3737
AMS3738A
AMS3739A
AMS3740A
AMS3750
ASTM D1763-76
MIL-C-47153(MI)
MIL-C-47163(MI)
MIL-C-47224(MI)
MIL-C-47233(MI)
MIL-C-47272(MI)
MIL-E-47231(MI)
MIL-HDBK-699A(MR)
MIL-I-46865(MI)
MIL-I-46877(1)(MI)
MIL-I-47151(MI)
MIL-M-0024041B(SHIPS)
MIL-P-43081B(MR)
MIL-P-47199(MI)
MIL-P-47298(2)(MI)
MIL-P-83455(USAF)
MIL-R-46092(1)(MI)
MIL-S-83384(USAF)

PREPREG
AMS3821
AMS3822
AMS3826
AMS3827
AMS3828
AMS3830
AMS3832
AMS3845
AMS3845/1
AMS3845/2
AMS3847
AMS3858A
AMS3867
AMS3867/1
AMS3867/2
AMS3867/3
AMS3894/1
AMS3894/2
AMS3894/3
AMS3894/4
AMS3894/5
AMS3894/6
AMS3894/7
AMS3894/8
AMS3894/9
AMS3894B
AMS3903
AMS3903/1
AMS3903/2
AMS3903/3
AMS3903/4
AMS3903/5
AMS3903/6
AMS3903/7
AMS3903/8

AMSAMS3906
AMS3906/1
AMS3906/2
AMS3906/3
AMS3906/4
AMS3906/5
AMS3906/6
AMS3906/7
L-P-383(1)
MIL-C-47257(MI)
MIL-G-46887(MI)
MIL-G-47024(MI)
MIL-G-47296(MI)
MIL-G-55636B(1)
MIL-G-83410(USAF)
MIL-HDBK-139(MU)
MIL-M-46861(2)(MI)
MIL-M-46862(MI)
MIL-P-16617B(OS)
MIL-P-43043B(MR)
MIL-P-46069(2)(MU)
MIL-R-2504B
MIL-R-25042B
MIL-R-25506B(1)(AS)
MIL-R-7575C(2)
MIL-R-83330
MIL-R-9300B
MIL-S-83474(USAF)

PROCESSING
MIL-HDBK-139(MU)
MIL-HDBK-149A
MIL-HDBK-700(MR)

PROPELLANT
MIL-C-20301
MIL-C-47171(MI)
MIL-HDBK-149A
MIL-L-47105(MI)
MIL-L-47106(MI)
MIL-L-51149(MI)
MIL-P-23942(1)(AS)
MIL-P-23943(1)(AS)
MIL-P-46122B(MR)
MIL-P-82658(1)(OS)
MIL-R-47252(MI)
MIL-R-82657(OS)
MIL-82664(OS)
MIL-R-83412A(USAF)

PROPYLENE
ANSI/ASTM D2853-70(1)
ASTM D2146-77

PUNCHING
ASTM D709-78

QPL
MIL-G-55636B(1)
MIL-HDBK-699A(MR)
MIL-M-14G
MIL-M-24325(2)(SHIPS)
MIL-M-24519(1)(NAVY)
MIL-P-997D(2)

QUARTZ
AMS3846
AMS3847
MIL-C-47272(MI)
MIL-M-47026(MI)
MIL-P-47134(MI)

RADAR APPLICATION
L-P-383(1)
MIL-P-25395A(ASG)
MIL-P-25421B
MIL-P-25518A(1)(ASG)
MIL-P-997D(2)
MIL-R-25506B(1)(AS)
MIL-R-7575C(2)
MIL-R-9300B

RADIO APPLICATION
ASTM D709-78
L-P-383(1)
MIL-P-25395A(ASG)
MIL-P-25421B
MIL-P-25518A(1)(ASG)
MIL-P-997D(2)
MIL-R-25506B(1)(AS)
MIL-R-7575C(2)
MIL-R-9300B

REINFORCEMENT
AMS3821
AMS3822
AMS3823B
AMS3824
AMS3825
AMS3826
AMS3827
AMS3828
AMS3835
AMS3835
AMS3837
AMS3842A
AMS3843A
AMS3845

AMS3845/1
AMS3845/2
AMS3846
AMS3847
AMS3858A
AMS3867
AMS3867/1
AMS3867/2
AMS3867/3
AMS3894/2
AMS3894/3
AMS3894/4
AMS3894/5
AMS3894/6
AMS3894/7
AMS3894/8
AMS3894/9
AMS3894B
AMS3902A
AMS3902A
AMS3903
AMS3903/1
AMS3903/2
AMS3903/3
AMS3903/4
AMS3903/5
AMS3903/6
AMS3903/7
AMS3903/8
AMS3906
AMS3906/1
AMS3906/2
AMS3906/3
AMS3906/4
AMS3906/5
AMS3906/6
AMS3906/7
ANSI/ASTM D2409-67(1)
ANSI/ASTM D2660-70(1)
ANSI/ASTM D2853-70(1)
ANSI/ASTM D3011-74(1)
ASTM D1636-75A
ASTM D1763-76
ASTM D2150-70(1977)
ASTM D2408-67(1972)
ASTM D2848-71
ASTM D2897-72
ASTM D2948-72
ASTM D3013-77
ASTM D3220-73
ASTM D3317-74A
ISO2113-1972
ISO2797-1974
L-P-383(1)
L-P-395C
MIL-C-47028(MI)
MIL-C-47083(MI)
MIL-F-46885(MI)
MIL-F-47018A(MI)
MIL-F-47078(MI)
MIL-F-47079(MI)
MIL-G-46887(MI)
MIL-G-47296(MI)
MIL-G-55636B(1)
MIL-G-83410(USAF)
MIL-HDBK-139(MU)
MIL-L-52696(ME)
MIL-M-24519(1)(NAVY)
MIL-M-46861(2)(MI)
MIL-M-46862(MI)
MIL-M-46891(MI)
MIL-M-47026(MI)
MIL-P-21347D
MIL-P-23943(1)(AS)
MIL-P-25395A(ASG)
MIL-P-25421B
MIL-P-25515C
MIL-P-43043B(MR)
MIL-P-46069(2)(MU)
MIL-P-46109C(MR)
MIL-P-46131B(MR)
MIL-P-46137A(MR)
MIL-P-46161A(MR)
MIL-P-46169A(MR)
MIL-P-46174(MR)
MIL-P-46892(MI)
MIL-P-47134(MI)
MIL-P-47253(MI)
MIL-P-81390
MIL-P-82650(OS)
MIL-R-46068(MU)
MIL-R-60346B
MIL-R-83330
MIL-S-83474(USAF)
MIL-Y-83371(USAF)

RIGID MATERIAL

AMS3734
AMS3735A
AMS3736
AMS3737
AMS3738A
AMS3739A
AMS3740A

ANSI/ASTM D1784-78
ANSI/ASTM D1788-78A
ANSI/ASTM D2953-71
ASTM D1431-78
ASTM D2146-77
L-P-1035A
L-P-399B
MIL-C-47027(MI)
MIL-C-47153(MI)
MIL-L-52696(ME)
MIL-M-19887A(SHIPS)
MIL-M-20693B
MIL-P-3409
MIL-P-46109C(MR)
MIL-R-46896(MI)
MIL-T-17171D(SHIPS)

ROCKET MOTOR
AMS3832
MIL-C-47171(MI)
MIL-G-46887(MI)
MIL-G-47024(MI)
MIL-G-47296(MI)
MIL-L-47105(MI)
MIL-L-47106(MI)
MIL-L-47108(MI)
MIL-L-47274(1)(MI)
MIL-M-46891(MI)
MIL-M-47026(MI)
MIL-P-23943(1)(AS)
MIL-P-47134(MI)
MIL-P-81255A
MIL-P-82650(OS)
MIL-R-475252(MI)
MIL-R-82664(OS)

ROD
AMS3650A
AMS3651C
MIL-M-14G
MIL-R-81090(Wep)

RODS
ASTM D709-78

ROOM-TEMPERATURE VULCANIZING
AMS3358
AMS3359
AMS3360
AMS3361
AMS3362A
AMS3363B
AMS3364A
AMS3365B
AMS3366B
AMS3370
AMS3371
MIL-C-47003(MI)
MIL-C-47163(MI)
MIL-HDBK-149A
MIL-HDBK-699A(MR)
MIL-P-47199(MI)
MIL-P-83455(USAF)
MIL-R-46092(1)(MI)
MIL-S-83384(USAF)

ROVING
AMS3828
AMS3832
ASTM D2150-70(1977)
ASTM D3317-74A
ISO2113-1972
ISO2797-1974
MIL-C-47257(MI)
MIL-G-46887(MI)
MIL-G-47024(MI)
MIL-G-47296(MI)
MIL-M-46861(2)(MI)
MIL-P-17549C(2)(SHIP)
MIL-P-23943(1)(AS)
MIL-P-82540(1)(OS)
MIL-P-82650(OS)
MIL-R-60346B
MIL-Y-83370A(USAF)
MIL-Y-83371(USAF)

SEALING MATERIAL
AMS3684
ARP765
MIL-C-82644(OS)
MIL-C-82644(OS)
MIL-E-47231(MI)
MIL-HDBK-139(MU)
MIL-HDBK-700(MR)
MIL-P-47298(2)(MI)
MIL-P-83455(USAF)
MIL-R-83322
MIL-R-83412A(USAF)

SELF-EXTINGUISHING MATERIAL
AMI12
AMS3821
AMS3827
ASTM D709-78
L-P-1035A
L-P-393A(2)
MIL-C-47153(MI)

MIL-C-47224(MI)
MIL-C-47233(MI)
MIL-I-46058C(4)
MIL-M-47137(MI)
MIL-N-18352A(NOrd)
MIL-P-3409
MIL-P-43038B(MR)
MIL-P-46115B(MR)
MIL-P-46120A(MRR)
MIL-P-46129A(MR)
MIL-P-46131B(MR)
MIL-P-46160(MR)
MIL-P-46161A(MR)
MIL-P-81390
MIL-P-81390
MIL-R-83330
MIL-S-83384(USAF)

SELF-LUBRICATING MATERIAL
MIL-M-20693B

SEMI-FLEXIBLE MATERIAL
MIL-C-47233(MI)

SHEET
AMS3650A
AMS3651C
AMS3842A
AMS3843A
AMS3858A
AMS3894/1
AMS3894/2
AMS3894/3
AMS3894/4
AMS3894/5
AMS3894/6
AMS3894/7
AMS3894/8
AMS3894/9
AMS3894B
AMS3906
AMS3906/1
AMS3906/2
AMS3906/3
AMS3906/4
AMS3906/5
AMS3906/6
AMS3906/7
ASTM D709-78
L-P-1035A
L-P-1183B
L-P-383(1)
MIL-C-15567(1)(OS)
MIL-G-55636B(1)
MIL-G-83410(USAF)
MIL-HDBK-139(MU)
MIL-M-14G
MIL-P-25421B
MIL-P-25515C
MIL-P-25515C
MIL-P-25770A(ASG)
MIL-P-46120A(MR)
MIL-P-46124A(MR)
MIL-P-46169A(MR)
MIL-P-47136(MI)
MIL-P-51431(EA)
MIL-R-81090(Wep)
MIL-R-83248(1)
MIL-R-83322
MIL-R-83485/1(USAF)
MIL-S-83474(USAF)
ZZ-R-765B(1)

SHOCK RESISTANT MATERIAL
ASTM D700-75
MIL-C-46866(MI)
MIL-C-47153(MI)
MIL-C-47163(MI)
MIL-C-47224(MI)
MIL-C-47233(MI)
MIL-C-47272(MI)
MIL-C-82644(OS)
MIL-M-47137(MI)
MIL-P-21347D

SILANE
MIL-P-82540(1)(OS)

SILICA
AMS3830
AMS3846
MIL-C-47028(MI)
MIL-C-47221(MI)
MIL-F-47018A(MI)
MIL-I-46865(MI)
MIL-M-47026(MI)

SILICONE
AMS3358
AMS3359
AMS3360
AMS3361
AMS3362A
AMS3363B
AMS3364A
AMS3368
AMS3369

AMS3370
AMS3371
AMS3372
ASTM D1418-79
ASTM D2000-77A
ASTM D709-78
MIL-C-47003(MI)
MIL-C-47016(MI)
MIL-C-47163(MI)
MIL-HDBK-149A
MIL-HDBK-700(MR)
MIL-I-46058C(4)
MIL-I-46877(1)(MI)
MIL-M-14G
MIL-P-25518A(1)(ASG)
MIL-P-47199(MI)
MIL-P-997D(2)
MIL-R-25506B(1)(AS)
MIL-R-46092(1)(MI)
MIL-S-83384(USAF)
ZZ-R-765B(1)

SPACE APPLICATION
MIL-HDBK-149A
MIL-R-83485/1(USAF)
MIL-S-83384(USAF)

STEAM-RESISTANT MATERIAL
L-P-385C
MIL-P-46115B(MR)
MIL-P-46129A(MR)
MIL-P-46131B(MR)
MIL-R-83285(USAF)

STRUCTURAL APPLICATION
AMS3823B
AMS3828
AMS3837
AMS3845
AMS3845/1
AMS3845/2
AMS3847
AMS3867
AMS3867/1
AMS3867/2
AMS3867/3
AMS3894/1
AMS3894/2
AMS3894/3
AMS3894/4
AMS3894/5
AMS3894/6
AMS3894/7
AMS3894/8
AMS3894/9
AMS3894B
AMS3902A
AMS3906/1
AMS3906/3
AMS3906/4
AMS3906/5
AMS3906/6
AMS3906/7
ASTM D2150-70(1977)
L-P-383(1)
MIL-C-47027(MI)
MIL-F-46885(MI)
MIL-F-47078(MI)
MIL-F-47079(MI)
MIL-G-46887(MI)
MIL-G-47024(MI)
MIL-G-47296(MI)
MIL-G-83410(USAF)
MIL-P-17549C(2)(SHIP)
MIL-P-23943(1)(AS)
MIL-P-25395A(ASG)
MIL-P-25421B
MIL-P-25515C
MIL-P-25515C
MIL-P-25518A(1)(ASG)
MIL-P-25770A(ASG)
MIL-P-43038B(MR)
MIL-P-46069(2)(MU)
MIL-P-46892(MI)
MIL-R-25042B
MIL-R-25042B
MIL-R-25506B(1)(AS)
MIL-R-46068(MU)
MIL-R-60346B
MIL-R-7575C(2)
MIL-R-83330
MIL-R-9299C
MIL-R-9300B
MIL-S-83474(USAF)

STYRENE
ANSI/ASTM D1788-78A
ASTM D1892-78
ASTM D703-78
MIL-R-82483A(OS)

STYRENE–ACRYLONITRILE
ANSI/ASTM D3011-74(1)
ASTM D1431-78
L-P-399B

STYRENE–BUTADIENE
ASTM D1418-79
ASTM D1892-78
L-P-398B
MIL-HDBK-149A
MIL-R-51209(MU)
MIL-S-21923(NORD)(2)

SURGICAL INSTRUMENTS
L-P-396B
MIL-P-46115B(MR)
MIL-P-46129A(MR)
MIL-P-46131B(MR)

TAPE
AMS3830
AMS3867
AMS3867/1
AMS3867/2
AMS3867/3
AMS3894/1
AMS3894/2
AMS3894/3
AMS3894/4
AMS3894/5
AMS3894/6
AMS3894/7
AMS3894/8
AMS3894/9
AMS3894B
AMS3906
AMS3906/1
AMS3906/2
AMS3906/3
AMS3906/4
AMS3906/5
AMS3906/6
AMS3906/7
BS 3412-1976
MIL-F-47018A(MI)
MIL-G-83410(USAF)
ZZ-R-765B(1)

TENSILE STRENGTH
AMS3894/2
AMS3894/3
AMS3894/4
AMS3894/6
AMS3894/7
AMS3903/1
AMS3903/2
AMS3903/3
AMS3903/4
AMS3903/5
AMS3903/6
AMS3903/7
AMS3903/8
ANSI/ASTM D2953-71
MIL-M-19887A(SHIPS)
MIL-P-22985B
MIL-R-60346B

TETRAFLUOROETHYLENE
ASTM D1418-79
ASTM D1457-78
ASTM D3159-78A

THERMAL INSULATION
AMS3830
MIL-C-47071(MI)
MIL-M-46891(MI)
MIL-M-47026(MI)
MIL-P-81255A

THERMOPLASTIC RESIN
AMI12
ANSI/ASTM D1755-78
ANSI/ASTM D1788-78A
ANSI/ASTM D2474-78
ANSI/ASTM D707-78
ANSI/ASTM D787-78
ASTM D1562-72
ASTM D1892-78
ASTM D3220-73
ASTM D3221-78
ASTM D706-78
ASTM D729-78
ISO1060-1975
ISO1872-1972(E)
J130
L-C-530B(1)
L-P-1041A
L-P-1183B
L-P-389A(1)
MIL-C-47027(MI)
MIL-HDBK-139(MU)
MIL-HDBK-700(MR)
MIL-M-24519(1)(NAVY)
MIL-P-3409
MIL-P-46120A(MR)
MIL-P-46122B(MR)
MIL-P-46133A(MR)
MIL-P-46160(MR)
MIL-P-46161A(MR)
MIL-P-47082(MI)
MIL-P-47253(MI)

THERMOSETTING RESIN
AMS3616
AMS3618
AMS3619
AMS3684
AMS3845
AMS3845/1
AMS3845/2
AMS3847
ANSI/ASTM D1201-62
ANSI/ASTM D705-62(19)
ASTM D1636-75A
ASTM D3013-77
ASTM D700-75
ASTM D704-62(1975)
ASTM D709-78
ISO2112-1977
L-C-530B(1)
MIL-C-47224(MI)
MIL-G-83410(USAF)
MIL-HDBK-139(MU)
MIL-HDBK-700(MR)
MIL-M-14G
MIL-M-24325(2)(SHIPS)
MIL-M-47137(MI)
MIL-P-16617B(OS)
MIL-P-23943(1)(AS)
MIL-P-25770A(ASG)
MIL-P-46892(MI)
MIL-P-997D(2)
MIL-R-21931A(OS)
MIL-R-25042B
MIL-R-25042B
MIL-R-25506B(1)(AS)
MIL-R-7575C(2)
MIL-R-9299C
MIL-R-9300B
MIL-S-83474(USAF)

TOW
MIL-G-83410(USAF)
MIL-Y-83371(USAF)

TRANSFER MOLDING
MIL-C-46866(MI)
MIL-C-47224(MI)
MIL-HDBK-139(MU)
MIL-HDBK-700(MR)
MIL-M-47137(MI)
MIL-P-46892(MI)

TRANSLUCENT MATERIAL
ANSI/ASTM D788-78A
ASTM D704-62(1975)
MIL-P-19735B
MIL-R-21607D(SH)

TRANSPARENT MATERIAL
AMS3368
AMS3370
L-P-1035A
L-P-380C
L-P-380C
L-P-396B
L-P-397C
MIL-C-47272(MI)
MIL-I-46058C(4)
MIL-I-46877(1)(MI)
MIL-P-19735B
MIL-S-83384(USAF)

TUBE
AMS3651C
ASTM D709-78
MIL-HDBK-699A(MR)
MIL-M-14G
MIL-R-81090(Wep)
ZZ-R-765B(1)

TYPE E GLASS
AMS3821
AMS3822
AMS3824
AMS3826
AMS3827
AMS3828
AMS3845
AMS3845/1
AMS3845/2
AMS3906/1
AMS3906/3
AMS3906/5
AMS3906/6
ANSI/ASTM D2409-67(1)
ANSI/ASTM D2660-70(1)
ASTM D2408-67(1972)
ASTM D2410-67(1972)
MIL-F-47018A(MI)
MIL-G-55636B(1)
MIL-P-46174(MR)
MIL-P-82650(OS)
MIL-R-60346B

TYPE S GLASS
AMS3832
AMS3837
AMS3906/2
AMS3906/4

AMS3906/7
MIL-R-60346B

ULTRAVIOLET-RESISTANT MATERIAL
L-P-380C
L-P-380C
MIL-P-46122B(MR)

UREA–FORMALDEHYDE
ANSI/ASTM D705-62(19)
ISO2112-1977
MIL-HDBK-700(MR)

URETHANE
ASTM D2000-77A
MIL-HDBK-700(MR)

VACUUM BAG MOULDING
AMS3821
AMS3822
AMS3826
AMS3827
AMS3906
MIL-HDBK-139(MU)
MIL-HDBK-700(MR)
MIL-R-21607D(SH)

VARNISH
MIL-R-15189A

VICAT SOFTENING POINT
ASTM D703-78
ISO1622-1975
L-P-396B

VINYL
ASTM D1418-79
MIL-HDBK-700(MR)

VINYL ACETATE
ANSI/ASTM D2474-78

VINYL CHLORIDE
ANSI/ASTM D1784-78
ANSI/ASTM D2287-78
ANSI/ASTM D2474-78
ASTM D729-78
ISO2798-1974

VINYL–SILANE
ANSI/ASTM D2409-67(1)

VINYLIDENE CHLORIDE
ANSI/ASTM D2474-78
ASTM D729-78
L-P-1041A

VISCOSITY
AMS3358
AMS3359
AMS3360
AMS3361
AMS3362A
AMS3363B
AMS3364A
AMS3365B
ISO/R1874-1971
MIL-I-46865(MI)
MIL-P-47199(MI)
MIL-R-82664(OS)
MIL-S-83384(USAF)

WEATHER-RESISTANT MATERIAL
AMS3858A
ANSI/ASTM D1248-74
ANSI/ASTM D2133-78
ANSI/ASTM D2133-78
ANSI/ASTM D2581-73
ANSI/ASTM D707-78
ASTM D1562-72
BS 3412-1976
L-C-530B(1)
L-P-380C
L-P-389A(1)
L-P-390C(1)
L-P-392A
L-P-394B
MIL-HDBK-700(MR)
MIL-M-20693B
MIL-P-46074B(MR)
MIL-P-47082(MI)
MIL-R-7575C(2)

YARN
AMS3835
ANSI/ASTM D2409-67(1)
ANSI/ASTM D2660-70(1)
ASTM D2408-67(1972)
ASTM D2410-67(1972)
ASTM D3317-74A
ISO2113-1972
MIL-G-83410(USAF)
MIL-Y-83370A(USAF)
MIL-Y-83371(USAF)

ZERO STRENGTH TIME
AMS3650A
L-P-385C
L-P-385C

APPENDIX I

Charts and Tables

1. Chemical Resistance Chart. Reprinted with permission of *Plastics World*, A Cahners Publication.
2. Engineering Thermoplastics—Properties Selection Guide. Reprinted with permission of Fiberfill.
3. How Plastics Are Derived. Reprinted with permission from *Plastics Technology*.
4. Temperature Conversion Chart.
5. Decimal Equivalent
6. Diameter and Areas of Circles.
7. Hardness Conversion Chart.
8. Measurement Conversions.
9. Thermal Conductivity Conversion.
10. Solvent Selection Guide.
11. Pressure Conversion Chart.
12. Properties Chart. Reprinted with permission from *Materials Engineering*, 1981; *Materials Selector*, Dec. 1980.

Guide to the effect Chemical Environments

PLASTIC MATERIAL / ENVIRONMENT	Aromatic solvents		Aliphatic solvents		Chlorinated solvents		Weak bases and salts		Strong bases		Strong acids		Strong oxidants		Esters and ketones		24-hr. Water absorption
temperature F	77	200	77	200	77	200	77	200	77	200	77	200	77	200	77	200	% change by weight
Acetals*	1-4	2-4	1	2	1-2	4	1-3	2-5	1-5	2-5	5	5	5	5	1	2-3	0.22 - 0.25
Acrylics	5	5	2	3	5	5	1	3	2	5	4	4-5	5	5	5	5	0.2 - 0.4
Acrylonitrile-Butadiene-Styrenes (ABS)	4	5	2	3-5	3-5	5	1	2-4	1	2-4	1-4	5	1-5	5	3-5	5	0.1 - 0.4
Aramids (aromatic polyamide)	1	1	1	1	1	1	2	3	4	5	3	4	2	5	1	2	0.6
Cellulose Acetates (CA)	2	3	2	3	3	4	2	3	3	5	3	5	3	5	5	5	2 - 7
Cellulose Acetate Butyrates (CAB)	4	5	1	3	3	4	2	4	3	5	3	5	3	5	5	5	0.9 - 2.0
Cellulose Acetate Propionates (CAP)	4	5	1	3	3	4	1	2	3	5	3	5	3	5	5	5	1.3 - 2.8
Diallyl Phthalates (DAP, filled)	1-2	2-4	2	3	2	4	2	3	2	4	1-2	2-3	2	4	3-4	4-5	0.2 - 0.7
Epoxies	1	2	1	2	1-2	3-4	1	1-2	1	2	2-3	3-4	4	4-5	2	3-4	0.01 - 0.10
Ethylene Copolymers (EVA) (Ethylene-Vinyl Acetates)	5	5	5	5	5	5	1	2	1	5	1	5	1	5	2	5	0.05 - 0.13
Fluorocarbons: Ethylene/Tetrafluoroethylene Copolymers (ETFE)	1	1	1	1	1	1	1	1	1	1	1	1	1	1	1	1	<0.03
Fluorocarbons: Fluorinated Ethylene Propylenes (FEP)	1	1	1	1	1	1	1	1	1	1	1	1	1	1	1	1	<0.01
Fluorocarbons: Perfluoroalkoxies (PFA)	1	1	1	1	1	1	1	1	1	1	1	1	1	1	1	1	<0.03
Fluorocarbons: Polychlorotrifluoroethylenes (CTFE)	1	1	1	1	3	4	1	1	1	1	1	1	1	1	1	1	0.01 - 0.10
Fluorocarbons: Polytetrafluoroethylenes (TFE)	1	1	1	1	1	1	1	1	1	1	1	1	1	1	1	1	0
Furans	1	1	1	1	1	1	2	2	2	2	1	1	5	5	1	1	0.01 - 0.20
Ionomers	2	4	1	4	4	4	1	4	1	4	2	4	1	5	1	4	0.1 - 1.4
Melamines (filled)	1	1	1	1	1	1	2	3	2	3	2	3	2	3	1	2	0.01 - 1.30
Nitriles (high barrier alloys of ABS or SAN)	1	4	1	2-4	1-4	2-5	1	2-4	1	2-4	2-5	5	3-5	5	1-5	5	0.2 - 0.5
Nylons	1	1	1	1	1	2	1	2	2	3	5	5	5	5	1	1	0.2 - 1.9
Phenolics (filled)	1	1	1	1	1	1	2	3	3	5	1	1	4	5	2	2	0.1 - 2.0
Polyallomers	2	4	2	4	4	5	1	1	1	1	1	3	1	4	1	3	<0.01

RATING CODE: 1. no effect or inert, **2.** slight effect, **3.** mild effect, **4.** softening or swelling, **5.** severe degradation.

MATERIALS

Some plastics incorporate a broad family of materials with widely differing properties in these environments. Special grades are also available with properties outside the ranges listed. For specific manufacturers, see PLASTICS WORLD Buyers Guide, Section 1.

ENVIRONMENT

Chemical effects in plastics are defined by conditions, which include stress cracking, etching or chemical attack (ASTM D-543), staining (ASTM D-2299) and swelling or solvating (ASTM D-543). These conditions are influenced by the time of exposure, temperature and concentrations or combination of these.

PLASTICS WORLD, 221 Columbus Ave., Boston, Mass. 02116 (617)536-7780

of
on Plastics

ENVIRONMENT / PLASTIC MATERIAL	Aromatic solvents		Aliphatic solvents		Chlorinated solvents		Weak bases and salts		Strong bases		Strong acids		Strong oxidants		Esters and ketones		24-hr. Water absorption
	temperature F																
	77	200	77	200	77	200	77	200	77	200	77	200	77	200	77	200	% change by weight
Polyamide-imides	1	1	1	1	2	3	1	1	3	4	2	3	2	3	1	1	0.22 - 0.28
Polyarylsulfones (PAS)	4	5	2	3	4	5	1	2	2	2	1	1	2	4	3	4	1.2 - 1.8
Polybutylenes (PB)	3	5	1	5	4	5	1	2	1	3	1	3	1	4	1	3	<0.01 - 0.3
Polycarbonates (PC)	5	5	1	1	5	5	1	5	5	5	1	1	1	1	5	5	0.15 - 0.35
Polyesters (thermoplastic)	2	5	1	3-5	3	5	1	3-4	2	5	3	4-5	2	3-5	2	3-4	0.06 - 0.09
Polyesters (thermoset-glass fiber filled)	1-3	3-5	2	3	2	4	2	3	3	5	2	3	2	4	3-4	4-5	0.01 - 2.50
Polyethylenes (LDPE-HDPE — low-density to high-density)	4	5	4	5	4	5	1	1	1	1	1-2	1-2	1-3	3-5	2	3	0.00 - 0.01
Polyethylenes (UHMWPE — ultra high molecular weight)	3	4	3	4	3	4	1	1	1	1	1	1	1	1	3	4	<0.01
Polyimides	1	1	1	1	1	1	2	3	4	5	3	4	2	5	1	1	0.3 - 0.4
Polyphenylene Oxides (PPO) (modified)	4	5	2	3	4	5	1	1	1	1	1	2	1	2	2	3	0.06 - 0.07
Polyphenylene Sulfides (PPS)	1	2	1	1	1	2	1	1	1	1	1	1	1	2	1	1	<0.05
Polyphenylsulfones	4	4	1	1	5	5	1	1	1	1	1	1	1	1	3	4	0.5
Polypropylenes (PP)	2	4	2	4	2-3	4-5	1	1	1	1	1	2-3	2-3	4-5	2	4	0.01 - 0.03
Polystyrenes (PS)	4	5	4	5	5	5	1	5	1	5	4	5	4	5	4	5	0.03 - 0.60
Polysulfones	4	4	1	1	5	5	1	1	1	1	1	1	1	1	3	4	0.2 - 0.3
Polyurethanes (PUR)	3	4	2	3	4	5	2-3	3-4	2-3	3-4	2-3	3-4	4	4	4	5	0.02 - 1.50
Polyvinyl Chlorides (PVC)	4	5	1	5	5	5	1	5	1	5	1	5	2	5	4	5	0.04 - 1.00
Polyvinyl Chlorides-Chlorinated (CPVC)	4	4	1	2	5	5	1	2	1	2	1	2	2	3	4	5	0.04 - 0.45
Polyvinylidene Fluorides (PVDF)	1	1	1	1	1	1	1	1	1	2	1	2	1	2	3	5	0.04
Silicones	4	4	2	3	4	5	1	2	4	5	3	4	4	5	2	4	0.1 - 0.2
Styrene Acrylonitriles (SAN)	4	5	3	4	3	5	1	3	1	3	1	3	3	4	4	5	0.20 - 0.35
Ureas (filled)	1	3	1	3	1	3	2	3	2	3	4	5	2	3	1	2	0.4 - 0.8
Vinyl Esters (glass fiber filled)	1	3	1-2	2-4	1-2	4	1	3	1	3	1	2	2	3	3-4	4-5	0.01 - 2.50

TEMPERATURE
In many cases the difference between the 77F and 200F ratings is simply a heat distortion effect with no degradation.

WATER ABSORPTION
In ordinary tap and salt water there is almost no chemical attack, however, ion-free water (such as distilled or deionized water) can show varying degrees of attack. Water absorption values have been included to quantify the most common effect of water.

*Widespread ratings exist since acetal copolymers are more inert than homopolymers in many "corrosive" environments.

2DVR77

Prepared by Plastics World and presented by Shell Chemical Co.

Plastalloy® Engineering Thermoplastics–Properties Selection Guide

					MECHANICAL								THERMAL				ELECTRICAL	
	Fiber Glass Content %/WT.	Specific Gravity	Water Absorption	Mold Shrinkage	Tensile Strength	Tensile Elongation	Tensile Modulus	Flexural Strength	Flexural Modulus	Impact Notched IZOD	Compressive Strength	Heat Distortion	Coefficient of Linear Thermal Expansion °F	Continuous Use Temperature	Flammability	Volume Resistivity	Dielectric Strength
Test Method		Grams/CC	%	IN./IN.	PSI	%	PSIx 10	PSIx 10^3	PSIx 10^5	FT./LB./IN.	PSIx 10^3	Temp. °F	IN./IN. x 10^5	°C	UL	OHM/CM	Volts/Mil.
		D-792	D-570	D-955	D-638	D-638	D-638	D-790	D-790	D-256	D-695	D-648	D-696	UL-Sub 94	UL-Sub 94	D-257	D-149
			@ 24 hrs.	1/8" Bar	@ 73°F	Break			@ 73°F	1/4" BAR		@ 66 PSI			1/4" BAR		
Styrene	**	1.06	.03	.005	6,700	2.2	4.6	14.0	4.5	0.2	14.0	184	3.8	45	HB	10^{16}	—
	20	1.20	.07	.001	11,000	1.0	11.0	15.5	9.6	1.0	16.1	200	2.3	50	HB	10^{16}	490
	30	1.28	.06	.001	13,500	1.0	13.0	17.0	12.2	1.0	17.4	210	2.0	50	HB	10^{16}	500
	40	1.38	.05	.0005	15,000	1.0	16.0	17.5	14.7	1.1	17.7	218	1.7	50	HB	10^{16}	500
SAN	**	1.08	.20	.004	10,500	3.0	5.6	15.0	5.5	0.5	15.0	200	3.6	50	HB	10^{16}	450
	20	1.22	.15	.001	13,000	2.0	12.5	18.7	11.0	1.0	19.5	205	2.2	60	HB	10^{16}	500
	30	1.30	.10	.0005	15,500	1.5	14.5	22.5	15.2	1.0	20.5	215	1.9	60	HB	10^{16}	500
	40	1.40	.09	.0005	17,200	1.5	18.0	23.4	18.0	1.0	21.5	220	1.6	60	HB	10^{16}	500
ABS	**	1.05	.30	.006	7,000	8.0	3.0	10.5	3.8	4.5	10.0	195	4.2	65	HB	10^{16}	410
	20	1.20	.15	.001	13,000	3.0	9.0	17.0	8.0	1.5	12.5	225	2.0	70	HB	10^{16}	480
	30	1.28	.13	.001	15,200	3.0	10.0	18.5	10.0	1.4	15.5	230	1.7	70	HB	10^{16}	480
	40	1.38	.10	.001	16,000	2.0	15.0	21.0	13.0	1.3	17.1	235	1.3	70	HB	10^{16}	480
ABS-FR (Flame Retardant)	**	1.22	.03	.007	5,800	5.1	3.5	12.0	3.3	4.0	7.5	220	4.3	70	V-O	10^{16}	465
	20	1.33	.03	.002	11,000	2.0	7.4	15.5	7.1	1.2	14.0	225	2.1	70	V-O	10^{16}	465
Polypropylene	**	.90	.01	.015	4,700	15.0	1.9	6.0	3.0	0.5	5.0	220	5.0	105	HB*	10^{16}	660
	20	1.04	.01	.004	8,500	3.0	5.5	9.0	6.0	0.8	6.0	300	2.5	105	HB*	10^{16}	500
	30	1.13	.03	.003	9,000	3.0	6.5	8.5	8.0	1.1	6.5	310	2.2	105	HB*	10^{16}	500
	40	1.22	.06	.003	10,000	2.0	7.5	9.0	10.0	1.3	7.0	320	2.0	105	HB*	10^{16}	500
Polypropylene (Glass Coupled)	**	.90	.01	.015	4,700	15.0	1.9	6.0	3.0	0.5	6.0	220	5.0	105	HB	10^{16}	660
	20	1.04	.01	.004	11,000	3.0	6.0	12.0	6.0	1.4	10.0	310	2.5	105	HB	10^{16}	500
	40	1.22	.05	.004	14,000	2.0	8.0	19.0	10.0	1.6	13.0	320	2.1	105	HB	10^{16}	500
Polyethylene	**	.96	.01	.020	4,300	9.0	1.5	5.5	2.2	1.3	4.0	160	6.0	85	HB	10^{16}	
	20	1.10	.03	.003	7,000	3.0	6.0	9.0	5.5	1.4	5.0	240	2.8	85	HB	10^{16}	500
	30	1.18	.02	.003	10,000	2.0	8.5	11.0	8.0	1.7	7.0	250	2.7	85	HB	10^{16}	500
	40	1.28	.02	.003	11,000	2.0	11.0	12.5	10.0	1.7	8.0	260	2.5	85	HB	10^{16}	500
Acetal	**	1.41	.03	.018	8,800	60.0	4.1	13.0	3.7	1.3	5.2	316	4.7	90	HB	10^{14}	480
	20	1.55	.05	.006	12,000	2.0	12.0	16.0	10.0	1.0	12.0	325	2.4	96	HB	10^{14}	480
	30	1.63	.06	.005	13,000	1.8	13.5	16.5	12.0	.8	12.0	325	2.2	96	HB	10^{14}	480

Modified PPO	**	1.06	.06	.006	9,600	60.0	3.5	13.5	3.6	5.0	16.0	260	3.3	90	HB	10^{17}	500
	20	1.20	.06	.002	15,000	2.0	10.0	18.0	7.5	1.6	16.0	295	2.0	90	HB	10^{17}	500
	30	1.29	.06	.002	17,000	2.0	12.5	20.0	11.0	1.8	18.0	320	1.4	90	HB	10^{17}	500
Modified PPO-FR (Flame Retardant)	**	1.06	.08	.006	9,600	60.0	3.5	13.5	3.6	5.0	16.0	260	3.3	90	V-O	10^{17}	500
	20	1.39	.06	.002	15,000	2.0	9.5	18.0	7.0	1.4	16.0	260	1.9	90	V-O	10^{17}	480
	30	1.40	.06	.001	16,000	2.0	12.0	20.0	10.5	1.5	17.0	270	1.3	90	V-O	10^{17}	480
Polyester	**	1.30	.08	.020	8,000	200.0	4.0	12.8	3.4	0.2	13.0	310	5.3	107	HB	10^{16}	460
	20	1.43	.08	.006	17,000	5.0	10.0	22.0	8.5	1.5	16.0	420	2.0	130	HB	10^{16}	500
	30	1.52	.06	.005	19,000	4.0	15.0	26.0	12.0	1.8	18.0	440	1.4	130	HB	10^{16}	500
	40	1.62	.04	.004	22,000	3.0	20.0	30.0	15.0	2.0	20.0	450	1.2	130	HB	10^{16}	500
Polyester-FR (Flame Retardant)	**	1.40	.06	.020	8,900	60.0	4.0	14.7	3.8	0.9	14.5	325	4.7	128	V-O	10^{15}	500
	30	1.66	.06	.002	19,000	3.0	16.0	25.5	13.0	1.3	18.0	410	1.2	130	V-O	10^{15}	500
Nylon 6	**	1.13	1.80	.013	11,800	200.0	4.0	15.0	4.0	1.0	13.0	365	4.6	95	V-2	10^{14}	400
	20	1.27	1.30	.004	18,500	3.0	10.0	23.0	8.0	1.5	21.5	420	2.3	102	HB	10^{14}	410
	30	1.37	1.10	.003	22,500	3.0	13.0	27.0	11.0	2.2	23.0	420	1.8	102	HB	10^{14}	420
	40	1.45	.90	.0025	26,800	2.0	14.0	30.0	13.0	3.0	23.0	420	1.6	102	HB	10^{14}	420
Nylon 6-FR (Flame Retardant)	**	1.30	.90	.012	12,300	60.0	4.2	16.0	4.0	1.0	13.0	370	3.6	95	V-O	10^{14}	400
	30	1.58	.85	.003	22,000	3.0	13.0	33.0	13.5	1.7	2.3	405	1.8	100	V-O	10^{14}	430
Nylon 6/6	**	1.14	1.50	.018	11,400	300.0	1.9	15.0	1.9	1.0	4.9	360	4.5	105	V-2	10^{14}	500
	20	1.28	1.00	.005	20,000	3.0	12.0	28.0	8.5	1.2	23.0	500	2.3	116	HB	10^{14}	500
	30	1.36	.90	.004	26,000	2.0	15.0	37.5	13.0	2.0	24.0	500	1.8	116	HB	10^{14}	500
	40	1.47	.60	.003	31,000	2.0	13.0	42.5	16.0	2.6	25.0	500	1.7	116	HB	10^{14}	500
Nylon 6/6-FR (Flame Retardant)	**	1.20	1.10	.018	9,700	35.0	1.9	13.0	1.75	1.0	4.9	400	4.5	105	V-O	10^{14}	400
	30	1.60	.90	.002	21,500	2.0	12.0	25.0	10.0	1.6	23.0	490	1.3	116	V-O	10^{14}	410
Nylon 612	**	1.07	.40	.010	8,800	150.0	2.9	12.5	2.9	1.0	11.0	300	5.0	100	V-2	10^{14}	400
	20	1.21	.22	.004	18,000	4.0	10.0	28.0	9.0	1.1	19.0	420	2.2	100	HB*	10^{14}	410
	30	1.30	.20	.0035	22,000	4.0	13.0	32.0	11.0	2.4	22.0	430	1.5	100	HB*	10^{14}	420
Polycarbonate	**	1.20	.10	.006	9,000	110.0	3.45	13.5	3.4	3.0	12.5	280	3.7	121	V-O	10^{16}	380
	10	1.27	.20	.003	13,000	5.0	7.0	16.0	6.0	2.0	18.0	300	1.6	127	V-O	10^{16}	440
	20	1.35	.20	.0025	16,000	5.0	9.0	20.0	8.0	2.2	20.0	300	1.5	127	V-O	10^{16}	480
	30	1.43	.10	.002	19,000	4.0	13.0	24.0	12.0	2.4	21.0	300	1.2	127	V-O	10^{16}	480
	40	1.52	.10	.0015	22,000	3.5	17.0	28.0	14.0	2.7	21.5	300	1.0	127	V-O	10^{16}	480
Polysulfone	**	1.24	.20	.007	10,200	75.0	3.6	15.4	3.9	0.6	14.0	345	3.1	140	V-O	10^{15}	425
	20	1.38	.20	.003	19,000	3.0	9.0	20.0	7.5	1.2	20.0	360	1.7	149	V-O	10^{15}	465
	30	1.46	.20	.003	21,500	3.0	12.0	22.5	10.0	1.4	22.5	365	1.4	149	V-O	10^{15}	465
	40	1.55	.10	.002	24,000	2.0	18.0	25.0	12.5	2.0	25.0	370	1.3	149	V-O	10^{15}	465
Polyphenylene Sulfide (PPS)	40	1.62	.02	.001	20,000	1.5	20.5	34.0	18.0	1.5	25.0	500	1.3	183	V-O	10^{16}	510

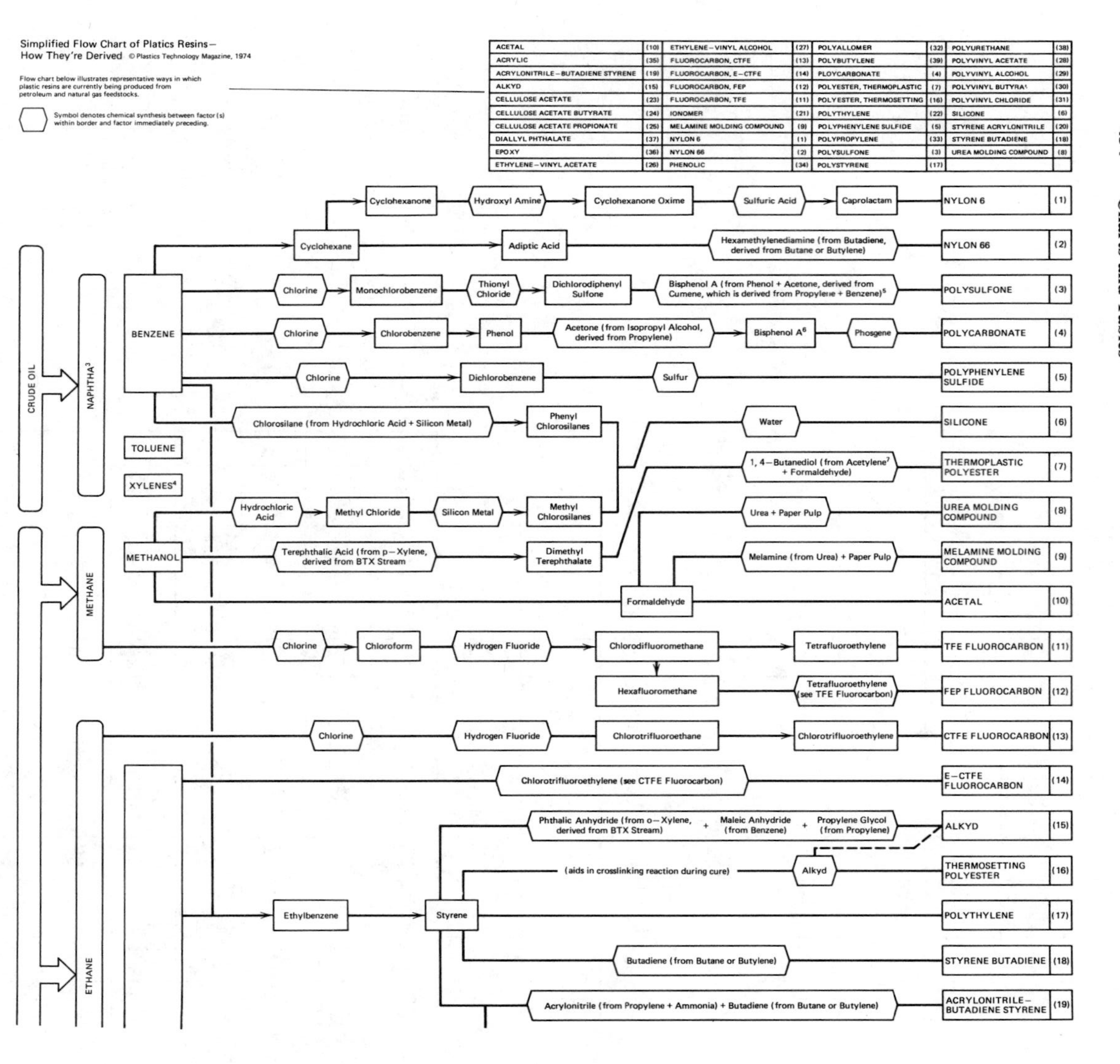

Simplified Flow Chart of Platics Resins–
How They're Derived © Plastics Technology Magazine, 1974
Flow chart below illustrates representative ways in which plastic resins are currently being produced from petroleum and natural gas feedstocks.
Symbol denotes chemical synthesis between factor(s) within border and factor immediately preceding.
ACETAL (10) ETHYLENE–VINYL ALCOHOL (27) POLYALLOMER (32) POLYURETHANE (38)
ACRYLIC (35) FLUOROCARBON, CTFE (13) POLYBUTYLENE (39) POLYVINYL ACETATE (28)
ACRYLONITRILE–BUTADIENE STYRENE (19) FLUOROCARBON, E–CTFE (14) PLOYCARBONATE (4) POLYVINYL ALCOHOL (29)
ALKYD (15) FLUOROCARBON, FEP (12) POLYESTER, THERMOPLASTIC (7) POLYVINYL BUTYRAL (30)
CELLULOSE ACETATE (23) FLUOROCARBON, TFE (11) POLYESTER, THERMOSETTING (16) POLYVINYL CHLORIDE (31)
CELLULOSE ACETATE BUTYRATE (24) IONOMER (21) POLYTHYLENE (22) SILICONE (6)
CELLULOSE ACETATE PROPIONATE (25) MELAMINE MOLDING COMPOUND (9) POLYPHENYLENE SULFIDE (5) STYRENE ACRYLONITRILE (20)
DIALLYL PHTHALATE (37) NYLON 6 (1) POLYPROPYLENE (33) STYRENE BUTADIENE (18)
EPOXY (36) NYLON 66 (2) POLYSULFONE (3) UREA MOLDING COMPOUND (8)
ETHYLENE–VINYL ACETATE (26) PHENOLIC (34) POLYSTYRENE (17)
CRUDE OIL
NAPHTHA³
METHANE
ETHANE
BENZENE
TOLUENE
XYLENES⁴
METHANOL
Cyclohexanone
Hydroxyl Amine
Cyclohexanone Oxime
Sulfuric Acid
Caprolactam
NYLON 6 (1)
Cyclohexane
Adiptic Acid
Hexamethylenediamine (from Butadiene, derived from Butane or Butylene)
NYLON 66 (2)
Chlorine
Monochlorobenzene
Thionyl Chloride
Dichlorodiphenyl Sulfone
Bisphenol A (from Phenol + Acetone, derived from Cumene, which is derived from Propylene + Benzene)⁵
POLYSULFONE (3)
Chlorine
Chlorobenzene
Phenol
Acetone (from Isopropyl Alcohol, derived from Propylene)
Bisphenol A⁶
Phosgene
POLYCARBONATE (4)
Chlorine
Dichlorobenzene
Sulfur
POLYPHENYLENE SULFIDE (5)
Chlorosilane (from Hydrochloric Acid + Silicon Metal)
Phenyl Chlorosilanes
Water
SILICONE (6)
1, 4–Butanediol (from Acetylene⁷ + Formaldehyde)
THERMOPLASTIC POLYESTER (7)
Hydrochloric Acid
Methyl Chloride
Silicon Metal
Methyl Chlorosilanes
Urea + Paper Pulp
UREA MOLDING COMPOUND (8)
Terephthalic Acid (from p–Xylene, derived from BTX Stream
Dimethyl Terephthalate
Melamine (from Urea) + Paper Pulp
MELAMINE MOLDING COMPOUND (9)
Formaldehyde
ACETAL (10)
Chlorine
Chloroform
Hydrogen Fluoride
Chlorodifluoromethane
Tetrafluoroethylene
TFE FLUOROCARBON (11)
Hexafluoromethane
Tetrafluoroethylene (see TFE Fluorocarbon)
FEP FLUOROCARBON (12)
Chlorine
Hydrogen Fluoride
Chlorotrifluoroethane
Chlorotrifluoroethylene
CTFE FLUOROCARBON (13)
Chlorotrifluoroethylene (see CTFE Fluorocarbon)
E–CTFE FLUOROCARBON (14)
Phthalic Anhydride (from o–Xylene, derived from BTX Stream) + Maleic Anhydride (from Benzene) + Propylene Glycol (from Propylene)
ALKYD (15)
(aids in crosslinking reaction during cure)
Alkyd
THERMOSETTING POLYESTER (16)
Ethylbenzene
Styrene
POLYTHYLENE (17)
Butadiene (from Butane or Butylene)
STYRENE BUTADIENE (18)
Acrylonitrile (from Propylene + Ammonia) + Butadiene (from Butane or Butylene)
ACRYLONITRILE–BUTADIENE STYRENE (19)

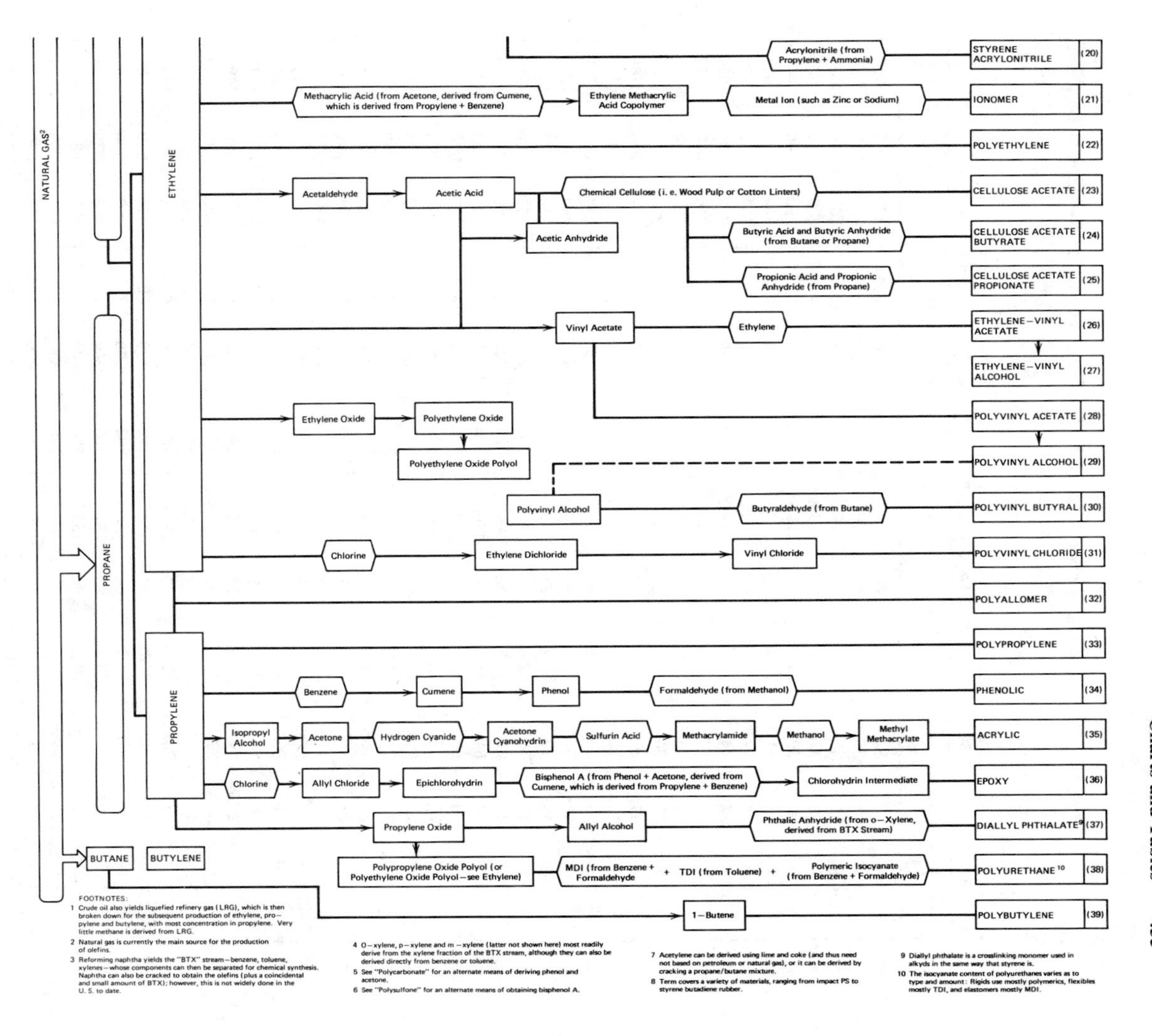

FOOTNOTES:

1 Crude oil also yields liquefied refinery gas (LRG), which is then broken down for the subsequent production of ethylene, propylene and butylene, with most concentration in propylene. Very little methane is derived from LRG.

2 Natural gas is currently the main source for the production of olefins.

3 Reforming naphtha yields the "BTX" stream—benzene, toluene, xylenes—whose components can then be separated for chemical synthesis. Naphtha can also be cracked to obtain the olefins (plus a coincidental and small amount of BTX); however, this is not widely done in the U. S. to date.

4 O–xylene, p–xylene and m–xylene (latter not shown here) most readily derive from the xylene fraction of the BTX stream, although they can also be derived directly from benzene or toluene.

5 See "Polycarbonate" for an alternate means of deriving phenol and acetone.

6 See "Polysulfone" for an alternate means of obtaining bisphenol A.

7 Acetylene can be derived using lime and coke (and thus need not based on petroleum or natural gas), or it can be derived by cracking a propane/butane mixture.

8 Term covers a variety of materials, ranging from impact PS to styrene butadiene rubber.

9 Diallyl phthalate is a crosslinking monomer used in alkyds in the same way that styrene is.

10 The isocyanate content of polyurethanes varies as to type and amount: Rigids use mostly polymerics, flexibles mostly TDI, and elastomers mostly MDI.

HOW TO CONVERT CENTIGRADE TEMPERATURE TO FAHRENHEIT AND VICE VERSA

Centigrade C° = 5/9 (F° − 32)   TEMPERATURE CONVERSION TABLES   To Fahrenheit F° = (9/5XC°) + 32

C.		F.	C.		F.	C.		F.	C.		F.
− 17.8	0	32	8.89	48	118.4	35.6	96	204.8	271	520	968
− 17.2	1	33.8	9.44	49	120.2	36.1	97	206.6	277	530	986
− 16.7	2	35.6	10.0	50	122.0	36.7	98	208.4	282	540	1004
− 16.1	3	37.4	10.6	51	123.8	37.2	99	210.2	288	550	1022
− 15.6	4	39.2	11.1	52	125.6	37.8	100	212.0	293	560	1040
− 15.0	5	41.0	11.7	53	127.4	38	100	212	299	570	1058
− 14.4	6	42.8	12.2	54	129.2	43	110	230	304	580	1076
− 13.9	7	44.6	12.8	55	131.0	49	120	248	310	590	1094
− 13.3	8	46.4	13.3	56	132.8	54	130	266	316	600	1112
− 12.8	9	48.2	13.9	57	134.6	60	140	284	321	610	1130
− 12.2	10	50.0	14.4	58	136.4	66	150	302	327	620	1148
− 11.7	11	51.8	15.0	59	138.2	71	160	320	332	630	1166
− 11.1	12	53.6	15.6	60	140.0	77	170	338	338	640	1184
− 10.6	13	55.4	16.1	61	141.8	82	180	356	343	650	1202
− 10.0	14	57.2	16.7	62	143.6	88	190	374	349	660	1220
− 9.44	15	59.0	17.2	63	145.4	93	200	392	354	670	1238
− 8.89	16	60.8	17.8	64	147.2	99	210	410	360	680	1256
− 8.33	17	62.6	18.3	65	149.0	100	212	413	366	690	1274
− 7.78	18	64.4	18.9	66	150.8	104	220	428	371	700	1292
− 7.22	19	66.2	19.4	67	152.6	110	230	446	377	710	1310
− 6.67	20	68.0	20.0	68	154.4	116	240	464	382	720	1328
− 6.11	21	69.8	20.6	69	156.2	121	250	482	388	730	1346
− 5.56	22	71.6	21.1	70	158.0	127	260	500	393	740	1364
− 5.00	23	73.4	21.7	71	159.8	132	270	518	399	750	1382
− 4.44	24	75.2	22.2	72	161.6	138	280	536	404	760	1400
− 3.89	25	77.0	22.8	73	163.4	143	290	554	410	770	1418
− 3.33	26	78.8	23.3	74	165.2	149	300	572	416	780	1436
− 2.78	27	80.6	23.9	75	167.0	154	310	590	421	790	1454
− 2.22	28	82.4	24.4	76	168.8	160	320	608	427	800	1472
− 1.67	29	84.2	25.0	77	170.6	166	330	626	432	810	1490
− 1.11	30	86.0	25.6	78	172.4	171	340	644	438	820	1508
− 0.56	31	87.8	26.1	79	174.2	177	350	662	443	830	1526
− 0	32	89.6	26.7	80	176.0	182	360	680	449	840	1544
0.56	33	91.4	27.2	81	177.8	188	370	698	454	850	1562
1.11	34	93.2	27.8	82	179.6	193	380	716	460	860	1580
1.67	35	95.0	28.3	83	181.4	199	390	734	466	870	1598
2.22	36	96.8	28.9	84	183.2	204	400	752	471	880	1616
2.78	37	98.6	29.4	85	185.0	210	410	770	477	890	1634
3.33	38	100.4	30.0	86	186.8	216	420	788	482	900	1652
3.89	39	102.2	30.6	87	188.6	221	430	806	488	910	1670
4.44	40	104.0	31.1	88	190.4	227	440	824	493	920	1688
5.00	41	105.8	31.7	89	192.2	232	450	842	499	930	1706
5.56	42	107.6	32.2	90	194.0	238	460	860	504	940	1724
6.11	43	109.4	32.8	91	195.8	243	470	878	510	950	1742
6.67	44	111.2	33.3	92	196.7	249	480	896	516	960	1760
7.22	45	113.0	33.9	93	199.4	254	490	914	521	970	1778
7.78	46	114.8	34.4	94	201.2	260	500	932	527	980	1796
8.33	47	116.6	35.0	95	203.0	266	510	950	532	990	1814

DECIMAL EQUIVALENTS OF FRACTIONS OF ONE INCH

1/64	0.015 625	17/64	0.265 625	33/64	0.515 625	49/64	0.765 625
1/32	0.031 250	9/32	0.281 250	17/32	0.531 250	25/32	0.781 250
3/64	0.046 875	19/64	0.296 875	35/64	0.546 875	51/64	0.796 875
1/16	0.062 500	5/16	0.312 500	9/16	0.562 500	13/16	0.812 500
5/64	0.078 125	21/64	0.328 125	37/64	0.578 125	53/64	0.828 125
3/32	0.093 750	11/32	0.343 750	19/32	0.593 750	27/32	0.843 750
7/64	0.109 375	23/64	0.359 375	39/64	0.609 375	55/64	0.859 375
1/8	0.125 000	3/8	0.375 000	5/8	0.625 000	7/8	0.875 000
9/64	0.140 625	25/64	0.390 625	41/64	0.640 625	57/64	0.890 625
5/32	0.156 250	13/32	0.406 250	21/32	0.656 250	29/32	0.906 250
11/64	0.171 875	27/64	0.421 875	43/64	0.671 875	59/64	0.921 875
3/16	0.187 500	7/16	0.437 500	11/16	0.687 500	15/16	0.937 500
13/64	0.203 125	29/64	0.453 125	45/64	0.703 125	61/64	0.953 125
7/32	0.218 750	15/32	0.468 750	23/32	0.718 750	31/32	0.968 750
15/64	0.234 375	31/64	0.484 375	47/64	0.734 375	63/64	0.984 375
1/4	0.250 000	1/2	0.500 000	3/4	0.750 000	1	1.000 000

DIAMETERS AND AREAS OF CIRCLES

Diam.	Area	Diam.	Area	Diam.	Area	Diam.	Area	Diam.	Area
1/64	0.00019	1/4	1.2272	1/2	9.6211	3/4	25.967	3/4	74.662
1/32	0.00077	5/16	1.3530	9/16	9.9678	13/16	26.535	7/8	76.589
3/64	0.00173	3/8	1.4849	5/8	10.321	7/8	27.109		
1/16	0.00307	7/16	1.6230	11/16	10.680	15/16	27.688	10	78.540
3/32	0.00690	1/2	1.7671	3/4	11.045			1/8	80.516
1/8	0.01227	9/16	1.9175	13/16	11.416	6	28.274	1/4	82.516
5/32	0.01917	5/8	2.0739	7/8	11.793	1/8	29.465	3/8	84.541
3/16	0.02761	11/16	2.2465	15/16	12.177	1/4	30.680	1/2	86.590
7/32	0.03758	3/4	2.4053			3/8	31.919	5/8	88.664
1/4	0.04909	13/16	2.5802	4	12.566	1/2	33.183	3/4	90.763
9/32	0.06213	7/8	2.7612	1/16	12.962	5/8	34.472	7/8	92.886
5/16	0.07670	15/16	2.9483	1/8	13.364	3/4	35.785		
11/32	0.09281			3/16	13.772	7/8	37.122	11	95.033
3/8	0.11045	2	3.1416	1/4	14.186			1/2	103.87
13/32	0.12962	1/16	3.3410	5/16	14.607	7	38.485		
7/16	0.15033	1/8	3.5466	3/8	15.033	1/8	39.871	12	113.10
15/32	0.17257	3/16	3.7583	7/16	15.466	1/4	41.282	1/2	122.72
1/2	0.19635	1/4	3.9761	1/2	15.904	3/8	42.718	13	132.73
17/32	0.22165	5/16	4.2000	9/16	16.349	1/2	44.179	1/2	143.14
9/16	0.24850	3/8	4.4301	5/8	16.800	5/8	45.664		
19/32	0.27688	7/16	4.6664	11/16	17.257	3/4	47.173	14	153.94
5/8	0.30680	1/2	4.9087	3/4	17.721	7/8	48.707	1/2	165.13
21/32	0.33824	9/16	5.1572	13/16	18.190				
11/16	0.37122	5/8	5.4119	7/8	18.665	8	50.265	15	176.71
23/32	0.40574	11/16	5.6727	15/16	19.147	1/8	51.849	1/2	188.69
3/4	0.44179	3/4	5.9396			1/4	53.456	16	201.06
25/32	0.47937	13/16	6.2126	5	19.635	3/8	55.088	1/2	213.82
13/16	0.51849	7/8	6.4918	1/16	20.129	1/2	56.745		
27/32	0.55914	15/16	6.7771	1/8	20.629	5/8	58.426	17	226.98
7/8	0.60132			3/16	21.125	3/4	60.132	1/2	240.53
29/32	0.64504	3	7.0686	1/4	21.648	7/8	61.862		
15/16	0.69029	1/16	7.3662	5/16	22.166			18	254.47
31/32	0.73708	1/8	7.6699	3/8	22.691	9	63.617	1/2	268.80
		3/16	7.9798	7/16	23.211	1/8	65.397	19	283.53
1	0.7854	1/4	8.2958	1/2	23.758	1/4	67.201	1/2	298.65
1/16	0.8866	5/16	8.6179	9/16	24.301	3/8	69.029		
1/8	0.9940	3/8	8.9462	5/8	24.850	1/2	70.882	20	314.16
3/16	1.1075	7/16	9.2806	11/16	25.406	5/8	72.760	1/2	330.06

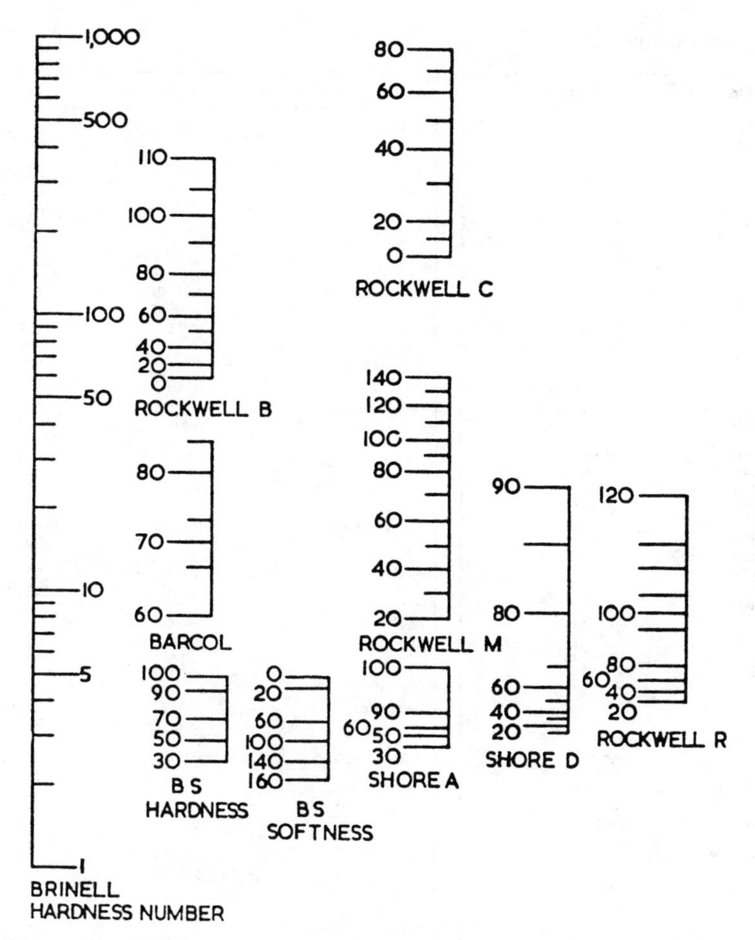

Approximate comparison of hardness scales.

Miscellaneous Measurement Conversions

Multiply This	By This	To Obtain This
acre	43,560	ft^2
	1.562×10^{-3}	sq mi
	160	sq rod
	0.4047	hectare
Angstrom unit	3.937×10^{-9}	in
	1×10^{-4}	μ
	0.1	mμ
	1×10^{-8}	cm
	1×10^{-10}	m
are	3.954	sq rod
	100	m^2
	119.6	yd^2
	0.02471	acre
atmosphere	760	mm Hg at 0°C
	29.92	in Hg at 0°C
	406.79	in H_2O at 4°C
	33.899	ft H_2O at 4°C
	407.16	in H_2O at 15°C
	33.93	ft H_2O at 15°C
	1.0333	kg/cm^2
	10.333	kg/m^2
	1.01325×10^6	$dyne/cm^2$
	14.696	lb/in^2
	2116.32	lb/ft^2
	1.0133	bar
bar	0.9869	atm
	750	mm Hg (0°C)
	10,197	kg/m^2
	1×10^6	$dyne/cm^2$
	14.50	lb/in^2
barrel	42	gal
Btu	777.98	ft-lb
	3.930×10^{-4}	hp-hr
	2.931×10^{-4}	kwh
	2.520×10^{-1}	cal (kg)
	107.6	kg-m
	1055	joule
Btu/sec	1055	watt
Btu/min	0.0236	hp
	17.58	watt
Btu/lb	0.55556	cal/gm
bushel	35.238	l
	0.3524	hectoliter
	4	peck
	1.2444	cu ft
calorie (gram)	3.968×10^{-3}	Btu
calorie/gm	1.8	Btu/lb
carat	0.2	g
cental	100	lb
centimeter	0.032808	ft
	0.393700	in
	10,000	μ
	1×10^8	Å
	0.01	m
	10	mm
centimeter of Hg(0°C)	0.01316	atm
	0.1934	lb/in^2
	27.845	lb/ft^2
	0.44604	ft H_2O at 4°C
	135.95	kg/m^2
	13332	$dyne/cm^2$
centimeter^2	100	mm^2
	1×10^{-4}	m^2
	0.1550	in^2
	0.00108	ft^2
	1.196×10^{-4}	yd^2
centimeter^3	0.061	in^3
	3.531×10^{-5}	ft^3
centimeters^3	1.3079×10^{-6}	yd^3
	1×10^{-6}	m^3
	0.99997	ml
	0.0338	oz (US fl)
	0.0351	oz (Brit. fl)
	10.567×10^{-4}	qt (US fl)
	8.7988×10^{-4}	qt (Brit. fl)
	2.6417×10^{-4}	gal (US)
	2.1997×10^{-4}	gal (Brit.)
centimeter/second	0.0328	ft/sec
	1.9685	ft/min
	0.06	m/min
	3.728×10^{-4}	mi/min
	0.02237	mi/hr
	0.03600	km/hr
centipoise	6.72×10^{-4}	lb/sec-ft
	3.60	kg/hr-m
chain (surveyors)	100	link
	66	ft
	20.117	m
cheval-vapeur	735.499	watt
	0.9863	hp
cord	128	ft^3
	3.625	stere
day	86400	sec
	1440	min
	0.143	wk
	0.0028	yr
decimeter	3.937	in
	0.328	ft
dram (dry)	0.0625	oz (wt)
	1.7718	g
dram (fluid)	3.6967	cc
	0.125	oz (fl)
dyne	1.020×10^{-3}	g
	2.248×10^{-6}	lb
dyne/centimeter^2	1×10^{-6}	bar
	10.197×10^{-4}	g/cm^2
	1.4504×10^{-5}	lb/in^2
ell	45	in
erg	9.4805×10^{10}	Btu
	1	dyne-cm
	7.376×10^{-8}	ft-lb
	2.388×10^{-11}	cal (kg)
fathom	6	ft
firkin	9	gal
foot	30.48	cm
	0.3048	m
foot H_2O(4°C)	0.0295	atm
	0.883	in Hg (0°C)
	2.419	cm Hg (0°C)
	0.4335	lb/in^2
	62.427	lb/ft^2
	304.79	kg/m^2
foot^2	0.111	yd^2
	3.587×10^{-8}	mi^2
	2.296×10^{-5}	acre
	0.0929	m^2
	929	cm^2
foot^3	1728	in^3
	0.037	yd^3
	28,316	cc
	0.02832	m^3
	28.316	l
	7.4805	gal (US)
	6.2288	gal (Brit.)

Miscellaneous Measurement Conversions . . . continued

Multiply This	By This	To Obtain This
$foot^3$ H_2O	62.42	lb (4°C)
	62.36	lb (15°C)
$foot^3$/min	0.4720	l/sec
	472	cc/sec
	448.831	gal/min
foot-pound	0.1383	kg-m
foot-pound/min	3.030×10^{-5}	hp
foot/min	0.01667	ft/sec
	0.0114	mph
	0.508	cm/sec
	0.005	m/sec
	0.3048	m/min
	0.0183	km/hr
foot/second	0.01136	mi/min
	0.6818	mph
	30.48	cm/sec
	18.288	m/min
	1.097	km/hr
furlong	660	ft
	220	yd
	40	rod
gallon (US)	0.8327	gal (Brit.)
	128	oz
	8	pt
	3785.4	cc
	3.785	l
	0.00379	m^3
	231	in^3
	0.1337	ft^3
	0.00495	yd^3
	8.3378	lb H_2O(60°F)
gallon (British)	1.2009	gal (US)
	4546	cc
	4.546	l
	277.419	in^3
	0.16054	ft^3
	10	lb H_2O(60°F)
	160	oz (Brit fl)
gallon/minute (US)	8.0208	ft^3/hr
	0.06308	l/sec
gill (US)	4	oz
grain	0.00229	oz (avoir.)
	0.0648	g
gram	0.0353	oz (avoir.)
	0.0022	lb
	15.432	grain
gram/$centimeter^2$	9.6784×10^{-4}	atm
	0.7356	mm Hg (0°C)
	0.0289	in Hg (0°C)
	0.3284	ft H_2O (60°F)
	0.0142	lb/in^2
	2.0482	lb/ft^2
	980.665	dyne/cm^2
	10	kg/m^2
gram/$centimeter^3$	62.428	lb/ft^3
	8.345	lb/gal (US)
	0.0361	lb/in^3
	1	g/ml
gram/liter	1000	ppm
	0.0624	lb/ft^3
gravity	32.174	ft/sec^2
	980.665	cm/sec^2
hand	4	in
hectare	2.471	acre
	100	are
	107,640	ft^2
	10,000	m^2
hogshead	63	gal
horsepower	42.418	Btu/min
	33,000	ft-lb/min
horsepower	550	ft-lb/sec
	0.7457	kw
	10.688	cal(kg)/min
	1.014	cheval-vapeur
hour	3600	sec
	0.04167	day
	0.0059	wk
hundredweight (cwt)	100	lb (short)
	112	lb (long)
inch	1000	mil
	0.083	ft
	0.02778	yd
	2.54	cm
	0.0254	m
	2.54×10^{8}	Å
inch Hg (0°C)	13.595	in H_2O
	1.133	ft H_2O
	0.0334	atm
	25.4	mm Hg
	34.5	g/cm^2
	345.3	kg/m^2
	33,864	dyne/cm^2
	0.4912	lb/in^2
	70.73	lb/ft^2
inch H_2O (4°C)	0.00245	atm
	0.07355	in Hg
	25.399	kg/m^2
	5.2022	lb/ft^2
	0.0361	lb/in^2
$inch^2$	6.4516	cm^2
	6.4516×10^{-4}	m^2
	0.0069	ft^2
	0.00077	yd^2
$inch^3$	16.387	cc
	5.7870×10^{-4}	ft^3
	1.6387×10^{-5}	m^3
	2.14335×10^{5}	yd^3
	0.003606	gal (Brit.)
	0.004329	gal (US)
	0.01639	l
	0.5541	oz (US fl)
	0.01488	qt (dry)
	0.01732	qt (US fl)
kilogram	35.274	oz (avoir)
	2.2046	lb
	9.842×10^{-4}	ton (long)
	0.0011	ton (short)
kilogram-meter	7.2330	ft-lb
kilogram/meter	0.67197	lb/ft
kilogram/$meter^2$	0.07356	mm Hg (0°C)
	0.00142	lb/in^2
	9.6777×10^{-5}	atm
	0.20482	lb/ft^2
kilogram/$meter^3$	0.06243	lb/ft^3
	0.001	g/cc
kiloliter	35.317	ft^3
kilometer	3280.8	ft
	0.53959	mi (naut)
	0.62137	mi
	1093.6	yd
kilometer/hour	27.7778	cm/sec
	54.68	ft/min
	0.9113	ft/sec
	0.5396	knot
	16.667	m/min
	0.27778	m/sec
$kilometer^2$	1.076×10^{7}	ft^2
	1.196×10^{6}	yd^2
	0.386	mi^2
	1×10^{6}	m^2
	247.1	acre
kilowatt	56.884	Btu/min
	1.3410	hp

Miscellaneous Measurement Conversions . . . continued

Multiply This	By This	To Obtain This
kilowatt-hour	3413	Btu
	1.3410	hp-hr
knot	51.479	cm/sec
	6080.2	ft/hr
	1.15155	mph
league	3	mi
light-year	9.4637×10^{12}	km
	5.88×10^{12}	mi
link	0.66	ft
liter	61.025	in^3
	0.035	ft^3
	0.2642	gal (US)
	33.814	oz (US fl)
	1.05668	qt (US fl)
	0.8799	qt (Brit fl)
liter/second	15.8507	gal/min
meter	1×10^{10}	Å
	3.2808	ft
	39.370	in
	5.3959×10^{-4}	mi (naut.)
	6.2137×10^{-4}	mi
	1.0936	yd
	1×10^{9}	μ
meter/minute	0.05468	ft/sec
	0.06	km/hr
	0.03728	mph
$meter^2$	0.01	are
	2.471×10^{-4}	acre
	10.7639	ft^2
	1550	in^2
	3.8610×10^{-7}	mi^2
	1.19598	yd^2
$meter^3$	35.314	ft^3
	61023	in^3
	1.3079	yd^3
	264.73	gal
	999.973	l
	1056.7	qt (US fl)
micron	1×10^{4}	Å
	1×10^{-4}	cm
	3.937×10^{-5}	in
mile (statute)	8	furlong
	63360	in
	1.60935	km
	1609.35	m
	0.8684	mi (naut)
	320	rod
	1760	yd
mile (nautical)	6080.2	ft
	1.85325	km
	1.1516	mi (statute)
miles/hour	44.704	cm/sec
	88	ft/min
	0.8684	knot
	26.82	m/min
milligram	3.5274×10^{-5}	oz (avoir)
	5.6438×10^{-4}	dram
	2.2046×10^{-6}	lb
millimeter	0.03937	in
	1000	μ
millimeter Hg (0°C)	1.316×10^{-3}	atm
	1333.22	$dyne/cm^2$
	1.3595	g/cm^2
	13.595	kg/m^2
	2.7845	lb/ft^2
millimicron	10	Å
	1×10^{-7}	cm
minute	6.9444×10^{-4}	day
	9.9206×10^{-5}	wk
month (mean calendar)	30.4202	day
	730.085	hr
month (mean calendar)	43,805	min
	2.6283×10^{6}	sec
ounce (avoirdupois)	16	dram
	437.5	grain
	0.91146	oz (troy)
	3.125×10^{-5}	ton (short)
	28.35	g
ounce (fluid)	29.5737	cc
	1.8047	in^3
	8	dram
	0.25	gill
	0.029573	l
	0.03125	qt
ounce (British fl)	28.413	cc
pace	2.5	ft
parts per million	0.0584	grain/gal
peck	537.6	in^3
	8.8096	l
	16	pint
pint (dry)	550.61	cc
	33.6003	in^3
	0.5506	l
pint (fluid)	473.179	cc
	28.875	in^3
	128	dram
	4	gill
	0.473168	l
pound	256	dram
	7000	grain
	453.5924	g
	0.45359	kg
	1.2153	lb (troy)
pound/foot	1.48816	kg/m
$pounds/foot^2$	4.7252×10^{-4}	atm
	4.7880×10^{-4}	bar
	478.78	$dyne/cm^2$
	0.48824	g/cm^2
	4.8824	kg/m^2
	0.35913	mm Hg (0°C)
	6.9445×10^{-3}	lb/in^2
$pound/in^2$	0.068046	atm
	70.307	g/cm^2
	703.07	kg/m^2
	51.715	mm Hg (0°C)
$pound/inch^3$	27.68	g/cc
$pound/foot^3$	0.016018	g/cc
	16.018	kg/m^3
	5.787×10^{-4}	lb/in^3
quart (dry)	0.03125	bushel
	1101.23	cc
	0.03889	ft^3
	1.1012	l
quart (liquid)	946.358	cc
	57.749	in^3
	0.03342	ft^3
	256	dram
	8	gill
	0.946333	l
quire	24	sheet
rod	0.25	chain
	16.5	feet
	0.025	furlong
	198	in
	25	link
	5.029216	m
	3.125×10^{-3}	mi
rood	0.25	acre
second	0.01667	min
	2.7778×10^{-4}	hr
	1.1574×10^{-5}	day

Miscellaneous Measurement Conversions . . . continued

Multiply This	By This	To Obtain This
slug	32.174	lb
	14.594	kg
span	9	in
stere	1	m^3
	999.973	l
stone	14	lb
	6.3503	kg
ton	20	cwt
	907.1846	kg
	0.89286	ton (long)
ton (long)	22.4	cwt
	1016.047	kg.
	1.12	ton (short)

Multiply This	By This	To Obtain This
ton (metric)	1000	kg
	2204.62	lb
watt	44.254	ft-lb/min
	$1.34.0 \times 10^{-3}$	hp
	3.41304	Btu/hr
week	168	hr
	10,080	min
	604,800	sec
yard	91.4402	cm
	5.68182×10^{-4}	mi
year	365.256	day
	8766.144	hr

Thermal Conductivity Values for Various Materials.

Values are shown in the following conductivity units:

(a) Btu per hr per sq ft per Deg F per inch

(b) Cal per sec per sq cm per Deg C per cm

Where values are expressed in other units, conversion to the desired units may be calculated by use of the following table:

		Cal/sec/ sq cm/°C/ cm	*Watt sq cm/°C/ cm*	*Btu/ sec/sq ft/ °F/in.*	*Btu/ hr/sq ft/ °F/in.*	*Btu/ hr/sq ft/ °F/ft*
Cal/sec/sq cm/°C/cm	×	1	4.186	0.8064	2903	241.9
Watt/sq cm/°C/cm	×	0.2389	1	0.1926	693.5	57.79
Btu/sec/sq ft/°F/in.	×	1.24	5.191	1	3600	300
Btu/hr/sq ft/°F/in.	×	0.004134	0.001442	0.0002778	1	0.08333
Btu/hr/sq ft/°F/ft	×	0.0003445	0.01730	0.003333	12	1

'Thermal conductivity expressed in any of the units in the left-hand column can be converted into any of the units in the headings of the columns by multiplying (×) by the number which is common to the row and column.

Thermal resistivity is the reciprocal of conductivity.

SOLVENT SELECTION GUIDE

Polymeric Material/Solvent	THF	Toluene	*N,N*-Dimethyl-formamide	Methylene Chloride	Chloro-form	2,2,2 Tri-fluoro-ethanol	1,1,1,3,3,3-Hexafluoro-2-propanol
ABS			◆				
Acrylics	◆		◆				
Alkyds	◆						◆
Amides/Imides			◆				
Asphalt	◆						
Butyl Rubber	◆	◆					
Cellulosics	◆		◆				
Epoxy	◆				◆		
Film-Forming Polyesters							◆
Glycerides	◆						
Natural Rubber		◆			◆		
Neoprene		◆					
Nonionic Surfactants	◆		◆				
Nylon						◆	◆
Phenolics	◆						
Polyacrylonitrile			◆				
Polybutadiene	◆	◆					
Polycarbonates	◆			◆	◆		
Polyesters/Unsaturated	◆			◆	◆		
Polyethylene*							
Polyethylene Oxide			◆				
Polyglycols	◆						
Polyisobutylene	◆	◆					
Polyisoprene	◆	◆					
Polyols	◆			◆	◆		
Polyphenylene Oxides	◆						
Polypropylene*							
Polystyrene	◆	◆	◆	◆	◆		
Polysulfones	◆				◆		
Polyurethane	◆		◆				
Polyvinyl acetate	◆						
PVC	◆						
Silicones		◆					
SBR	◆	◆					

PRESSURE CONVERSION CHART

BY FACTOR TO OBTAIN →

MULTIPLY GIVEN NUMBER OF ↓

GIVEN	lb/in^2	in H_2O (at +39.2°F)	cm H_2O (at +4°C)	in Hg (at +32°F)	mm Hg (Torr) (at 0°C)	dyne/cm^2 (1μ bar)	newton/m^2 (PASCAL)	kgm/cm^2	bar	atm. (A_n)	lb/ft^2	ft H_2O (at +39.2°F)
lb/in^2	1.0000	2.7680×10^{1}	7.0308×10^{1}	2.0360	5.1715×10^{1}	6.8948×10^{4}	6.8948×10^{3}	7.0306×10^{-2}	6.8947×10^{-2}	6.8045×10^{-2}	1.4400×10^{2}	2.3067
in H_2O (at +39.2°F)	3.6127×10^{-2}	1.0000	2.5400	7.3554×10^{-2}	1.8683	2.4908×10^{3}	2.4908×10^{2}	2.5399×10^{-3}	2.4908×10^{-3}	2.4582×10^{-3}	5.2022	8.3333×10^{-2}
cm H_2O (at +4°C)	1.4223×10^{-2}	0.3937	1.0000	2.8958×10^{-2}	0.7355	9.8064×10^{2}	9.8064×10^{1}	9.9997×10^{-4}	9.8064×10^{-4}	9.6781×10^{-4}	2.0481	3.2808×10^{-2}
in Hg (at +32°F)	4.9116×10^{-1}	1.3596×10^{1}	3.4532×10^{1}	1.0000	2.5400×10^{1}	3.3864×10^{4}	3.3864×10^{3}	3.4532×10^{-2}	3.3864×10^{-2}	3.3421×10^{-2}	7.0727×10^{1}	1.1330
mm Hg (Torr) (at 0°C)	1.9337×10^{-2}	5.3525×10^{-1}	1.3595	3.9370×10^{-2}	1.0000	1.3332×10^{3}	1.3332×10^{2}	1.3595×10^{-3}	1.3332×10^{-3}	1.3158×10^{-3}	2.7845	4.4605×10^{-2}
dyne/cm^2 (1μ bar)	1.4504×10^{-5}	4.0147×10^{-4}	1.0197×10^{-3}	2.9530×10^{-5}	7.5006×10^{-4}	1.0000	1.0000×10^{-1}	1.0197×10^{-6}	1.0000×10^{-6}	9.8692×10^{-7}	2.0886×10^{-3}	3.3456×10^{-5}
newton/m^2 (PASCAL)	1.4504×10^{-4}	4.0147×10^{-3}	1.0197×10^{-2}	2.9530×10^{-4}	7.5006×10^{-3}	1.0000×10^{1}	1.0000	1.0197×10^{-5}	1.0000×10^{-5}	9.8692×10^{-6}	2.0885×10^{-2}	3.3456×10^{-4}
kgm/cm^2	1.4224×10^{1}	3.9371×10^{2}	1.00003×10^{3}	2.8959×10^{1}	7.3556×10^{2}	9.8060×10^{5}	9.8060×10^{4}	1.0000	9.8060×10^{-1}	9.678×10^{-1}	2.0482×10^{3}	3.2809×10^{1}
bar	1.4504×10^{1}	4.0147×10^{2}	1.0197×10^{3}	2.9530×10^{1}	7.5006×10^{2}	1.0000×10^{6}	1.0000×10^{5}	1.0197	1.0000	9.8692×10^{-1}	2.0885×10^{3}	3.3456×10^{1}
atm. (A_n)	1.4696×10^{1}	4.0679×10^{2}	1.0333×10^{3}	2.9921×10^{1}	7.6000×10^{2}	1.0133×10^{6}	1.0133×10^{5}	1.0332	1.0133	1.0000	2.1162×10^{3}	3.3900×10^{1}
lb/ft^2	6.9445×10^{-3}	1.9223×10^{-1}	4.882×10^{-1}	1.4139×10^{-2}	3.591×10^{-1}	4.7880×10^{2}	4.7880×10^{1}	4.8824×10^{-4}	4.7880×10^{-4}	4.7254×10^{-4}	1.0000	1.6019×10^{-2}
ft H_2O (at +39.2°F)	4.3352×10^{-1}	1.2000×10^{1}	3.0480×10^{1}	8.826×10^{-1}	2.2419×10^{1}	2.9890×10^{4}	2.9890×10^{3}	3.0479×10^{-2}	2.9890×10^{-2}	2.9499×10^{-2}	6.2427×10^{1}	1.0000

Tensile yield strength[a] 10^3 psi (MPa)

Material ↓	High	Low
Cobalt & its alloys	290 (1999)	26 (179)
Low alloy hardening steels; wrought, quenched & tempered	288 (1986)	76 (524)
Stainless steels, standard martensitic grades; wrought, heat treated	275 (1896)	60 (414)
Rhenium	270 (1862)	—
Ultra-high strength steels; wrought, heat treated	270 (1862)	170 (1172)
Stainless steels, age hardenable; wrought, aged	237 (1634)	105 (724)
Nickel & its alloys	230 (1586)	10 (69)
Stainless steels, specialty grades; wrought, 60% cold worked	226 (1558)	102 (703)
Tungsten	220 (1517)	—
Molybdenum & its alloys	210 (1448)	82 (565)
Titanium & its alloys	191 (1317)	27 (186)
Carbon steels, wrought; normalized, quenched & tempered	188 (1296)	58 (400)
Low alloy carburizing steels; wrought, quenched & tempered	178 (1227)	62 (427)
Nickel-base superalloys	172 (1186)	40 (276)
Alloy steels, cast; quenched & tempered	170 (1172)	112 (772)
Stainless steels; cast	165 (1138)	31 (214)
Tantalum & its alloys	168 (1089)	48 (331)
Steel P/M parts; heat treated	154 (1062)	75 (517)
Ductile (nodular) irons, cast	150 (1034)	40 (276)
Copper casting alloys[b]	140 (965)	9 (62)
Stainless steels, standard austenitic grades; wrought, cold worked	140 (965)	75 (517)
Columbium & its alloys	135 (931)	35 (241)
Iron-base superalloys; cast, wrought	134 (924)	40 (276)
Cobalt base superalloys, wrought	116 (800)	35 (241)
Bronzes, wrought[b]	114 (786)	14 (97)
Heat treated low alloy constructional steels; wrought, mill heat treated	110 (758)	90 (621)
High copper alloys, wrought[b]	110 (758)	9 (62)
Stainless steels, standard martensitic grades; wrought, annealed	105 (724)	25 (172)
Cobalt base superalloys, cast	100 (689)	75 (517)
Heat treated carbon constructional steels; wrought, mill heat treated	100 (690)	42 (290)
Hafnium	96 (662)	32 (221)
Brasses, wrought[b]	92.5 (638)	10 (69)
Aluminum alloys, 7000 series	91 (627)	14 (97)
Alloy steels, cast; normalized & tempered	91 (627)	38 (262)
Copper-nickel-zincs, wrought[b]	90 (620)	18 (124)
Copper nickels, wrought[b]	85 (586)	13 (90)
Malleable irons, pearlitic grades; cast	80 (552)	45 (310)
High strength low alloy steels; wrought, as-rolled	80 (552)	42 (290)
Stainless steels, specialty grades; wrought, annealed	80 (552)	27 186)
Stainless steels, standard ferritic grades; wrought, cold worked	80 (552)	45 (310)
Carbon steels, wrought; carburized, quenched & tempered	77 (531)	46 (317)
Carbon steel, cast; quenched & tempered	75 (517)	—
Stainless steel (410) P/M parts; heat treated	75 (517)	—
Steel P/M parts; as-sintered	75 (517)	30 (207)
Coppers, wrought[b]	72 (496)	10 (69)
Aluminum alloys, 2000 series	66 (455)	10 (69)
Ductile (nodular) austenitic irons, cast	65 (448)	28 (193)
Zinc foundry alloys	64 (441)	30 (207)
Zinc alloys, wrought	61 (421)	23 (159)
Stainless steels, standard ferritic grades; wrought, annealed	60 (414)	35 (241)
Aluminum alloys, 5000 series	59 (407)	6 (41)
Aluminum alloys, 6000 series	55 (379)	7 (48)
Aluminum casting alloys	55 (379)	8 (55)
Carbon steels, cast; normalized & tempered	55 (379)	48 (331)
Stainless steels, standard austenitic grades; wrought, annealed	55 (379)	30 (207)
Stainless steel P/M parts, as-sintered	54 (372)	40 (276)
Rare earths	53 (365)	9.5 (66)
Zirconium & its alloys	53 (365)	15 (103)
Depleted uranium	50 (345)	35 (241)
Aluminum alloys, 4000 series	46 (317)	—
Thorium	45 (310)	26 (179)
Magnesium alloys, wrought	44 (303)	13 (90)
Silver	44 (303)	8 (55)
Carbon steels, cast; normalized	42 (290)	38 (262)
Beryllium & its alloys	40 (276)	5 (34)
Aluminum alloys, 3000 series	36 (248)	6 (41)
Carbon steel, cast; annealed	35 (241)	—
Malleable ferritic cast irons	35 (241)	32 (221)
Palladium	30 (207)	5 (34)
Gold	30 (207)	—
Magnesium alloys, cast	30 (207)	12 (83)
Polyimides, reinf	28 (193)	5 (34)
Platinum	27 (186)	2 (14)
Iron P/M parts; as-sintered	26 (179)	11 (76)
Aluminum alloys, 1000 series	24 (165)	4 (28)
Polyphenylene sulfide, 40% gl reinf	21 (145)	—
Polysulfone, 30-40% gl reinf	19 (131)	17 (117)
Acetal, copolymer, 25% gl reinf	18.5 (128)	—
Styrene acrylonitrile, 30% gl reinf	18 (124)	—
Phenylene oxide based resins, 20-30% gl reinf	17 (117)	14.5 (100)
Poly (amide-imide)	17 (117)	13.3 (92)
Polystyrene, 30% gl reinf	14 (97)	—
Zinc die-casting alloys	14 (96)	—
Polyimides, unreinf	13 (90)	7.5 (52)
Nylons, general purpose	12.6 (87)	7.1 (49)
Polyether sulfone	12.2 (84)	—
Polyphenylene sulfide, unreinf	11 (76)	—
Polysulfone, unreinf	10.2 (70)	—
Acetal, homopolymer, unreinf	10 (69)	—
Nylon, mineral reinf	10 (69)	9 (62)
Polypropylene, gl reinf	10 (69)	6 (41)
Polystyrene, general purpose	10 (69)	5.0 (34)
Phenylene oxide based resins, unreinf	9.6 (66)	7.8 (54)
Acetal, copolymer, unreinf	8.8 (61)	—
ABS/Polycarbonate	8.0 (55)	—
Lead & its alloys[b]	8 (55)	1.6 (11)
Polyaryl sulfone	8 (55)	—
ABS/Polysulfone (polyaryl ether)	7.5 (52)	—
Acrylic/PVC	7.0 (48)	6.5 (45)

Material ↓	High	Low
Tin & its alloys	6.6 (45)	1.3 (7)
ABS/PVC, rigid	6.0 (41)	—
Polystyrene, impact grades	6.0 (41)	2.8 (19)
Polypropylene, general purpose	5.2 (36)	4.8 (33)
ABS/Polyurethane	4.5 (31)	3.7(26)
Polypropylene, high impact	4.3 (30)	2.8 (19)

[a] At 0.2% offset for metals, unless otherwise noted; tensile strength at yield for plastics, per ASTM D638. [b] At 0.5% offset.

Specific strength,[a] 10^3 in. (10^3 m)

Material ↓	High	Low
Graphite-epoxy, 56-60 v/o graphite	3645 (92)	98 (2.5)
Boron-epoxy, 55 v/o boron	2829 (72)	177 (4.5)
Polyesters, thermoset pultrusions	1429 (36)	345 (8.8)
Titanium & its alloys	1043 (26.5)	171 (4.3)
Stainless steels, standard martensitic grades; wrought, heat treated	982 (24.9)	214 (5.44)
Ultra-high strength steels; wrought, heat treated	954 (24.2)	601 (15.3)
Aluminum alloys, 7000 series	892 (22.6)	144 (3.6)
Cobalt & its alloys	878 (22.3)	89 (2.3)
Stainless steels, age hardenable; wrought, aged	842 (21.4)	373 (9.47)
Nickel & its alloys	687 (17.5)	36 (0.9)
Magnesium alloys, wrought	667 (17.0)	268 (6.8)
Carbon steels; wrought; normalized, quenched & tempered	665 (16.9)	205 (5.21)
Aluminum alloys, 2000 series	647 (16.4)	101 (2.6)
Vinylidene chloride copolymer, oriented	645 (16)	242 (6.1)
Aluminum alloys, 5000 series	602 (15.3)	62 (1.6)
Alloy steels, cast; quenched & tempered	600 (15.2)	396 (10.1)
Ductile (nodular) irons, cast	592 (15.0)	158 (4.01)
Aluminum alloys, 6000 series	561 (14.2)	72 (1.8)

Material ↓	High	Low
Aluminum casting alloys	539 (13.7)	86 (2.2)
Beryllium & its alloys	533 (13.5)	75 (1.9)
Nylons, 30% gl reinf	521 (13)	396 (10)
Nickel-base superalloy	518 (13.1)	143 (3.6)
Titanium carbide-base cermets	515 (13)	—
Stainless steels, standard austenitic grades; wrought, cold worked	495 (12.6)	265 (6.73)
Aluminum alloys, 4000 series	474 (12.0)	—
Polyester, thermoplastic, PET, 45 & 30% glass reinf	459 (11.7)	288 (7.3)
Iron-base superalloys; cast, wrought	457 (11.6)	137 (3.48)
Magnesium alloys, cast	456 (11.6)	185 (4.7)
Copper casting alloys	433 (11.0)	33 (0.84)
Molybdenum & its alloys	420 (10.7)	221 (5.6)
Polycarbonate, 40 & 20% gl reinf	418 (11)	327 (8.3)
Columbium & its alloys	394 (10.0)	115 (2.9)
Stainless steels, standard martensitic grades; wrought, annealed	375 (9.53)	89 (2.26)
Styrene acrylonitrile, 30% gl reinf	367 (9.3)	—
Aluminum alloys, 3000 series	363 (9.22)	61 (1.5)
Bronzes, wrought	355 (9.0)	54 (1.4)
Rhenium	355 (9.0)	—
Cobalt base superalloys	351 (8.9)	116 (2.9)
Plastic foams, rigid, integral skin, reinf	351 (8.9)	105 (2.7)
High copper alloys, wrought	341 (8.7)	30 (0.76)
Polyester, thermoplastic, PBT, 40 & 15% glass reinf	328 (8.3)	245 (6.2)
Alloy steels, cast; normalized & tempered	322 (8.18)	134 (3.40)
Acetal, copolymer; 25% gl reinf	319 (8.1)	—
Plastic foams; rigid, integral skin, unreinf	319 (8.1)	48 (1.2)
Tungsten	314 (7.9)	—
Styrene acrylonitrile, unreinf	312 (7.9)	247 (6.3)
Nylons, unreinf	307 (7.8)	173 (4.4)
Brasses, wrought	300 (7.6)	34 (0.86)
Malleable pearlitic cast irons	300 (7.62)	169 (4.29)
Polystyrene, 30% gl reinf	298 (7.6)	—
Stainless steels, standard ferritic grades; wrought, cold worked	291 (7.39)	164 (4.17)
Copper-nickel-zincs, wrought	285 (7.2)	57 (1.4)
Carbon steels, wrought; carburized, quenched & tempered	272 (6.91)	163 (4.14)
Polyester, thermoplastic, PBT/PET blind, 30 ½ 15% gl reinf	267 (6.8)	231 (5.9)
Carbon steel, cast; quenched & tempered	265 (6.73)	—
Polyaryl sulfone	265 (6.7)	—
Copper nickels, wrought	263 (6.7)	41 (1.0)
Polystyrene, general purpose	263 (6.7)	132 (3.4)
Tantalum & its alloys	263 (6.7)	5 (0.13)
Acrylics, mldgs	253 (6.4)	133 (3.4)
Boron carbide	247 (6.3)	—
Polyphenylene oxide based resins, unreinf	246 (6.2)	200 (5.1)
Aluminum alloys, 1000 series	245 (6.2)	41 (1.04)
Ductile (nodular) austenitic irons, cast	238 (6.05)	102 (2.59)
Tungsten carbide-base cermets	236 (6)	—
Acrylic, cast	235 (6.0)	141 (3.6)
Polyacrylate, unfilled	233 (5.9)	—
Silicone	224 (5.7)	—
Coppers, wrought	223 (5.7)	31 (0.8)
Zirconium & its alloys	221 (5.6)	64 (1.62)
Polycarbonate, unreinf	218 (5.5)	193 (4.9)
Stainless steels, standard ferritic grades; wrought, annealed	218 (5.54)	127 (3.23)
Alumina ceramic	214 (5.4)	—
Polyphenylene sulfide, unreinf	208 (5.3)	—
Hafnium	204 (5.2)	68 (1.72)
Polyurethane plastics	198 (5.0)	106 (2.7)
Ethyl cellulose	195 (5.0)	73 (1.9)
Carbon steels, cast; normalized & tempered	194 (4.92)	170 (4.32)
Stainless steels, standard austenitic grades; wrought, annealed	194 (4.93)	106 (2.69)
Acetal, homopolymer, standard	192 (4.9)	—
ABS/Polycarbonate	191 (4.9)	—
Polyester, thermoplastic, PBT, 45 & 35% glass/mineral reinf	187 (4.75)	170 (4.3)
Polyaryl ether	183 (4.6)	—
Fluorocarbon, ETFE & ECTFE; gl reinf	179 (4.5)	—
Cellulose acetate	174 (4.4)	65 (1.7)
Acetal, copolymer; standard	173 (4.4)	—
Silicon carbide	172 (4.4)	—
Polypropylene, general purpose	169 (4.3)	148 (3.8)
Chlorinated polyvinyl chloride	162 (4.1)	135 (3.4)
Polyester, thermoplastic, PBT, 30 & 10% mineral filled	161 (4.1)	145 (3.6)
Cellulose nitrate	160 (4.1)	140 (3.6)
PVC, PVC-acetate, rigid	160 (4.1)	110 (2.8)
Cellulose acetate butyrate	160 (4.1)	70 (1.8)
Polyethylene, high molecular weights	159 (4.0)	—
Polystyrene, impact grades	156 (4.0)	86 (2.2)
Acetal homopolymer; 20% gl reinf	152 (3.9)	—
Mullite	151 (3.8)	—
Cellulose acetate propionate	149 (3.8)	92 (2.3)
Rare earths	149 (3.8)	88 (2.2)
Carbon steels, cast; normalized	148 (3.8)	134 (3.40)
Silicon nitride	148 (3.8)	—
Magnesia	147 (3.7)	—
ABS/PVC	143 (3.6)	62 (1.6)
Beryllia	139 (3.5)	—
Acrylic/PVC	137 (3.5)	116 (2.9)
Natural rubber	136 (3.5)	—
Fluorocarbon, PVF_2	134 (3.4)	113 (2.9)
Malleable irons, ferritic grades; cast	134 (3.40)	123 (3.12)
Polymethylpentene	133 (3.4)	—
Synthetic isoprene	132 (3.4)	—
Vinylidene chloride copolymer, unoriented	129 (3.3)	65 (1.7)
Polyethylene, high density	128 (3.3)	84 (2.1)
Allyl diglycol carbonate	125 (3.2)	104 (2.6)
Urethane, polyether and polyester	125 (3.2)	—
Nitrile	124.7 (3.2)	—
Carbon steel, cast; annealed	124 (3.15)	—
Chromium carbide-base cermets	124 (3.1)	—
Propylene ethylene	121 (3.1)	—
Ionomer	118 (3.0)	—
Silver	116 (3.0)	—
Thermoplastic elastomers	112.5 (2.9)	—
Boron nitride	112 (2.8)	—
Polypropylene, high impact	108 (2.7)	92 (2.3)
Thorium	107 (2.7)	61.9 (1.57)
Ethylene propylene, ethylene propylene diene	107 (2.7)	—
Silica	104 (2.6)	—
Ethyl butene	103 (2.6)	—
Styrene butadiene	102.9 (2.6)	—
Isobutylene-isoprene	100 (2.5)	—
Forsterite	95 (2.4)	—
Polysulfide	94 (2.4)	—
Chloroprene	89 (2.3)	—
Polybutadiene	88 (2.2)	—
Cordierite	87 (2.2)	—
Zirconia	86 (2.2)	—
Chlorinated polyethylene	83 (2.1)	—
Ethylene vinyl acetate	82 (2.1)	41 (1.0)
Fluorocarbon, PTFE	82 (2.1)	32 (0.8)
Polybutylenes, homopolymer	76 (1.9)	67 (1.7)
Polyethylene, low density	76 (1.9)	27 (0.7)
Fluorocarbon, ETFE & ECTFE; unreinf	74 (1.9)	65 (1.7)
Fluorocarbon, PTFCE	74 (1.9)	60 (1.5)
Depleted uranium	73 (1.85)	51.2 (1.3)
Polyethylene, medium density	71 (1.8)	59 (1.5)

Material ↓	High	Low
Ethylene ethyl acrylate	71 (1.8)	47 (1.2)
Palladium	69.1 (1.75)	11.5 (0.292)
Polybutylenes, copolymer	68 (1.7)	28 (0.7)
Steatite	68 (1.7)	—
Polyacrylate	62.5 (1.6)	—
Epichlorohydrin	60 (1.5)	—
Chlorosulfonated polyethylene	58 (1.5)	—
Zirconia	58 (1.5)	—
Fluorocarbon, PFA	55 (1.4)	—
Propylene oxide	54.1 (1.4)	—
Fluorocarbon, FEP	45 (1.1)	32 (0.8)
Fluorocarbon	43 (1.1)	—
Gold	42 (1.07)	—
Platinum	34.8 (0.88)	2.58 (0.065)
Tin & its alloys	23 (0.6)	5 (0.13)
Lead & its alloys	20 (0.51)	4 (0.1)

* Strength-weight ratio determined by dividing tensile yield strength of metals and ultimate tensile strength of nonmetallics by density.

Modulus of elasticity in tension,

10^6 psi (10^4 MPa)

Material ↓	High	Low
Silicon carbide	95 (65.5)	13 (8.96)
Tungsten carbide-base cermets	94.3 (65)	61.6 (42.5)
Tungsten carbide	94 (64.8)	65 (44.8)
Osmium	80 (55.1)	—
Iridium	79 (54.5)	—
Titanium, zirconium, hafnium borides	73 (50.3)	71 (48.95)
Ruthenium	68 (46.9)	—
Rhenium	68 (46.9)	—
Boron carbide	65 (44.8)	42 (28.96)
Boron	64 (44.1)	—
Tungsten	59.0 (40.6)	—
Beryllia	58 (39.9)	39 (26.9)
Titanium carbide-base cermets	57 (39.3)	42 (28.96)
Rhodium	55 (37.9)	—
Titanium carbide	55 (37.9)	36 (24.8)
Molybdenum & its alloys	53 (36.5)	46 (31.7)
Tantalum carbide	53 (36.5)	—
Magnesia	50 (34.5)	35 (24.1)
Alumina ceramic	50 (34.5)	30 (20.7)
Columbium carbide	49 (33.8)	—
Beryllium carbide	46 (31.7)	30 (20.7)
Chromium	42 (28.9)	—
Beryllium & its alloys	42.0 (28.9)	27.0 (18.6)
Graphite-epoxy composites	40 (27.6)	20 (13.4)
Cobalt base superalloys	36.0 (24.8)	29.0 (19.9)
Zirconia	35 (24.1)	23 (15.8)
Nickel & its alloys	34.0 (23.4)	19.0 (13.1)
Cobalt & its alloys	33.6 (23.1)	30.0 (20.7)
Nickel-base superalloys	33.5 (23.1)	18.3 (12.6)
Iron-base superalloys; cast & wrought	31 (21.4)	28 (19.3)
Silicon nitride	31 (21.4)	9 (6.2)
Alloy steels; cast	30 (20.7)	29 (20)
Boron-epoxy composites	30 (20.7)	—
Carbon steels; cast	30 (20.7)	—
Carbon steels, carburizing grades; wrought	30 (20.7)	29 (20)
Carbon steels, hardening grades; wrought	30 (20.7)	29 (20)
Depleted uranium	30 (20.7)	20 (13.8)
Stainless steels, age hardenable; wrought	30 (20.7)	28 (19.3)
Stainless steels, specialty grades; wrought	30 (20.7)	27 (18.6)
Ultra-high strength steels; wrought	30 (20.7)	27 (18.6)
Stainless steels; cast	29 (20)	24 (16.5)
Stainless steels, standard austenitic grades; wrought	29 (20)	28 (19.3)
Stainless steels, standard ferritic grades; wrought	29 (20)	—
Stainless steels, standard martensitic grades; wrought	29 (20)	—
Boron-aluminum composites	28 (19.3)	—
Malleable irons, pearlitic grades; cast	28 (19.3)	26 (17.9)
Tantalum & its alloys	27.0 (18.6)	21.0 (14.4)
Ductile (nodular) irons; cast	25 (17.2)	22 (15.2)
Malleable ferritic cast irons,	25 (17.2)	—
Platinum	25 (17.2)	—
Gray irons; cast	24 (16.5)	9.6 (6.62)
Copper nickels, wrought	22.0 (15.1)	18.0 (12.4)
Mullite	21 (14.5)	—
Zircon	21 (14.5)	—
Ductile (nodular) austenitic irons; cast	20 (13.8)	13 (8.96)
Hafnium	20 (13.8)	—
Copper casting alloys	19.3 (13.3)	11.0 (7.6)
Vanadium	19 (13.1)	18 (12.4)
High copper alloys, wrought	19.0 (13.1)	17.0 (11.7)
Coppers, wrought	18.7 (12.9)	17.0 (11.7)
Titanium & its alloys	18.5 (12.7)	11.0 (7.6)
Copper-nickel-zincs, wrought	18.0 (12.4)	18.0 (12.4)
Palladium	18.0 (12.4)	—
Brasses, wrought	18.0 (12.4)	15.0 (10.3)
Bronzes, wrought	17.5 (12.0)	16.0 (11.0)
Polycrystalline glass	17.3 (11.9)	12.5 (8.6)
Columbium & its alloys	16.0 (11.0)	11.5 (7.9)
Silicon	15.5 (10.7)	—
Zirconium & its alloys	14.0 (9.6)	13.8 (9.5)
Zinc alloys, wrought	14.0 (9.6)	6.2 (4.3)
Rare earths	12.2 (8.4)	2.2 (1.5)
Gold	12.0 (8.2)	—
Aluminum alloys, 4000 series	11.4 (7.9)	—
Silver	11.0 (7.6)	—
Boron nitride	11 (7.6)	7 (4.8)
Aluminum alloys, 2000 series	10.8 (7.4)	10.2 (7.0)
Silica	10.5 (7.24)	—
Aluminum alloys, 7000 series	10.4 (7.2)	10.3 (7.1)
Aluminum alloys, 5000 series	10.3 (7.1)	10.0 (6.9)
Thorium	10.3 (7.1)	—
Aluminum alloys, 1000 series	10.0 (6.9)	10.0 (6.9)
Aluminum alloys, 3000 series	10.0 (6.9)	10.0 (6.9)
Aluminum alloys, 6000 series	10.0 (6.9)	10.0 (6.9)
Thorium	10.0 (6.9)	—
Tin & its alloys	7.7 (5.3)	6.0 (4.1)
Cordierite	7 (4.8)	—
Magnesium alloys, wrought	6.5 (4.5)	6.0 (4.1)
Magnesium alloys, cast	6.5 (4.5)	6.5 (4.5)
Polyesters, thermoset, pultrusions, general purpose	6.0 (4.14)	2.3 (1.59)
Epoxy, gl laminates	5.8 (4.0)	3.3 (2.28)
Glass fiber-epoxy composites	5 (3.44)	—
Bismuth	4.6 (3.17)	—
Polyimides, gl reinf	4.5 (3.1)	—
Carbon graphite	4.0 (2.75)	0.6 (0.41)
Graphite, pyrolytic	4.0 (2.75)	—
Phenolics; reinf	3.3 (2.28)	0.35 (0.24)
Alkyds	2.9 (2.0)	1.9 (1.31)
Graphite, recrystallized	2.7-1.5 (1.86-1.03)	0.85-0.8 (0.59-0.55)
Hickory (shag bark)	2.2 (1.51)	—
Locust (black)	2.1 (1.44)	—
Polyester, thermoplastic, PET, 45 & 30% glass reinf	2.1 (1.45)	1.3 (0.90)
Birch (yellow)	2.0 (1.37)	—
Douglas fir (coast type)	2.0 (1.37)	—
Lead & its alloys	2.0 (1.38)	—
Pine (long needle, ponderosa)	2.0 (1.37)	1.3 (0.89)
Polyesters, thermoset, reinf mldgs	2.0 (1.38)	1.2 (0.83)
Ash (white)	1.8 (1.24)	—
Graphite, gen pur	1.8 (1.24)	0.5 (0.34)
Maple (sugar)	1.8 (1.24)	—
Oak (red, white)	1.8 (1.24)	—
Styrene acrylonitrile; 30% gl reinf	1.8 (1.24)	—
Beech	1.7 (1.17)	—
Carbon & graphite, fibrous reinf	1.8 (1.24)	0.3 (0.2)

Material ↓	High	Low
Graphite, premium	1.7 (1.17)	0.7 (0.48)
Walnut (black)	1.7 (1.17)	—
Polycarbonate, 40% gl reinf	1.7 (1.17)	0.86 (0.59)
Spruce (sitka)	1.6 (1.10)	—
Poplar (yellow)	1.6 (1.10)	—
Carbon, petroleum coke base	1.6 (1.10)	2.3 (1.58)
Indium	1.57 (1.08)	—
Basswood	1.5 (1.03)	—
Elm (rock)	1.5 (1.03)	—
Polysulfone, 30-40% gl reinf	1.5 (1.03)	1.1 (0.76)
Cypress (Southern bald)	1.4 (0.96)	—
Nylons; 30% gl reinf	1.4 (0.97)	1.0 (0.69)
Polyester, thermoplastic, PBT, 40 & 15% glass reinf	1.4 (0.97)	0.8 (0.55)
Cedar (Port Orford)	1.3 (0.89)	—
Cottonwood (black)	1.3 (0.89)	—
Phenylene oxide based resins; 20-30% gl reinf	1.3 (0.90)	0.93 (0.64)
Redwood (virgin)	1.3 (0.89)	—
Acetal, copolymer; 25% gl reinf	1.25 (0.86)	—
Carbon, anthracite coal base	1.2 (0.82)	0.6 (0.41)
Diallyl phthalates, reinf	1.2 (0.83)	0.6 (0.41)
Fir (balsam)	1.2 (0.83)	—
Hemlock (Eastern, Western)	1.2 (0.83)	1.5 (1.03)
Pine (Eastern white)	1.2 (0.83)	—
Polybutadienes	1.2 (0.83)	0.4 (0.28)
Polystyrene, 30% gl reinf	1.2 (0.83)	—
Polyphenylene sulfide, 40% gl reinf	1.12 (0.77)	—
Fluorocarbon, ETFE & ECTFE; gl reinf	1.1 (0.76)	—
Melamines, cellulose electrical	1.1 (0.76)	1.0 (0.69)
Cedar (Eastern red)	0.9 (0.62)	—
Polyimides, unreinf	0.70 (0.48)	0.45 (0.31)
Polyesters, thermoset, cast, rigid	0.65 (0.45)	0.15 (0.10)
Acetal, homopolymer; unreinf	0.52 (0.36)	—
Acrylics, cast, general purpose	0.50 (0.34)	0.35 (0.24)
Acrylics, mldgs	0.50 (0.34)	0.23 (0.16)
Nylon, mineral reinf	0.5 (0.34)	—
Polystyrene, general purpose	0.50 (0.34)	0.46 (0.32)
Styrene acrylonitrile; unreinf	0.50 (0.34)	0.40 (0.28)
Nylons; general purpose	0.48 (0.33)	0.28 (0.19)
Polyphenylene sulfide, unreinf	0.48 (0.33)	—
Polystyrene, impact grades	0.47 (0.32)	0.15 (0.10)
Epoxies, cast	0.45 (0.31)	0.05 (0.03)
Polycarbonate, unreinf	0.45 (0.31)	0.34 (0.23)
Acrylonitrile butadiene styrene (ABS)	0.42 (0.29)	0.29 (0.20)
Acetal, copolymer; unreinf	0.41 (0.28)	—
Phenylene oxide based resins; unreinf	0.38 (0.26)	0.36 (0.25)
ABS/Polycarbonate	0.37 (0.26)	—
Acrylic/PVC	0.37 (0.26)	0.34 (0.23)
Polyaryl sulfone	0.37 (0.26)	—
Polysulfone; unreinf	0.36 (0.25)	—
Polyether sulfone	0.35 (0.24)	—
ABS/PVC, rigid	0.33 (0.23)	—
ABS/Polysulfone (Polyaryl ether)	0.32 (0.22)	—
Allyl diglycol carbonate	0.30 (0.21)	—
Fluorocarbon, PTFCE	0.30 (0.21)	0.19 (0.13)
Fluorocarbon, ETFE & ECTFE; unreinf	0.24 (0.17)	—
ABS/Polyurethane	0.22 (0.15)	0.16 (0.11)
Polypropylene, general purpose	0.22 (0.15)	0.16 (0.11)
Polymethylpentene	0.21 (0.14)	—
Fluorocarbon, PVF	0.2 (0.14)	0.17 (0.12)
Vinylidene chloride copolymer, oriented	0.20 (0.138)	—
Polypropylene, high impact	0.13 (0.09)	—
Polyethylene, high molecular weight	0.1 (0.069)	—
Fluorocarbon, FEP	0.07 (0.05)	0.05 (0.03)
Fluorocarbon, PTFE	0.07 (0.05)	0.04 (0.03)
Vinylidene chloride copolymer, unoriented	0.07 (0.048)	—
Polybutylene, homopolymer	0.036 (0.025)	0.034 (0.023)
Polybutylene, copolymer	0.034 (0.023)	0.012 (0.008)
Polyacrylate, unfilled	0.29 (0.20)	—
Polyethylenes, low density	0.027 (0.019)	0.020 (0.014)
PVC, PVC-acetate, non-rigid	0.003 (0.0021)	0.0004 (0.00027)

Specific stiffness, 10^6 in. (10^5 m)

Material ↓	High	Low
Silicon carbide	819 (208)	—
Boron	792 (201)	—
Boron carbide	714 (181)	—
Graphite-epoxy composites	701 (178)	351 (89)
Beryllium & its alloys	560 (142)	403 (102)
Beryllia	537 (136)	—
Columbium & its alloys	466.4 (118.4)	377.0 (95.7)
Boron-epoxy composites	410 (104)	—
Magnesia	388 (99)	—
Beryllium carbide	357 (91)	—
Alumina ceramic	357 (91)	—
Boron-aluminum composites	280 (71)	—
Silicon nitride	270 (69)	—
Titanium carbide-base cermets	219 (56)	—
Silicon	184 (47)	—
Glass, polycrystalline	184 (47)	—
Mullite	176 (45)	—
Columbium carbide	173 (44)	—
Gold	171.9 (43.6)	—
Tungsten carbide-base cermets	171 (43)	—
Chromium	162 (41.1)	—
Titanium, zirconium, hafnium borides	159 (40)	—
Ruthenium	154.2 (39.2)	—
Tungsten carbide	152 (39)	—
Zirconia	150 (38)	—
Boron nitride	138 (35)	—
Molybdenum & its alloys	124.3 (31.5)	106 (26.9)
Magnesium alloys; wrought	123.7 (31.5)	98.5 (25.1)
Rhodium	123 (31.2)	—
Aluminum alloys, 4000 series	117.5 (29.8)	—
Cobalt base superalloys	109.1 (27.7)	96.6 (24.5)
Silica	109 (28)	—
Stainless steels, age hardenable; wrought	107 (27.2)	99 (25.1)
Aluminum alloys, 7000 series	106.2 (27.0)	101.9 (25.8)
Alloy steels; cast	106 (26.9)	102 (25.9)
Carbon steels, carburizing grades; wrought	106 (26.9)	102 (25.9)
Carbon steels; cast	106 (26.9)	—
Carbon steels, hardening grades; wrought	106 (26.9)	103 (26.2)
Iron-base superalloys; cast, wrought	106 (26.9)	96 (24.4)
Ultra-high strength steels; wrought	106 (26.9)	95 (24.1)
Aluminum alloys, 2000 series	105.8 (26.8)	103.0 (26.2)
Aluminum alloys, 5000 series	105.2 (26.7)	105.1 (26.6)
Malleable irons, pearlitic grades; cast	105 (27.7)	98 (24.9)
Stainless steels, standard ferritic grades; wrought	105 (26.7)	—
Nickel-base superalloys	104.0 (26.4)	65.3 (16.6)
Stainless steels, standard martensitic grades; wrought	104 (26.4)	—
Aluminum alloys, 6000 series	103.1 (26.2)	102.0 (25.9)
Cobalt & its alloys	103.1 (26.2)	101.8 (25.8)
Aluminum alloys, 1000 series	102.0 (25.9)	—
Aluminum alloys, 3000 series	102.0 (25.9)	101.0 (25.6)
Nickel & its alloys	101.8 (25.8)	67.4 (17.1)
Titanium & its alloys	101.1 (25.7)	69.6 (17.7)
Magnesium alloys; cast	100 (25.4)	98.5 (25.1)
Ductile (nodular) irons; cast	98 (24.9)	87 (22.1)
Tantalum carbide	98 (24.9)	—
Osmium	97.6 (24.8)	—
Iridium	97.1 (24.7)	—
Zirconia	97 (24.7)	—
Malleable irons, ferritic grades; cast	96 (24.4)	—
Gray irons; cast	92 (23.4)	37 (9.4)
Rhenium	89.4 (31.6)	—
Vanadium	85.9 (21.8)	81.4 (20.6)
Tungsten	84.3 (21.4)	—
Cordierite	78 (20)	—
Ductile (nodular) austenitic irons; cast	73 (18.5)	48 (12.2)
Glass fiber-epoxy composites	67 (17)	—

Material ↓	High	Low
Bronzes, wrought	61.5 (15.6)	54.5 (13.8)
Copper casting alloys	59.7 (15.1)	40.8 (10.3)
High copper alloys; wrought	58.8 (14.9)	57.0 (14.4)
Zirconium & its alloys	58.7 (14.9)	58.3 (14.8)
Brasses, wrought	58.4 (14.8)	50.6 (12.8)
Coppers, wrought	57.8 (14.7)	52.9 (13.4)
Copper-nickel-zincs, wrought	57.3 (14.5)	56.9 (14.4)
Copper nickels, wrought	56.0 (14.2)	55.7 (14.1)
Lead & its alloys	48.7 (12.3)	—
Tantalum & its alloys	45.0 (11.4)	34.7 (8.8)
Depleted uranium	43.9 (11.1)	29.3 (7.44)
Hafnium	42.5 (10.8)	—
Palladium	41.4 (10.5)	—
Styrene acrylonitrile; 30% gl reinf	37 (9.4)	—
Polyester, thermoplastic, PET 45 & 30% glass reinf.	34.4 (8.7)	23.2 (5.9)
Rare earths	34.4 (8.7)	20.3 (5.15)
Polyesters, thermoset, reinf mldgs	34 (8.6)	20 (5.1)
Platinum	32.2 (8.17)	—
Polycarbonate, 40 & 20% gl reinf	31.0 (7.9)	17.6 (4.5)
Nylons; 30% gl reinf	29 (7.4)	21 (5.3)
Silver	29.0 (7.3)	—
Phenylene oxide based resins; 20-30% gl reinf	28 (7.1)	21 (5.3)
Polysulfone; 30-4% gl reinf	28 (7.1)	21 (5.3)
Tin & its alloys	26.6 (6.7)	22.9 (5.8)
Polystyrene; 30% gl reinf	26 (6.6)	—
Polyester, thermoplastic, PBT, 40 & 15% glass reinf	24.1 (6.1)	15.7 (4.0)
Thorium	24.4 (6.2)	—
Acetal, copolymer; 25% gl reinf	22 (5.6)	—
Polyphenylene sulfide; 40% gl reinf	19 (4.8)	—
Fluorocarbon, ETFE & ECTFE; gl reinf	16.4 (4.2)	—
Polyimides, unreinf	13.5 (3.4)	8.7 (2.2)
Polystyrene, general purpose	13 (3.3)	12 (3.0)
Styrene acrylonitrile	13 (3.3)	10.4 (2.6)
Bismuth	12.9 (3.27)	—
Polystyrene, impact grades	12.2 (3.1)	3.9 (0.99)
Acrylics, mldgs	12.0 (3.1)	5.5 (1.4)
Acrylics, cast, general purpose	11.8 (3.0)	8.2 (2.1)
Polycarbonate, unreinf	10.2 (2.6)	7.7 (1.95)
Acrylonitrile butadiene styrene (ABS)	10.1 (2.6)	7.0 (1.8)
Acetal, homopolymer; unreinf	10 (2.5)	—
Polyphenylene sulfide; unreinf	10 (2.5)	—
Phenylene oxide based resins; unreinf	9.7 (2.5)	9.2 (2.3)
Nylon; mineral reinf	9.4 (2.4)	—
ABS/polycarbonate	8.6 (2.2)	—
Acetal, copolymer; unreinf	8.0 (2.0)	—
Polysulfone; unreinf	8.0 (2.0)	—
ABS/PVC, rigid	7.8 (2.0)	—
Acrylic/PVC	7.8 (2.0)	7.2 (1.8)
Polyaryl sulfone	7.6 (1.9)	—
Polyether sulfone	7.1 (1.8)	—
Polypropylene, general purpose	6.8 (1.7)	4.9 (1.2)
Polyacrylate, unfilled	6.74 (1.7)	—
Allyl diglycol carbonate	6.3 (1.6)	—
Indium	5.94 (1.5)	—
ABS/polyurethane	5.7 (1.5)	4.2 (1.1)
Polypropylene, high impact	4 (1.0)	—
Fluorocarbon, ETFE & ECTFE; unreinf	3.9 (0.99)	—
Fluorocarbon, PTFCE	3.9 (0.99)	2.5 (0.64)
Fluorocarbon, PVF	3.1 (0.79)	2.7 (0.69)
Polyethylene, high molecular weight	2.9 (0.74)	—
Vinylidene chloride copolymer, unoriented	1.13 (0.29)	—
Polybutylene, homopolymer	1.09 (0.28)	1.03 (0.26)
Polybutylene, copolymer	1.05 (0.27)	0.37 (0.09)
Fluorocarbon, FEP	0.90 (0.23)	0.65 (0.17)
Fluorocarbon, PTFE	0.88 (0.22)	0.5 (0.13)
Polyethylene, low density	0.82 (0.21)	0.61 (0.15)

Hardness of metals, Brinell

Material ↓	High	Low
White & alloy irons; cast	700	130
Osmium	670	300
Low alloy steels, wrought; normalized, quenched & tempered	627	202
Stainless steels, wrought martensitics hardened & tempered	580	180
Rhenium	555	331
Molybdenum & its alloys	555	179
Nickel & its alloys	534	75
Stainless steels, cast	470	130
Tungsten	443	330
Low alloy steels, wrought; carburized, quenched & tempered	429	212
Copper casting alloys	415	35
Alloy steels, cast; quenched & tempered	401	262
Rhodium	401	100
Iridium	351	200
Gray irons; cast	350	140
Ruthenium	350	200
Nickel-base superalloys	341	302
Titanium & its alloys	331	—
Ductile (nodular) irons; cast	300	140
Hafnium	285	277
Malleable irons, pearlitic grades; cast	269	163
Stainless steels, standard martensitic grades, wrought; annealed	260	150
Ductile (nodular) austenitic cast irons	240	130
Tantalum & its alloys	237	—
Alloy steels, cast; normalized & tempered	217	137
High strength low alloy steels, 42,000-65,000 psi yld str, wrought; as rolled	190	149
Depleted uranium	187	—
Zirconium & its alloys	179	112
Stainless steels, standard austenitic grades, wrought; annealed	170	143
Aluminum alloys, 7000 series	160	60
Malleable ferritic cast irons	156	110
Aluminum casting alloys	145	40
Aluminum alloys, 2000 series	135	45
Zinc foundry alloys	125	85
Aluminum alloys, 4000 series	120	—
Aluminum alloys, 6000 series	120	25
Palladium	118	40
Columbium & its alloys	114	—
Platinum	106	40
Aluminum alloys, 5000 series	105	28
Zinc die-casting alloys	91	82
Silver	90	26
Magnesium alloys, wrought	82	46
Magnesium alloys, cast	80	50
Aluminum alloys, 3000 series	77	28
Rare earths	77	17
Gold	66	25
Aluminum alloys, 1000 series	44	19
Tin & its alloys	29	5
Lead & its alloys	17	4.7
Indium	0.9	—

Hardness of plastics, Rockwell R

Material ↓	High	Low
Melamine, cellulose electrical	125	115
Polyphenylene sulfide; unreinf	124	—
Polyphenylene sulfide; 40% gl reinf	123	—
Chlorinated polyvinyl chloride	122	117
Nylon, mineral reinf	121	119
Polyester, thermoplastic, PBT; unreinf	120	117
Polyester, thermoplastic, PET, 45 & 30% glass reinf	120	—
Polyester, thermoplastic, PBT/PET blend, 30 ½ 15% gl. reinf	120	119
Cellulose acetate	120	49
Nylons, general purpose	120	111
PVC, PVC-acetate; rigid	120	110
Polysulfone; unreinf	120	—

Material ↓	High	Low
Phenylene oxide based resins; unreinf	119	115
Polyester, thermoplastic; PBT, 45 & 35% glass/mineral reinf	119	114
Polyester, thermoplastic, PBT; 30% gl reinf	119	118
Polyester, thermoplastic, PBT, 30 & 10% mineral filled	118	112
ABS/Polycarbonate	117	—
ABS/Polysulfone (polyaryl ether)	117	—
Acrylonitrile butadiene styrene (ABS)	115	75
Fluorocarbon, PTFCE	115	110
Polypropylene; gl reinf	115	90
Cellulose acetate butyrate	114	23
Fluorocarbon, PVF$_2$	110	—
Cellulose acetate propionate	109	57
ABS/PVC	102	—
Polypropylene, general purpose	100	80
Fluorocarbon, ETFE & ETCFE, unreinf	95	—
ABS/Polyurethane	82	70
Fluorocarbon, PTFE	55	35

Hardness of rubber & elastomers, Shore A

Material ↓	High	Low
Thermoplastic elastomers	100	35
Urethane, polyether and polyester	100	80
Styrene butadiene	100	40
Natural rubber	100	30
Chlorinated polyethylene	95	40
Chloroprene	95	40
Chlorosulfonated polyethylene	95	50
Silicone	90	25
Polyacrylate	90	40
Nitrile	90	20
Fluorocarbon	90	65
Ethylene propylene, ethylene propylene diene	90	30
Epichlorohydrin	80	40
Isobutylene-isoprene	80	40
Polybutadiene	80	45
Polysulfide	80	20
Propylene oxide	80	40
Synthetic isoprene	80	40

Hardness of ceramics, Mohs

Material ↓	High	Low
Mullite	18	14
Columbium carbide	10	9
Beryllium carbide	9+	—
Tantalum carbide	9+	—
Beryllia	9	—
Alumina	9	—
Titanium carbide	9	8
Zirconium carbide	9	8
Zirconia	8	7
Zircon	8	—
Steatite	7.5	—
Thoria	7	—
Magnesia	6	—
Porcelain enamel	6	3½
Calcia	6.0	3.3
Boron and aluminum nitrides	2	—

Maximum service temperature of nonmetallics,[a] F (K)

Material ↓	High	Low
Thoria	4890* (2972)	—
Beryllia	4350* (2672)	—
Calcia	4350 (2672)	—
Magnesia	4350* (2672)	—
Zirconia	4350* (2672)	—
Silicon carbide	4200 (2589)	—
Boron carbide	4100 (2533)	—
Alumina ceramic	3540* (2222)	—
Mullite	3200 (2033)	—
Forsterite	1832 (1273)	—
Steatite	1832 (1273)	—
Mica, phlogopite	1800 (1255)	1400 (1033)
Mica, ceramoplastic	1500 (1116)	700 (644)
Porcelain enamel	1500 (1088)	700 (644)
Mica, natural muscovite	1110 (872)	—
Silicone coatings	1000 (811)	—
Tungsten carbide flame sprayed	1000 (811)	—
Mica, glass bonded	700 (644)	—
Polyimides	600 (589)	—
Silicone elastomer	600 (589)	—
Silicone plastics, reinf	600 (589)	—
Epoxies, cycloaliphatic, unreinf	550 (561)	480 (522)
Fluorocarbon, PTFE	550 (561)	—
Phenolics, reinf	550 (561)	220 (378)
Fluorocarbon coatings	550 (561)	—
Glass, soda lime (tempered)	550 (561)	—
Epoxies	500 (533)	250 (394)
Epoxy novolacs, unreinf	500 (533)	450 (505)
Fluorocarbon elastomers	500 (511)	—
Glass, borosilicate (tempered)	500 (533)	—
Polyaryl sulfone	500 (533)	—
Polyphenylene sulfide	500 (533)	—
Alkyds, reinf	450 (505)	300 (422)
Felt, TFE fluorocarbon	450 (505)	—
Melamines, reinf	430 (494)	170 (350)
Diallyl phthalates	400 (478)	300 (422)
Epoxy amine coatings	400(477)	—
Fluorocarbon, FEP	400 (478)	—
Fluorocarbon, CTFE	390 (472)	350 (450)
Polyacrylate elastomer	375 (464)	—
Felt, nylon	350 (450)	—
Felt, polyester	350 (450)	—
Phenolic coatings	350 (450)	—
Polyesters, thermoplastic, PBT & PTMT	350 (450)	—
Polyesters, thermoset, reinf	350 (450)	300 (422)
Polysulfones	350 (450)	340 (444)
Ethylene propylene, ethylene propylene diene	325 (436)	—
Chlorosulfonated polyethylene elastomer	325 (436)	—
Chloronated polyethylene elastomers	300 (422)	—
Epichlorohydrin elastomers	300 (422)	—
Fluorocarbon, ETFE & ECTFE	300 (422)	—
Fluorocarbon, PVF$_2$	300 (422)	—
Isobutylene-isoprene elastomers	300 (422)	—
Nitrile elastomers	300 (422)	—
Nylons	300 (422)	175 (353)
Polypropylene, general purpose	300 (422)	225 (380)
Polyurethane coatings	300 (422)	—
Propylene oxide elastomers	300 (422)	—
Thermoplastic elastomers	300 (422)	—
Chlorinated polyether	290 (416)	—
Polymethylpentene	275 (408)	—
Felt, polypropylene	250 (394)	—
Polyaryl ether	250 (394)	—
Polycarbonate	250 (394)	—

Material ↓	High	Low
Polyethylene, high density	250 (394)	175 (353)
Polypropylene, high impact	250 (394)	200 (366)
Polysulfide elastomers	250 (394)	—
Polyurethane plastics, unreinf	250 (394)	190 (361)
Urethane, polyether and polyester	250 (394)	—
Chloroprene elastomers	240 (389)	—
Propylene ethylene	240 (389)	190 (361)
Acrylics	230 (383)	125 (325)
Acrylonitrile butadiene styrene	230 (383)	130 (328)
Chlorinated polyvinyl chloride	230 (383)	—
Felt, rayon viscose	225 (380)	—
Polybutylene	225 (380)	—
Cellulose acetate	220 (378)	140 (333)
Cellulose butyrate	220 (378)	140 (333)
Cellulose propionate	220 (378)	155 (341)
Phenylene oxide based resins	220 (378)	175 (353)
ABS/Polycarbonate	220 (378)	—
Acetal, copolymer	212 (373)	—
Allyl diglycol carbonate	212 (373)	—
Polyethylene, low density	212 (373)	180 (355)
Alkyd coatings	200 (366)	—
Chloroprene coatings	200 (366)	—
Polybutadiene elastomers	200 (366)	—
Polyester coatings	200 (366)	—
Styrene acrylonitrile	200 (366)	140 (333)
Acetal, homopolymer	195 (364)	—
Ethylene ethyl acrylate	190 (361)	—
Polyurethane plastics	190 (361)	—
Ethyl cellulose	185 (358)	115 (319)
Acrylic coatings	180 (355)	—
Ionomer	180 (355)	160 (344)
Natural rubber	180 (355)	—
Nitrocellulose coatings	180 (355)	—
Styrene butadiene elastomers	180 (355)	—
Synthetic isoprene	180 (355)	—
Polystyrenes	175 (353)	140 (333)
Polyvinyl chloride	175 (353)	140 (333)
Ureas, cellulose reinf	170 (350)	—
Polyvinyl coatings	150 (339)	—
Cellulose nitrate	140 (333)	—

[a] At zero stress.

Heat deflection temperature of plastics,[a]

F (K)

Material ↓	High	Low
Silicone plastics	>900(>755)	—
Polyimides	680 (633)	582 (588)
Poly (amide-imide)	545 (558)	520 (544)
Epoxy, cycloaliphatic diepoxides, cast, rigid	525 (547)	300 (422)
Polyaryl sulfone	525 (547)	—
Polybutadienes	500 (533)	—
Nylons; 30% gl reinf	495 (530)	420 (489)
Polyester, thermoplastic, PET, 45 & 35% glass reinf	442 (501)	428 (493)
Epoxy novolacs	425 (491)	422 (300)
Polyphenylene sulfide; 40% gl reinf	425 (491)	—
Polyester, thermoplastic, PBT, 40 & 15% glass reinf	405 (480)	375 (464)
Alkyds	>400(>478)	350 (450)
Epoxy; mineral gl reinf	400 (478)	340 (444)
Melamines	400 (478)	265 (403)
Polyesters, thermoset, cast rigid	400 (478)	120 (322)
Polyesters, thermoset, reinf mldgs	400 (478)	375 (464)
Polyesters, thermoset, pultrusions	400 (478)	—
Polyether sulfone	400 (478)	—
Plastic foams, rigid, integral skin; gl reinf	390 (472)	162 (345)
Polyester, thermoplastic, PBT, 45 & 35% glass/mineral reinf	390 (472)	340 (444)
Polyester, thermoplastic, PBT, 30 & 10% mineral filled	385 (469)	150 (339)
Polyester, thermoplastic PBT/PET blend, 30 & 15% gl reinf	380 (466)	320 (433)
Polysulfones	365 (458)	345 (447)
Vinylidene chloride copolymer	339 (150)	328 (130)
Polyacrylate, unfilled	320 (433)	—
Acetals	325 (436)	212 (373)
Phenylene oxide based resins, 20-30% gl reinf	310 (428)	282 (412)
ABS/Polysulfone	300 (422)	—
Nylon; mineral reinf	300 (422)	—
Polypropylene; gl reinf	300 (422)	250 (394)
Polycarbonates	295 (419)	260 (400)
Fluorocarbon, ETFE & ECTFE; gl reinf	285 (414)	—
Polyphenylene sulfide, unreinf	278 (410)	—
Plastic foams, rigid, integral skin; unreinf	>270(>405)	94 (307)
Polyphenylene oxide based resins; unreinf	265 (403)	212 (373)
Acrylonitrile butadiene styrene (ABS)	245 (391)	180 (355)
Chlorinated polyvinyl chloride	234 (385)	202 (368)
Fluorocarbon, PVF_2	232 (384)	—
Epoxy, cast, rigid	230 (383)	—
ABS/Polycarbonate	220 (378)	—
Fluorocarbon, PTFE; ceramic reinf	220 (378)	170 (350)
Nylons, general purpose	220 (378)	155 (341)
Polystyrenes	220 (378)	210 (372)
Styrene acrylonitrile	220 (378)	210 (372)
ABS/Polyurethane	207 (370)	201 (367)
Cellulose acetate butyrate	196 (364)	118 (321)
Cellulose acetate	188 (360)	117 (320)
Acrylic/PVC	185 (358)	160 (344)
Fluorocarbon, PTFCE	178 (354)	151 (339)
Cellulose acetate propionate	173 (351)	129 (327)
Fluorocarbon, ETFE & ECTFE; unreinf	170 (350)	160 (344)
Epoxy, cast, flexible	155 (341)	90 (305)
Polybutylene, homopolymer	140 (333)	120 (322)
Polypropylene, general purpose	140 (333)	135 (330)
Polypropylene, high impact	140 (333)	120 (322)
Polyester, thermoplastic, PBT unreinf	130 (328)	123 (324)

[a] At 264 psi (1.82 MPa) stress.

Specific heat, Btu/lb·F (J/kg·K)

Material ↓	High	Low
Polyester, thermoset, cast, rigid	0.55 (2301)	0.30 (1255)
Polyethylenes, high density	0.55 (2301)	0.46 (1925)
Polyethylenes, low & medium density	0.55 (2301)	0.53 (2218)
Nylons	0.5 (2092)	0.3 (1255)
Epoxies, cast, rigid	0.5 (2092)	0.4 (1673)
Polypropylene, high impact	0.48 (2008)	0.45 (1883)
Beryllium & its alloys	0.45 (1883)	—
Propylene, general purpose	0.45 (1883)	—
Cellulose acetate	0.42 (1757)	0.3 (1255)
Cellulose acetate butyrate	0.4 (1674)	0.3 (1255)
Phenolics, reinf	0.40 (1674)	0.28 (1172)
Cellulose acetate propionate	0.4 (1674)	0.3 (1255)
ABS/Polysulfone (polyaryl ether)	0.35 (1464)	—
Acetals	0.35 (1464)	—
Acrylics	0.35 (1464)	0.33 (1381)
Polyester, thermoset; reinf mldgs	0.35 (1464)	0.20 (837)
Polystyrene, general purpose	0.35 (1464)	0.30 (1255)
Polystyrene, impact grade	0.35 (1464)	0.30 (1255)
Wrought magnesium alloys	0.346 (1448)	0.245 (1025)
Silicon carbide, (0-2550 F; 255-1672 K)	0.34 (1422)	0.285 (1192)
Beryllium carbide (85-210 F; 303-372 K)	0.334 (1397)	—
Fluorocarbon, PVF_2	0.33 (1381)	—
Styrene acrylonitrile	0.33 (1381)	—
Vinylidene chloride copolymer	0.32 (1339)	—
Polyimides, unreinf	0.31 (1297)	0.27 (1130)
Boron	0.307 (1284)	—
ABS/Polycarbonate	0.3 (1255)	—
Allyl diglycol carbonate	0.3 (1255)	—
Polycarbonate; unreinf	0.30 (1255)	—
Acrylic/PVC	0.29 (1213)	—

Material ↓	High	Low
Fluorocarbon, FEP	0.28 (1172)	—
Polyesters, thermoset, pultrusions	0.28 (1172)	0.24 (1004)
Polyimides, gl reinf	0.27 (1130)	—
Magnesia (400 F; 477 K)	0.26 (1088)	—
Polyphenylene sulfide	0.26 (1088)	—
Polystyrene; 30% gl reinf	0.26 (1088)	—
Beryllia (200 F; 366 K)	0.25 (1046)	—
Fluorocarbon, PTFE	0.25 (1046)	—
Magnesium alloys, cast	0.245 (1025)	0.245 (1025)
Polysulfone	0.24 (1004)	—
Aluminum alloys, 6000 series	0.23 (962)	0.23 (962)
Aluminum alloys, 7000 series	0.23 (962)	0.23 (962)
Aluminum alloys, 5000 series	0.23 (962)	0.22 (920)
Aluminum alloys, 2000 series	0.23 (962)	0.22 (920)
Aluminum alloys, 3000 series	0.22 (920)	0.22 (920)
Boron carbide	0.22 (920)	—
Aluminum alloys, 1000 series	0.22 (920)	—
Fluorocarbon, PTFCE	0.22 (920)	—
Calcia	0.21 (879)	0.075 (314)
Nickel-base superalloys	0.20 (837)	0.106 (443)
Forsterite	0.2 (837)	—
Alumina ceramic	0.19 (795)	—
Polycrystalline glass	0.190 (795)	0.18 (753)
Boron and aluminum nitrides	0.17 (711)	—
Zirconia	0.16 (669)	—
Silica	0.16 (669)	—
Silicon	0.1597 (668.2)	—
Titanium & its alloys	0.155 (648)	0.110 (460)
Silicon nitride	0.15 (627)	—
Titanium, zirconium, hafnium borides	0.15 (627)	0.05 (209)
Cobalt & its alloys	0.14 (585)	—
Nickel & its alloys	0.14 (585)	0.091 (381)
Stainless steel, cast	0.14 (586)	0.11 (460)
Ductile (nodular) irons; cast	0.13 (544)	—
Gray irons; cast	0.13 (544)	—
Titanium carbide	0.13 (544)	—
Zinc foundry alloys	0.125 (523)	0.104 (410)
Cobalt base superalloys	0.12 (502)	0.09 (376)
Iron-base superalloys; cast, wrought	0.12 (502)	0.10 (418)
Specialty stainless steels; wrought	0.12 (502)	0.10 (418)
Stainless steels, wrought austenitics	0.12 (502)	—
Vanadium	0.12 (502.1)	—
Zircon	0.12 (502)	—
Indium	0.117 (489)	—
Manganese	0.114 (476.9)	—
Alloy steels, cast	0.11 (460)	0.10 (418)
Carbon steels, cast	0.11 (460)	0.10 (418)
Carbon steels, hardening grades; wrought	0.11 (460)	0.10 (418)
Stainless steels, standard and ferritic & martensitic grades wrought	0.11 (460)	—
Chromium	0.1068 (446.8)	—
High copper alloys, wrought	0.10 (418)	0.09 (376)
Bronzes, wrought	0.10 (418)	0.09 (376)
Copper, casting alloys	0.10 (418)	0.09 (376)
Zinc die-casting alloys	0.10 (418)	0.10 (418)
Zinc, wrought alloys	0.096 (402)	0.094 (393)
Coppers, wrought	0.092 (385)	0.092 (385)
Brasses, wrought	0.09 (376)	0.09 (376)
Copper-nickel-zinc, wrought	0.09 (376)	0.09 (376)
Copper-nickel, wrought	0.09 (376)	0.09 (376)
Zirconium carbide (400 F; 477 K)	0.08 (334)	—
Columbium carbide (400 F; 477 K)	0.08 (334)	—
Zirconium & its alloys	0.0659 (276)	—
Molybdenum & its alloys	0.065 (272)	—
Columbium and its alloys	0.065 (272)	0.060 (251)
Rhodium	0.059 (247)	—
Palladium	0.058 (243)	—
Ruthenium	0.057 (238)	—
Silver	0.056 (234)	—
Tungsten carbide +15% cobalt flame sprayed coatings	0.056 (234)	—
Tin & its alloys	0.054 (226)	—
Tungsten carbide + 9% cobalt flame sprayed coatings	0.048 (201)	—
Tungsten carbide (400 F; 477 K)	0.04 (167)	—
Tantalum carbide	0.04 (167)	—
Tantalum & its alloys	0.036 (151)	—
Rhenium	0.035 (146)	—
Hafnium	0.035 (146)	—
Lead & its alloys	0.032 (134)	0.031 (130)
Gold	0.031 (130)	—
Tungsten	0.034 (142)	—
Iridium	0.031 (130)	—
Platinum	0.031 (130)	—
Osmium	0.031 (130)	—
Thorium	0.03 (126)	—
Depleted uranium	0.03 (126)	—
Bismuth	0.0294 (123.0)	—
Thorium	0.0282 (117.9)	—

Thermal conductivity, Btu·ft/hr·sq ft·F (W/m·K)

Material ↓	High	Low
Silver plating	244 (422)	—
Silver (212 F, 373 K)	242 (419)	—
Copper, casting alloys	226 (391)	16 (28)
Coppers, wrought	226 (391)	112 (194)
Copper plating	222 (384)	—
High copper alloys, wrought	218 (377)	62 (107)
Graphite, pyrolytic	215 (372.1)	108 (186.9)
Gold (212 F, 373 K)	172 (298)	—
Gold plating	169 (292)	—
Graphite recrystallized	140-90 (242.3-155.8)	80-39 (138.5-67.5)
Aluminum alloys, 1000 series	135 (234)	128 (222)
Brasses, wrought	135 (234)	15 (26)
Aluminum alloys, 6000 series	125 (216)	99 (171)
Beryllium & its alloys (212 F, 373 K)	123 (213)	87 (151)
Bronzes, wrought	120 (208)	20 (35)
Aluminum alloys, 5000 series	116 (201)	67.4 (117)
Aluminum alloys, 2000 series	111 (192)	82.5 (143)
Aluminum alloys, 3000 series	111 (192)	93.8 (162)
Boron nitride	105 (182)	—
Tungsten (212 F, 373 K)	96.6 (167)	—
Beryllia (2190 F; 1472 K)	95.2 (164)	—
Graphite, premium	95 (164.4)	65 (112.5)
Graphite, gen pure	94 (162.7)	65 (112.5)
Aluminum casting alloys	92.5 (160)	51 (88)
Molybdenum & its alloys (212 F, 393K)	84.5 (146)	—
Magnesium alloys; wrought	79 (137)	29 (50)
Zinc foundry alloys	72.5 (125)	66.3 (115.1)
Aluminum alloys, 7000 series	70 (121)	70 (121)
Carbon-graphite	66 (114.2)	18 (31.2)
Zinc die-casting alloys (158-284 F, 343-413 K)	65.5 (113)	62.9 (109)
Zinc plating	64.2 (111.1)	—
Zinc, wrought alloys	62.2 (108)	60.5 (105)
Magnesium alloys, cast	58 (100)	24 (42)
Tungsten carbide	50.8 (88)	16.5 (28.5)
Tungsten carbide-base cermets	50.1 (86.7)	25.7 (44)
Rhodium (212 F, 393 K)	50 (87)	—
Carbon & graphite, fibrous reinf	50 (86.5)	2 (3.5)
Nickel & its alloys	49.6 (86)	5.67 (10)
Silicon	48.4 (83.7)	—
Rhenium (212 F, 373 K)	43.7 (76)	—
Platinum (212 F, 373 K)	42 (73)	—
Indium	41.1 (71.1)	—
Palladium (212 F, 373 K)	41 (71)	—
Tin & its alloys	37 (64)	—
Nickel plating	34.4 (59.5)	—
Iridium (212 F, 373 K)	34 (59)	—
Columbium & its alloys (212 F, 373 K)	31.5 (55)	—
Tantalum & its alloys (212 F, 373 K)	31.5 (55)	—
Copper nickels, wrought	31 (54)	17 (29)
Gray irons; cast (212 F, 373 K)	30 (51.9)	25 (43.3)
Malleable irons; cast	29.5 (51.1)	—
Ultrahigh strength steels	29 (50.2)	17 (29.4)

Material ↓	High	Low
Carbon & alloy steels; cast (212 F, 373 K)	27 (46.7)	—
Carbon steels, carburizing & hardening grades; wrought (212 F, 373 K)	27 (46.7)	—
Nickel-base superalloys (1600 F, 1144 K)	27 (47)	5.25 (9)
Copper-nickel-zincs, wrought	26 (45)	17 (29)
Silicon carbide (2200 F, 1478 K)	25 (43)	9 (15)
Thorium (392 F; 473 K)	21.8 (37.7)	—
Stainless steels, standard martensitic grades; wrought (212 F, 373 K)	21.2 (36.7)	11.7 (20.2)
Rare earths (82 F, 301 K)	21.0 (36)	0.023 (0.040)
Lead plating	20.1 (34.8)	—
Alumina flame sprayed coating	20 (34.6)	19 (32.9)
Ductile (nodular) irons; cast (212 F, 373 K)	20 (34.6)	18 (31)
Lead & its alloys (212 F, 373 K)	19.6 (34)	16 (28)
Silicon nitride	19 (33)	8 (14)
Cr_2O_3 flame sprayed coating	18 (31.1)	—
Vanadium (212 F; 373 K)	17.9 (31.0)	—
Depleted uranium (212 F, 373 K)	17.2 (30)	—
Cobalt & its alloys (1300 F, 978 K)	16.6 (29)	5.2 (9)
Boron carbide (2200 F, 1478 K)	16 (27.7)	—
Cobalt base superalloys, cast (1300 F, 978 K)	15.8 (27)	11.9 (20.6)
Stainless steels, standard ferritic grades; wrought (212 F; 373 K)	15.6 (27)	13.8 (23.9)
Stainless steels; cast (212 F; 373 K)	15.2 (26.3)	7.5 (13)
Iron-base superalloys; wrought & cast (1100-1200 F; 866-922 K)	15 (26)	10.5 (18.2)
Specialty stainless steels; wrought (212 F; 373 K)	14.4 (24.9)	6.5 (11.2)
Cobalt base superalloys, wrought (1300 F, 978 K)	13.1 (22.6)	11.0 (19.0)
Tantalum carbide	12.8 (22)	—
Beryllium carbide (68-795 F; 243-697 K)	12.1 (20.9)	—
Stainless steels, age hardenable; wrought (212 F; 373 K)	12.1 (20.9)	8.7 (15.1)
Zirconium carbide	11.9 (20.5)	—
Titanium carbide	11.9 (20.5)	9.9 (17)
Titanium & its alloys	11.5 (20)	3.9 (7)
Carbon, anthracite coal base	11 (19.0)	7 (12.1)
Zirconium & its alloys (212 F, 373 K)	9.6 (17)	8.1 (14)
Stainless steels, standard austenitic grades; wrought (212 F; 373 K)	9.4 (16.3)	8.2 (14.2)
Porcelain enamel	9.0 (15.5)	6.0 (3.46)
Columbium carbide	8.23 (14.2)	—
Zirconium oxide flame sprayed coating	8 (13.8)	7 (12.1)
Cadmium plating	5.3 (9.17)	—
Tungsten carbide + 9% cobalt flame sprayed coating	5.3 (9.2)	—
Tungsten carbide + 15% cobalt flame sprayed coating	5.3 (9.2)	—
Carbon, petroleum coke base	5 (8.7)	3 (5.2)
Bismuth	4.83 (8.37)	—
Calcia (1830 F; 1272 K) 9% pososity	4.12 (7.13)	—
Zircon	3.61 (6.2)	2.88 (4.9)
Forsterite	2.40 (4.1)	1.94 (3.3)
Cordierite	2.40 (4.1)	0.97 (1.7)
Polycrystalline glass	1.95 (3.4)	1.13 (1.9)
Steatite	1.94 (3.3)	1.45 (2.5)
Wood comp board, hardboard (tempered)	1.50 (2.6)	1.10 (1.65)
Magnesia (2190 F; 1472 K; 22% porosity)	1.47 (2.5)	—
Mullite	1.47 (2.5)	1.38 (2.4)
Wood comp board, particleboard (medium density)	1.0 (0.40)	—

Material ↓	High	Low
Epoxies, molded	0.87 (0.5)	0.1 (0.17)
Silica, 0.9% porosity	0.80 (1.4)	—
Wood comp board, medium density	0.60 (1.04)	0.50 (0.87)
Polyimides, unreinf	0.57 (0.99)	0.21 (0.36)
Thermoplastic polyester, PBT, 30% carbon reinf	0.542 (0.938)	—
Zirconia (2190 F; 1472 K), 28% porosity	0.53 (0.9)	—
Silicone plastics	0.50 (0.87)	0.075 (0.13)
Wood comp board, structural insulating	0.45 (0.78)	0.27 (0.47)
Polyesters, thermoset, pultrusions	0.42 (0.73)	0.33 (0.57)
Mica, phlogophite	0.4 (0.69)	0.3 (0.51)
Mica, natural muscovite	0.36 (0.62)	0.25 (0.43)
Felt, flurocarbon (fine)	0.35 (0.605)	—
Phenolics, molding grades	0.309 (0.535)	0.116 (0.201)
Epoxies, cast, rigid	0.3 (0.52)	0.1 (0.17)
Polyimide, gl reinf	0.299 (0.517)	—
Mica, ceramoplastic	0.29 (0.50)	—
Nylons, 30% gl reinf	0.29 (0.50)	0.1 (0.17)
Melamine, gl fiber	0.28 (0.48)	—
Mica, glass bonded	0.24 (0.42)	—
Felt, polyester	0.24 (0.42)	0.175 (0.302)
Felt, rayon viscose (medium)	0.23 (0.40)	0.19 (0.33)
Acrylonitrile butadiene styrene	0.20 (0.35)	0.08 (0.14)
Melamine, cellulose electrical	0.20 (0.35)	0.17 (0.29)
Cellulose acetate	0.19 (0.33)	0.10 (0.17)
Cellulose acetate butyrate	0.19 (0.33)	0.10 (0.17)
Cellulose acetate propionate	0.19 (0.33)	0.10 (0.17)
Polyethylenes, low-high density & high molecular weight	0.19 (0.33)	—
Felt, nylon (fine)	0.183 (0.31)	—
Polyester, thermoplastic, PET, 45 & 30% glass reinf	0.183 (0.31)	—
ABS/Polysulfone	0.173 (0.299)	—
Polyphenylene sulfide; unreinf	0.167 (0.289)	—
Acetal, copolymer	0.16 (0.28)	—
Polysulfone, unreinf	0.15 (0.26)	—
Fluorocarbon, PTFCE	0.145 (0.251)	—
Nitrile	0.143 (0.247)	—
Polypropylene, high impact	0.143 (0.247)	—
Styrene butadiene	0.143 (0.247)	—
Fluorocarbon, PTFE	0.14 (0.24)	—
Fluorocarbon, PVF_2	0.14 (0.24)	—
Nylons, general purpose	0.14 (0.24)	0.1 (0.17)
Polyesters, thermoset, reinf high strength mldgs	0.14 (0.24)	0.11 (0.19)
Acetal, homopolymer, unreinf	0.13 (0.22)	—
Acrylics	0.13 (0.22)	0.12 (0.21)
Fluorocarbon	0.13 (0.22)	—
Silicone	0.13 (0.22)	—
Polycarbonate, 40% gl reinf	0.128 (0.222)	—
Polyester, thermoplastic, PBT, 45 & 35% glass/mineral reinf	0.128 (0.222)	0.100 (0.17)
Phenylene oxide based resins, unreinf	0.125 (0.216)	0.092 (0.159)
Polybutylene, homopolymer	0.125 (0.216)	—
Allyl diglycol carbonate	0.12 (0.21)	—
Fluorocarbon, FEP	0.12 (0.21)	—
Polycarbonate, unreinf	0.118 (0.204)	0.113 (0.196)
Polystyrene; 30% gl reinf	0.117 (0.202)	—
Polypropylene, general purpose	0.113 (0.196)	0.10 (0.17)
Chloroprene	0.112 (0.19)	—
Thermoplastic polyester, PBT; unreinf & 30% gl reinf	0.108 (0.187)	0.092 (0.159)
PVC, PVC-acetate	0.10 (0.17)	0.07 (0.12)
Phenylene oxide based resins, 20 & 30% gl fiber reinf	0.096 (0.166)	0.092 (0.159)
Polyaryl sulfone	0.092 (0.159)	—
Polystyrene, general purpose	0.090 (0.156)	0.058 (0.10)
Polystyrene, impact grades	0.090 (0.156)	0.024 (0.042)
Thermoplastic elastomers	0.087 (0.15)	—
Acrylic/PVC	0.084 (0.145)	—
Natural rubber	0.082 (0.14)	—
Synthetic isoprene	0.082 (0.14)	—
ABS/Polycarbonate	0.079 (0.137)	—
Plastic foams, rigid, no surface skin	0.077 (0.133)	0.009 (0.016)
Chlorosulfonated polyethylene	0.065 (0.112)	—
Isobutylene-isoprene	0.053 (0.09)	—

Material ↓	High	Low
Vinylidene chloride copolymer	0.053 (0.092)	—
Felt, wool	0.03 (0.52)	—
Felt, polypropylene (fine)	0.0216 (0.0375)	—
Thoria (2190 F; 1472 K), 17% porosity	0.0 (0.0)	—

Coefficient of thermal expansion,[a]

10^{-6} in./in./F (10^{-6} m/m/K)

Material ↓	High	Low
Polyethylene, medium & high density	167 (301)	83 (149)
ABS/polyurethane	121 (218)	116 (209)
Fluorocarbon, PFA	111 (200)	67 (121)
Polyethylenes, low density	110 (198)	89 (160)
Fluorocarbon, FEP	105 (189)	83 (149)
Cellulose acetate	90 (162)	44 (79)
Cellulose acetate butyrate	90 (162)	60 (108)
Cellulose acetate propionate	90 (162)	60 (108)
Vinylidene chloride copolymer	88 (158)	—
Fluorocarbon, PVF_2	85 (153)	—
Acrylonitrile butadiene styrene (ABS)	72 (130)	16 (29)
Polybutylene, homopolymer	71 (128)	—
Phenylene oxide based resins; unreinf	68 (38)	59 (33)
Plastic foams, rigid, integral skin type; unreinf[b]	67 (121)	50 (90)
Acrylics	60 (108)	30 (54)
Allyl diglycol carbonate	60 (108)	—
Polypropylene, high impact	59 (106)	40 (72)
Polypropylene, general purpose	58 (104)	38 (68)
Polyester, thermoset, cast, rigid	56 (101)	39 (70)
Polystyrene, impact grades	56 (101)	22 (40)
Fluorocarbon, PTFE	55 (99)	—
Diallyl phthalates	50 (90)	22 (40)
Epoxy, cast, flexible	50 (90)	30 (54)
Nylons, general purpose	50 (90)	45 (81)
Nylon; mineral reinf	50 (90)	27 (49)
Polyester, thermoplastic, PBT/PET blend, 30 & 15% gl reinf	50 (90)	25 (45)
Silicone plastics	50 (90)	25 (45)
Polyester, thermoplastic, PBT, 30 & 10% mineral filled	48 (86)	42 (76)
Polystyrene, general purpose	48 (86)	33 (59)
Acetal, copolymer; unreinf	47 (85)	—
Acetals; 20-25% gl reinf	47 (85)	20 (36)
ABS/PVC, rigid	46 (83)	—
Acetals, homopolymer; unreinf	45 (81)	—
Acrylic/PVC	44 (79)	35 (63)
Polyacrylate, unfilled	40 (72)	35 (63)
Polyester, thermoplastic, PBT, unreinf	40 (72)	—
Fluorocarbon, PTFCE	39 (70)	—
Chlorinated polyvinyl chloride	38 (68)	—
Polycarbonate; unreinf	38 (68)	18 (32)
Polyesters, thermoplastic, PBT, 45 & 35% glass/mineral reinf	38 (68)	28 (50)
ABS/polycarbonate	37 (67)	—
Styrene acrylonitrile; unreinf	37 (67)	36 (65)
Polyester, thermoplastic, PBT, 40 & 15% glass reinf	35 (63)	28 (50)
Epoxy, cast, rigid	33 (59)	—
Polysulfone; unreinf	31 (56)	—
Alkyds	30 (54)	10 (18)
Polyphenylene sulfide; unreinf	30 (54)	—
Melamine, cellulose electrical	28 (50)	11 (20)
Polyimides; unreinf	28 (50)	25 (45)
Polyarylsulfone	26 (47)	—
Nylons; 30% gl reinf	25 (45)	12 (22)
Poly(amide-imide), high impact	24 (43)	—
Phenolics, mldg grades, general purpose	23 (41)	12 (22)
Polyimide; 15% graphite reinf	23 (41)	—
Polyphenylene sulfide	22 (40)	—
Magnesium alloys, wrought (68-750 F, 293-672 K)	21.8 (39.2)	14.0 (25.2)
Epoxy; mineral gl reinf	20 (36)	10 (18)
Fluorocarbon, PTFE; ceramic reinf	20 (36)	17 (31)
Phenylene oxide based resins; 20-30% gl reinf	20 (36)	14 (25)
Polyester, thermoplastic, PET, 45 & 30% glass reinf	20 (36)	13 (23.4)
Zinc alloys, wrought (68-212 F, 293-373 K)	19.3 (34.7)	6.0 (10.8)
Poly(amide-imide), high modulus	19 (34)	—
Polyester, thermoset; gl reinf high strength mlds	19 (34)	13 (23)
Phenolics, impact mldg grades	18 (32)	6.7 (12)
Polystyrene; 30% gl reinf	18 (32)	—
Fluorocarbon, ETFE & ECTFE; gl reinf	17 (31)	—
Lead & its alloys (68-212 F, 293-373 K)	16.3 (29.3)	16.0 (28.8)
Polysulfone; 30-40% gl reinf	16 (29)	12 (22)
Styrene acrylonitrile; 30% gl reinf	16 (29)	—
Zinc foundry alloys	15.5 (26.9)	12.9 (23.2)
Zinc die-casting alloys (68-212 F, 293-373 K)	15.2 (27.4)	—
Phenolics, heat resistant mldg grades	15 (27)	7.8 (14)
Fluorocarbon, ETFE & ECTFE; unreinf	14 (25)	—
Magnesium alloys, cast	14 (25)	—
Aluminum casting alloys (68-212 F, 293-373 K)	13.7 (24.7)	9.0 (16.2)
Aluminum alloys, 5000 series (68-212 F, 293-373 K)	13.4 (24.1)	13.1 (23.6)
Aluminum alloys, 3000 series (68-212 F, 293-373 K)	13.3 (23.9)	12.9 (23.2)
Aluminum alloys, 2000 series (68-212 F, 293-373 K)	13.2 (23.8)	12.4 (22.3)
Aluminum alloys, 1000 series (68-212 F, 293-373 K)	13.2 (23.8)	13.1 (23.6)
Aluminum alloys, 6000 series (68-212 F, 293-373 K)	13.1 (23.6)	12.9 (23.2)
Aluminum alloys, 7000 series (68-212 F, 293-373 K)	13.1 (23.6)	13.0 (23.4)
Tin & its alloys (32-212 F, 273-373 K)	13.0 (23.4)	—
Manganese (68 F, 293 K)	12.2 (22.0)	—
Copper casting alloys (68-572 F, 293-573 K)	12.0 (21.6)	9.0 (16.2)
Brasses, wrought (68-572 F, 293-573 K)	11.8 (21.2)	10.0 (18.0)
Bronzes, wrought (68-572 F, 293-573 K)	11.8 (21.2)	8.3 (14.9)
Silver (32-212 F, 273-373 K)	10.9 (19.6)	—
Aluminum alloys, 4000 series (68-212 F, 293-373 K)	10.8 (19.4)	—
Iron-base superalloys; wrought & cast (70-1500 F, 294-1089 K)	10.7 (19.3)	9.4 (16.9)
White & alloy cast irons (70°F, 294°K)	10.7 (19.3)	4.5 (8.1)
Specialty stainless steels; wrought (70-212 F, 294-373 K)	10.5 (18.9)	4.8 (8.6)
Ductile (nodular) austenitic irons; cast (70-400 F, 294-478 K)	10.4 (18.7)	7.0 (12.6)
Stainless steels; cast (70-1000 F, 294-811 K)	10.4 (18.7)	6.4 (11.5)
Stainless steels, standard, austenitic grades; wrought (32-212 F, 273-373 K)	10.4 (18.7)	8.3 (14.9)
Cobalt & its alloys (70-1800 F, 294-1255 K)	9.9 (17.8)	6.8 (12.2)
Nickel-base superalloys (70-200 F, 294-366 K)	9.89 (17.8)	5.92 (10.6)
Cobalt base superalloys; cast (70-1800 F, 294-1255 K)	9.8 (17.6)	8.7 (15.7)
High copper alloys; wrought (68-572 F, 293-573 K)	9.8 (17.6)	9.0 (16.2)
Coppers; wrought (68-572 F, 293-573 K)	9.8 (17.6)	9.3 (16.7)
Nickel & its alloys	9.6 (17.2)	6.2 (11.2)
Copper nickels; wrought (68-572 F, 293-573 K)	9.5 (17.1)	9.0 (16.2)
Carbon-graphite	9.4 (16.9)	1.0 (1.8)
Cobalt base superalloys; wrought (70-1800 F, 294-1255 K)	9.4 (16.9)	9.0 (16.2)

Material ↓	High	Low
Copper-nickel-zincs; wrought (68-572 F, 293-573 K)	9.3 (16.7)	9.0 (16.2)
Polycarbonate; 40% gl reinf	9.3 (16.7)	—
Beryllium & its alloys (70 F, 294 K)	9.0 (16.2)	6.4 (11.5)
Carbon steels, carburizing grades; wrought (70-1200 F, 294-922 K)	8.4 (15.1)	—
Alloy steels; cast (70-1200 F, 294-922 K)	8.3 (14.9)	8.0 (14.4)
Carbon steels; cast (70-1200 F, 294-922 K)	8.3 (14.9)	—
Carbon steels, hardening grades; wrought (70-1200 F, 294-922 K)	8.3 (14.9)	7.5 (13.5)
Melamine; gl reinf	8.2 (14.8)	—
Stainless steels, age hardenable; wrought (70-212 F, 294-373 K)	8.2 (14.8)	5.3 (9.5)
Gold (68 F, 293 K)	7.9 (14.2)	—
Magnesia (68-2550 F, 293-1672 K)	7.78 (14)	—
Depleted uranium (70 F, 294 K)	7.7 (13.9)	—
Malleable irons, pearlitic grades; cast (68-212 F, 293-373 K)	7.5 (13.5)	—
Titanium carbide-base cermets (68-1200 F, 293-922 K)	7.5 (13.5)	4.3 (7.7)
Bismuth	7.39 (13.3)	—
Calcia (70-1000 F, 294-811 K)	7.0 (12.6)	—
Thorium (70 F, 294 K)	6.94 (12.5)	—
Gray irons; cast (32-212 F, 273-373 K)	6.8 (12.2)	6.0 (10.8)
Carbon & Graphite, fibrous reinf	6.7 (12.1)	1.8 (3.2)
Ductile (nodular) irons; cast (70-400 F, 294-478 K)	6.6 (11.9)	—
Stainless steels, standard ferritic grades; wrought (32-212 F, 273-373 K)	6.6 (11.9)	5.2 (9.4)
Palladium (68 F, 293 K)	6.5 (11.7)	—
Stainless steels, standard martensitic grades; wrought (32-212 F, 273-373 K)	6.2 (11.2)	5.5 (9.9)
Titanium & its alloys (68-1000 F, 293-811 K)	6.0 (10.8)	4.5 (8.1)
Chromium carbide-base cermets (68-576 F, 293-575 K)	6.0 (10.8)	—
Malleable irons, ferritic grades; cast (68-212 F, 293-373 K)	5.9 (10.6)	—
Beryllium carbide (77-1472 F, 298-1073 K)	5.8 (10.4)	—
Beryllia (68-2550 F, 293-1672 K)	5.28 (9.5)	—
Thoria (68-2550 F, 293-1672 K)	5.28 (9.5)	—
Ruthenium (68 F, 293 K)	5.1 (9.2)	—
Polyester, thermoplastic, PBT; 30% carbon reinf	5 (9)	—
Polyester, thermoset, pultrusions	5 (9)	3 (5.4)
Vanadium (70 F, 294 K)	4.96 (8.94)	—
Platinum (68 F, 293 K)	4.9 (8.8)	—
Titanium, zirconium, hafnium borides (70-4000 F, 294-2478 K)	4.8 (8.6)	4.2 (7.5)
Forsterite (68-212 F, 293-373 K)	4.72 (8.5)	—
Rhodium (68 F, 293 K)	4.7 (8.49)	—
Glass, soda lime	4.7 (8.46)	—
Boron (68-1380 F, 293-1022 K)	4.61 (8.3)	—
Tantalum carbide (77-1472 F, 298-1073 K)	4.6 (8.3)	—
Graphite, recrystallized (70 F, 294 K)	4.5 (8.1)	0.4 (0.09)
Alumina ceramic (77-1830 F, 298-1272 K)	4.3 (7.74)	—
Boron nitride (70-1800 F, 294-1255 K)	4.17 (7.5)	—
Titanium carbide (77-1472 F, 298-1073 K)	4.1 (7.4)	3.7 (6.6)
Tungsten carbide	4.1 (7.4)	2.5 (4.5)
Steatite (68-212 F, 293-373 K)	3.99 (7.2)	3.33 (5.9)
Tungsten carbide-base cermets (68-1200 F, 293-922 K)	3.9 (7)	2.5 (4.5)
Columbium & its alloys; wrought (70 F, 294 K)	3.82 (6.9)	3.80 (6.8)
Iridium	3.8 (6.8)	—
Zirconium carbide (77-1472 F, 298-1073 K)	3.7 (6.6)	—
Rhenium (70 F, 294 K)	3.7 (6.7)	—
Osmium	3.6 (6.5)	—
Zirconium & its alloys (212 F, 373 K)	3.6 (6.5)	3.1 (5.6)
Chromium (68 F, 293 K)	3.4 (6.2)	—
Hafnium (70 F, 294 K)	3.4 (6.2)	—
Glass, polycrystalline (77-570 F, 298-572 K)	3.2 (5.7)	0.2 (0.36)
Zirconia (68-2190 F, 293-1472 K)[c]	3.1 (5.6)	—
Mullite (68-212 F, 293-373 K)	3.0 (5.4)	2.7 (4.8)
Molybdenum & its alloys (70 F, 294 K)	2.7 (4.9)	—
Silicon	2.6 (4.67)	—
Tungsten (70 F, 294 K)	2.5 (4.5)	—
Silicon carbide (0-2550 F, 255-1672 K)	2.4 (4.3)	2.17 (3.9)
Graphite, pyrolitic (70 F, 294 K)	2.2 (4.0)	1.1 (2.0)
Cordierite (68-212 F, 293-373 K)	2.08 (3.7)	—
Zircon (68-212 F, 293-373 K)	1.84 (3.3)	1.31 (2.3)
Glass, borosilicate	1.83 (3.2)	—
Boron carbide (0-2550 F, 255-1672 K)	1.73 (3.1)	—
Carbon, anthracite coal base (70 F, 294 K)	1.5 (2.7)	1.3 (2.3)
Carbon, petroleum coke base (70 F, 294 K)	1.5 (2.7)	1.3 (2.3)
Silicon nitride (70-1800 F, 294-1255 K)	1.37 (2.4)	—
Graphite, premium (70 F, 294 K)	0.5 (0.9)	0.963 (0.113)
Glass, 96% silica	0.44 (0.79)	—
Glass, fused silica (quartz)	0.31 (0.56)	—
Silica (68-2280 F, 293-1522 K)	0.28 (0.5)	—
Mica, phlogopite	0.27 (0.49)	0.144 (0.26)
Mica, natural muscovite	0.18 (0.32)	—
Mica, ceramoplastic	0.067 (0.12)	0.062 (0.11)
Mica, glass bondad	0.058 (0.10)	—
Graphite, general purpose	0.064 (0.115)	0.055 (0.099)

[a] For −22 to 86 F (243-303 K) for plastics; as indicated for other materials. [b] Polyethylene, polypropylene and polystyrene types.
[c] Depends on degree of stabilization.

Dielectric strength of nonmetallics,[a]

v/mil (10^6 v/m)

Material ↓	High	Low
Fluorocarbon, PFA	2000 (78.7)	—
Mica (step by step, 1/8 in.)	2000 (78.7)	1000 (39.37)
Chlorinated polyvinyl chloride	1500 (59.1)	1220 (48.0)
PVC, PVC-acetate, rigid	1400 (55.1)	725 (28.5)
Ionomer	1000 (39.4)	—
Polyesters, thermoplastic, PBT & PTMT, unreinf	750 (29.5)	540 (21.2)
Polyester, thermoplastic, PBT; 30% gl reinf	750 (29.5)	—
Polymethylpentenes	700 (27.6)	—
Acrylic/PVC	670 (26.37)	430 (16.9)
Polyallomer	650 (25.6)	500 (19.7)
Polypropylene, general purpose	650 (25.6)	—
Polypropylene, high impact	650 (25.6)	450 (17.7)
Polystyrene, impact grades	650 (25.6)	300 (11.8)
ABS/PVC, rigid	600 (23.6)	—
Cellulose acetate	600 (23.6)	250 (9.8)
Fluorocarbon, PTFCE	600 (23.6)	530 (20.9)
Fluorocarbon, FEP	600 (23.6)	500 (19.7)
Polybutadienes	600 (23.6)	400 (15.7)
Polyphenylene sulfide, unreinf	595 (23.4)	—
Acetals, 20-25% gl reinf	580 (22.8)	500 (19.7)
Polyimides, unreinf	560 (22.0)	310 (12.2)
Ethylene ethyl acrylate	550 (21.7)	—
Acrylics	530 (20.9)	400 (15.7)
Ethylene vinyl acetate	525 (20.7)	—
Styrene acrylonitrile; 30% gl reinf	515 (20.3)	—
Polystyrene, general purpose	>500(>19.7)	—
Acetals, unreinf	500 (19.7)	—

Material ↓	High	Low
Fluorocarbon, PTFE	500 (19.7)	400 (15.7)
Mica, glass bonded	500 (19.7)	—
Phenylene oxide based resins, unreinf	500 (19.7)	400 (15.7)
Porcelain enamel	500 (19.7)	400 (15.7)
Styrene acrylonitrile, unreinf	500 (19.7)	400 (15.7)
Fluorocarbon, ETFE & ECTFE; unreinf	490 (19.3)	—
Polyphenylene sulfide	490 (19.3)	—
Polyethylenes, low-high density & high molecular weight	480 (18.9)	—
Polysulfone, 30 & 40% gl reinf	480 (18.9)	—
Polypropylene; gl reinf	475 (18.7)	317 (12.5)
Nylons, general purpose	470 (18.5)	385 (15.2)
Cellulose acetate propionate	450 (17.7)	300 (11.8)
Diallyl phthalates; reinf	450 (17.7)	350 (13.8)
Nylons; 30% gl reinf	450 (17.7)	400 (15.7)
Polycarbonate, unreinf	450 (17.7)	380 (15.0)
Polycarbonate, 40% gl reinf	450 (17.7)	—
Poly (amide-imide)	440 (17.3)	430 (16.9)
ABS/Polysulfone (polyaryl ether)	430 (16.9)	—
Phenolics	425 (16.7)	200 (7.8)
Polysulfone, unreinf	425 (16.7)	—
Acrylonitrile butadiene styrene	415 (16.3)	300 (11.8)
Cellulose acetate butyrate	400 (15.7)	250 (9.8)
Ceramoplastic	400 (15.7)	270 (10.6)
Melamines, cellulose electrical	400 (15.7)	350 (13.8)
Phenolics, heat resistant mldg grades	400 (15.7)	210 (8.3)
Phenolics, impact mldg grades	400 (15.7)	300 (11.8)
Polystyrene, 30% gl reinf	396 (15.6)	—
Alkyds	350 (13.8)	290 (11.4)
Polyaryl sulfone	350 (13.8)	—
Polycrystalline glass	350 (13.78)	250 (9.84)
Phenolics, general purpose mldg grades	340 (13.4)	234 (9.2)
Alumina ceramic	300 (11.81)	200 (7.87)
Mullite	300 (11.81)	—
Allyl diglycol carbonate	290 (11.4)	—
Steatite	280 (11.02)	145 (5.71)
Fluorocarbon, PVF_2	260 (10.2)	—
Zircon	250 (9.84)	—
Forsterite	250 (9.84)	—
Cordierite	230 (9.06)	140 (5.51)

[a] Short term.

Dielectric constant of nonmetallics[a]

Material ↓	High	Low
Alumina ceramic	10.0	8.0
Zircon	10.0	8.0
Silicon nitride	9.4	—
Mica	8.7-5.4	—
Melamines	7.9	5.2
Fluorocarbon, PVF_2	7.5	—
Polycrystalline glass	7.13, 10^3 cps (10^3Hz)	5.62
Cellulose acetate	7.0	3.2
Mullite	7.0	6.5
Phenolics	7.0	4.0
Ceramoplastic	6.9	6.8
Alkyds	6.8	4.5
Mica, glass bonded	6.7	—
Forsterite	6.5	6.2
Steatite	6.3, 6 x 10^1 cps (6 x 10^1 Hz)	5.9
Cordierite	6.23	4.02
Cellulose acetate butyrate	6.2	3.2
Polyesters, thermoset, cast, flexible	6.1	3.7
Nylons; 30% gl reinf	5.4	3.5
Epoxies	5.2	2.78
Epoxy novolacs	5.1	4.3
Boron nitride	4.8	4.1
Polyesters, thermoset, general purpose reinf mldgs	4.75	4.55
Glass, borosilicate	4.6	—
Polyesters, thermoset, cast, rigid	4.4	2.8
Silicone plastics	4.3	3.4
Acetal, homopolymer, 20% gl reinf	4.0	—
Polystyrene, impact grades	4.0	2.4
Acetal, copolymer, 25% gl reinf	3.9	—
Poly (amide-imide)	3.9	3.8
Polyimides; unreinf	3.9	3.5
Polyphenylene sulfide; 40% gl reinf	3.9	—
Allyl diglycol carbonate	3.8	3.5
Glass, 96% silica	3.8	—
Glass, fused silica (quartz)	3.8	—
Nylons, general purpose	3.8	3.5
Acetals, unreinf	3.7	—
Cellulose acetate propionate	3.7	3.4
Polyaryl sulfone	3.7	—
Polyester, thermoplastic, PBT, 30% gl reinf.	3.7	—
ABS/Polycarbonate	3.6	3.2
Polyether sulfone	3.5	—
Polycarbonate; 40% gl reinf	3.48	—
Acrylic/PVC	3.44	3.06
Polysulfone, 30 & 40% gl reinf	3.4	—
Styrene acrylonitrile, 30% gl reinf	3.4	—
Polybutadienes	3.3	—
Acrylonitrile butadiene styrene (ABS)	3.2	2.4
Polyesters, thermoplastic, PBT & PTMT, unreinf	3.2	3.1
Polyphenylene sulfide, unreinf	3.2	—
Polypropylenes	3.2	2.0
ABS/Polysulfone (polyaryl ether)	3.10	—
Polycarbonate; unreinf	3.1	3.0
Polysulfone; unreinf	3.0	—
Styrene acrylonitrile	3.0	2.6
Polystyrene; 30% gl reinf	3.0	—
Phenylene oxide based resins; 20 & 30% gl reinf	2.9	—
Acrylics	2.9	2.5
Phenylene oxide based resins, unreinf	2.7	2.6
Polystyrene, general purpose	2.7	2.5
Fluorocarbon, ETFE & ECTFE; unreinf	2.5	—
Flurocarbon, PTFCE	2.4	2.3
Polybutylene, copolymer	2.25	2.18
Polybutylene, homopolymer	2.25	—
Fluorocarbon, FEP	2.1	—
Fluorocarbon, PFA	2.1	—
Fluorocarbon, PTFCE	2.1	—
Polymethylpentenes	2.1	—
Polyethylene, cellular foam	1.84	1.05
Plastic foams, rigid, no surface skin	1.1	2.0

[a] At 10^6 cyc/sec (10^6 Hz) unless otherwise indicated.

Plastics – Including Reinforced & Filled Types

ABS (Acrylonitrile Butadiene Styrene)

Type →		Medium Impact	Very High Impact	Low Temp Impact	Heat Resistant	Self-extinguishing (No PVC)	Transparent
PHYSICAL PROPERTIES	ASTM						
Specific gravity	D792	1.05-1.07	1.01-1.06	1.02-1.04	1.06-1.08	1.21	1.07
Refractive index, nD	D542	—	—	—	—	—	1.538
Transparency (visible light), %	—	Opaque	Opaque	Opaque	Opaque	Opaque	85
MECHANICAL PROPERTIES							
Tensile	D638						
Strength, 1000 psi (yield)		6.3-8.0	4.5-6.0	4.0-6.0	7.0-8.0	6.0	5.6
Elongation, % (ultimate)		5-20	20-50	30-200	20	20	—
Modulus, 1000 psi		330-400	200-310	200-310	350-420	320	290
Flexural	D790						
Strength, 1000 psi		9.9-11.8	6.0-9.8	5-8	11.0-12.0	10.0	9.95
Modulus, 1000 psi		350-400	200-320	200-320	350-420	330	300
Compressive str, 1,000 psi (2% offset)	D695	10.5-11.0	—	—	9.3-11.0	—	7.0
Impact strength							
Izod (notched), lb/in.	D256	2.0-4.0	5.0-7.0	6-10	2.0-4.0	4.0	5.3
Gardner, ft-lb		—	—	—	—	—	—
Hardness							
Rockwell	PD785	R108-115	R85-105	R75-95	R107-116	R100	R100
Shore	PD785	—	—	—	—	—	—
Barcol		—	—	—	—	—	—
Bending fatigue str (Woehler), 1,000 psi at 10^7 cycles	D671	—	—	—	—	—	—
Abrasion res (Taber, CS-17 wheel), mg/1000 cycles	D1044	—	—	—	—	—	83.8
Coef of static friction	D1894						
Against self		—	—	—	—	—	0.501
Against steel		—	—	—	—	—	0.281
ELECTRICAL PROPERTIES							
Vol res, ohm-cm	D257	2.7×10^{16}	$1\text{-}4 \times 10^{16}$	$1\text{-}4 \times 10^{16}$	$1\text{-}5 \times 10^{16}$	—	—
Dielec str, v/mil (dry)	D149						
Short-time		385	300-375	300-415	360-400	—	—
Dielec constant (dry)	D150						
60 Hz		2.8-3.2	2.8-3.5	2.5-3.5	2.7-3.5	—	—
1 MHz		2.75-3.0	2.4-3.0	2.4-3.0	2.8-3.2	—	—
Dissip factor (dry)	D150						
60 Hz		0.003-0.006	0.005-0.010	0.005-0.01	0.030-0.040	—	—
1 MHz	D495	0.008-0.009	0.008-0.016	0.008-0.016	0.005-0.015	—	—
Arc res, sec		—	—	—	—	—	—
THERMAL PROPERTIES							
Ther cond, Btu/hr/sq ft/°F/in. (Cenco-Fitch)	C177	0.96—2.16	0.12—1.68	0.96—1.68	1.44—2.40	—	—
Coef of ther exp, 10^{-5}/°F	D696	3.2-4.8	5.0-6.0	5.0-6.0	3.7-5.1	3.7-6.3	0.46
Specific heat, Btu/lb/°F	—					—	—
Heat deflection temp °F	D648						
At 66 psi		—	—	—	—	198	195
At 264 psi		185-223	180-218	185-224	220-245	180	183
Brittleness temp, °F	D746	—	—	—	—	—	—
Max temp for continuous use (no load), °F	—	—	—	—	—	150	—
APPLICABLE PROCESSING METHODS		Injection mldg, extrusion, thermofrmg, blow mldg, foam mldg.					Injection mldg.
CHEMICAL RESISTANCE		Highly res to aqueous acids, alkalis, salts. Res to concentrated phosphoric and hydrochloric acids, alcohols, and animal, vegetable and mineral oils. Disintegrated by concentrated sulfuric or nitric acids. Soluble in esters, ketones, ethylene dichloride					
USES		Pipe, appliance housings, housewares, lawn and garden equipment, chrome plated parts, highway safety devices, extruded profiles, shoe heels, fume hoods and ducts, toys, office equipment; also available as formable sheet for such uses as cases, luggage, refrigerator linings					

[1] For a bar 2½ x ½ x ¼ in.

Acetals

Material →		Homopolymer[c]			Copolymer[d]		
		Standard	20% Glass Reinforced	22% TFE Filled	Standard	25% Glass Coupled	High Flow
PHYSICAL PROPERTIES	ASTM						
Specific Gravity	D792	1.425	1.56	1.54	1.410	1.59	1.410
Ther Cond, Btu/hr/sq ft/°F/in.	—	1.56	—	—	1.92	—	—
Coef of Ther Exp, 10^{-5} per °F	D696	4.5	2.0-4.5	4.5	4.7	2.2-4.7	4.7
Specific Heat, Btu/lb/°F	—	0.35	—	—	0.35	—	0.35
Refractive Index, n_D	D542	Opaque	Opaque	Opaque	transulcent	translucent	translucent
Water Absorption (24 hr), %	D570	0.25	0.25	0.20	0.22	0.29	0.22
MECHANICAL PROPERTIES							
Tensile Strength, 1000 psi	D638						
Break	—	10.0	8.5	6.9	—	16	—
Yield	—	10.0	—	—	8.8	—	8.8
Elongation, %	D638			12			
Ultimate	—	25	7	—	60-75	2-3	40
Yield	—	12	—	—	12	—	12
Mod of Elast in Ten, 10^5 psi	D638	5.2	—	—	4.1	12.0	4.1
Flex Strength, 1000 psi	D790	14.1	—	—	13	24	13
Mod of Elast in Flex, 10^5 psi	D790	4.1	8.8	4.0	3.75	10.5	3.75
Impact Str (Izod, notched), ft-lb/in.	D638	1.4	0.8	0.7	1.3-1.5	1.1	1.0
Compr Str (1%), 1000 psi	D695	5.2	5.2	4.5	4.5	—	4.5
Hardness (Rockwell)	D785	M94	M90	M78	M80	—	M80
Coef of Static Frict (against steel)	—	0.1-0.3	0.1-0.3	0.05-0.15	0.15	—	0.15
Abr Res (Taber, CS-17), mg/1000 cyc	D1044	14-20	33	9	14	—	14
ELECTRICAL PROPERTIES							
Volume Resistivity, ohm-cm	D257	1×10^{15}	5×10^{14}	—	1×10^{14}	0.8×10^{14}	1.0×10^{14}
Dielectric str (short-time), v/mil[e]	D149	500	500	—	500	600	500
Dielectric constant	D150						
60 Hz		3.7	4.0	—	3.7 @ 100	4.12 @ 100	3.7 @ 100
1 MHz		3.7	4.0	—	3.7	4.04	3.7
Dissipation factor	D150						
60 Hz		0.0048	0.0047	—	0.001 @ 100	0.003 @ 100	0.001 @ 100
1 MHz		0.0048	0.0036	—	0.006	0.006-0.007	0.006
Arc resistance, sec	D495	129[b]	188	—	240	142	240
HEAT RESISTANCE							
Max Rec Service Temp, F	—	195	195	195	220	220	220
Deflection Temp, F	D648						
66 psi		338	345	329	316	—	316
264 psi		255	3.5	212	230	322	230
APPLICABLE PROCESSING METHODS		Injection molding, extrusion, rotational molding, blow molding			Injection molding, extrusion, rotational molding, blow molding, foam molding		
CHEMICAL RESISTANCE		Exc res to most organic solvents, incl aliphatic, aromatic hydrocarbons. Not rec for strong acids and alkalis.	Same as standard homopolymer	Same as standard homopolymer	Excellent res to strong alkalis. Most organic solvents including alcohols, ketones, esters, aliphatic and aromatic hydrocarbons and glycols do not seriously alter properties. Not recommended for use in strong mineral acids and oxidizing reagents.		
USES		Appliance parts, gears, bushings, aerosol bottles, auto, plumbing, textile, consumer uses	Same as homopolymer. Where high stiffness and dimensional stability are required	Same as homopolymer. Where low friction and high resistance to wear are required	Appliance parts, gears, bushings, aerosol bottles, auto, plumbing, textile machinery, consumer products	Same as standard grade where high stiffness and greater thermal stability are required.	Same as standard grade, especially consumer products requiring faster cycles.

a 10% deformation. b 15-mil specimen. c "Delrin" is most common tradename. d "Celcon" is most common tradename. e 0.090-in. thick specimen
f in air

Acrylics

Type →		Cast Sheets, Rods		Moldings		
		General Purpose Type I[a]	General Purpose Type II[a]	Grades 5, 6, 8[b]	High Impact Grade	Modified (XT acrylic)
PHYSICAL PROPERTIES	ASTM					
Specific Gravity	D792	1.17-1.19	1.18-1.20	1.18-1.19	1.12-1.16	1.10-1.12
Ther Cond, Btu/hr/sq ft/°F/in.		1.44	1.44	1.44	1.44	1.56
Coef of Ther Exp, 10^{-5} per °F	D696	4.5	4.5	3-4	4-6	4.4-4.5
Spec Ht, Btu/lb/°F	—	0.35	0.35	0.35	0.34	0.33
Refractive Index	D542	1.485-1.500	1.485-1.495	1.489-1.493	—	1.51
Transmittance (luminous, 0.125 in.), %	D791	91-92	91-92	>92	—	86-88
Haze, %	D672	1-2	1-2	<3	—	—
Water Absorption (24 hr), %	D570	0.3-0.4	0.2-0.4	0.3-0.4	0.2-0.3	0.3
MECHANICAL PROPERTIES						
Mod of Elast in Tension, 10^5 psi	D638	3.5-4.5	4.0-5.0	3.5-5.0	2.3-3.3	3.7-4.3
Ten Str, 1000 psi	D638	6-9	8-10	9.5-10.5	5.5-8.0	7.0-8.0
Elong (in 2 in.), %	D638	2-7	2-7	3-5	>25	12-30
Hardness (Rockwell)	D785	M80-90	M96-102	M80-103	L60-94	M45-68
Impact Str (Izod notched), ft-lb/in.	D256	0.4	0.4	0.2-0.4	0.8-2.3	1-2
Mod of Elast in Flex, 10^5 psi	D790	3.5-4.5	4.0-5.0	3.5-5.0	2.8-3.6	3.5-4.0
Flex Str, 1000 psi	D790	12-14	15-17	15-16	8.7-12.0	11-13
Compr Yld Str (0.1% offset), 1000 psi	D695	12-14	14-18	14.5-17	7.3-12.0	9.5-11.5
ELECTRICAL PROPERTIES						
Vol Res, ohm-cm	D257	$>10^{15}$	$>10^{15}$	$>10^{14}$	2.0×10^{16}	$>10^{16}$
Dielec Str (short time), v/mil	D149	450-530	450-500	400	400-500	400-500
Dielec Const						
60 Hz	D150	3.5-4.5	3.5-4.5	3.5-3.9	3.5-3.9	—
1 MHz	D150	2.7-3.2	2.7-3.2	2.7-2.9	2.5-3.0	2.78-2.86
Dissip Factor						
60 Hz	D150	0.05-0.06	0.05-0.06	0.04-0.06	0.03-0.04	0.026-0.029
1 MHz	D150	0.02-0.03	0.02-0.03	0.02-0.03	0.01-0.02	0.022-0.025
Arc Resistance, sec	......	No track	No track	No track	No track	No track
APPLICABLE PROCESSING METHODS		Thermoforming, casting		Injection molding, extrusion, thermoforming, blow molding		Injection molding, extrusion, thermoforming, blow molding
HEAT RESISTANCE						
Max Recommended Svc Temp, F		140-160	180-200	155-190	—	160
Heat Dist Temp, F		150-180	190-225	166-250[d]	169-205	195 (264 ps)
CHEMICAL RESISTANCE		Resists weak alkalis, acids and aliphatic hydrocarbons. Attacked by esters, ketones, aromatic hydrocarbons, chlorinated hydrocarbons and concentrated acids				
USES		Transparent aircraft enclosures, radio and television parts, lighting, drafting equipment, signs		Decorative and functional automotive parts, reflectors, protective goggle lenses, radio and television parts, appliances	Shoe heels, control knobs, business machine and piano keys, pump parts, sprinkler heads, tool handles	Packaging, lenses containers, shields

[a] ASTM D702. [b] Range includes typical values for Grades 5, 6, and 8, and may be superior to minimum or maximum requirements for these grades as detailed in ASTM D788. [c] Cenco-Fitch. [d] D788 specified values for Grades 5, 6, and 8: 149 F, 162 F, 183 F respectively.

Alkyds and Thermoset Carbonate

Material →		Allyl Diglycol Carbonate	Alkyds			
			Putty (encapsulating)	Rope (general purpose)	Granular (high speed molding)	Glass Reinforced (heavy duty parts)
PHYSICAL PROPERTIES	ASTM					
Specific Gravity	D792	1.32	2.05–2.15	2.20–2.22	2.21–2.24	2.02–2.10
Ther Cond, Btu/hr/sq ft/°F/in.	—	1.45	4.2—7.2	4.2—7.2	4.2—7.2	2.4—3.6
Coef of Ther Exp, per °F	D696	6 x 10^{-5}	1-3 x 10^{-5}	1-3 x 10^{-5}	1-3 x 10^{-5}	1-3 x 10^{-5}
Specific Heat, Btu/lb/°F	—	0.3	—	—	—	—
Water Absorption (24 hr), %	D570	0.20	0.10–0.15	0.05–0.08	0.08–0.12	0.07–0.10
Transparency (visible light), %	—	89–92	Opaque	Opaque	Opaque	Opaque
Refractive Index, n_D	D542	1.50	—	—	—	—
MECHANICAL PROPERTIES						
Tensile Strength, 1000 psi	D638	5–6	4–5	7–8	3–4	5–9
Tensile Mod, 10^5 psi	D638	3.0	20–27	19–20	24–29	20–25
Elongation, %	D638		—	—	—	—
Impact Str (Izod notched), ft-lb/in.	D256	0.2–0.4	0.25–0.35	2.2	0.30–0.35	8–12
Flex Strength, 1000 psi	D790	—	8–11	19–20	7–10	12–17
Mod of Elast in Flex, 10^5 psi	D790	2.5–3.3	—	22–27	22–27	22–28
Compr Strength, 1000 psi	D690	22.5	20–25	28	16–20	24–30
Hardness (Barcol)	D785	M95–M100 (Rockwell)	60–70	70–75	60–70	70–80
ELECTRICAL PROPERTIES						
Volume Resistivity, ohm-cm	D257	4 x 10^{14}	10^{14}	10^{14}	1 x 10^{14}–1 x 10^{15}	10^{14}
Dielectric Str (step-by-step), v/mil	D149	290	300–350	290	300–350	300–350
Dielectric Constant	D150					
60 Hz		4.4	5.4–5.9	7.4	5.7–6.3	5.2–6.0
1 MHz		3.5-3.8	4.5–4.7	6.8	4.8–5.1	4.5–5.0
Dissipation Factor	D150					
60 Hz		0.03-0.04	0.030–0.045	0.019	0.030–0.040	0.02–0.03
1 MHz		0.1-0.2	0.016–0.020	0.023	0.017–0.020	0.015–0.022
Arc Resistance, sec	D495	185	180	180	180	180
APPLICABLE PROCESSING METHODS		Casting	Injection molding, compression molding, transfer molding			
HEAT RESISTANCE						
Max Rec Service Temp, F	—	212	250	300	300	300
Deflection Temp (264 psi), F	D648	—	350–400	>400	350–400	>400
CHEMICAL RESISTANCE		Resists nearly all solvents including acetone, benzene and gasoline, and practically all chemicals except highly oxiding acids	Resistant to weak acids; attacked by alkalis; practically unattacked by organic liquids such as alcohols, hydrocarbons and fatty acids			
USES		Aircraft windows, lenses, marine glazing, vending machine windows, slides, watch crystals, safety windows	Encapsulation of resistors, coils and small electronic parts	Molding of tube bases and sockets, connectors, tuning devices, electrical instrument parts, switches and relays. Parts for transformers, motor controllers and automotive ignition systems		Heavy duty circuit breaker and switchgear, stand-off insulators, electrical motor brush holders and end plates

Alloys-ABS/Polycarbonate, ABS/PVC, Acrylic/PVC, ABS/Polysulfone, ABS/Polyurethane

Material ➧		ABS/ Polycarbonate	ABS/PVC (rigid)	Acylic/ PVC	ABS/Poly-sulfone (Polyaryl ether)	ABS/Poly-urethane
PHYSICAL PROPERTIES	ASTM					
Specific Gravity	D792	1.19	1.21	1.28-1.35	1.14	1.04-1.07
Refractive Index, nD	D542	—	—	—	—	—
Transparency (visible light), %	D1003	Opaque	Opaque	Translucent to opaque	Opaque	Opaque
MECHANICAL PROPERTIES						
Tensile	D638					
Str (yield), 1,000 psi		8.2	5.5-7.5	5.2-6.8	7.5	3.15-3.85
Elong (ult), %		—	—	75-150	25-90	120-200
Modulus, 1,000 psi		376	290	310-370	320	100-180
Flexural	D790					
Str, 1,000 psi		13-14	9-11	9.7-11	11	1.7-5.5
Modulus, 1,000 psi		352-400	300-430	250-425	300	100-170
Compressive str, 1,000 psi (2% offset)	D695	—	7.4	8.4	—	3.6-4.8
Impact str						
Izod (notched), ft-lb/in.	D256	10.3-10.5	2.0-13.0	1.5-15	8.0-10.0	9.7-12.0
Gardner, lb/in.		—	—	—	—	—
Hardness						
Rockwell	D785	R117	R102	R110-105	R117	R70-82
Shore	D785	—	—	—	—	—
Bending fatigue str (Woehler), 1,000 psi	D671	—	—	—	—	—
Abrasion res (Taber, CS-17 wheel, 1000g)	D1044					
mg/100 cycles		—	—	40	—	30-35
Coef of friction	D1894					
Against self		—	—	—	0.28	—
Against steel		—	—	—	0.31	—
ELECTRICAL PROPERTIES						
Vol res, ohm-cm	D257	$2.2\text{-}4 \times 10^{16}$	—	1.6×10^{15}	1.5×10^{16}	—
Dielec str, v/mil (dry)	D149					
Short-time		1250-1550	600	430-670	430	—
Step-by-step		—	—	647	—	—
Dielec constant (dry)	D150					
60 Hz		3.08	—	3.33-3.86	3.14	—
1 MHz		3.2-3.6	—	3.06-3.44	3.10	—
Dissip factor (dry)	D150					
60 Hz		0.019-0.021	—	0.016	0.006	—
1 MHz		0.020	—	0.017	0.007	—
Arc res, sec	D495	—	—	42-80	180	—
THERMAL PROPERTIES						
Ther cond, Btu/hr/sq ft/°F/in.	C177	0.95	—	1.01	2.07	—
Coef of ther exp, 10^{-5}/°F	D696	37	4.6	3.5-4.4	—	11.6-12.1
Specific heat, Btu/lb/°F		0.3	—	0.293	0.35	—
Heat deflection, °F	D648					
At 66 psi		238	—	177-205	320	214-216
At 264 psi		220-252	161-210	138-195	300	183-199
Brittleness temp, °F	D746	—	—	−46	—	—
Max temp for continuous use (no load), F	—	190-200	—	—	—	—
CHEMICAL AND ENVIRONMENTAL RESISTANCE						
Water absorption, %	D570					
In 24 hr		—	0.12	0.06-0.09	—	0.35-0.41
Saturation		—	—	—	—	—
Weathering		Slight effect	Slight effect	Resist	Embrittles	Slight effect
Acids						
Weak		Resist	Resist	Resist	Resist	Resist
Strong		Attack	Attack	Attack	Resist	Attack
Alkalis						
Weak		Resist	Resist	Resist	Resist	Resist
Strong		Resist	Resist	Resist	Resist	Resist
Organic solvents		Attack	Attack	Attack	Attack	Attack
Fuel		—	—	—	—	Attack
Oil and grease		—	—	—	—	—
METHODS OF PROCESSING		Inj mldg, extrusion	Inj mldg, extrusion, thermoforming	Extrusion, thermo-forming, inj mldg	Inj mldg, extru-sion	Inj mldg, extrusion, thermo-forming

Cellulose Acetate

ASTM Grade[a]		H6	H4	H2	MH	MS	S2
PHYSICAL PROPERTIES	ASTM						
Specific Gravity	D792	—	1.29-1.31	1.25-1.31	1.24-1.31	1.23-1.30	1.22-1.30
Ther Cond, Btu/hr/sq ft/°F/in.	C177	1.20—2.28	1.20—2.28	1.20—2.28	1.20—2.28	1.20—2.28	1.20—2.28
Coef of Ther Exp, 10^{-5} per °F	D696	4.4-9.0	4.4-9.0	4.4-9.0	4.4-9.0	4.4-9.0	4.4-9.0
Refractive Index	D542	1.46-1.50	1.46-1.50	1.46-1.50	1.46-1.50	1.46-1.50	1.46-1.50
Spec Ht, Btu/lb/°F	—	0.3-0.42	0.3-0.42	0.3-0.42	0.3-0.42	0.3-0.42	0.3-0.42
Luminous Transmittance, %	D791	75-90	75-90	80-90	80-90	80-90	80-95
Haze, %	D672	2-15	2-15	2-10	2-10	2-10	2-8
Water Absorption (24 hr), %	D570	—	1.7-2.7	1.7-2.7	1.8-4.0	2.1-4.0	2.3-4.0
MECHANICAL PROPERTIES							
Ten Str at Fracture, 1000 psi	D638	—	6.4-7.8	5.3-7.2	4.6-6.3	3.3-5.3	3.0-4.4
Impact Str (Izod), ft-lb/in. of notch	D256	—	1.1-3.1	1.5-3.9	2.0-4.9	2.4-6.5	3.2-6.8
Modulus of Elast in Flex, 10^5 psi	D747	—	2.0-2.55	1.50-2.35	1.50-2.15	1.25-1.90	1.05-1.65
Flex Str at Yield, 1000 psi	D790	—	8.1-11.15	6.0-10.0	4.4-8.65	3.8-7.1	3.5-5.7
Heat Deflection Temp, °F, 264 psi		—	145-188	120-172	128-155	123-141	117-129
ELECTRICAL PROPERTIES							
Vol Res, ohm-cm	D257	10^{10}-10^{13}	10^{10}-10^{13}	10^{10}-10^{13}	10^{10}-10^{13}	10^{10}-10^{13}	10^{10}-10^{13}
Dielec Str (short-time), v/mil	D149	250-600	250-600	250-600	250-600	250-600	250-600
Dielec Const, 1 MHz	D150	3.2-7.0	3.2-7.0	3.2-7.0	3.2-7.0	3.2-7.0	3.2-7.0
Dissip Factor, 1 MHz	D150	—	0.01-0.10	0.01-0.10	0.01-0.10	0.01-0.10	0.01-0.10
CHEMICAL RESISTANCE		Unattacked by water, salt water solutions, white gasoline, oleic acide, 5% acetic acid and dilute sulfuric acid. Decomposed by 30% sulfuric, 10% nitric and 10% hydrochloric acids, sodium hydroxide, 10% ammonium hydroxide. Dissolved by acetone, ethyl acetate.					
USES		Film, tape, blister packaging, appliance and optical parts, handles, toys, novelties, buttons, tags.					

[a] According to ASTM D706-74.

Cellulose Acetate Butyrate and Cellulose Acetate Propionate

ASTM Grade[a]		Cellulose Acetate Butyrate			Cellulose Acetate Propionate		
		H4	MH	S2	H6	H4	H2
PHYSICAL PROPERTIES	ASTM						
Specific Gravity	D792	1.22	1.18-1.20	1.15-1.18	1.22	1.20-1.21	1.19
Ther Cond, Btu/hr/sq ft/°F/ft	C177	0.10-0.19	0.10-0.19	0.10-0.19	0.10-0.19	0.10-0.19	0.10-0.19
Coef of Ther Exp, 10^{-5} per °F	D696	(6-9) x 10^{-5}	(6-9) x 10^{-5}	(6-9) x 10^{-5}	(6-9) x 10^{-5}	(6-9) x 10^{-5}	(6-9) x 10^{-5}
Refractive Index	D543	1.46-1.49	1.46-1.49	1.46-1.49	1.46-1.49	1.46-1.49	1.46-1.49
Spec Ht, Btu/lb/°F	—	0.3-0.4	0.3-0.4	0.3-0.4	0.3-0.4	0.3-0.4	0.3-0.4
Luminous Transmittance, %	D791	75-92	80-92	85-95	80-92	80-92	80-92
Haze, %	D672	2-5	2-5	2-5	2-5	2-5	2-5
Water Absorption (24 hr), %	D570	2.0	1.3-1.6	0.9-1.3	1.6-2.0	1.3-1.8	1.6
MECHANICAL PROPERTIES							
Ten Str at Fracture, 1000 psi	D638	6.9	5.0-6.0	3.0-4.0	6.7-7.1	5.6-6.7	4.3-5.8
Impact Str (Izod), ft-lb/in. of notch	D256	1.2-1.4	3.2-4.8	6.8-10.0	1.0-2.0	1.5-3.0	2.0-7.5
Modulus of Elast in Flex, 10^5 psi	D747	2.7-2.9	1.9-2.2	1.1-1.4	2.9-3.4	2.45-3.0	1.9-2.6
Flex Str at Yield, 1000 psi	D790	9.0	5.6-6.7	2.5-4.0	8.0-8.9	6.5-8.4	4.8-7.2
Heat Deflection Temp, °F, 264 psi		196	146-160	118-130	193-202	167-181	141-164
ELECTRICAL PROPERTIES							
Vol Res, ohm-cm	D257	10^{11}-10^{14}	10^{11}-10^{14}	10^{11}-10^{14}	10^{11}-10^{14}	10^{11}-10^{14}	10^{11}-10^{14}
Dielec Str (short-time), v/mil	D149	250-400	250-400	250-400	300-450	300-450	300-450
Dielec Const, 1 MHz	D150	3.2-6.2	3.2-6.2	3.2-6.2	3.4-3.7	3.4-3.7	3.4-3.7
Dissip Factor, 1 MHz	D150	0.02-0.05	0.02-0.05	0.02-0.05	0.02-0.05	0.02-0.05	0.02-0.05
CHEMICAL RESISTANCE		Unaffected by 3% sulfuric, 5% acetic, 10% hydrochloric and oleic acids; discolored by 10% nitric acid. Unaffected by 1% sodium hydroxide and 2% sodium carbonate; slightly softened by 10% sodium hydroxide and discolored by 10% ammonium hydroxide. Unaffected by white gasoline, but swollen or dissolved by ethyl alcohol, acetone, ethyl acetate, ethylene dichloride, carbon tetrachloride and toluene. Unaffected by water, salt water and 3% hydrogen peroxide.					
USES		Outdoor signs, handles, pipes, pens, telephones, steering wheels, blister packaging, knobs					

[a] According to ASTM D707 and D1562, respectively.

Diallyl Phthalates

Type ➧		Orlon-Filled	Dacron-Filled	Asbestos-Filled	Glass Fiber-Filled
PHYSICAL PROPERTIES	ASTM				
Specific Gravity	D792	1.31-1.35	1.40-1.65	1.50-1.96	1.55-1.85
Coef of Ther Exp, per °F	D696	5.0 x 10^{-5}	5.2 x 10^{-5}	4.0 x 10^{-5}	2.2-2.6 x 10^{-5}
Water Abs (122 F. 48 hr), %	–	0.2-0.5	0.2-0.5	0.4-0.7	0.2-0.4
MECHANICAL PROPERTIES					
Mod of Elast in Tension, psi[a]	D638	6 x 10^5	–	12 x 10^5	–
Ten Str, psi	D638	4500-6000	4600-5500	4000-6500	5500-9500
Hardness (Rockwell)	D785	M108	–	M107	M108
Impact Str (Izod notched), ft-lb/in.	D256	0.5-1.2	1.7-4.5	0.30-0.50	0.5-15.0
Flex Str, 1000 psi	D790	7.5-10.5	9-11.5	8-10	10-18
Compr Str, 1000 psi	D695	20-25	20-30	18-25	25
ELECTRICAL PROPERTIES					
Dielec Str, v/mil					
Short Time (dry)	D149	400	376-390	350-450	350-430
Short Time (wet[b])	D149	375	360-391	300-400	300-420
Step-by-Step (dry)	D149	350	350-374	300-400	300-420
Step-by-Step (wet[b])	D149	325	350-361	250-350	275-420
Dielec Breakdown, kv					
Short Time (dry)	–	65-75	65	55-80	63-70
Short Time (wet[b])	–	60-65	60	55	45-65
Step-by-Step (dry)	–	55-60	60	38-70	55-65
Step-by-Step (wet[b])	–	46-60	55	39-60	45-65
Dissip Factor[c]					
Dry	D150	0.023, 0.015	0.008, 0.015	0.05, 0.03	0.01, 0.015
Wet[d]	D150	0.045, 0.040	0.009, 0.017	0.154, 0.050	0.012, 0.020
Dielec Const[e]					
Dry	D150	3.9, 3.3	3.8, 3.6	5.2, 4.5	4.5, 4.2
Wet[d]	D150	4.1, 3.4	3.9, 3.7	6.5, 4.8	4.6, 4.4
Vol Res, megohm-cm[d]	D257	60,000-6,000,000	100-25,000	100-5000	10,000-50,000
Surface Res, megohms[d]	D257	25,000-2,500,000	500-25,000	100-5000	10,000-100,000
Arc Resistance, sec	D495	85-115	105-125	125-140	125-135
APPLICABLE PROCESSING METHODS		Injection molding, compression molding, transfer molding, layup molding			
HEAT RESISTANCE					
Max Recommended Svc Temp, F	–	300	300-370	350-450	400-450
Heat Dist Temp, F	D648	240-266	270-290	300-350	350-500
CHEMICAL RESISTANCE		Unaffected by weak acids and alkalis and organic solvents. Slightly affected by strong acids and alkalis			
USES		Molding compounds—connectors, potentiometers, plugboards, housings, appliance fixtures, resistors, insulators, etc. Prepregs—radomes, aircraft leading edges, housings, nose cones, air ducts, etc. Laminates—decorative sheets for surfacing real and grain-printed wood and fabrics, etc.			

[a] Conditioned 48 hr at 122 F. [b] Tested after 48-hr immersion in water at 122 F. [c] Values given for frequencies of 1 kc and 1 mc, in that order. [d] Conditioned 30 days at 100% RH and 158 F. [e] Flame-resistant type is available. [f] 480 hr, 257 F.

Epoxies

Type →		Standard Epoxies (Bisphenol A)			Epoxy Novolacs		Cycloaliphatic
		Cast, Rigid[a]	Cast Flexible[b]	Molded[c]	Cast, Rigid[d]	Molded[e]	Cast, Rigid[f]
PHYSICAL PROPERTIES	ASTM						
Specific Gravity	D792	1.15	1.14-1.18	1.80-2.0	1.24	1.7	1.22
Ther Cond, Btu/hr/sq ft/°F/in.	C177	1.2—3.6	—	1.2—6.0	—	—	—
Coef of Ther Exp, 10^{-5} per °F	D696	3.3	3.5	1-2	—	1.7-2.2	1.6-3.0
Specific Heat, Btu/lb/°F	—	0.4-0.5	—	—	—	—	—
Water Absorp (24 hr), %	D570	0.1-0.2	0.4-1.0	0.3-0.8	—	0.11-0.2	0.1-0.7[g]
Transparency (visible light), %	—	90	85	Opaque	—	Opaque	—
Refractive Index, n_D	D542	1.61	1.61	—	—	—	—
MECHANICAL PROPERTIES							
Tensile Str, 1000 psi	D638	9.5-11.5	1.4-7.6	8-11	8-12	5.2-5.3	9.6-12.0
Tensile Mod, 10^5 psi	D638	4.5	0.5-2.5	—	4-5	—	4.8-5.0
Elongation, %	D638	4.4	1.5-60	—	2-5	—	2.2-4.8
Impact Str (Izod notched), ft-lb/in.	D256	0.2-0.5	0.3-2.0	0.4-0.5	0.5	0.3-0.5	—
Flexural Str, 1000 psi	D790	14-18	1.2-12.7	19-22	11-16	10-12	12-13
Mod of Elast in Flex, 10^5 psi	D790	4.5-5.4	0.36-3.9	15-25	4-5	—	4.4-4.8
Compr Strength, 1000 psi	D695	16.5-24	—	34-38	17-19	22-26	30-50
Hardness (Rockwell)	D785	106M	50-100M	75-80 (Barcol)	107-112[h]	94-96D[i]	—
ELECTRICAL PROPERTIES							
Vol Resist, ohm-cm	D257	6.1×10^{15}	9.1×10^{8}—6.7×10^{9}	$1\text{-}5 \times 10^{15}$	2.10×10^{14}	$1.4\text{-}5.5 \times 10^{14}$	$>10^{16}$
Dielectric Str (step-by-step), v/mil	D149	>400	400-410	360-400	—	280-400	444 (short time)
Dielectric Constant	D150						
60 Hz		4.02	4.43-4.79	4.4-5.4	3.96-4.02	4.7-5.7	3.34-3.39
1 MHz		3.42	2.78-3.52	4.1-4.6	3.53-3.58	4.3-4.8	—
Dissipation Factor	D150						
60 Hz		0.0074	0.0048-0.0380	0.011-0.018	0.0855-0.0074	0.0071-0.025	0.001-0.007
1 MHz		0.032	0.0369-0.0622	0.013-0.020	0.029-0.028	—	—
Arc Resistance, sec	D495	100	75-98	135-190	—	180-185	120
APPLICABLE PROCESSING METHODS		Casting	Casting	Injection, compression and transfer molding	Casting	Injection, compression, transfer molding,	Casting
HEAT RESISTANCE							
Max Rec Svc Temp. F	—	175-190	100-125	<400	450	450-500	450-500
Heat Deflection Temp (264 psi), F	D648	230	90-155	340-400	300-400	300-425	300-525
CHEMICAL RESISTANCE		Highly res to water and bases; less res to acids and oxidizing agents			Res water and strong alkalies, more res to acid and oxidizing agents than standard epoxies		Highly res to weather
USES		Potting and encapsulation of electronic components, precision castings, tools and dies, patching compounds		Electrical moldings, such as condensers, switch plates, connector plugs, resistor bobbins and wirewound resistors, molded coils, relay assemblies	Impregnation and potting requiring high heat res; adhesives	Electrical and electronic encapsulation designed for high temp	Encapsulation, impregnation and potting req outstanding arc and tracking res

[a] 13 phr of 11.1 A curing agent. [b] 30-80 phr of flexible curing agent. [c] Mineral-glass reinforced. [d] 28 phr methylene dianiline: cure—16 hr at 130 F, 2 hr at 257 F, 2 hr at 347 F. [e] Mineral-filled proprietary compounds. [f] 12 phr of hexahydrophthalic and anhydride, 12 phr sodium alcoholate accelerator; cure-24 phr at 250 F and post cure of 3 phr at 400 F. [g] 1 hr at 212 F. [h] 1 mc. [i] Durometer.

Fluorocarbons

Type ➛	ASTM	Polytrifluoro-chloroethylene (PTFCE)	Polytetrafluoro-ethylene (PTFE)	Ceramic-Reinforced (PTFE)	Fluorinated Ethylene Propylene (FEP)	Polyvinylidene-fluoride (PVF_2)	ETFE & ECTFE Std.	ETFE & ECTFE Glass reinf.	PFA
PHYSICAL PROPERTIES	ASTM								
Specific Gravity	D792	2.10-2.15	2.1-2.3	2.2-2.4	2.14-2.17	1.77	1.68	1.86	2.12-2.17
Ther Cond, Btu/hr/sq ft/°F/in.	•	1.74	1.68	—	1.44	1.68	—	—	—
Coef of Ther Exp, per °F x 10^{-5}	D696	3.99	7.5-8.4	5.5	5.3-10.7	8.5	5-7.8	0.9-1.8	5.2-11.5
Refractive Index	D542	1.43	1.35	—	1.34	1.42	1.44	—	—
Specific Heat, Btu/lb/°F		0.22	0.25	—	0.28	0.33	—	—	—
Transmittance (luminous), %	D791	80-92	—	—	—	—	0.7-0.8	—	—
Water Absorption (24 hr), %	D570	0.00	0.01	>.2	0.01	0.03	—	0.01	0.03
MECHANICAL PROPERTIES									
Mod of Elast in Compr, psi	D638	1.8 x 10^{5}	0.70-0.90 x 10^{5}	1.5-2.0 x 10^{5}	0.6-0.8 x 10^{5}	1.7-2 x 10^{5}	—	—	—
Mod of Elast in Tension, psi	D638	1.9-3.0 x 10^{5}	0.38-0.65 x 10^{5}	1.5-2.0 x 10^{5}	0.5-0.7 x 10^{5}	1.7-2 x 10^{5}	2.4 x 10^{5}	11 x 10^{5}	—
Ten Str, 1000 psi	D638	4.6-5.7	2.5-6.5	.75-2.5	2.5-3.5	7.2-8.6	4.0-4.5	12.0	4.3
Elongation (in 2 in.), %	D638	125-175	250-350	10-200	250-330	200-300	150-200WIA	9	200
Hardness (Rockwell)	D785	R110-115	52D	R35-55	58D	R110	R95	—	D60
Abrasion Res, gm/cycle	[d]	0.0080	—	—	—	0.0006-0.0012	0.005	—	—
Impact Str (Izod notched), ft-lb/in.	D256	3.50-3.62	2.5-4.0	—	No break	3.8	No break	7	—
Mod of Elast in Flexure, psi	D747	2.0-2.5 x 10^{5}	0.6 x 10^{5}	4.64 x 10^{5}	0.8 x 10^{5}	—	2.4 x 10^{5}	9.5 x 10^{5}	1.0 x 10^{5}
Flex Str (0.1% offset), 1000 psi	D790	3.5	—	—	—	2	—	15	—
Compr Str (0.1% offset), 1000 psi	D695	2.0	0.7-1.8	1.4-1.8	1.6	12.8-14.2	—	—	—
ELECTRICAL PROPERTIES									
Volume Resistivity, ohm-cm	D257	10^{18}	>10^{18}	10^{15}	>2 x 10^{18}	5 x 10^{14}	10^{16}	—	>10^{18}
Dielec Str (short time), v/mil	D149	530-600	400-500	300-400	500-600	260	490	—	2000
Dielectric Constant									
60 Hz	D150	2.6-2.7	2.1	2.9-3.6	2.1	10.0	2.6	—	2.1
1 MHz	D150	2.30-2.37	2.1	2.9-3.6	2.1	7.5	2.5	—	2.1
Dissipation Factor									
60 Hz	D150	0.02	0.0002	.0005-.0015	0.0003	0.050	0.0007	—	0.0002
1 MHz	D150	0.007-0.010	0.0002	.0005-.0015	0.0003	0.184	0.009	—	0.0003
Arc Resistance, sec		>360	>200	—	>165	50-60	135	—	—
APPLICABLE PROCESSING METHODS		Compression mldg, isotactic pressing	Compression molding, isostactic molding		Inject molding, extrus compres molding	Inject molding, extrus, compres molding	Inject, rotational, blow & transfer molding; extrus; thermoforming; foam		Injection mldg, extrusion, blow mldg
HEAT RESISTANCE									
Max Rec Svc Temp, F		380	550	450-500	400	340	300-355	—	500
Heat Dist Temp, F									
66 psi	D648	196-291	—	350-480	—	300	220-240	—	—
264 psi	D648	151-178	—	170-220	—	232	160-170	285	—
CHEMICAL RESISTANCE		High res to corrosive chemical & most organic solvent	Inert to most chemicals and solvents with exception of alkali metals. Halogenated solvents at high temperatures and pressure have some effect			Res to most acids and bases except fuming sulfuric	Res to most acids & bases	Res to most acids & bases	Same as PTFE
USES		Chemical pipes pump parts, cables, tank linings, connectors, connector inserts, valve diaphragms, insulation	Chemical pipes, valves and liners, gaskets, packings, pump bearings and impellers, electrical equip, anti-adhesive coatings	Bearings, bushings, wear surfaces, elec insulators, gaskets, packings, valve seats in corrosive conditions	Electronic instruments, valve linings, laminates, corrosion resistant and non-adhesive coatings	Seals; chemical pipe and fittings, gaskets; electrical jackets and primary insulation; finishes			PTFE uses req more ease of processing

[a] A proprietary material consisting of polyetrafluorocthylene and special constituents designed to improve TFE's mechanical and thermal properties while retaining its electrical and chemical characteristics. [b] Range covers properties for compounds containing from 10-25% glass. [c] Cenco-Fitch. [d] Federal Spec L-P-406A No. 1092.1. [e] From 73 to 500 F.

Nylons

Type →	ASTM	Type 6: General Purpose[a]	Type 6: Glass Fiber (30%) Reinforced[a]	Type 6: Cast	Type 6: Flexible[a] Copolymers	Type 11	Type 12	Transparent
PHYSICAL PROPERTIES	ASTM							
Specific Gravity	D792	1.14	1.37	1.15	1.12-1.14	1.04	1.01	1.06-1.12
Ther Cond., Btu/hr/sq ft °F/in.	—	1.2	1.2-1.7	1.2-1.7	—	1.5	1.7	—
Coef of Ther Exp., 10^{-5} per °F	D696	4.8	1.2	4.4	—	5.5	7.2	2.8-4.31
Specific Heat, Btu/lb/°F.	—	0.4	—	0.4	—	0.58	0.28	0.327
Refractive Index, n_D	D542	—	—	—	—	—	—	1.535-1.566
Water Absorption (24 hr), %	D570	1.7-1.8	1.3	0.6	0.8-1.4	0.4	0.25	0.41
MECHANICAL PROPERTIES								
Tensile Strength (2 in./min)[d]	D638							
Ultimate		9.5-12.5	21.23	12.8	7.5-10.0	—	7.1-8.5	—
Yield		8.5-12.5	—	12.8	7.5-10.0	8.5	5.5-6.5	9.8-10.7
Elongation (2 in./min), %	D638							
Ultimate		30-220	2.4	20	200-320	100-120	120-350	75-150
Yield		—	—	5	—	—	5.8	9
Mod of Elast in Tension, 10^5 psi	D638	—	10-12	5.4	—	1.78-1.85	1.7-2.1	3.1-4.05
Flex Strength, 1000 psi	D790	Unbreak.	26-34	16.5	3.4-16.4	—	—	12.3-13.26
Mod of Elast in Flex, 10^5 psi	D790	1.4-3.7	10-12	5.05	0.92-3.2	1.51	—	2.9-3.86
Impact Str (Izod notched), ft-lb/in.	D256	0.8-1.2	3-2.3	1.2	1.5-19	3.3-3.6	1.2-4.2	1.1
Compr Strength (1% offset)[d]	D695	9.7	19-20	14	—	—	—	3.39
Fatigue Str[h], 1000 psi	D671							
10^4 cyc		—	7.0[g]	—	—	—	—	—
10^5 cyc		—	6.0	—	—	—	—	—
10^7 cyc		—	5.7	—	—	—	—	—
Hardness (Rockwell)	D785	R118-R120	R121	R116	R72-R119	R100-R108	R106	M89-M93
Coef of Dyn Frict	—	—	—	0.32[c]	—	—	—	—
Abrasion Res (Taber, CS-17), mg/1000 cycles	D1044	5	—	2.7	—	—	—	3.2-21
ELECTRICAL PROPERTIES								
Volume Resistivity, ohm-cm	D257	4.5×10^{13}	$2.8\text{-}10^{11}$ to 1.5×10^{15}	2.6×10^{14}	—	2×10^{13}	$10^{14}\text{-}10^{15}$	$>5 \times 10^{15}$
Dielectric Str (short time), v/mil	D149	385	400-450	380	440	425	840	371-670
Dielectric Constant	D150							
60 Hz		4.0-5.3	4.6-5.6	4.0	3.2-4.0	3.3(10^3cps)	3.6(10^3cps)	3.5-3.99
1 MHz		3.6-3.8	3.9-5.4	3.3	3.0-3.6	—	—	3.3
Dissipation Factor	D150							
60 Hz		0.06-0.014	0.022-0.008	0.015	0.007-0.010	0.03	0.04(10^3cps)	0.001-0.028
1 MHz		0.03-0.04	0.019-0.015	0.05	0.010-0.015	0.02	—	0.019
Arc Resistance, sec	D495	—	92-81	—	—	—	—	120
HEAT RESISTANCE								
Deflection Temp, F	D648							
66 psi		360	425-428	420	260-350	302	—	285
264 psi		155-160	420-419	410	115-130	131	—	256
APPLICABLE PROCESSING METHODS		Injection molding, extrusion, rotational mldg, blow mldg		Rotational molding	Extrusion, inj mldg		Extrusion, inj mldg	Extr, inj, blow, compr mldg
CHEMICAL RESISTANCE		Resists esters, ketones, alkalis, weak acids, alcohols and common solvents. Not resistant to conc mineral acids		Res most organic chem, such as alcohols, ketones, hydrocarbons and chlor solv. Att by str acids. phenols, str agents	Res esters, ketones, alkalis, weak acids, alcohols and common solvents. Not res to conc mineral acids	Res alkalis, petroleum products and common organic solvents. Not res to phenols and conc acids and oxidants	Res alkalis, petroleum products and common organic solvents. Not res to phenols and conc acids and oxidants	Res weak acids, alkali, strong alkali, oil, greases. Attacked by strong acids, alcohols and org solvents
USES		Bearings, gears, bushings, coils, rod, tubing, tape	Gen pur type 6 parts req stiffness, dimen stab	Bearings, wearplates, bushings, gears, rollers, stock shapes	Parts requiring high impact strength or flexibility	Elec insulation and other nylon uses where low mois absorp is needed	Filament, rod, tubing, sheet, moldings req dim stability and low moist absorp	Lenses, containers, gauges, fuel tanks, processing equipment housings

[a]Dry, as-molded properties unless otherwise noted. [b]Non-cross-linked, can be cross-linked. [c]Dynamic, no lubrication, nylon to steel. [d]1000 psi [e]0.4 self-ext to slow burn. [f]Heat stabilized [g]Moisture conditioned to 50% relative humidity. [h]Cyclic failure stress at 1800 cyc./min.

Nylons

Type →		6/6 Nylon: General Purpose Molding[a]	6/6 Nylon: Glass Fiber Reinforced[b]	6/6 Nylon: Glass Fiber, Molybdenum Disulfide Filled[c]	6/6 Nylon: General Purpose Extrusion[d]	6/6 Nylon: High impact[ai]	6/9 Nylon[a]	6/12 Nylon	Mineral reinf nylon
PHYSICAL PROPERTIES	ASTM								
Specific Gravity	D792	1.13-1.15,—	1.37, 1.47	1.37—1.41	1.13, 1.15	1.09	1.07-1.09	1.06-1.08	1.47
Ther Cond, Btu/hr/sq ft/°F/in	—	1.7, —	1.5, 3.3	—	1.7, —	—	1.5	1.5	—
Coef of Ther Exp, 10^5 per °F	D696	4.5, —	2.1, 1.4	1.75	—	—	8.3	5.0	2.7-5.0
Specific Heat, Btu/lb/°F	—	0.3-0.5	—	—	0.3-0.5	—	0.3-0.5	0.3-0.5	—
Refractive Index, n_D	D542	Transluc	Opaque	Opaque	Opaque	Opaque	Translucent	Translucent	Opaque
Water Absorption (24 hr), %	D570	1.5, —	0.9, 0.8	0.5-0.7	1.5	—	0.48	0.4	0.5-0.8
Coef of Static Frict (against self)	—	0.04-0.13, —	—	—	—	—	—	—	0.23
MECHANICAL PROPERTIES									
Tensile Strength, 1000 psi	D638								
Ultimate		11.8, 11.2	25, 30	19-22	12.6, 8.6	7.8, 6.0	5.9, 5.0	8.8, 8.8	—
Yield		11.8, 8.5	—	—	12.6, 8.6	—	8.5, 6.5	8.8, 7.4	9-10
Elongation, %	D638								
Ultimate		60, 300	1.8, 2.2	3	90, 240	40, 210	125, 300	150, 340	10-25
Yield		5, 25	—	—	5, 30	—	10, 10	7, 40	—
Mod of Elast in Tension, 10^5 psi	D638	4.75, 3.85	14, 20	—	—	—	2.75, 1.5	—	5.0
Flex Strength, 1000 psi	D790	Unbreak.	26, 35	26-28	—	—	11, 8	—	12-16
Mod of Elast in Flex, 10^5 psi	D790	410, 175	10, 18	11-13	4.1, 1.75	2.55, 1.25	3.0, 1.6	2.9, 1.8	3.3-6.0
Imp Str (Izod notched), ft-lb/in	D638	1.0, 2.0	2.5, 3.4[e]	—	1.3, —	15, 15-25	1.1, 2.0	1.0, 1.4	1.0-1.5
Compr Strength (1%), 1000 psi	D695	4.9, —	20, 24	—	4.9 (1%), —	1.9	—	2.4, —	—
Fatigue Str.[f] 1000 psi	D671								
10^4 cyc		6.5, 3.4[g]	8.0, 9.0[h]	—	—	—	—	—	—
10^5 cyc		5.9, 3.2[g]	6.5, 7.3[h]	—	—	—	—	—	—
10^6 cyc		5.3, 3.1[g]	6.0, 7.0[h]	—	—	—	—	—	—
10^7 cyc		5.2, 3.1[g]	5.9, 7.0[h]	—	—	—	—	—	—
Hardness (Rockwell)	D785	R118, R108	E60, E80	M95-100	R118-108	R112	R111	R114, —	R119-121
Abrasion Res (Taber CS-17, 1000g), mg/1000 cycles	D1044	3-5, 6-8	—	—	—, 3-5	—	—	—, 5.7	12-30
ELECTRICAL PROPERTIES									
Volume Resistivity, ohm-cm	D257	10^{14}-10^{15}	5.5×10^{15}, 2.6×10^{15},	—	10^{15}	10^{14}, 10^{13}	3.3×10^{15} 2.6×10^{13}	10^{15}, 10^{12}	10^{16}
Dielectric Str (short time), v/mil	D149	385	400, 480	300-400	—	390, 330	540, 560	—	280-485
Dielectric Constant	D150								
60 Hz		4.0, —	4.0, 4.4	—	—	3.2, 5.5	3.6, 5.4	4.0, 6.0	—
1 MHz		3.6, —	3.5, 4.1	—	—	3.1, 3.9	3.2, 3.4	3.5, 4.0	—
Dissipation Factor	D150								
60Hz		0.014, 0.04	0.018, 0.009	—	—	0.013	0.02, 0.09	—	—
1 MHz		0.04, —	0.017, 0.018	—	—	0.017	0.02, 0.02	0.02, 0.03	—
Arc Resistance sec	D495	120	148, 100	135	120	72, 77	—	—	115
HEAT RESISTANCE									
Max Rec Service Temp, F	—	250-300[d]	250-300[d]	250-300[d]	250-300[d]	—	225-275[d]	350	—
Deflection Temp, F	D648								
66 psi		470	507, 509	—	470	420	330	330	400
264 psi		220	495, 500	—	220	160	140	180	300
CHEMICAL RESISTANCE			Inert to most organic chemicals such as esters, ketones, alcohols and hydrocarbons. Resist alkalis and salt solutions, but att by phenols, formic acid, strong mineral acids and strong oxidizing agents						
APPLICABLE PROCESSING METHODS		Injection molding			Extrusion	Injection molding, extrusion		Injection molding, blow molding, extrusion	Injection molding
USES		Bearings, gears, bushings, coil forms, brush backs, rod, tubing		Mech parts where lubrication is undesirable or diff	Tubing, rod, pipe, sheeting, laminations	Protective helmets, tool handles and housings	Jacketing for wire and cable, special molded parts		Elec housings and mech. parts

[a]Where two values are given, first is for dry, as-molded material, and second for moisture equilibrium in air; single value pertains to dry material unless otherwise noted. [b]First value for 30% glass fiber and second for 40%. All values at moisture equilibrium. [c]30% glass fiber. [d]Heat stabilized for max heat resistance. [e]¼ in. [f]cyclic failure stress at 1800 cyc/min. [g]Second value is for material moisture conditioned to 50% relative humidity. [h]Values for material moisture conditioned to 50% relative humidity. [i]Zytel ST-801 (DuPont)

Phenolics

Type and Filler →		General—Woodflour and Flock	Shock—Paper, Flock or Pulp	High Shock—Chopped Fabric or Cord	Very High Shock—Glass Fiber	Arc Resistant—Mineral	Rubber Phenolic—Woodflour or Flock	Rubber Phenolic—Chopped Fabric	Rubber Phenolic—Asbestos
PHYSICAL PROPERTIES	ASTM								
Specific Gravity	D792	1.32-1.46	1.34-1.46	1.36-1.43	1.75-1.90	1.6-3.0	1.24-1.35	1.30-1.35	1.60-1.65
Ther Cond, Btu/hr/sq.ft/°F/ft	C177	0.097-0.3	0.1-0.16	0.097-0.170	0.20	0.24-0.34	0.12	0.05	0.04
Coef of Ther Exp 10^{-5} per °F	D696	1.66-2.50	1.6-2.3	1.60-2.22	0.88	—	0.83-2.20	1.7	2.2
Spec Ht, Btu/lb/°F		0.35-0.40	—	0.30-0.35	0.28-0.32	0.28-0.32	0.33	—	—
Water Absorption (24 hr), %	D570	0.3-0.8	0.4-1.5	0.4-1.75	0.1-1.0	0.2	0.5-2.0	0.5-2.0	0.10-0.50
MECHANICAL PROPERTIES									
Mod of Elast in Tension, 10^5 psi	D638	8-13	8-12	9-14	30-33	10-30	4-6	3.5-6	5-9
Ten Str, 1000 psi	D638, D651	5.0-8.5	5.0-8.5	5-9	5-10	6	4.5-9	3.5	4
Elong (in 2 in.), %	D638	0.4-0.8	—	0.37-0.57	0.2	—	0.75-2.25	—	—
Hardness (Rockwell)	D785	E85-100	E85-95	E80-90	E50-70	E80-90	M40-90	M57	M50
Impact Str (Izod notched), ft-lb/in.	D256	0.24-0.50	0.4-1.0	0.6-8.0	10-33	0.32	0.34-1.0	2.0-2.3	0.3-0.4
Mod of Elast in Flex, 10^5 psi	D790	8-12	8-12	9-13	30-33	10-30	4-6	3.5	5.0
Flex Str, 1000 psi	D790	8.5-12	8.0-11.5	8-15	10-45	10	7-12	7	7
Compr Str, 1000 psi	D695	22-36	24-35	15-30	17-30	20	12-20	10-15	10-20
ELECTRICAL PROPERTIES									
Vol Res, ohm-cm	D257	10^9-10^{13}	1-50 × 10^{11}	10^{10}	7-10 × 10^{12}	6 × 10^{12}	10^8-10^{11}	10^{11}	10^{11}
Dielec Str (short time), v/mil	D149	200-425	250-350	200-350	200-370	380	250-375	250	350
Dielec Const									
60 Hz	D150	5.0-9.0	5.6-11.0	6.5-15.0	7.1-7.2	7.4	9-16	15	15
1 MHz	D150	4.0-7.0	4.5-7.0	4.5-7.0	4.6-6.6	5.0	5	5	5
Dissip Factor									
60 Hz	D150	0.05-0.30	0.08-0.35	0.08-0.45	0.02-0.03	0.13-0.16	0.15-0.60	0.5	0.15
1 MHz	D150	0.03-0.07	0.03-0.07	0.03-0.09	0.02	0.10	0.1-0.2	0.09	0.13
Arc Resistance, sec	D495	5-60	5-60	5-60	60	180	7-20	10-20	5-20
APPLICABLE PROCESSING METHODS		Injection molding, compression molding, transfer molding				Injection molding, compression molding, transfer molding, foam molding, reinforced layup molding, laminating			
HEAT RESISTANCE									
Max Rec Svc Temp, F		300-350	300	250-300	350-450	400	212-300	212-225	225-360
Deflection Temp, F	D648	260-360	290-340	250-340	600	335	220-270	220-280	250-300
CHEMICAL RESISTANCE		Severely attacked by strong acids and strong alkalis. Effects of dilute acids, alkalis and organic solvents vary with the reagent. Chemical resistance varies with the particular formulation and not all materials of a type are equally resistant							
USES		Mechanical applications include pulleys, wheels, motor housings, handles. Electrical uses include coil forms, ignition parts, condenser housings, fuse blocks, instrument panels. Thermal applications include handles, appliance connector plugs. Chemical uses include photographic development tanks, rayon spinning buckets and parts, milking machine cups. Decorative uses include radio and television cabinets, handles, knobs, buttons							

Plastic Foams--Rigid (structural, integral skin type)

Type ▸ Density, lb/ft³ ▸		ABS 50	Polycarbonate 50	Polypropylene 37.5	Ph ox[b] based 50	Polyurethane 34.5	Polyester 56
	ASTM						
Ther Cond. Btu/hr/sq ft/°F/in.	D2326	0.58-2.1	0.09	—	0.084	—	—
Coef of Ther. per °F × 10^{-5}	D696	—	2.5 × 10^{-5}	—	3.8 × 10^{-5}	—	—
Water Absorption, % vol	C272	0.4-0.6	—	—	—	—	—
Heat Deflection (264 psi), F	D648	170	270	140	205	150	380
Max Rec Service Temp, F	—	180	—	—	—	—	—
Tensile Str, psi	D1623	2700	4800	2100	3300	—	11,000
Ultimate Ten Elong, %	D1623	—	—	—	—	—	—
Mod of Elast in Tension, 1000 psi	D1623	—	—	—	—	—	1000
Compr Str, psi (10%)	D1621	1000	7500	—	5500	1600	—
Mod of Elast in Compr, 1000 psi	D1621	—	—	—	—	—	—
Flex Str, psi	D790	3700	9500	3200	5000	4200	20,000
Mod of Elast in Flex, 1000 psi	D790	125	290	120	220	150	—
Shear Str, psi	C273	—	—	—	—	—	—
Mod of Elast in Shear, 1000 psi	C273	—	—	—	—	—	—
Hardness (Shore D)	—	—	—	62	—	70	—
Impact Str (Izod, unnotched), ft-lb	—	—	—	1.25	—	—	—

[a]Samples 0.25 in. thick [b]Phenylene oxide based resin.

Phenylene Oxides, Polysulfones, Polyarylsulfone

Material →		Phenylene Oxides (Noryl)			Polysulfones		Polyphenyl-sulfone
		SE-100	SE-1	Glass Fiber Reinforced[a]	Standard	Glass Fiber Reinforced[b]	
PHYSICAL PROPERTIES	ASTM						
Specific Gravity	D792	1.10	1.06	1.21, 1.27	1.24	1.41, 1.55	1.29
Ther Cond, Btu/hr/sq ft/°F/in.	C177	1.10	1.5	1.15, 1.1	1.8	—	2
Coef of Ther Exp, 10^{-5} per °F	D696	3.8	3.3	2.0, 1.4	3.1	1.6, 1.2	3.1
Specific Heat, Btu/lb/°F	—	—	—	—	0.24	—	0.28
Refractive Index, n_D	D542	Opaque	Opaque	Opaque	1.63	Opaque	—
Water Absorption (24 hr), %	D570	0.07	0.07	0.06	0.22	0.22, 0.18	—
MECHANICAL PROPERTIES							
Tensile Strength, 1000 psi	D638						
Ultimate		—	—	—	—	—	—
Yield		7.8	9.6	14.5, 17.0	10.2	17, 19	10.4
Elongation, %	—						
Ultimate		50	60	4-6	50-100	—	60
Yield		—	—	—	5.6	2, 1.6	7
Mod of Elast in Tension, 10^5 psi	D638	3.8	3.55	9.25, 13.3	3.6	10.9, 14.9	3.1
Flex strength, 1000 psi	D790	12.8	13.5	20.5, 22	15.4	25, 28	12.4
Mod of Elast in Flex, 10^5 psi	D790	3.6	3.6	7.4, 10.4	3.9	12, 15.5	3.3
Impact Str (Izod notched), ft-lb/in.	D638	5.0	5.0	2.3	1.3	1.8, 2.0	1.2
Compr Strength, 1000 psi	D695	12	16.4	17.6, 17.9	13.9	—	14.4
Fatigue Strength[d], 1000 psi	D671						
10^4 cyc		—	—	—	—	14, —	—
10^5 cyc		—	—	—	—	6.5, —	—
10^6 cyc		—	—	—	—	5.0, —	—
10^7 cyc		—	—	—	—	4.5, —	—
Hardness (Rockwell)	D785	R115	R119	L106, L108	R120	M84	M83
Coef of Static Frict (against self)	—	—	—	—	0.67	—	—
Abrasion Res (Taber, CS-17), mg/1000 cycles	—	100	20	35	20	—	9.4
ELECTRICAL PROPERTIES							
Volume Resistivity ohm-cm	D257	10^{17}	10^{17}	10^{17}	5×10^{16}	10^{17}	3.5×10^{15}
Dielectric Str (short time), v/mil	D149	400(1/8 in.)	500(1/8 in.)	1020(1/32 in.)	425	480	371 (1/8 in.)
Dielectric Constant	D150						
60 Hz		2.65	2.69	2.93	3.06	3.55	3.44
1 MHz		2.64	2.68	2.92	3.03	3.41	3.45
Dissipation Factor	D150						
60 Hz		0.0007	0.0007	0.0009	0.0008	0.0019	0.00058
1 MHz		0.0024	0.0024	0.0015	0.0034	0.0049	0.00764
Arc Resistance, sec	D495	75	75	120	122	114	41
HEAT RESISTANCE							
Max Rec Service Temp, F		—	212	—	340	350	375
Deflection Temp, F	D648						
66 psi		230	279	293, 317	358	389	—
264 psi		212	265	282, 310	345	365	400
APPLICABLE PROCESSING METHODS		Injection molding, extrusion, thermo forming, foam molding			Injection molding, extrusion, thermoforming		Inj. molding, extrusion
CHEMICAL RESISTANCE		Excellent res to aqueous media such as detergents and weak and strong acids and bases even at elevated temperatures. Many halogenated and aromatic hydrocarbons will soften or partially dissolve these materials			Res to inorganic acids, alkalis and aliphatic hydrocarbons; partially soluble, or swells in ketones and aromatic hydrocarbons; soluble in chlorinated hydrocarbons		Res to mineral acids, alkali, salt solutions, detergents, hydrocarbon oils, chlorinated hydrocarbons
USES		Automotive dashboards & electrical connectors; appliance & business machine housings, cabinets, consoles & covers; coffee brewers & dispensers; pump & plumbing parts; valves; tape cartridge platforms; coil assemblies; bus bar insulators; switch housings; terminal blocks; tuner bars; light fixtures			Coil bobbins, switches, terminal blocks, battery cases, connectors, circuit carriers, sockets, tube bases, medical instrumentation, printed circuit boards, appliance components, microwave cookware, food processing equipment		Aircraft, electrical parts, with high impact and heat resistance; transparency if required

[a] Where two values are given, first applies to 20% glass fiber and second to 30%, otherwise same value applies to both. [b] Where two values are given, first applies to 30% glass fiber and second to 40%, otherwise same value applies to both. [c] 10% deformation. [d] Cyclic failure stress at 1800 cyc/min.

Materials and Type →		Polyarylate[c]	Polybutadienes	Polybutylenes	
				Copolymer	Homopolymer
PHYSICAL PROPERTIES	ASTM				
Specific gravity	D792	1.2	1.6-2.0	0.894-0.910	0.910-0.915
Refractive Index, n_D	D542	1.6	—	—	150
Transparency (visible light), %	D1003	—	Opaque	Translucent	Translucent
MECHANICAL PROPERTIES					
Tensile Strength (yld), 1000 psi	D638	10	5-12	0.9-2.2	2.2-2.5
Elongation (ult), %		50	—	400-500	300-400
Tensile Modulus, 1000 psi	D790	290	400-1200	12-34	34-36
Flexural Strength 1000 psi		11-15	7-21	—	—
Flexural Modulus, 1000psi		310-320	700-1800	—	—
Compressive str[a], 1000 psi	D695	12.9	11-20	—	—
Fatigue str[b], 1000 psi	D671				
10^4		—	—	—	—
10^5		—	—	—	—
10^6		—	—	—	—
10^7		—	—	—	—
Impact strength					
Izod (notched), lb/in.	D256	4.0-5.5	1-10	—	—
Gardner, ft-lb/in.		—	1-4	—	—
Hardness					
Rockwell	D785	—	E30-70	—	—
Shore	D785	—	—	34-50D	53-60D
ELECTRICAL PROPERTIES					
Vol res, ohm-cm	D257	2×10^{16}	1.5×10^{15}	—	—
Dielec str, v/mil (dry)	D149				
Short time		400	400-600	—	
Dielec constant	D150				
60 Hz		3.08	3.3	2.18-2.25	2.25
1 MHz		2.96	3.3	2.18-2.25	2.25
Dissip factor	D150				
60 Hz		0.001-0.002	0.009	0.0002	0.0002
1 MHz		0.02	0.003	0.0002	0.0002
Arc res, sec (tungsten electrodes)	D495	—	—	—	—
THERMAL PROPERTIES					
Ther cond, Btu/hr/sq ft/°F/in.	C177	—	—	—	1.5
Coef of therm exp, 10^{-5}/°F	D696	3.5-4	—	—	7.1
Specific heat, Btu/lb/°F		—	—	—	—
Heat deflection temp, F	D648				
At 66 psi		345	—	—	215-235
At 264 psi		320	500	—	120-140
Brittleness temp, F	D746	—	—	−5 to −35	−5
Max temp cont use (no load), F	—	265[d] (300)[e]	350-500	—	225
CHEMICAL AND ENVIRON RES	D570				
Water absorption, %					
In 24 hr		0.2	0.10	0.01	0.01
Weathering		—	Discolors	—	Embrittles
Acids					
Weak		—	Resists	Resists	Resists
Strong		—	Resists	—	Resists
Alkali					
Weak		—	Resists	Resists	Resists
Strong		—	Resists	Resists	Resists
Organic solvents		—	Resists	—	Slight swell
Fuels		—	Resists	—	Slight swell
Oil and grease		—	Resists	—	Slight swell
METHODS OF PROCESSING		Injection mldg, extrusion	Calendering, casting, extrusion; compression, injection, rotational, transfer mldg	Extrusion casting, thermoforming, injection, rotational, compression mldg	Extrusion, casting, thermoforming; injection, rotational, compression mldg
USES		Outdoor applications, solar energy collectors, appliances, snap-fit connectors, spring clips, tinted glazing	Wire and cable jackets, foot wear, floor tiles, gaskets, tires, sealants, adhesive	Polymer additives and blends; hot melt adhesives	Water, gas, chemical pipe and fittings; pressure vessels; drum liners; construction sheeting; packaging film

[a]2% offset. [b]Cyclic failure stress at 1800 cycles/min. [c]Aromatic polyester. [d]Continuous. [e]Intermittent.

Polycarbonates and Polymethyl Pentenes

Materials and Type →		Polycarbonates				Polymethyl pentenes
		Unfilled	20% gl reinf	30% gl reinf	40% gl reinf	
PHYSICAL PROPERTIES	ASTM					
Specific gravity	D792	1.19-1.22	1.35	1.43	1.52	0.83
Refractive Index, n_D	D542	1.586	—	—	—	1.465
Transparency (visible light), %	D1003	85-89	Opaque	Opaque	Opaque	90
MECHANICAL PROPERTIES						
Tensile Strength (yld), 1000 psi	D638	8.5-9.0	16 (ult)	19 (ult)	23 (ult)	4
Elongation (ult), %		90-115	4-6	3-5	3-5	15
Tensile Modulus, 1000 psi	D790	325-340	860	1250	1680	210
Flexural Strength, 1000 psi		12-13.5	19	23	27	—
Flexural Modulus, 1000psi		310-345	800	1100	1400	140-200
Compressive str[a], 1000 psi	D695	10-12.5	16	18	21	—
Fatigue str[b], 1000 psi	D671					
10^4		5.4	—	—	14.5	—
10^5		2.0	—	—	8.7	—
10^6		1.0	—	—	6.1	—
10^7		0.8	—	—	6.0	—
Impact strength						
Izod (notched), lb/in.	D256	12-16	2.0	2.0	2.5	0.8
Gardner, ft-lb/in.		—	—	—	—	—
Hardness						
Rockwell	D785	M68-74	R122 (M91)	R120 (M92)	M93	L67-74
Shore	D785	—	—	—	—	—
ELECTRICAL PROPERTIES						
Vol res, ohm-cm	D257	8.2×10^{16}	$>10^{16}$	$>10^{16}$	4×10^{16}	10^{16}
Dielec str, v/mil (dry)	D149					
Short time		380-425	490	475	450	700
Dielec constant	D150					
60 Hz		3.01-3.17	3.17	3.35	3.53	2.12
1 MHz		2.96-3.05	3.13	3.31	3.48	2.12
Dissip factor	D150					
60 Hz		0.0009	0.0009	0.0011	0.0013	—
1 MHz		0.010	0.0073	0.007	0.0067	—
Arc res, sec (tungsten electrodes)	D495	120	120	120	120	—
THERMAL PROPERTIES						
Ther cond, Btu/hr/sq ft/°F/in.	C177	1.35-1.41	1.47	1.50	1.53	—
Coef of therm exp. 10^{-5}/°F	D696	3.75	1.49	1.21	0.93	—
Specific heat, Btu/lb/°F		0.30	—	—	—	—
Heat deflection temp, F	D648					
At 66 psi		280-290	300	305	310	—
At 264 psi		260-288	295	295	295	—
Brittleness temp, F	D746	−200	—	—	—	—
Max temp cont use (no load), F	—	—	—	—	—	250-320
CHEMICAL AND ENVIRON RES	D570					
Water absorption, %						
In 24 hr		0.15	0.16	0.14	0.12	0.01
Weathering		Discolors	—	—	To none	None
Acids						
Weak		Resists	Resists	Resists	Resists	Attacks
Strong		Attacks	Attacks	Attacks	Attacks	Resists
Alkali						
Weak		Res-attacks	Res-attacks	Res-attacks	Res-attacks	Resists
Strong		Attacks	Attacks	Attacks	Attacks	Resists
Organic solvents		Attacks	Attacks	Attacks	Attacks	Attacks
Fuels		Attacks	Attacks	Attacks	Attacks	Resists
Oil and grease		Resists	Resists	Resists	Resists	Resists
METHODS OF PROCESSING		Blow, foam, injection, rotational mldg; extrusion, thermoforming	Injection mldg			Blow mldg, extrusion, injection mldg, thermoforming
USES		Electrical parts, portable tool housings, light globes, lenses, sports goods, glazing sheet, impellers, automotive parts, body armor				Laboratory and medical ware light diffusers, lenses reflectors, vending machines parts, packaging

[a] 2% offset. [b] Cyclic failure stress at 1800 cyc/min

Polyesters—Thermosets

Type →		Cast Polyester		Reinforced Polyester Moldings			Pultrusions	
		Rigid	Flexible	High Strength (glass fibers)	Heat and Chemical Resistant (asbestos)	Sheet Molding Compounds, General Purpose	General purpose	High performance
PHYSICAL PROPERTIES	ASTM							
Specific Gravity	D792	1.12-1.46	1.06-1.25	1.8-12.0	1.5-1.75	1.65-1.80	1.61	1.94
Ther Cond, Btu/hr/sq ft/°F/in.	—	1.20—1.44	—	1.32-1.68	—	—	4	5
Coef of Ther Exp, per °F	D696	$3.9\text{-}5.6 \times 10^{-5}$	—	$13\text{-}19 \times 10^{-6}$	—	—	5×10^{-6}	3×10^{-6}
Specific Heat, Btu/lb/°F	—	0.30-0.55	—	0.25-0.35	—	0.20-0.25	0.28	0.24
Water Absorption (24 hr), %	D570	0.20-0.60	0.12-2.5	0.5-0.75	0.25-0.50	0.15-0.25	0.75	0.75
Transparency (visible light), %	—	—	—	Opaque	Opaque	Opaque	Opaque	Opaque
Refractive Index, n_D	D542	1.53-1.58	1.50-1.57	—	—	—	—	—
MECHANICAL PROPERTIES								
Tensile Strength, 1000 psi	D638	4-10	1-8	5-10	4-6	15-17	20[a]	100[a]
Tensile Modulus, 10^5 psi	D638	1.5-6.5	0.001-0.10	16-20	12-15	15-20	23[a]	60[a]
Elongation, %	D638	1.7-2.6	25-300	0.3-0.5	—	—	—	—
Impact Strength (Izod notched), ft/lb/in.	D256	0.18-0.40	4.0	1-10	0.45-1.0	5-15	18[a]	18[a]
Flexural Strength, 1000 psi	D790	14-18	4-16	6-26	10-13	26-32	30[a]	—
Mod of Elast in Flex, 10^5 psi	D790	1-9	0.001-0.39	15-25	—	15-18	20[a]	—
Compr Strength, 1000 psi	D690	12-37	1-17	20-26	20-25	22-36	—	—
Hardness (Barcol)	D785	35-50	6-40	60-80	40-70	45-60	50[a]	50[a]
ELECTRICAL PROPERTIES								
Volume Resistivity, ohm-cm	D257	10^{13}	10^{12}	1×10^{12} 1×10^{13}	1×10^{12}- 1×10^{13}	6.4×10^{15}- 2.2×10^{16}	—	—
Dielectric Str (step-by-step), v/mil	D149	300-400	300-400	200-400	350	400-440	—	—
Dielectric Constant	D150							
60 Hz		2.8-4.4	3.18-7.0	—	—	4.62-5.0	4.5(1)	4.5(1)
1 MHz		2.8-4.4	3.7-6.1	—	—	4.55-4.75	—	—
Dissipation Factor	D150							
60 Hz		0.003-0.04	0.01-0.18	—	—	0.0087-0.04	0.03(1)	0.03(1)
1 MHz		0.006-0.04	0.02-0.06	—	—	0.0086-0.022	—	—
Arc Resistance, sec	D495	115-135	125-145	130-170	—	130-180	80	80
APPLICABLE PROCESSING METHODS		Casting		Layup molding, laminating, compression molding, transfer molding, injection molding		Matched metal die compression molding	Pultrusion	
HEAT RESISTANCE								
Max Rec Svc Temp, F	—	250-300	150-250	250-400	300	300	250-300	300
Heat Deflection Temp (264 psi), F	D648	120-400	—	400	375-400	375-400	400	400
CHEMICAL RESISTANCE		Slightly to heavily attacked by strong acids; attacked by strong alkalis, ketones and solvents		Good to exc res to weak acids, organic solvents, weak alkalis; good res to strong acids; poor to fair res to strong alkalis	Good to exc res to weak acids, organic solvents and weak alkalis; good res to strong acids; air to good res to strong alkalis	Good to exc res to weak acids, organic solvents and weak alkalis; good res to strong acids; poor to fair res to strong alkalis	Good to exc res to weak acids, organic solvents, weak alkalies good res to strong acids; poor to fair res to strong alkalies	
USES		Electrical components, buttons, decorative architectural uses	Flooring, tooling encapsulants, buttons and shields	Chairs, housings, covers, trays molded panels, bezels, motor shrouds, electrical parts, fans, helmets			Elec components, corrosion res construction, high-str-to-wt mech parts	

[a] Longitudinal direction.

Polyester — Thermoplastic

Type→		Polybutylene Terephthalate (PBT)			
		Unreinforced	Unreinforced, flame-retardant	Glass-reinforced[d]	Glass-reinforced[e] flame-retardant
PHYSICAL PROPERTIES	ASTM				
Specific Gravity	D792	1.31	1.41-1.43	1.42-1.6	1.53-1.7
Mold Shrinkage, in./in. × 10^{-3}	—	17-23	15-23	2-6[b](5-8)[c]	2-8[b](8-10)[c]
Water absorption (24 hr), %	D570	0.08-0.09	0.07-0.09	0.05-0.07	0.05-0.07
Coef of Friction, self	D1894	0.17	0.12-0.16	0.16-0.17	0.12-0.24
metal		0.13	0.10-0.14	0.12-0.14	0.10-0.16
MECHANICAL PROPERTIES					
Tensile Strength, 1000 psi	D638	7.5-8.5	8	12.5-19	13-19.5
Elongation, %	D638	5-300	15-80	2-6	1-6
Elast Mod in Tension, 10^5 psi	D638	—	—	8-14	11-19
Flex Strength, 1000 psi	D790	12-12.4	12.8-14.7	21-30	21-29.9
Elast Mod in Flex, 10^5 psi	D790	3.3-3.6	3.7-4.0	6.75-13	7.3-15
Impact Str (notched Izod), ft-lb/in.	D256	0.7-1.0	0.5-0.8	0.9-2.2	1.0-1.5
Compr Str, 1000 psi	D695	13	14.5	15-18	15-22
Shear Strength, 1000 psi	D732	7.7	7.7	8-8.9	8-9.7
Hardness (Rockwell)	D785	R117-M75	R120-M81	R118	R119-M94
Abrasion Res[a], mg/1000 cyc	D1044	9	14	12-19	17-22
ELECTRICAL PROPERTIES					
Volume Resistivity, 10^{15} ohm-cm	D257	1.0-40	1.0-40	1.0-32	4-36
Dielectric Str (short time), v/mil[h]	D149	400-420	350-470	450-560	410-570
Dielectric constant: 100 Hz	D150	3.2-3.3	3.2-3.3	3.6-3.8	3.6-3.9
1 MHz		3.1	3.1	3.4-3.7	3.0-3.7
Dissipation Factor: 100 Hz	D150	0.002	0.002-0.003	0.0015-0.002	0.002-0.0064
1 MHz		0.02-0.03	0.02	0.02	0.014-0.020
Arc Resistance, sec	D495	75-190	63-110	125-146	85-130
THERMAL PROPERTIES					
Max Rec Service Temp, °F	—	—	—	—	—
Heat Deflection Temp, °F @ 66 psi	D648	310-324	325-354	413-442	409-435
@ 264 psi		123-130	135-160	375-405	360-406
Ther Cond, Btu·ft/hr·ft²·°F	C177	1.1	1.2	1.2-1.54	1.2-1.63
Coef of Ther Exp, 10^{-5} in./in. · °F	D635	4.0	3.6	1.3-2.5[b](4.3-5.4)[c]	1.3-1.4[b](4-5.4)[c]
Specific Heat, Btu/lb·°F	C351	0.36-0.55	—	0.3	0.22-0.23
Flammability:					
Rate of Burning, in./min	D635	0.7-1.0	0	0.8-1.3	0
Flammability Rating	UL 94	HB	V-0[f]	HB[g]	V-0[f]
Limiting Oxygen Index, %	D2863	20	33	20-20.5	28.5-33
APPLICABLE PROCESSING METHODS		Injection molding, extrusion			
CHEMICAL RESISTANCE		Resistant to aliphatic hydrocarbons, gasoline, carbon tetrachloride, perchloroethylene, oils, fats, alcohols, glycols, ethers, high molecular weight esters and ketones, dilute acids and bases, detergents, most aqueous salt sol, at ambient temp. Attacked by strong acids and bases. Resistant to potable water at ambient temp. Prolonged use in water above 125 F not recommended. Swollen by ethylene dichloride, low molecular weight ketones and substituted aromatic compounds			
USES		Gears, bearings, valves, pump parts, fittings, rollers, cams, bushings, electronic parts, textile machinery parts, tape cassettes, fasteners			

Properties for injection moldings are at room temperature. Where property ranges are indicated, values apply to a variety of grades. a Taber CS-17 wheel. b Parallel to flow direction. c Perpendicular to flow direction. d 15 to 40% glass reinforcement. e 15 to 30% glass reinf. f @ 0.028 and 1/32 in. thick. g @ 0.028. to 0.058 in. thick. h 1/8 in.

Polyester — Thermoplastic

Type→		PBT (cont'd)			PET[h]	PBT/PET
		Mineral-filled[a]	Mineral/glass-reinforced[b]	High-Impact	Glass-reinforced[i]	Glass-reinforced[j]
PHYSICAL PROPERTIES	ASTM					
Specific Gravity	D792	1.36-1.47	1.55-1.86	1.25-1.47	1.56-1.69	1.43-1.66
Mold Shrinkage, in./in. × 10^{-3}	—	3-22	3-8[c] (5-10)[d]	3-17[c] (17-21)[d]	2[c] (8-9)[d]	4-8
Water absorption (24 hr), %	D570	0.06-0.10	0.04-0.08	0.08-0.10	0.05-0.06	0.06-0.07
Coef of Friction, self	D1894	0.18-0.25	0.23-0.27	—	—	—
metal		0.22-0.27	0.25-0.30	—	—	—
MECHANICAL PROPERTIES						
Tensile Strength, 1000 psi	D638	7.1-9	9.5-12.5	5.8-13	16.1-28	12-16
Elongation, %	D638	3-150	3-4	3.5-270	2.1-2.8	3
Elast Mod in Tension, 10^5 psi	D638	—	—	—	12.8-21.1	—
Flex Stength, 1000 psi	D790	12.5-18	16-20	7-19.5	21.3-41.0	20-26
Elast Mod in Flex, 10^5 psi	D790	3.9-12.4	8.5-12	2.2-9.2	13.0-20.0	6.5-10
Impact Str (notched Izod), ft-lb/in.	D256	1.0-1.9	0.7-1.0	3.4-18	1.2-2.4	0.7-1.5
Compr Str, 1000 psi	D695	9.3-13.6	11-19	—	13.3-26	15-21.8
Shear Strength, 1000 psi	D732	6.5-7.5	6.8-8	5.3	11.5-12.5	7.4-8.9
Hardness (Rockwell)	D785	R112-M57	R82-M75	R113-M42	M80-R120	R119-R120
Abrasion Res[m], mg/1000 cyc	D1044	20-37	25-55	22-37	14.9-21	14-19
ELECTRICAL PROPERTIES						
Volume Resistivity, ohm-cm × 10^{15}	D257	>1-69	>1-19	>1-3.8	0.1	0.02-9
Dielectric Str (short time), v/mil[n]	D149	390-600	500-600	350-400	403-550	470-530
Dielectric Constant: 100 Hz	D150	3.1-4.3	3.7-4.0	3.2-3.64	4.21-4.22[k]	3.6
1MHz		2.9-3.9	3.5-3.7	3.0-3.43	3.5-3.9	3.5
Dissipation Factor: 100 Hz	D150	0.002	0.0012-0.0025	0.002-0.004	0.0024-0.0096	0.0018
1 MHz		0.02	0.014-0.02	0.02-0.03	0.011-0.014	0.016
Arc Resistance, sec	D495	125-130	101-129	129-146	81-126	68-136
THERMAL PROPERTIES						
Max Rec Service Temp, °F	—	—	—	—	—	—
Heat Deflection Temp, °F @ 66 psi	D648	300-420	410-420	250-379	—	400-430
@ 264 psi		150-385	340-390	125-374	428-442	320-380
Ther Cond, Btu · ft/hr · ft² · °F	C177	—	—	—	2.2	—
Coef of Ther Exp, in./in. · °F × 10^{-5}	D696	3.4-4.5[c] (5.0)[d]	1.5-2.3[c] (4-5.1)[d]	—	1.3-2.0	3.1-5.5[c] (1.8-4.4)[d]
Specific Heat, Btu/lb · °F	C351	—	—	—	—	—
Flammability						
Rate of Burning, in./min	D635	—	—	—	—	—
Flammability Rating	UL 94	HB, V-2[e]	V-0[f]	HB[g]	HB, V-0, V-2	HB, V-0, V-2
Limiting Oxygen Index, %	D2863	—	—	—	20-30	—

APPLICABLE PROCESSING METHODS: Injection molding, extrusion

CHEMICAL RESISTANCE: Resistant to aliphatic hydrocarbons, gasoline, carbon tetrachloride, perchloroethylene, oils, fats, alcohols, glycols, ethers, high molecular weight esters and ketones, dilute acids and bases, detergents, most aqueous salt sol, at ambient temp. Attacked by strong acids and bases. Resistant to potable water at ambient temp. Prolonged use in water above 125 F not recommended. Swollen by ethylene dichloride, low molecular weight ketones and substituted aromatic compounds

USES: Gears, bearings, valves, pump fittings, rollers, cams, bushings, electronic parts, textile machinery parts, tape cassettes, fasteners

Properties for injection moldings are at room temperature unless otherwise noted. Where property ranges are indicated, values are for a variety of plastics in each category. a 10 — 30% mineral filled. b 35 — 45% total reinforcement. c Parallel to flow direction. d Perpendicular to flow direction. e @ 0.058 in. thick. f @ 0.032 in. thick. g @ 0.028 in. thick. h Polyethylene terephthalate. i 30 — 45% glass-reinforced. j 15 — 30% glass-reinforced. k 60 Hz. m Tabor CS-17 Wheel. n 1/8 in.

Polyethylenes

Type ➡		Type I—Lower Density (0.910-0.925)			Type II—Medium Density (0.926-0.940)	
		Melt Index 0.3-3.6	Melt Index 6-26	Melt Index 200	Melt Index 20	Melt Index 1.0-4.0
PHYSICAL PROPERTIES	ASTM					
Specific Gravity	D792	0.910–0.925	0.918–0.925	0.910	0.930	0.930–0.940
Ther Cond, Btu/hr/sq ft/°F/in.	C177	2.28	2.28	2.28	2.28	2.28
Coef of Ther Exp, 10^{-5} per °F	D696	8.9–11.0	8.9–11.0	11.0	8.3–16.7	8.3–16.7
Refractive Index	D542	1.51	1.51	1.51	1.51	1.51
Spec Ht, Btu/lb/°F		0.53–0.55	0.53–0.55	0.53–0.55	0.53–0.55	0.53–0.55
Water Absorption (24 hr), %	D570	<0.01	<0.01	<0.01	<0.01	<0.01
MECHANICAL PROPERTIES						
Mod of Elast in Tension, 10^5 psi	D638	0.21–0.27	0.20–0.24	—	—	—
Ten Str, 1000 psi	D412	1.4–2.5	1.4–2.0	0.9–1.1	2.0	3.3
Elong, %	D412	500–725	125–675	80–100	200	800
Hardness (Shore)	D785	C73, D50–52	C73, D47–53	D45	D55	D 59
Impact Str (Izod), ft-lb/in. notch	D256	—	—	—	—	—
Brittleness Temp, F		<—94	<—4	<14	<—148	<—105
Mod of Elast in Flex, 10^3 psi	D747	13–27	12–30	10	35-50	65
Shear Str, 1000 psi		1.6–1.85	1.4–1.7	1	—	—
ELECTRICAL PROPERTIES						
Vol Res, ohm-cm	D257	10^{17}–10^{19}	10^{17}–10^{19}	10^{17}–10^{19}	>10^{15}	>10^{15}
Dielec Str (short time), v/mil	D149	480	480	480	480	480
Dielec Const	D150	2.3	2.3	2.3	2.3	2.3
Dissip Factor	D150	<0.0005	<0.0005	<0.0005	<0.0005	<0.0005
APPLICABLE PROCESSING METHODS		Injection molding, extrusion, rotational molding, blow molding, foam molding			Injection molding, extrusion, thermoforming, rotational molding, blow molding	
HEAT RESISTANCE						
Vicat Softening Point, F		176–201	176–201	—	215-230	220-239
CHEMICAL RESISTANCE		Excellent resistance to acids and alkalis at normal temperature, except oxidizing acids such as nitric, chlorosulfonic and fuming sulfuric. Below 122 F, insoluble in organic solvents; at higher temperatures, soluble to varying degrees in hydrocarbons and halogenated hydrocarbons, but insoluble in more polar liquids. Generally, a higher melt index material has greater solubility				
USES		Injection moldings: kitchen utilityware, toys, process tank liners,closures, packages, sealing rings, battery parts. Blow moldings: squeeze bottles for packaging, containers for drugs. Film: wrapping materials for food, clothes, other items. Wire and cable: high frequency insulation, jacketing. Pipe: chemicals handling, irrigation systems, natural gas transmission				

Type ➡		Type III—Higher Density (0.941-0.965)			High Molecular Weight
		Melt Index 0.2-0.9	Melt Index 0.1-12.0	Melt Index 1.5-15	
PHYSICAL PROPERTIES	ASTM				
Specific Gravity	D792	0.953-0.962	0.963	0.96	0.94
Ther Cond, Btu/hr/sq ft/°F/in.	C177	2.28	2.28	2.28	2.28
Coef of Ther Exp, 10^{-5} per°F	D696	8.3–16.7	8.3–16.7	8.3–16.7	—
Refractive Index	D542	1.54	1 54	1.54	—
Spec Ht., Btu/lb/°F		0.46–0.55	0.46–0.55	0.46–0.55	—
Water Absorption (24 hr), %	D570	<0.01	<0.01	<0.01	<0.01
Flammability, ipm	D635	1.0	1.0	1.0	1.0
MECHANICAL PROPERTIES					
Mod of Elast in Tension, 10^5 psi	D638	—	—	—	1.0
Ten Str, 1000 psi	D 882/638	3.2/4.2	3.4/3.8	3.2/4.3	5.4
Ultimate/Elongation,%	D 882/638	700-1000	600-900	500-900	400
Hardness (Shore)	D935	D68–70	D60–70	D68–70	60-65

Polyethylenes

Type →		Type III—Higher Density (0.941-0.965)			
		Melt Index 0.2-0.9	Melt Index 0.1-12.0	Melt Index 1.5-15	High Molecular Weight
MECHANICAL PROPERTIES (Cont'd)					
Impact Str (Izod), ft-lb/in. notch	D256	1.8	1.2-2.0	1.2-2 5	>20
Brittleness Temp, F	D746	−106 to −180	<−105	<−105	<−100
Mod of Elast in Flex, 10^3 psi	D747	130–150	90–125	140	75
ELECTRICAL PROPERTIES					
Vol Res, ohm-cm	D257	$>10^{15}$	$>10^{15}$	$>10^{15}$	$>10^{15}$
Dielec Str (short time), v/mil	D149	480	480	480	480
Dielec Const	D150	2.3	2 3	2 3	2 3
Dissip Factor	D150	<0.0005	<0.0005	<0.0005	<0.0005
APPLICABLE PROCESSING METHODS		Injection molding, extrusion, thermoforming, rotational molding, blow molding, foam molding			Extrusion, compression molding, and special injection molding
HEAT RESISTANCE					
Vicat Softening Point, F		258-266	255	260	—
CHEMICAL RESISTANCE		Same basic chemical resistance as Types I and II, but better resistance to some specific chemicals			
USES		Refrigerator parts, packaging, structural housing panels, pipe, defroster and heater ducts, sterilizable housewares and hospital equipment, hoops, battery parts, blow molded containers and parts, film wrapping materials, wire and cable insulation, and chemical resistant pipe			

[a]Powder. [b] Flow molding.

Olefin Copolymers

Type →		EEA (Ethylene Ethyl Acrylate)	EVA (Ethylene Vinyl Acetate)	Ethylene-Butene	Propylene-Ethylene	Ionomer	Poly-allomer
PHYSICAL PROPERTIES	ASTM						
Specific Gravity	D792	0.93	0.94	0.95	0.91	0.94	0.898-0.904
Tensile Impact, ft-lb/in.^{2}	D1822	500	690	—	150	400	—
Tensile Str, 100 psi	D638	2.0	3.6	3.5	4	4	30-43
Izod Impact Str (notched), ft-lb/in.	D256	—	—	0.4	1.1	9-14	1.5
Hardness, Shore D	D785	35	36	65	—	60	—
Elongation (in 1 in.), %	D638	650	650	20	—	450	300-400
Flexural Modulus, psi	D790	—	—	165	140	—	0.7-1.3 x 10^5
ELECTRICAL PROPERTIES							
Vol Res, 10^{15} ohm-cm	D257	2.4	0.15	—	—	10	$>10^{16}$
Dielec Str (short time), v/mil	D149	550	525	—	—	1000	500-650
Dielec Const, 60 Hz	D150	2.8	3.16	—	—	2.4	2.3
Dissip Factor, 60 Hz	D150	0.001	0.003	—	—	0.003	>0.0005
THERMAL PROPERTIES							
Heat Deflection Temp (66 psi), F	D648	—	—	—	104	105	122-133[a]
Brittleness Temp, F	D746	−155	−148	−35	—	−160	—
Softening Point, Vicat	D1525	147	147	243	—	162	—
CHEMICAL RESISTANCE		Res most weak mineral acids, alkalis; att by chlorinated hydrocarbons, straight-chain paraffinic solvents, benzene		Generally satisfactory res to chemicals; good res to stress corrosion		Res to organic solvents; att by oxidizing acids	Res to strong alk, weak acids, organic solvents; att slowly by acid reagents
USES		Tubing, seals, bushings, dampening pads, rug backing, elec insulation, floor mats, other flexible items		Packaging, appliances, furniture, wire and cable insulation, rigid parts		Skin packaging; coated substrates; clear, flexible parts	Molded parts req toughness and good hinge properties

[a] For 264 psi.

Polyimides, Poly (amide-imide)

Material →		Polyimides			Poly(amide-imide)[g]	
		Unreinforced	15% Graphite	Glass Reinf	High Modulus	High Impact
PHYSICAL PROPERTIES	ASTM					
Specific Gravity	D792	1.43	1.51	1.90	1.45	1.40
Ther Cond, Btu/hr/sq ft/°F/in.	C177	2.5-6.8	6.0	3.59[b]	2.5	1.7
Coef of Ther Exp, per °F x 10^{-5}	D696	2.5-2.8	2.3	0.8	1.5	2.0
Specific Heat, Btu/lb/°F	—	0.27-0.31[b, c]	—	0.27[b, c]	—	—
Refractive Index, n_D	D542	Opaque	Opaque	Opaque	Opaque	Opaque
Water Absorption (24 hr), %	D570	0.24-0.47	0.19	0.2	0.22	0.28
Coef of Friction	—	—	—	—	—	—
MECHANICAL PROPERTIES						
Tensile Strength, 1000 psi	D638					
Ultimate		—	—	—	19.6	26.9
Yield		7.5-13	5-9	28	—	—
Elongation, %	D638					
Ultimate		<1-8	1.2-5	<1	6	12
Yield		—	—	—	—	—
Mod of Elast in Tension, 10^5 psi	D638	4.7-7.5	5.4-6	45	9.2	7.3
Flex Strength, 1000 psi	D790	11-17	6.6-15	56	26.4	30.7
Mod of Elast in Flex, 10^5 psi	D790	4.5-7.0	5.0-5.4	38.4	9.2	6.6
Impact Str (Izod notched), ft-lb/in.	D638	0.5[d]-1.0	0.5[d]	17[d]	1.1	2.5
Compr Strength, 1000 psi	D695	27.4-40	32	42	30	4.0
Hardness (Rockwell)	D785	97-99M	88M	114E	109 M	78E
Abrasion Res (Taber, CS-10 wheel), mg/ 1000 cycles, gm loss	D1044	0.080[a]	—	20	—	—
ELECTRICAL PROPERTIES						
Volume Resistivity, ohm-cm	D257	10^{15}-10^{16}	—	9.2 x 10^{15}	—	1.2 × 10^{17}
Dielectric Str (short time), v/mil	D149	310-560	—	300	—	600
Dielectric Constant	D150					
60 Hz		3.6-4.1	—	4.84	—	—
1 MHz		3.5-3.9	—	4.74	—	4.0
Dissipation Factor	D150					
60 Hz		0.002-0.003	—	0.0034	—	—
1 MHz		0.004-0.011	—	0.0055	—	0.009
Arc Resistance, sec	D495	152-230	—	50-180	—	125
HEAT RESISTANCE						
Max Rec Service Temp, F	—	500-800	800	500	500	500
Deflection Temp, F	D648					
66 psi		—	—	—	—	—
264 psi		582-680	680	660	525	525
APPLICABLE PROCESSING METHODS		Compression molding, laminating, filament winding, machining, sintering			Compression molding, injection molding, extrusions	
CHEMICAL RESISTANCE		Resist polar and nonpolar organic solvents, dilute acids and bases			Resists most acids, oils, fuels, greases, alcohol and organic solvents	
USES		Bushings, valve seats and high temp mechanical parts		High temp bearing uses, such as jet engine components	Bushings, valve seats, grinding wheels and other high temperature mechanical parts	

[a] Extruded sheet. [b] G. E. test. [c] Cal/gm/°C. [d] ASTM D256. [e] In./1000 hr. dry, 10,000 PV. [f] This grade is an electrical conductor. [g]bearing grade (graphite), unreinforced.

Polypropylene, Polyphenylene Sulfide, Polyether Sulfone

Material ➧		Polypropylene				Polyphenylene Sulfide		Poly-ether sulfone
		General Purpose	High Impact	Glass Reinforced	Flame Retardant	Glass and Mineral Filled[c]	40% Glass Reinforced	
PHYSICAL PROPERTIES	ASTM							
Specific Gravity	D792	0.900-0.910	0.900-0.910	1.04-1.22	1.2	1.8-2.1	1.64	1.37
Ther Cond, Btu/hr/sq ft/°F/in.	—	1.21-1.36	1.72	—	—	—	2.0	1.9
Coef of Ther Exp, 10^{-5} per °F	D696	3.8-5.8	4.0-5.9	1.6-2.4	—	—	2.2	3.1
Specific Heat, Btu/lb/°F	—	0.45	0.45-0.48	—	—	—	0.25	—
Refractive Index, n_D	D542	Transl-op	Trans-op	Opaque	Opaque	Opaque	Opaque	Transparent
Water Ab (24 hr), %	D570	<0.01-0.03	<0.01-0.02	0.02-0.05	0.02-0.03	0.03	0.05	0.43
MECHANICAL PROPERTIES								
Tensile Strength, 1000 psi	D638, C							
Maximum		4.8-5.5	—	—	—	—	—	—
Yield		4.8-5.2	2.8-4.3	6-10	3.6-4.2	11-13	19.5	12.2
Elongation, %	—							
Break		30->200	30->200	2-4	3-15	0.6-0.7	1	—
Yield		9-15	7-13	—	—	—	—	40-80
Mod of Elast in Ten, 10^5 psi	D638, B	1.6-2.2	1.3	812	1.5-2.4	—	11.2	3.54
Flex Yld Strength, 1000 psi	D790, B	6-7	4.1	8-11	—	17-21	29	18.65
Mod of Elast in Flex, 10^5 psi	D790, B	1.7-2.5	1.0-2.0	4-8.2	1.9-6.1	18-21	17	3.73
Impact Str (Izod notched), ft-lb/in.	D256	0.4-2.2	1.5-12	0.5-2	2.2	0.5-1.0	1.4	1.6
Compr Yld Str, 1000 psi	D695	5.5-6.5	4.4	6.5-7	—	16-17	21	—
Fatigue Str[a], 1000 psi	D671							
10^4 cyc		—	—	5.5[b]	—	—	—	—
10^5 cyc		—	—	4.5[b]	—	—	—	—
10^7 cyc		—	—	4.5[b]	—	—	—	—
Hardness (Rockwell)	D785	R80-R100	R28-95	R90-R115	R60-R105	R120/121	R123	M88
ELECTRICAL PROPERTIES								
Vol Res, ohm-cm	D257	$>10^{17}$	10^{17}	1.7×10^{16}	4×10^{16}-10^{17}	1-3×10^{15}	4.5×10^{16}	10^{17}-10^{18}
Diel Str (short time), v/mil	D149	650(125 mil)	450-650	317-475	485-700	340-521	450	400
Dielec Const	D150							
60 Hz		2.20-2.28	2.20-2.28	2.3-2.5	2.46-2.79	3-11	3.79	3.5
1 MHz		2.23-2.24	2.23-2.27	2-2.5	2.45-2.70	3.22 4.6-6.6	3.88	3.5
Dissip Factor	D150							
60 Hz		0.00005-0.0007	<0.0016	0.002	0.0007-0.017	0.0004	0.0037	0.001
1 MHz		0.0002-0.0003	0.0002-0.0003	0.003	0.0006-0.003	0.0007	0.0041	0.006
Arc Res, sec	D495	125-136	123-140	73-77	15-40	182-200	34	116
HEAT RESISTANCE								
Max Rec Svc Temp, F	—	230	—	250	205	—	500	400
Deflection Temp, F	D648							
66 psi		205-230	190-235	275-310	245-280	—	—	—
264 psi		135-140	120-140	250-300	155	>500	485	400
APPLICABLE PROCESSING METHODS		Extrusion, injection mldg, thermoforming, rotational molding, blow mldg, coating, foam mldg				Injection mldg, coating, compression mldg		Inj mldg, extrusion
CHEMICAL RESISTANCE		Res to most acids, alkalis and saline solutions, even at higher temp; res to higher aliphatic solvents and polar substances. Above 175 F, soluble in such aromatic substances as toluene and xylene, and chlorinated hydrocarbons				Exc res to org solv below 375 F. Unaff by strong alkalis or aqueous inorg salt solns		Res most inorg reagents, strong acids & alkalis, org chem. Att by conc oxid acids, some org solv, chlor hydrocarbons
USES		Hospital ware, housewares, applicances, radio and TV housings, film fibers	Luggage seating packaging, housings, auto parts, containers, wire coating	Housings, shrouds, cases, panels and mechanical parts	Electrical uses to meet UL requirements, housings and shields	Corrosion resistant pump components, valves and pipe; auto., elect. comp.	Pump vanes, valve parts, gaskets, fuel cells and parts req chem res at high temps	Engine fittings, hi-temp elec parts, appliance components

[a]Cyclic failure stress at 1800 cyc/min. [b]30% glass-reinforced [c]Properties are ranges of standard and colored grades.

Polystyrenes

Type →		Polystyrenes				Styrene Acrylonitrile (SAN)	Glass Fiber (30%) Reinforced SAN
		General Purpose	Medium Impact	High Impact	Glass Fiber (30%) Reinf		
PHYSICAL PROPERTIES	ASTM						
Specific Gravity	D792	1.04	1.04-1.07	1.04-1.07	1.29	1.04-1.07	1.35
Ther Cond, Btu/hr/sq ft/°F/in.	—	0.70—1.08	0.29—1.08	0.29—1.08	1.40	—	—
Coef of Ther Exp, 10^{-5} per °F	D696	3.3–4.8	3.3–4.7	2.2–5.6	1.8	3.6–3.7	1.6
Specific Heat, Btu/lb/°F	—	0.30–0.35	0.30–0.35	0.30–0.35	0.256	0.33	—
Refractive Index, n_D	D542	1.60	Opaque	Opaque	Opaque	1.565–1.569	Opaque
Water Absorption (24 hr), %	D570	0.03-0.2	0.03-0.09	0.05-0.22	0.07	0.20-0.35	0.15
MECHANICAL PROPERTIES							
Tensile Strength, 1000 psi	D638						
Ultimate		5.0-10	6.0	3.3–5.1	14	9.5-12.0	18
Yield		5.0-10	6.0	2.8–5.3	14	—	18
Elongation, %	D638						
Ultimate		1.0-2.3	3.0-40	30-40	1.1	0.5-3.7	1.4
Yield		1.0-2.3	1.2–3.0	1.5-2.0	1.1	—	1.4
Mod of Elast in Tension, 10^5 psi	D638	4.6–5.0	3.9–4.7	1.50-3.80	12.1	4.0-5.0	17.5
Flex Strength, 1000 psi	D790	10-15	—	—	17	—	22
Mod of Elast in Flex, 10^5 psi	D790	4–5	3.5–5.0	2.3–4.0	12	—	14.5
Impact Str (Izod notched), ft-lb/in.	D638	0.2–0.4	0.5–0.7	0.8–1.8	2.5	0.30–0.45	3.0
Compr Strength, 1000 psi	D695	11.5–16.0	4–9	4–9	19	—	2.3
Fatigue Strength[a], 1000 psi	D671						
10^4 cyc		—	—	—	8.0	—	—
10^5 cyc		—	—	—	7.0	—	—
10^6 cyc		—	—	—	6.0	—	—
10^7 cyc		—	—	—	5.0	—	—
Hardness (Rockwell)	D785	M72	M47–65	M3–43	M85–95	M80-85	M90–100
Abrasion Res (Taber), mg/1000 cycles	—	—	—	—	164	—	—
ELECTRICAL PROPERTIES							
Volume Resistivity, ohm-cm	D257	$>10^{16}$	$>10^{16}$	$>10^{16}$	3.6×10^{16}	$>10^{16}$	4.4×10^{16}
Dielectric Str (short time), v/mil	D149	>500	>425	300–650	396	400–500	515
Dielectric Constant	D150						
60 Hz		2.45–2.65	2.45–4.75	2.45–4.75	3.1	2.6–3.4	3.5
1 MHz		2.45–2.65	2.4–3.8	2.5–4.0	3.0	2.6–3.02	3.4
Dissipation Factor	D150						
60 Hz		0.0001–0.0003	0.0004–0.002	0.0004–0.002	0.005	>0.006	0.005
1 MHz		0.0001–0.0005	0.0004–0.002	0.0004–0.002	0.002	0.007-0.010	0.009
Arc Resistance, sec	D495	60-135	20-135	20–100	28	100–150	65
HEAT RESISTANCE							
Max Rec Service Temp, F	—	160–205	125–165	125–165	190–200	175–190	—
Deflection Temp, F	D648						
66 psi		—	—	—	230	—	230
264 psi		220 max	210 max	210 max	220	210–220	220
APPLICABLE PROCESSING METHODS		Injection molding, extrusion, thermoforming, rotational molding, blow molding, foam molding					
CHEMICAL RESISTANCE		Res alkalis, salts, low alcohols, glycols and water. Fair res to mineral chemicals and vegetable oils. Not res to aromatic and chlorinated hydrocarbons			No effect by weak acids, strong acids; att by oxid acids; no effect by weak alkalis; att slowly by str alkalis; soluble in aromatic and chlor hydrocarbons	Res to alkalis and acids, animal and vegetable oils, soaps, detergents and household chemicals	No effect by weak acids, alkalis, strong acids; att by oxid acids, str alkalis; soluble in ketones, esters, some chlor hydrocarbons
USES		Thin parts, long flow parts, toys, appliances, containers, film, monofilaments and housewares	Radio cabinets, toys, containers, packaging and closures	Containers, cups, lids, large thin wall parts, auto parts, TV cabinets, trays and appliance housings	Auto dashboard skeletons, camera housings and frames, tape reels, fan blades	Kitchenware, closures, film, containers, lenses, battery cases	Camera housings and frames, auto bezels, electrical components, handles, auto panels

[a] Cyclic failure stress at 1800 cyc/min.

Polyvinyl Chloride and Copolymers

Type →		Polyvinyl Chloride, Polyvinyl Chloride Acetate			Vinylidene Chloride[a] Copolymer	Chlorinated Polyvinyl Chloride
		Nonrigid—General	Nonrigid—Electrical	Rigid—Normal Impact		
PHYSICAL PROPERTIES	ASTM					
Specific Gravity	D792	1.20-1.55	1.16-1.40	1.32-1.58	1.68-1.75	1.49-1.58
Ther Cond, Btu/hr/sq ft/°F/in.	C177	0.84—1.20	0.84—1.20	0.84—1.20	0.64	—
Coef of Ther Exp, 10^{-5} per °F	D696	—	—	2.7-3.3	8.78	3.8
Refractive Index	D542	—	—	—	1.60-1.63	—
Spec Ht, Btu/lb/°F	—	—	—	—	0.32	—
Water Absorption (24 hr), %	D570	0.2-1.0	0.40-0.75	0.03-0.40	>0.1	0.02-0.15
MECHANICAL PROPERTIES						
Mod of Elast in Tension, 10^5 psi	D412	0.004-0.03	0.01-0.03	3.5-4.0[b]	0.7-2.0	—
Ten Str, 1000 psi	D412	1-3.5	2-3.2	5.5-8	4-8, 15-40	—
Elong (in 2 in.), %	D638	200-450	220-360	1-10	15-25, 20-30	—
Hardness (Rockwell)	D785	—	—	R110-120	M50-65	R117-122
Hardness (Shore)	D676	A50-100	A78-100	D70-85	>A95	—
Impact Str (Izod notched), ft-lb/in	D256	Variable	Variable	0.5-10	2-8, 0.053	1.0-3.0
Mod of Elast in Flex, psi	D790	—	—	3.8-5.9 × 10^5	—	3.8-4.50
100% Modulus, psi	—	600-2800	600-2800	—	—	
Flex Str, 1000 psi	D790	—	—	11-16	15-17, flexible	14.5-17
Compr Str, 1000 psi	D695	—	—	11-12	—	—
Compr Yld Str, 1000 psi	D695	—	—	10-11	75-85	—
Cold Flex Temp, F	D1043	−70 to 0	−7 to +20	—	—	—
Cold Bend Temp, F	—	−40 to −4	−49 to −4	—	—	—
ELECTRICAL PROPERTIES						
Vol Res, ohm-cm	D257	1-700 x 10^{12}	4-300 x 10^{11}	10^{14}->10^{16}	10^{14}-10^{16}	10^{15}
Dielec Str (short time), v/mil	D149	—	24-500	725-1400	—	1220-1500
Dielec Const (60 Hz)	D150	5.5-9.1	6.0-8.0	2.3-3.7	3-5	3.08
Dissip Factor (60 Hz)	D150	0.05-0.15	0.08-0.11	0.020-0.03	0.03-0.15	0.019-0.021
Loss Factor (60 Hz)	D150	—	1.0-1.2	0.030-0.072	—	—
APPLICABLE PROCESSING METHODS		Injection molding, extrusion, thermoforming, blow molding, foam molding, slush molding, calendering			Extrusion, calendering	Calendering, compression molding, extrusion, injection molding, thermoforming
HEAT RESISTANCE						
Max Rec Svc Temp, F	—	150-220	140-220	150-165	170-212	230
Heat Dist Temp, F						
66 psi	D648	—	—	170-185	190-210	215-247
264 psi	D648	—	—	140-170	130-150	202-234
CHEMICAL RESISTANCE		Generally resistant to alkalis and weak acids. Moderately to not resist to strong acids. Not resistant to keotones and esters; aromatic hydrocarbons produce swelling			Excellent to all acids and most common alkalis[c]	Res to acids, alkalis, oil, grease and most organic solvents
USES		Parts made by molding, high speed extrusion, calendering. Blown extruded film. Vacuum cleaner parts, handlebar grips, doll parts, hair curlers, safety goggle cups, grommets, toy tires, garden hose, and protective garments	Parts made by calendering, extrusion. Insulation and jacketing for: communication and low-tension power wire and cable, building wiring, appliance and machine tool cords, and switchboard cable	Parts made by calendering, laminating, molding, extrusion. Fume hoods and ducts, storage tanks, chemical piping, plating tanks, phonograph records. Sheets and shapes for decorative panels, other building uses	Extrusions: gasket rods, valve seats, flexible chemical tubing and pipe, tape for wrapping joints, chemical conveyor belts. Moldings: spray-gun handles, acid dippers, parts for rayon producing equipment	Hoods, ducts, exterior bldg components, pipe.

[a] Where two values or ranges are given, they represent unoriented and oriented forms, respectively. [b] Modulus of elasticity in compression. [c] Unaffected by aliphati and aromatic hydrocarbons, alcohols, esters, etc. [d] Barrel temperature. [e] Stock temperature.

Rubber and Elastomers

Rubber — Molded, Extruded

Type →	Natural Isoprene	Synthetic Isoprene	Polybutadiene	Styrene-butadiene	Isobutylene Isoprene	Nitrile	Chloroprene
ASTM Designation	**NR**	**IR**	**BR**	**SBR**	**IIR**	**NBR**	**CR**
PHYSICAL PROPERTIES							
Specific Gravity	0.92-0.93	0.93	0.91	0.94	0.92	0.98	1.23-1.25
Ther Cond, Btu/hr/sq ft/F/ft	0.082	0.082	—	0.143	0.053	0.143	0.11
Coef of Ther Exp (cubical), 10^{-5} per °F	37	37	37.5	37	32	39	34
Colorability	—	—	—	Good	Good	Excellent	Fair
MECHANICAL PROPERTIES							
Hardness, Durometer	30A-100A	30A-100A	45A-80A	30A-90D	30A-100A	30A-100A	30A-95A
Ten Str, 1000 psi	3.5-4.5	2.5	2.5	2.5-3.0	>2.0	1.0-3.5	0.5-3.5
Modulus (100%), psi	—	—	300-1500	300-1500	50-500	490	100-3000
Elongation, %	500-700	300-750	450	450-500	300-800	400-600	100-800
Compression Set, Method B, %	10-30	—	10-30	5-30	25[a]	5-20	20-60[b]
Resilience, %							
Yerzley (ASTM 945)	80	—	50-90	20-90	30	—	50-80
Rebound (Bashore)	—	—	—	10-60	—	—	50-80
Hysteresis Resistance	Excellent	Excellent	Good	Fair-Good	—	—	Very good
Flex Cracking Resistance	—	—	Excellent	Good	—	—	Very good
Tear Resistance	Excellent	Good	Good	Fair	Good	Good	Good
Abrasion Resistance	Excellent	Excellent	Excellent	Excellent	Good	Excellent	Excellent
Impact Resistance	Excellent	Excellent	Good	Excellent	Good	Good	Excellent
ELECTRICAL PROPERTIES							
Vol Res, ohm-cm	—	—	—	5.0-8.4 x 10^{13}	2.0 x 10^{16}	3.5 x 10^{10}	2.0 x 10^{13}
Dielectric Str, v/mil	400-600	—	400-600	600-800	600-900	250	400-600
Dielectric Constant							
60 Hz	—	—	—	—	2.31	—	8.0
1 MHz	2.9	—	3.3	—	2.25	—	6.7
THERMAL PROPERTIES							
Service Temperature, F							
Min for Cont Use	−70	−60	−150	−75	−50	−60	−60
Max for Cont Use	250	180	200	250	300	300	225
ENVIRONMENTAL RESISTANCE							
Ozone	Poor	Poor	Poor	Poor	Excellent	Poor	Very good
Oxidation	Good	Good	Good	Good	Excellent	Fair-Good	Very good
Weathering	Fair	Fair	Fair	Fair	Excellent	Good	Very good
Water	Excellent	Excellent	Excellent	Excellent	Excellent	Excellent	Good
Radiation	Fair-Good	Fair-Good	Poor	Good	Poor	Fair-Good	Good
Alkalies	Fair-Good	Fair-Good	Fair-Good	Fair-Good	Excellent	Fair-Good	Excellent
Aliphatic Hydrocarbons	Poor	Poor	Poor	Poor	Fair	Excellent	Good
Aromatic Hydrocarbons	Poor	Poor	Poor	Poor	Fair-Good	Good	Fair
Halogenated Hydrocarbons	Poor	Poor	Poor	Poor	Poor	Poor	Poor
Alcohol	Good	Good	Good	Good	Very good	Very good	Good
Animal, Vegetable Oils	Poor-Good	Poor-Good	Poor-Good	Poor	Good	Excellent	Good
Acids							
Dilute	Fair-Good	Fair-Good	Fair-Good	Fair-Good	Excellent	Excellent	Excellent
Concentrated	Fair-Good	Fair-Good	Fair-Good	Fair-Good	Good	Excellent	Fair-Good
Synthetic Lubricants (diester)	Poor-Fair	Poor-Fair	Poor-Fair	Poor	Fair	Fair-Good	Poor
Hydraulic Fluids							
Silicates	Poor-Good	Poor-Good	Poor-Good	Poor-Good	Fair	Fair	Poor-Good
Phosphates	Poor-Good	Poor-Good	Poor-Good	Poor-Good	Good	Poor	Poor
Permeability to Gases	Low	Low	Low	Low	Very low	Very low	Low-Medium
Limiting Oxygen Index	—	—	—	—	18-19	17-20	38-45
USES	Pneumatic tires, tubes; power transmission belts; gaskets; shock absorption, seals against air, moisture, sound, dirt; sponge stock; heels, soles	Same as natural rubber	Pneumatic tires; heels, soles; gaskets, seals, belting, sponge stocks; used in blends with other rubbers for better resilience, abrasion resistance, low temp properties	Same as natural rubber and polybutadiene	Truck and auto tire inner tubes; curing bags for tire vulcanization and molding; steam hose, diaphragms, flexible elec insul, shock, vibration absorption	Gasoline, chemical, oil hose; tubing, gaskets, seals, O-rings; heels, soles; conveyor belting, printing goods and binders for friction materials	Wire and cable; belts, hose, extruded goods, coatings; molded and sheet goods; adhesives, automotive gaskets, seals; petroleum, chemical tank linings

[a] 70 hr at 257 F.

Rubber and Elastomers — Molded, Extruded

Type	Polysulfide	Ethylene Propylene	Chlorinated Polyethylene	Chlorosulfonated Polyethylene	Ethylene/ acrylic	Epichlorohydrin	Polynorbornene
ASTM Designation	PTR	EPM, EPDM	CM	CSM		CO, ECO	
PHYSICAL PROPERTIES							
Specific Gravity	1.35	0.86	1.16-1.32	1.11-1.28	1.08-1.12	1.27-1.49	0.96
Ther Cond, Btu·ft/hr·sq ft·F	—	0.15	—	0.065	—	—	—
Coef of Ther Exp (cubical), 10^{-5} per F	—	32	—	27	—	—	—
Colorability	Fair	Good-Excellent	Excellent	Excellent	—	Good	Excellent
MECHANICAL PROPERTIES							
Hardness, Durometer	20A-80A	30A-90A	50A-95A	40A-95A	64A	30A-95A	15-100A
Ten Str, 1000 psi	0.5-1.5	0.5-3.5	0.9-3.0	0.5-3.5	1.95	2-3	1.0-4.0
Modulus (100%), psi	—	100-3000	700-2200	100-3000	800[f]	150-2000	100-1500
Elongation, %	210-450	100-700	100-700	100-700	450	200-800	100-600
Compression Set, Method B, %	29-38	20-60[b]	5-30	35-80[a]	—	20[b]	10-600
Resilience, %							
Yerzley (ASTM 945)	—	40-75	—	30-70	—	50-80	—
Rebound (Bashore)	—	40-75	15-40	30-70	20	45-75	—
Hysteresis Resistance	—	Good	Good	Fair-Good	—	Good	Excellent
Flex Cracking Resistance	—	Very good	Excellent	Very good	Excellent	Very good	Good
Tear Resistance	Poor-Fair	Fair-Good	Good	Fair	Excellent	Good	Good
Abrasion Resistance	Poor-Fair	Good-Excellent	Excellent	Excellent	Excellent	Fair-Good	Excellent
Impact Resistance	Poor-Fair	Very good	Excellent	Very good	—	Good	Excellent
ELECTRICAL PROPERTIES							
Vol Res, ohm-cm	5×10^{13}	2×10^{16} - 1×10^{17}	—	1×10^{14}	1.9×10^{12}	—	—
Dielectric Str, v/mil	—	500-1000	—	650	7.30	—	—
Dielectric Constant							
60 Hz	7.3[c]	2.25-3.0	—	7.0	—	—	—
1 MHz	6.8[d]	2.2-2.85	—	6.0[e]	—	—	—
THERMAL PROPERTIES							
Service Temperature, F							
Min for Cont Use	-50	-70	-60	-50	−30	-15 to -80	−60
Max for Cont Use	250	350	300	275	400	325	250
ENVIRONMENTAL RESISTANCE							
Ozone	Excellent	Outstanding	Outstanding	Outstanding	Outstanding	Excellent	Poor
Oxidation	Excellent	Excellent	Outstanding	Outstanding	Excellent	Excellent	Good
Weathering	Excellent	Outstanding	Good	—	Excellent	Excellent	Fair
Water	Good	Excellent	Good	Good	Excellent	Good	Excellent
Radiation	Fair	Good	Excellent	Very good	—	Poor	Fair-Good
Alkalies	Good	Good-Excellent	Excellent	Excellent	Excellent	Good	Fair-Good
Aliphatic Hydrocarbons	Excellent	Poor-Good	Excellent	Good	Good	Excellent	Poor
Aromatic Hydrocarbons	Excellent	Poor	Poor	Fair	Good	Very good	Poor
Halogenated Hydrocarbons	Fair-Good	Fair-Poor	—	Poor	Good	Good	Poor
Alcohol	Very good	Poor-Good	Excellent	Good	Fair	Good	Good
Animal, Vegetable Oils	Excellent	Fair	Very good	Good	Excellent	Excellent	Fair-Good
Acids							
Dilute	Good	Excellent	Outstanding	Excellent	Excellent	Good	Fair-Good
Concentrated	Poor	Fair-Good	Excellent	Very good	Poor	Poor	Fair-Good
Synthetic Lubricants (diester)	Good	Fair-Good	Poor	Poor	—	Fair-Good	Good
Hydraulic Fluids							
Silicates	Poor-Good	Fair-Good	Good	Poor	Good	Very good	Fair-Good
Phosphates	Poor-Fair	Good-Excellent	Good	Poor	Good	Poor-Fair	Fair-Good
Permeability to Gases	Very low	Medium	Low	Low	—	Low	Very low
Limiting Oxygen Index	—	10-20	30-35	30-36	48[g]	25-33	
USES	Seals, gaskets, diaphragms, valve seat disks, flexible mountings, hose in contact with solvents, balloons, boats, life vests, rafts	Elec insul, jacketing; footwear, sponge, proofed fabrics; auto weather strip, hose, belts; auto, appliance parts; parts req outstanding ozone, heat res	Hose, tubing, belting, molded goods	Flex chemical and petroleum tube and hose, rolls, tank linings, high temp belts; wire and cable; shoe soles and heels; flooring; building products	Automotive ignition wire jackets, spark plug boots, coolant and power steering hose, motor mounts, timing belts, transmission seals	Diaphragms, print rolls, belts, oil seals, molded mechanical goods, gaskets, hose for petroleum handling; low temp parts	Gaskets, body and engine mount, shock absorbing sponges, heels, soles, electrical and appliance parts, low-temperature parts

[a]22 hr at 212 F. [b]70 hr at 212 F. [c]1 KC. [d]1 MC. [e]Estimated. [f]At 200%. [g]Mineral filled compounds.

Rubber and Elastomers — Molded, Extruded

Type →	Polyacrylate	Silicone	Fluorosilicone	Fluorocarbon	Polyurethanes	Propylene Oxide
ASTM Designation	ACM, ANM	VMQ	FVMQ	FPM	AU, EU	PO
PHYSICAL PROPERTIES						
Specific Gravity	1.09	1.1-1.6	1.4	1.4-1.95	1.02-1.25	1.01
Ther Cond, Btu · ft/hr · sq ft · F	—	0.13	0.13	0.06-1.3	0.09-0.10	—
Coef of Ther Exp (cubical), 10^{-5} per °F	—	45	45	—	5-25	—
Colorability	Good	Excellent	Very good	Very good	Good-Excellent	Good
MECHANICAL PROPERTIES						
Hardness, Durometer	40A-90A	20A-90A	40A-70A	55A-95A	10A-80D	40A-80A
Ten Str, 1000 psi	1.8-2.0	1.5	<2.0	<2.0	0.8-8.0	> 2
Modulus (100%), psi	100-1500	—	—	200-2000	25-5000	—
Elongation, %	100-400	100-800	200-400	150-450	250-800	500-670
Compression Set, Method B, %	10-60	10	—	20-25[a]	10-45[b]	—
Resilience, %						
Yerzley (ASTM 945)	—	30-60	—	40-70	5-75	—
Rebound (Bashore)	—	—	—	40-70	20-65	—
Hysteresis Resistance	—	Fair-Good	Good	Good	Fair-Good	Very good
Flex Cracking Resistance	Fair	Fair-Excellent	Good	Good	Good-Excellent	Very good
Tear Resistance	Fair-Good	Fair	Fair	Fair-Very good	Outstanding	Excellent
Abrasion Resistance	Good	Poor	Poor	Good	Exc-Outst	Good
Impact Resistance	Poor	Poor-Good	Fair	Good	Exc-Outst	Excellent
ELECTRICAL PROPERTIES						
Vol Res, ohm-cm	7×10^{12}	1×10^{14}-1×10^{16}	—	2×10^{13}	0.3×10^{10}-4.7×10^{13}	—
Dielectric Str, v/mil	800	400-700	—	500	330-700	—
Dielectric Constant						
60 Hz	—	2.95-4.0	—	5.0-10.0[d]	4.7-9.53	—
1 MHz	—	2.95-4.0	—	—	5.9-8.51	—
THERMAL PROPERTIES						
Service Temperature, F						
Min for Cont Use	-40	-178	-90	-50	-65	-80
Max for Cont Use	400	600	400	>550	250	< 250
ENVIRONMENTAL RESISTANCE						
Ozone	Excellent	Excellent	Excellent	Outstanding	Excellent	Excellent
Oxidation	Excellent	Excellent	Excellent	Outstanding	Excellent	Excellent
Weathering	Excellent	Excellent	Excellent	Excellent	Good	Excellent
Water	Fair-Poor	Excellent	Excellent	Good	Good-Excellent	Excellent
Radiation	Fair	Fair-Good	Fair-Excellent	Fair-Good	Good-Excellent	Poor
Alkalies	Poor	Poor-Fair	Very good	Fair-Good	Poor-Fair	Very good-Excellent
Aliphatic Hydrocarbons	Excellent	Poor-Good	Excellent	Excellent	Excellent	Fair
Aromatic Hydrocarbons	Poor	Poor-Good	Excellent	Excellent	Fair-Good	Poor-Fair
Halogenated Hydrocarbons	Poor	Fair	Excellent	Good	Poor-Fair	Poor-Fair
Alcohol	Poor	Fair	Good-Excellent	Fair-Good	Poor-Good	Fair-Good
Animal, Vegetable Oils	Very good	Good	Good	Excellent	Excellent	Good
Acids						
Dilute	Fair	Very good	Excellent	Fair-Excellent	Fair	Fair
Concentrated	Fair	Good	Excellent	Fair-Excellent	Poor	Poor
Synthetic Lubricants (diester)	Good	Good	Outstanding	Good-Excellent	Poor-Good	Fair-Good
Hydraulic Fluids						
Silicates	Good	Poor	Outstanding	Excellent	Fair	Fair-Good
Phosphates	Poor	Good	Poor	Excellent	Poor	Poor
Permeability to Gases	Medium	High	Medium	Low	Medium	High
Limiting Oxygen Index	—	20-30	—	50-100	15-20	—
USES	Oil hose, search lt gaskets, white or pastel colored goods, auto gaskets, O-rings (especially for res to extreme pressure and to lubricants, oils containing sulfur)	High and low temperature electrical insulation seals, gaskets, diaphragms, duct work, O-rings, tubing for food and medical uses	Parts req res to high temp solv or oils; seals, gaskets, O-rings	O-rings, brke seals, shft seals, gaskets, hose and ducting connectors, diaphragms, carburetor needle tips, lined valves, packings, roll coverings	Fork lift truck and airplne tail wheels; back-up wheels for turbine blade grinders; spinning cots for glass fiber, hydraulic accum; heels; rolls, gaskets, seals, mechanical goods	Electrical insulation, molded mechanical goods

[a] 22 hr at 392 F. [b] 22 hr at 158 F. [c] 158 F. [d] 1000 cycles.

Thermoplastic Elastomers

Types	Olefinics	Styrenics	Polyester Urethanes	Polyether Urethanes	Copolyether Ester
PHYSICAL PROPERTIES					
Specific Gravity	0.84-1.07	0.9-1.2	1.18-1.23	1.10-1.20	1.17-1.25
Ther Cond, Btu/hr/sq ft/F/ft	0.08-0.09	0.09	0.14-0.19	0.09-0.17	0.154
Coef of Ther Exp (cubical), 10^{-5} per °F	6-12	7.2-7.7	—	5.4-9.5	36
Colorability	Excellent	Excellent	Excellent	Excellent	Excellent
MECHANICAL PROPERTIES					
Hardness, Durometer	35A-95A	45A-95A	70A-80D	80A-65D	40D-72D
Ten Str, 1000 psi	0.65-4.46	0.5-3.2	5.0-8.0	3.0-8.0	3.7-8.7
Modulus (100%), psi	200-2200	100-800	650-5400	650-3400	1100-2000
Elongation, %	50-1000	600-800	390-750	350-650	350-450
Compression Set, Method B, %	45-92	32-59	27-50	25-90	56-60
Resilience, %					
Yerzley (ASTM 945)	—	71-75	—	—	—
Rebound (Bashore)	35-50	—	35	25-50	40-60
Hysteresis Resistance	Fair	—	Fair	Fair-Good	Good
Flex Cracking Resistance	Fair	—	Good-Excellent	Excellent	Excellent
Tear Resistance	Good-Excellent	Good	Good-Excellent	Exc-Outstanding	Excellent
Abrasion Resistance	Fair-Good	—	Exc-Outstanding	Exc-Outstanding	Excellent
Impact Resistance	Exc-Outstanding	Excellent	Excellent	Exc-Outstanding	Excellent
ELECTRICAL PROPERTIES					
Vol. Res, ohm-cm	1×10^{16}	2×10^{16}	2.1×10^{12}-4.7×10^{13}	0.88×10^{12}-11×10^{12}	2.25×10^{13} -4.5×10^{14}
Dielectric Str, v/mil	600	400-510	330-460	440-730	525-900
Dielectric Constant					
60 Hz	2.41	2.5	5.75-6.34	6.0	—
1 MHz	2.41	2.5	4.53-5.15	4.21	4.6
THERMAL PROPERTIES					
Service Temperature, F					
Min for Cont Use	−60	−65	−65	−65	−60
Max for Cont Use	250	300	212	180	300
ENVIRONMENTAL RESISTANCE					
Ozone	Good-Excellent	Excellent	Excellent	Excellent	Excellent
Oxidation	Good	—	Excellent	Good-Excellent	Excellent
Weathering	Exc-Outstanding	—	Good	Good	Good
Water	Excellent	Good	Good	Good-Excellent	Excellent
Radiation	Good	—	Good	Good-Excellent	Excellent
Alkalies	Good-Excellent	Good	Poor-Fair	Poor-Fair	Good-Excellent
Aliphatic Hydrocarbons	Fair-Good	Poor	Excellent	Excellent	Good-Excellent
Aromatic Hydrocarbons	Poor	Poor	Fair-Good	Fair-Good	Excellent
Halogenated Hydrocarbons	Poor	Poor	Fair	Poor-Fair	Poor
Alcohol	Fair-Good	Good	Poor-Fair	Fair-Good	Good
Animal, Vegetable Oils	Fair-Good	—	Excellent	Excellent	Excellent
Acids					
Dilute	Good-Excellent	Good	Fair	Fair	Good
Concentrated	Good-Excellent	Fair	Poor	Poor	Poor
Synthetic Lubricants (diester)	Good	—	Poor	Poor-Good	Excellent
Hydraulic Fluids					
Silicates	Fair-Good	—	Poor	Fair	Excellent
Phosphates	Fair-Good	—	Poor	Poor	Good
Permeability to Gases	Low	—	Medium	Medium	Medium
Limiting Oxygen Index	17-20	—	—	15-20	20[b]
USES	Cable jacketing, sheeting, tubing, blown film, tires, gaskets, seals	Tubing, sheeting, disposable medical products, sealants, shoe soles, asphalt modification	Tubing, film, seals, gaskets, shoe soles, automotive fascia, O-rings, diaphragms, casters	Tubing, film, cable jacketing, seals, gaskets, diaphragms, O-rings, bladders, wheels, casters	Tubing, hose, film, gears, wire and cable jacketing, gaskets, belting, sheeting, all terrain vehicle tracks

[a] 10^3 cycles [b] 30 with additives

Index

Abbreviations, polymer materials, 375, 376
Abrasion resistance, 73, 377
Absolute viscosity, *see* Dynamic viscosity
Accelerated weathering tests, 129-140
  effects of carbon-arc exposure, 130
  effects of fluorescent UV lamp exposure, 129
  effects of xenon-arc exposure, 134
  limitations, 135
Acceptance sampling, 325-334
Acetic acid immersion test, 266
Acetone immersion test, 265
Air cannon impact test, 70
Amorphous plastics, 147, 377
Analytical tests, 239-247
ANSI, 359
AOQL sampling plans, 334
Apparent density, 190, 243, 377
Apparent melt viscosity, 173
AQL sampling plan, 328
Arapahoe smoke test, 220
Arc resistance, 120, 377
  table, 121
ASTM, 360
Attributes control charts, 324

Bacterial resistance, 142
Ball impact tester, 69
Ball rebound test, 296
Barcol hardness, 81
Birefringence, 152, 275, 310, 377
Bottle drop impact tester, 71
Boyle's law, 284
Brittle fracture, 50, 377
Brittleness temperature, 110, 378
Brookfield viscometer, 195, 378
Bruceton staircase method, 64
Bubble viscometer, 195, 378
Bulk density, 243, 377
Bulk factor, 190, 377
Buoyancy factor, 288
Burst strength tests, 260-264, 378
  long term burst strength test, 261
  quick burst strength test, 260

Calibrated solvent test, 234
Calorimetric measurements, 184
Capillary flow analysis, 254
Capillary rheometer, 171, 378
Carbon arc lamps, 130
Cellular plastics, *see* Foam plastics
Charpy impact test, 56, 378
Charts and tables, 429-487
Chemical properties, 231-238
Chemical resistance, 231
Chemical resistance chart, 430-431
Chip impact test, 57
Chroma, 156, 378
Chromatogram, 181, 182
Closed cell, 284
Coefficient of thermal expansion, 106, 378
Color, 154
Colorimeter, 156, 379
Compressive strength, 28, 264, 378
Compressometer, *see* Deflectometer
Conditioning, 249, 379
Conditioning procedures, 249-251
Consumers' risk, 327
Conversion chart, 440-443
Copolymer, 277
Copper wire test, 276
Couplants, 352
Crack initiation energy, 49
Crack propagation energy, 49
Crazing, 234
Creep, 15, 31, 379
Creep curve, 31
Creep data, 40-45
  interpretation and applications, 34
Creep modulus, 37, 379
Creep properties, 30-46
Creep strain, 32, 35, 36
Critical stress, 234
Crush test, 264
Crystalline plastics, 147
Crystallinity, 52, 89, 147, 379
Cup flow test, 193, 379
Cup viscosity test, 259, 379
Cyclic loading, 74

Decimal equivalents chart, 437
Deflectometer, 25, 30
Density, 241, 379
Density gradient technique, 241
Dielectric breakdown, 114
Dielectric constant, 117, 380
  table, 118
Dielectric strength, 114, 379
  table, 115

Differential scanning calorimetry, 184, 380
Dilatometer, 109
Dilute solution viscosity, 174-178
  applications, 177
  limitations, 177
  measurements, 174
Dissipation factor, 117, 380
Drop impact test, 63, 380
Ductile failure, 50
Durometer hardness, 81, 380
Dynamic viscosity, 174

Electrical properties, 113-126
Elongation, 7, 18, 380
Endothermic, 184
End product testing, 266
Environmental stress cracking resistance, 235, 380
Environmental test chamber, 22
Exothermic, 184, 196
Extensometer, 18, 380

Failure analysis, 307-316, 380
  identification analysis, 309
  microtoming, 310
  stress, 310
  visual, 309
Failures, types of:
  chemical, 308
  environmental, 308
  mechanical, 308
  thermal, 308
Falling dart impact tester, 64
Fatigue endurance limit, 75
Fatigue life, 75
Fatigue resistance, 74
FDA, 360
Fiberglass orientation, effects of, 20
Flammability, 199-230, 381
  cellular plastics, 216
  flexible plastics, 205
  self-supporting plastics, 206
Flammability requirements, 227, 229
Flammability standards, agencies regulating, 228
Flash ignition temperature, 208
Flaw detection, 351, 353
Flexible foam test methods, 291-298
  air flow test, 293
  constant deflection compression test, 292
  compressions, load deflection, 293
  density, 294
  dry heat, 293
  fatigue, 293
  load deflection test, 292
  resilience, 296
  steam autoclave, 291
  tear resistance, 294
  tension test, 295
Flexural creep, 32
Flexural fatigue test, 75
Flexural modulus, 26, 381
  effect of temperature, 28
Flexural properties, 23-28
  factors affecting, 27
Flexural strength, 23, 381
Flow point, 257
Flow tests, 190
  factors affecting, 192
Foam plastics, 283-305, 381
Foam properties chart, 299-303
Frequency generator, 352
Fungal resistance, 142
Fusion point, 253
Fusion test, 253

Gelation point, 197, 381
Gel permeation chromatography, 178-183
Gel time, 196, 381
Gel time meter, 197
Glossary, 377-388
Glossimeter, 163, 381
Guarded hot plate, 103, 106

Hardness, tests, 79-84
  table, 79
Hardness scales comparison chart, 439
Haze, 150, 381
Hazemeter, 150
Heat deflection temperature, 90, 382
  limitations, 92
  test variables, 92
Heat distortion temperature, *see* Heat deflection temperature
Heat resistance (long term) test, 96
High rate impact test, 67
High rate tension test, 61
Homopolymer, 277
Hooke's law, 8, 382
Hoop stress, 261, 382
Hue, 156, 382
Hue, value/chroma chart, 157
Hydrostatic design stress, 262
Hydrostatic pressure tester, 262, 263
Hygroscopic, 382

Identification analysis, 271-281
Identification of plastics materials, 277-281
  thermoplastics:
    ABS, 277
    acetal, 277
    acrylic, 278
    cellulose acetate, 278
    cellulose acetate butyrate, 278
    cellulose propionate, 278
    fluorocarbons, 278
    nylons, 278
    polycarbonate, 279

polyester, 279
polyethylene, 279
polyphenylene oxide, 280
polystyrene, 279
polysulfone, 280
polyurethane, 280
PVC, 279
thermosets:
diallyl pthalate, 280
epoxy, 280
melamine formaldehyde, 281
phenol formaldehyde, 280
polyester, 281
silicones, 281
urea formaldehyde, 281
Ignitability, 208
Ignition furnace, 209
Ignition properties, 208-210
Ignition temperature determination, 209
Immersion test, 232
Impact properties, 49-71
Impact resistance, 49
Impact strength, 49, 382
factors affecting, 50
Impact tests, type of, 52
Incandescence resistance test, 206
Incoherent light, 152
Index of refraction, *see* Refractive index
Inherent viscosity, 176, 382
Instrumented impact testing, 66
Intrinsic viscosity, 177, 382
Isochronous stress-strain curves, 36, 382
Izod impact test, 54, 383

Kinematic viscosity, 174, 195, 259

Lightness, 156
Light scattering, 177
Load-energy-time curve, 67
LTPD sampling plans, 334
Luminous transmittance, 150, 383

Material characterization tests, 165-198
for thermosets, 189
Maxwell model, 9
Mechanical properties, 7-8
Melt index test, 166
factors affecting, 167
Melting point determination test, 274
Fisher-Johns method, 274
Kofler method, 275
Metamerism, 160, 383
Microbial growth resistance testing, 142-143
bacterial resistance test, 142
fungal resistance test, 142
limitations, 143
Microoganisms, 128
Microtoming, 313

Modulus of elasticity, 8, 383
Moisture analysis, 244
Mold control, 336
Molding index, 193
Molecular orientation, 21, 27, 51, 89
Molecular weight, 178, 383
Molecular weight distribution, 165, 178, 383

NBS, 361
NBS smoke test, 219
NEMA, 361
Newtonian viscosity, 174, 384
NFPA, 361
Non-destructive testing, 351-358
Non-Newtonian viscosity, 174
Notching machine, 55
Notch sensitivity, 50, 384
NSF, 362

Open cell content, 284
Operating characteristics curve, 325-326
Optical properties, 147-164
Osmotic pressure, 177
OSU release rate test, 222
Oxygen index, 211, 384
Oxygen index test, 211-214
factors affecting, 213

Particle size, 246
Particle size determination test, *see* Sieve analysis test
Particle size distribution, 246
Peak exothermic temperature, 196, 384
Pendulum impact tester, 54
Pendulum impact tests, 53-57
effect of test variables, 57
limitations, 57
Permeability, 290
Permeance, 290
Permittivity, *see* Dielectric constant
Photoelastic, 152, 310
Photoelasticity, 152, 384
Piezoelectric material, 352
PLASTEC, 362
Plasticizer absorption tests, 254-258
burette method, 257
centrifuge force method, 258
torque rheometer method, 257
Plastics identification chart, 272-273
Plunger impact tester, 69
Polarizing, 152
Polyaxial stress, 235
Polyelectrolyte, 178
Polymerization, 165, 384
Porosity, 284
Pourability, 290
Pressure conversion chart, 446
Pressure rating (pipe), 263

Process control charts, 319
Process quality control, 335
Producers' risk, 327
Product liability, 347
Product quality control, 335
Properties chart, 447-487
Proportional limit, 8, 384
Pulse-echo technique, 353
Pycnometer, 241, 285

Quality, 317
Quality assurance manual, 339
Quality control, 317-345
  documentation, 338
  process, 335
  product, 335
  raw material, 334
  visual standards, 335
  workmanship standards, 336
Quality control system, 334

Radiant heat energy, 214
Radiant panel test, 214, 221
Radiation pyrometer, 214
Rate-of-rise test, 297
Raw material quality control, 334
Reduced viscosity, 176
Refractive index, 148, 384
Relative thermal indices, 97-102
  definition, 97
  table, 98
Relative viscosity, 175, 385
Residual stress, 92, 152, 266, 311
Resonance technique, 355
Rheological measurements, 172
Rheometer, 171, 385
Rigid foam test methods, 283-290
  cell size, 284
  compressive properties, 284
  density, 283
  dielectric constant, 290
  dimensional stability, 287
  dissipation factor, 290
  flammability, 290
  flexural properties, 287
  open cell content, 284
  shear properties, 287
  tensile properties, 286
  water absorption, 288
  water vapor transmission, 289
  weathering properties, 290
Rockwell hardness, 79, 385
Rotational rheology, 195

Safety standards organizations, 391-393
Sampling plans, 327
Sampling tables, 329
Saturation, *see* Chroma
Secant modulus, 8, 385
Self-ignition temperature, 208
Sensitizing agent, 235
Setchkin apparatus, 209
Shear rate, 173, 385
Shear strength, 71, 385
Shear stress, 173, 385
Shrinkage voids, 315
Sieve analysis test, 246
Smoke density test, 218
Smoke generation tests, 218-222
S-N curve, 75
Solubility test, 275
Solvent selection guide, 445
Solvent stress-cracking resistance, 233
SPE, 363
Specification, 2-25, 399-428
Specific gravity test, 239, 276, 385
Specific viscosity, 176
Spectrophotometer, 152, 158, 385
Specular gloss, 162, 385
SPI, 363
Spiral flow test, 192, 385
Stain resistance test, 232
Standard, 2-5
Standard laboratory atmosphere, 249
Statistical quality control, 318-334
Strain, 7, 385
Strain gauge, 312
Stress, 7, 385
Stress analysis, 152, 154, 310-313
  brittle coating method, 311
  chemical method, 312
  photoelastic method, 310
  strain gauge method, 312
Stress concentration, 50, 66, 153, 386
Stress cracking, 233
Stress optical sensitivity, 152, 386
Stress relaxation, 14, 46
Stress-strain diagram, 8, 385
Sulfide staining, 233
Surface burning characteristics, 214
Surface resistance, 119
Surging, 260, 386

Taber abraser, 73
Temperature conversion chart, 436
Tensile creep, 32
Tensile fatigue test, 76
Tensile impact test, 58
Tensile modulus, 18
Tensile strength, 16, 386
  effect of strain rate, 23
  effect of temperature, 23
Tensile tests, 15-23
  factors affecting, 21
Terepolymer, 277
Test equipment manufacturers (index), 367-374

Testing, reasons for, 1
Testing laboratories, 397-398
Testing organizations, 359-365
Thermal analysis, 183-189
Thermal conductivity, 102, 386
  table, 104
Thermal conductivity conversion chart, 444
Thermal expansion, 106
  table, 107
Thermal index, 97
Thermal properties, 89-112
Thermal stress cracking, 235
Thermogravimetric analysis, 185, 386
Thermomechanical analysis, 187, 386
Thickness measurement, 351
Torque rheometer, 253
Torque rheometer test, 253-258
Torsion pendulum test, 93, 386
Toss factor, 53, 56
Toughness, 49, 386
Tracking, 120
Trade names, 389-390
Trade publications, 395-396
Transducer, 352
Translucency, 147
Transmission technique, 355
Transparency, 147
Tunnel test, 214
TVI drying test, 245-246

UL 94 flammability tests, 222-227
  factors affecting, 227
UL requirements (electrical), 122
UL temperature index, 97
Ultimate strength, 8
Ultrasonic testing, 351, 387
Ultraviolet radiation, 127, 387
Underwriters laboratories, 364

Value, 156
Variables control chart, 319
Vicat softening temperature, 92, 387
Viscosity, 174, 387
Viscosity tests (thermosets), 193
Visual standards, 335
Voids, 50
Volume resistance, 119
Volume resistivity, 119
  table, 120

Water absorption, 243, 387
Weathering properties, 127-145, 387
  effects of:
    microorganisms, 128
    moisture, 128
    oxygen, 128
    thermal energy, 128
    UV radiation, 127
Weld line, 50
Whiteness index, 159
Workmanship standards, 336

Xenon arc lamps, 134

Yellow card, 100
Yellowness index, 159, 387
Yielding, 50
Yield point, 8, 387
Yield strength, 8, 387
Young's modulus, 8, 387

Zahn viscosity cup, 259, 388